LUMINESCENCE - AN OUTLOOK ON THE PHENOMENA AND THEIR APPLICATIONS

Edited by **Jagannathan Thirumalai**

Luminescence - An Outlook on the Phenomena and their Applications
http://dx.doi.org/10.5772/62517
Edited by Jagannathan Thirumalai

Contributors
Jagannathan Thirumalai, Seshadri Meruva, Anjos V, Bell M.J.V, Hirobumi Suzuki, Hiroaki Matsui, Mokhotswa Dhlamini, Farida Selim, Pooneh Saadatkia, Chris Varney, Prasada Rao Talakonda, Ken Munyikwa, Manoj Kumar Mahata, Hans Hofsäss, Ulrich Vetter, Yuichi Oba, Jose Rui Lima Paitio, Alfredo Morales-Sánchez, Santiago Cabañas Tay, Liliana Palacios, Mariano Aceves-Mijares, Antonio Coyopol Solis, Sergio Alfonso Pérez García, Liliana Licea Jiménez, Carlos Domínguez Horna, Evgeniy Lipatov, Dmitrii Genin, Denis Grigor'ev, Victor Tarasenko, Yuriy Tokarev, Olga Mashukova, Carlos Ruvalcaba

Published by InTech
Janeza Trdine 9, 51000 Rijeka, Croatia

Publishing Process Manager Iva Lipovic
Technical Editor SPi Global
Cover InTech Design team

First published November, 2016
Printed in Croatia

Additional hard copies can be obtained from orders@intechopen.com

Luminescence - An Outlook on the Phenomena and their Applications, Edited by Jagannathan Thirumalai
p. cm.
Print ISBN 978-953-51-2762-8
Online ISBN 978-953-51-2763-5

For EU product safety concerns:
IN TECH d.o.o., Prolaz Marije Krucifikse Kozulić 3, 51000 Rijeka, Croatia,
info@intechopen.com or visit our website at intechopen.com.

Meet the editor

Dr. Jagannathan Thirumalai received his PhD from Alagappa University, Karaikudi in 2010. He was also awarded the Post-doctoral Fellowship from Pohang University of Science and Technology (POSTECH), Republic of Korea, in 2013. Currently, he is working as assistant professor of physics, B.S. Abdur Rahman University, Chennai. His research interests focus on luminescence, self-assembled nanomaterials and thin film opto-electronic devices. He has published more than 50 SCOPUS/ISI indexed papers and 8 book chapters and is a member of several national and international societies. Currently, he is acting as a principal investigator for a funded project towards the application of luminescence-based thin film opto-electronic devices, funded by the Science and Engineering Research Board (SERB), India. As an expert in opto-electronics and nanotechnology area, he has been invited as an external and internal examiner to MSc and PhD theses and a reviewer for international and national journals.

Contents

Preface

Luminescence is more enthralling and significant towards mankind since the prehistoric times and is named after new material that defined them, for example the occurrence of luminescence through the aurora borealis, luminescent wood, glow worms, putrid fish and meat (in the course of Middle Ages and past). The contemporary age is known as the lumen ("bright" or "radiant") age and/or materials age. Luminescence has been implemented in ground-breaking research, especially in the sub-nanoscale, which studies the properties of luminescence on both excitation and emission and is widely distributed in physical, chemical, material, medical and biological sciences.

The growth of innovative luminescence technologies virtually relies on the usage of existing physico-chemical and biological materials. The origin of modern industrialized technologies generates employments and superior living atmospheres. In the outline, learning about luminescent materials may lead to innovative applications in the scientific community, which can improve the quality of life expectancy rather than just following scientific inquisitiveness. At present modern luminescent materials are in developing stage and are at the forefront of physical, chemical and medicinal sciences with promising scope.

Luminescence has a wide range of applications in everyday life, starting from conventional fluorescent lighting to digital radiography in the field of magnetic resonance imaging (MRI), electronic portal imaging device (EPID), light-emitting diodes (LEDs), solid-state lasers, luminescent solar concentrators and other electrical and electronic equipment. For examining the structure and dynamics of substance or living systems at a molecular or supramolecular level, luminescence is a powerful tool. Currently, fluorescence is used in various fields, such as genetics, cell and molecular biology, biochemistry, microbiology, bioinformatics, biometrics, forensics, flow cytometry, medical diagnostics, nanomaterials, DNA sequencing, etc.

The purpose of this book is to offer timely and in-depth coverage of designated advancements in luminescent materials by the contributors. Also, researchers from various fields are working in the field of luminescence throughout the world. This book deals with the many elements of knowledge necessary to comprehend, both quantitatively and qualitatively, the cutting-edge information in luminescent materials with an overview about the introduction of luminescence, the types of luminescence with material examples, field of applications, excitation and de-excitation processes of atom/molecules, dynamics, rare-earth ions, photon down-/up-converting materials, luminescence dating, lifetime, bioluminescence microscopical perspectives and prospects, some topical instrumentation and its various technological applications for wide range of luminescent materials. The chapters of this book have been contributed by esteemed researchers in the area and cover the frontier areas of research and developments in the field of biomedical and materials science technology. This book is envisioned for students and researchers wishing to gain a deeper understanding about the fundamental concepts that are of concern in the relevant fields, and it is our anticipation that the readers will find this book useful.

This book consists of 14 chapters that have been divided into 4 sections. Section one includes five chapters on the luminescent materials and their associated phenomena. This section deals with the introduction of luminescence, the types of luminescence with material examples, charge-transfer transitions and the new development of high temperature composites based on cadmium/zinc oxides, rare-earth ions-doped oxides, diamond for application in various semiconductor devices, polarized PL of non-polar ZnO layers and their QW structures in terms of the crystal symmetry and the in-plane lattice strain using PLD technique, crystal structure and morphological analysis, excitation-intensity effect on nanostructures/thin films with different morphologies, diamond-based LED device, the dynamics of energy transfer and persistent luminescence associated with novel luminescence applications.

Section two includes three chapters on the photo-physical properties and their emerging applications which deal with the review of rare-earth-doped up-conversion materials, silicon-rich dielectric materials (nitride and oxide) and rare-earth ions in glassy hosts for desired applications such as biomedical application, surface modification, high luminescent emission (resistive switching) in metal oxide semiconductor technology and luminescent solar concentrators.

Section three consists of three chapters on the thermoluminescence dating: from theory to applications which include discussion on the advanced thermoluminescence setup, current and future trends in luminescence dating, lithium borate/zinc tellurite glasses towards the application of the evaluation of trap levels and the characterization of donors and acceptors in semiconductors and dielectrics (wide band gap materials), earth sciences and archaeology and radiation dosimetry.

Section four comprises three chapters on the bioluminescence perspectives and prospects which include discussion on the design and studies with bioluminescence microscopy, cell biology, single live-cell analysis, interaction between bioluminescence and specific photoreceptor adaptations in fishes to detect the biological light (metamorphosis of eye and luminous tissues), the Black Sea ctenophores-aliens as an index of their physiological state and three-dimensional imaging with dissimilar functionalizations specific for anticipated applications, for instance, biomedical and industrial applications.

Finally, I will never forget that my first steps in the field of luminescence were guided by Dr. R. Jagannathan and Professor R. Chandramohan; our friendly collaboration for many years was very fruitful. I would like to thank all the contributors of the chapters in this book for their tremendous efforts in constructing outstanding work. Last but not least, I would like to express my sincere gratitude to Ms. Iva Lipović, publishing process manager, for the effective communication and assistance during the preparation of this book.

Jagannathan Thirumalai
B.S. Abdur Rahman University,
India

Section 1

Luminescent Materials and their Associated Phenomena

Chapter 1

The Impact of Luminescence in Technological Scale

Jagannathan Thirumalai

Additional information is available at the end of the chapter

http://dx.doi.org/10.5772/64625

1. A epigrammatic testimony of luminescence

From the prehistoric times, the term 'luminescence' is more fascinating towards mankind. One can simply look at the logically occurring luminescence through the aurora borealis, luminescent wood, glow worms, putrid fish and meat [1]. The effect was wearing a veil in secrecy and illustrated consequently in the Middle Ages and past. The most primitive printed report of a solid-state luminescent material originated from a Chinese text that was published in the Song dynasty (960–1279 A.D.), quite referred to a book (never recovered) from the period 140–88 B.C. It narrates a painting picture of a cow munch grass in an outside field. In the darkness, the cow would be seen repose within a shelter [1–3]. Perhaps, the first man-made ink was exploited using a persistent phosphor material. Harvey [3] dispenses a tremendous description of these untimely interpretations far beyond the purview of the current reassess. In general, the name phosphorus is mentioned only for the chemical element, whereas specific microcrystalline luminescent materials are referred as phosphors. The first artificial phosphor exemplified in Western literature dates from 1603. Then, the Italian alchemist and shoemaker Vincenzio Cascariolo's phosphor (1870) manifest was the earliest commercially available phosphor, referred to as "Balmain's paint," a barium sulphide preparation. Phosphors (light-bearing materials) are optical transducers that yield luminescence when the material is suitably excited. The idiom 'luminescence' (the Greek translation of *lucifer*, means light bearer) was first initiated by the German physicist, Eilhardt Wiedemann, in 1888, to facilitate the discrimination among the emission of light (luminescence) from thermally excited substances/molecules under suitable excitation devoid of escalating their average kinetic energy.

After 1900, the modern period luminescence experimentations were started on the inspirations of promising quantum mechanics approaches [4–6]. During the antediluvian 1900s, the progress of quantum theory bestowed a concrete evidence on theoretical groundwork about the enormous accretion of spectroscophical facts. A comprehensive understanding of luminescent emission led from quantum theory, which voluntarily elucidating prior interpretation and consenting predictions of innovative occurrence. Subsequently, during the period of 1920–1930, theoretical concepts of luminescence are very well implicit among researchers, and it was documented to facilitate luminescence spectroscopy is intrinsically added novel perceptive than absorption spectroscopy. A minimum of five independent luminescence properties are able to be estimated which are the features of a testing sample module [4–6]:

1. Emission intensity by monitoring the excitation wavelength.
2. Excitation intensity by monitoring the emission wavelength.
3. Decay time of the excited state.
4. Emission of polarization.
5. Quantum yield.
6. Anisotropy.

As an assessment, the merely alternative variable calculated using absorption spectroscopy is the transmission spectra (Beer-Lambert law). However, most studies concern the activity of luminescence concepts in the prediction of innovative occurrences as summarized in **Table 1**.

Type	Examples	Applications
Fluorescence	Lignum nephriticum ('kidneywood'), aragonite, and so on (in all the below cited luminescence types) [7]	Display devices, fluorescent hydrogels, biomarkers
Phosphorescence	Eu^{2+}-doped strontium silicate-aluminate and so on (in all the below cited luminescence types) [8]	Traffic signals, phosphorescent paint ('Leuchtgelb')
Photoluminescence	Halophosphate (fluoro-or chloro-apatite): $Ca_5(PO_4)_3(F,CI):Sb^{3+}$, Mn^{2+} [9]	Fluorescent lamp
Radioluminescence	Paint with radium, gaseous tritium light source (GTLS) [10]	Wristwatch faces, gun sights, nuclear reactors and radioisotopes
Cathodoluminescence	$Ca_3Gd_7(PO_4)(SiO_4)_5O_2$: Ce^{3+}, Tb^{3+} and Mn^{2+} [11]	Cathode ray tube, monitors, field emission device
Electroluminescence	Zn(S,Se): Cu^+, ZnS: Cu^+ [12]	LED, EL displays
Thermoluminescence	ZnS: Mn^{2+}, Radioactive irradiation, quartz [13, 14]	Archaeology, dating of burnt flint, pressure gauge temperature
Chemiluminescence	Oxidation of luminol, fluorescein, rhodamines, coumarins, oxazines [15]	Analytical chemistry
Bioluminescence	Green fluorescent protein [16]	Cell tracking, fast-acting biocides
Candoluminescence/ Pyroluminescence	Zinc oxide and cerium oxide or thorium dioxide, trimethyl borate, alkali metals and alkali earth metals [17, 18]	Gas mantles or limelight
Galvanoluminescence	Electrolysis of sodium bromide (NaBr) [19]	Fabrication of electrolytic cell
Sonoluminescence	Collapse of gas-filled bubbles in a liquid [20]	Bomb-resistant baggage container for wide body aircraft

Type	Examples	Applications
Mechanoluminescence/ Mechanochromic luminescence	$CaZnOS:Mn^{2+}$ and $CaZr(PO_4)_2:Eu^{2+}$ [21]	Mechanical stress in industrial plants, structures and living bodies
Triboluminescence/ Fractoluminescence	$ZnS:Mn^{2+}$ [22, 23]	Diamond, quartz, emission of electromagnetic radiation (EMR)—sensors/smart materials
Crystaloluminescence	NaCl [24]	Image intensification techniques (spatial, temporal and spectral)
Injection luminescence	LED [25]	Basic research
Negative luminescence	InSb, (Hg,Cd)Te, Ge and InAs [26]	Electronic device

Table 1. Different types of luminescence, with material examples, and field of applications.

Luminescence is a process having a wide range of applications in everyday life, starting from the conventional fluorescent lighting they extend to digital radiography in the field of magnetic resonance imaging (MRI) [27], electronic portal imaging device (EPID) [28], light-emitting diodes (LEDs) [29, 30], solid-state lasers [31], luminescent solar concentrators [32] and many/much other electrical and electronic equipment employ luminescent materials. Recently, electroluminescent display that shows promise for making flexible electroluminescent flat panel display (FEL-FPD) technology [33] is emerging worldwide; it also provides an excellent platform for a foundation for a no-compromise hang-on-the wall TV. In the field of biochemistry and biophysics, the fluorescence spectroscopy and time-resolved fluorescence are deemed as the first and foremost research equipment and this prominence has transformed and expanded nowadays with modern spectroscophical equipment. Currently, fluorescence as one of the foremost tactics was meticulously utilized in dissimilar areas of biochemistry, cell and molecular biology, genetics, bioinformatics, microbiology, biometrics, forensics, flow cytometry, medical diagnostics, nanomaterials, DNA sequencing, etc. The usage of fluorescence proves a dramatic growth in cellular and molecular imaging. Fluorescence imaging should be able to disclose the localized analysis of intra-cellular molecules, every so often at the stage of the detection of single molecule [34].

2. Technological advancements in the science of luminescence spectroscopy

All and sundry is having numerous astonishing moments to have a high regard for the spectacular engagement in recreation of luminosity, the consequence and the good organization of the assistance offered through optical devices to expand our prospect, in addition to reward for its ensnared defects to make ourselves with optical illusions. The well-equipped spectroscophical techniques possess broad accessibility by means of ease procedure, selectivity, sensitivity, accuracy, speed and precision [6, 9, 34]. The novel applications of fluorescence have proffered innovative technological advancements over few decades and these technological features were rapidly implemented for ground-breaking research. It is pointed out that two-photon or multi-photon excitation and multi-photon microscopy is one of the important technologies by employing the fluorescence mechanism [35–38]. By two-photon absorption process fluorophores can be excited by means of femtosecond pump-pulse lasers with regular pulse width. These lasers have turned out to be simple to utilize and are equipped with microscopes in the recent days. **Table 2** summarizes some major innovative technological advancement associated with the science of luminescence activities in a broad spectrum.

Instrumentation	References
Time-resolved fluorescence spectroscopy	[6]
Transient-absorption spectroscopy (flash spectroscopy)	[39]
Time-resolved infrared spectroscopy	[40]
Time-resolved two-photon photoelectron (2PPE) spectroscopy (or) time-resolved photoemission spectroscopy (or) laser-based angle-resolved photoemission spectroscopy	[41–44]
Fluorescence lifetime imaging spectroscopy	[45]
Fluorescence correlation spectroscopy	[46, 47]
Single-molecule fluorescence spectroscopy	[48]
Fluorescence microscopy (epi-fluorescence, confocal)	[49]
Two-photon excitation fluorescence microscopy	[50]
Near-field scanning optical microscopy (or) optical stethoscopy	[51]

Table 2. Different types of luminescence spectroscophical instrumentation.

In fluorescence microscopy, the controlled excitation from the phenomenon of two-photon excitation has created a prevalent employability. Only through the focal plane of a microscope the image processing could be achieved through multi-photon excitation process [49]. This is a major benefit, since fluorescence images may get deformed from fluorescence process from top and bottom of the focal plane. There is no definite phase fluorescence so as to reduce the dissimilarity in non-confocal fluorescence microscopy; as a result, the images are obtained with good resolution. Such images are currently being achieved in numerous research laboratories.

Recently, a variety of scientific themes in association to the perspective of analytical advancements in luminescence spectroscopy and luminescence-based imaging in the field of earth sciences and related disciplines were discussed in detail [52]. Cathodoluminescence (CL) spectroscopy can be employed to detect and differentiate diverse generation of minerals or mineral by its variable CL colours or as an efficient technique on behalf of spatially resolved analysis of point/lattice defects (e.g. radiation-induced defects or vacancies, or broken bonds induced from electron defects) in solids by using the CL spectral measurements [53]. A new approach where fluorescence methods combined with modern chemo-metric approaches, such as bio-specific and other sensors, shows significant potential in the detection of cultural heritage and its degradation, explosives, residues and their components using time-resolved photoluminescence spectroscopy (TRPL) and fluorescence lifetime imaging (FLIM) [54]. Similarly, the total reflection x-ray fluorescence (TXRF) spectrometry is an energy-dispersive x-ray method that is employed for determining the elemental and chemical analysis (in stainless steel metal release) and is also suitable for small-sample analyses like airborne silver nanoparticles (NPs) from fabrics [55].

Thus, the invention of modern luminescence technology-oriented spectroscophical tools employed with multi-photon excitation/emission is one of the most important mechanisms that encompasses with radiative energy transfer, energy transfer by resonant exchange, energy transfer by spatial process, energy exchange by spin coupling, energy transfer by non-resonant processes and so on, which involved during photophysical processes even in a molecular level. As a result, the up-to-date activity in luminescence-based spectroscophical instrumentation has been correlated to expand our prospects towards new ideas in the field of biological science, physical chemistry, food science, pharmacology, nanotechnology, photovoltaics/solar cells, LEDs and displays, environmental science and so on.

In connection to the aforementioned aspects, the proper evaluation of environmental risks pertinent to recent experimental standards with reference to technological perspectives based on the growth inhibition caused by the chemical substances require necessary qualitative assessment such as the assessment of mechanism articulating toxicity. Therefore, it is affirmed that this assessment is need to be developed for building improvement towards ecological preservation and to deep evasion against human health.

3. Conclusion

As discussed above, luminescence is not only well conceived, but a pioneer across the globe with innovative scientific developments; however, facts also demonstrate that it has been and will prolong to be imperative towards ground-breaking research against novel applications for the societal cause. The most important worldwide challenges amongst the major noteworthy progress are in diverse fields of biochemistry, cell, molecular biology, genetics, bioinformatics, microbiology, bioinformatics, biometrics, forensics, flow cytometry, medical diagnostics and the addition of nanotechnology. The dispute of novel spectroscophical/microscopical innovation comprises interdisciplinary areas that must continue to be improved for these innovative global developments in spectral imaging, fluorescence lifetime, time-correlated single-photon counting, kinetic chemical reaction rates, singlet-triplet dynamics, visual implants, non-invasive optical biopsy and neurology. Thus, studies on inimitable luminescence technological surroundings might provide an insights about atoms/molecules that may perhaps turn out to be the future harbingers of green energy in the upcoming scenario.

Acknowledgements

This work was partially supported by the Department of Science and Technology, Government of India (SR/FTP/PS-135/2011).

The authors apologize for inadvertent omission of any pertinent references.

Conflict of interest

The authors declare that there is no conflict of interests regarding the publication of this paper.

Author details

Jagannathan Thirumalai

Address all correspondence to: thirumalaijg@gmail.com; jthirumalai@bsauniv.ac.in

Department of Physics, B. S. Abdur Rahman University, Vandalur, Chennai, Tamil Nadu, India

References

[1] Yen W. M., Weber M. J. Inorganic Phosphors: Compositions, Preparation and Optical Properties. 1st ed. Boca Raton, FL, USA: CRC Press; 2004. 456 p. DOI: 10.1201/9780203506325.

[2] Yen W. M., Shionoya S., Yamamoto H. Phosphor Handbook. 2nd ed. Boca Raton, FL, USA: CRC Press; 2007. 1057 p. DOI: 0-8493-3564-7.

[3] Harvey E. N. A History of Luminescence from the Earliest Times until 1900. 1st ed. Philadelphia, PA: American Philosophical Society; 1957. 774 p. DOI: 00390708.

[4] Goldberg M. C., Weiner E. R. The science of luminescence. In: Goldberg M. C., editor. Luminescence Applications in Biological, Chemical, Environmental, and Hydrological Sciences. 193rd Meeting. Denver, Colorado: ACS Symposium Series; Washington, DC: American Chemical Society; Vol. 383; December 30, 1989. p. 1–22. DOI: 10.1021/bk-1989-0383.ch001.

[5] Widemann E. U. Fluorescenz und phosphorescenz I. Abhandlung(On Fluorescence and Phosphorescence). Ann. Phys. 1888;270(7):446–463. DOI: 10.1002/p.18882700703.

[6] Valeur B., Berberan-Santos M. N., editors. Molecular Fluorescence: Principles and Applications. 2nd ed. Germany: Wiley-VCH Verlag GmbH & Co. KGaA; 2001. p. 1–399. DOI: 10.1002/9783527650002.

[7] Acuna A. U., Amat-Guerri F. Early history of solution fluorescence: the lignum nephriticum of Nicolas Monardes. In: Wolfbeis O. S., (Series editor) Berberan-Santos M. N. (Vol. editors). Fluorescence of Supermolecules, Polymers, and Nanosystems. 4th ed. of Springer Series on Fluorescence. New York, NY, USA: Springer Science & Business Media; 2007. p. 3–20. DOI: 10.1007/4242_2007_006.

[8] Sophya Preethi K. R., Lu C.-H., Thirumalai J., Jagannathan R., Natarajan T. S., Nayak N. U., Radhakrishna R., Jayachandran M., Trivedi D. C. $SrAl_4O_7$: Eu^{2+} nanocrystals:

synthesis and fluorescence properties . J. Phys. D: Appl. Phys. 2004;37(19):2664–2669. DOI: 10.1088/0022-3727/37/19/009.

[9] McCarthy G. J. The Rare Earths in Modern Science and Technology. 2nd ed. New York: Springer Science & Business Media; 2012. 647 p. DOI: 10.1007/978-1-4613-3054-7.

[10] Bower K. E., Barbanel Y. A., Shreter Y. G., Bohnert G. W., editors. Polymers, Phosphors, and Voltaics for Radioisotope Microbatteries. 1st ed. Boca Raton, FL: CRC Press; 2002. 504 p. DOI: 10.1201/9781420041392.ch4.

[11] Zhang Y., Li G., Geng D., Shang M., Peng C., Lin J. Color-tunable emission and energy transfer in $Ca_3Gd_7(PO_4)(SiO_4)_5O_2$:$Ce^{3+}$/$Tb^{3+}$/$Mn^{2+}$ phosphors. Inorg. Chem. 2012;51(21): 11655–11664. DOI: 10.1021/ic3015578.

[12] Warkentin M., Bridges F., Carter S. A., Anderson M. Electroluminescence materials ZnS:Cu,Cl and ZnS:Cu,Mn,Cl studied by EXAFS spectroscopy. Phys. Rev. B. 2007;75(7): 075301–075309. DOI: 10.1103/PhysRevB.75.075301.

[13] Zahedifar M., Taghavinia N., Aminpour M. Synthesis and Thermoluminescence of ZnS:Mn^{2+} Nanoparticles. In: Salamin Y. I., Hamdan N. M., Al-Awadhi H., Jisrawi N. M., Tabets N., editors. AIP Conference Proceedings; 10–12 April 2007; Sharjah (United Arab Emirates). Melville, NY: AIP Publishing; 2007. p. 128–132. DOI: 10.1063/1.2776701.

[14] McKeevers S. W. S., editor. Thermoluminescence of Solids. Volume 3 of Cambridge Solid State Science Series. Cambridge, England: Cambridge University Press; 1988. 392 p. DOI: 9780521368117.

[15] Abel B., Odukoya B., Mohammed M., Aslan K. Enhancement of the chemiluminescence response of enzymatic reactions by plasmonic surfaces for biosensing applications. Nano. Biomed. Eng. 2015;7(3):92–101. DOI: 10.5101/nbe.v7i3.p92-101.

[16] Tsien R. Y. The green fluorescent protein. Annu. Rev. Biochem. 1998;67(1):509–44. DOI: 10.1146/annurev.biochem.67.1.509.

[17] Minchin L. T. Luminescence of oxides under flame excitation. Trans. Faraday Soc. 1939;35(1):163–170. DOI: 10.1039/TF9393500163.

[18] Kitsinelis S., Kitsineliss S., editors. Light Sources. Basics of Lighting Technologies and Applications. 2nd ed. Boca Raton, FL, USA: CRC Press; 2015. 295 p. DOI: 10.1201/b18456-13.

[19] Sullivan R. R., Dufford R. T. A further study of galvano-luminescence. J. Opt. Soc. Am. 1931;21(8):513–523. DOI: 10.1364/JOSA.21.000513.

[20] McNamara W. B., III, Didenko Y. T., Suslick K. S. Sonoluminescence temperatures during multi-bubble cavitation. Nature. 1999;401(October):772–775. DOI: 10.1038/44536.

[21] Zhang J.-C., Xu C.-N., Wang X. S., Long Y.-Z. Novel elastico-mechanoluminescence materials CaZnOS:Mn^{2+} and CaZr$(PO_4)_2$:Eu^{2+}. J. Adv. Dielectrics. 2014;4(3):1430003–1430007. DOI: 10.1142/S2010135X14300035.

[22] Clegg W., Bourhill G., Sage I. Hexakis(antipyrine-O)terbium(III) triiodide at 160 K: confirmation of a centrosymmetric structure for a brilliantly triboluminescent complex. Acta Cryst. 2002;E58(4):m159–m161. DOI: 10.1107/S1600536802005093.

[23] Chmel A., Shcherbakov I. Fractoluminescence from brittle and ductile homogeneous solids. J. Lumin. 2014;153(September):85–89. DOI: 10.1016/j.jlumin.2014.03.005.

[24] Alexander A. J. Deep ultraviolet and visible crystalloluminescence of sodium chloride. J. Chem. Phys. 2012;136(6):064512. DOI: 10.1063/1.3684548.

[25] Nakamura S., Chichibu S. F., editors. Introduction to Nitride Semiconductor Blue Lasers and Light Emitting Diodes. Illustrated ed. Boca Raton, FL, USA: CRC Press; 2000. 386 p. DOI: 10.1.1.398.4326.

[26] Malyutenko V. K. Negative luminescence in semiconductors: a retrospective view. Phys. E. 2004;20(3–4):553–557. DOI: 10.1016/j.physe.2003.09.008.

[27] Roux S., Bazzi R., Rivière C., Lux F., Perriatand P., Tillement O. Rare earth nanomaterials in magnetic resonance imaging. In: Yang T. T. T., editor. Rare Earth Nanotechnology. 1st ed. Boca Raton, FL, USA: CRC Press; 2012. 262 p. DOI: 978-981-4364-20-1.

[28] Yeboah C., Pistorius S. Monte Carlo studies of the exit photon spectra and dose to a metal/phosphor portal imaging screen. Med. Phys. 2000;27(2):330–339. DOI: 10.1118/1.598835.

[29] Ma M., Mont F. W., Yan X., Cho J., Fred S. E., Kim G. B., Sone C. S. Effects of the refractive index of the encapsulant on the light-extraction efficiency of light-emitting diodes . Opt. Exp. 2011;19(5):A1135–A1140. DOI: 10.1364/OE.19.0A1135.

[30] Krishnan R., Thirumalai J. Synthesis and up/down conversion luminescence properties of $Na_{0.5}R_{0.5}MoO_4$:Ln^{3+} (R^{3+} = La, Gd) (Ln^{3+} = Eu, Tb, Dy, Yb/Er) thin phosphor films grown by pulsed laser deposition technique. RSC Adv. 2014;4(109):64258–64266. DOI: 10.1039/C4RA11274A.

[31] Barbet A., Paul A., Gallinelli T., Balembois F., Blanchot J.-P., Forget S., Chénais S., Druon F., Georges P. Light-emitting diode pumped luminescent concentrators: a new opportunity for low-cost solid-state lasers. Optica. 2016;3(5):465–468. DOI: 10.1364/OPTICA.3.000465.

[32] Meinardi F., McDaniel H., Carulli F., Colombo A., Velizhanin K. A., Makarov N. S., Simonutti R., Klimov V. I., Brovelli S. Highly efficient large-area colourless luminescent solar concentrators using heavy-metal-free colloidal quantum dots. Nat. Nanotechnol. 2015;10(10):878–885. DOI: 10.1038/nnano.2015.178.

[33] Crawford G. P., editor. Flexible Flat Panel Displays. Wiley Series in Display Technology. Hoboken, NJ: John Wiley & Sons; 2005. 556 p. DOI: 10.1002/0470870508.ch1.

[34] Lakowicz J. R., editor. Principles of Fluorescence Spectroscopy. 3rd ed. New York: Springer; 2006. XXVI, 954 p. DOI: 10.1007/978-0-387-46312-4.

[35] Justel T., Nikol H., Ronda C. New developments in the field of luminescent materials for lighting and displays. Angew. Chem. Int. Ed., . 1998;37(22):3084. DOI: 10.1002/(SICI)1521-3773.

[36] Diaspro A., editor. Confocal and Two-Photon Microscopy, Foundations, Applications, and Advances. 1st ed. New York: Wiley-Liss; 2002. 580 p. DOI: 978-0471409205.

[37] Masters B. R., Thompson B. J., editors. Selected Papers on Multiphoton Excitation Microscopy. 1st ed. Washington, DC: SPIE Optical Engineering Press Bellingham; 2003. 694 p. DOI: 9780819447487.

[38] Zipfel W. R., Williams R. M., Webb W. W. Nonlinear magic: multiphoton microscopy in the biosciences. Nat. Biotechnol. 2003;21(11):1369–1377. DOI: 10.1038/nbt899 .

[39] Burda C., El-Sayed M. A. High-density femtosecond transient absorption spectroscopy of semiconductor nanoparticles. A tool to investigate surface quality. Pure Appl. Chem. 2000;72(1–2):165–177. DOI: 10.1351/pac200072010165.

[40] Muhammad S., Moncho S., Brothers E. N., Darensbourg M. Y., Darensbourg D. J., Bengali A. A. Time resolved infrared spectroscopy: kinetic studies of weakly binding ligands in an iron–iron hydrogenase model compound. Inorg. Chem. 2012;51(13):7362–7369. DOI: 10.1021/ic300785z.

[41] Stolow A., Bragg A. E., Neumark D. M. Femtosecond time-resolved photoelectron spectroscopy. Chem. Rev. 2004;104(4):1719–1758. DOI: 10.1021/cr020683w.

[42] Weinelt M. Time-resolved two-photon photoemission from metal surfaces. J. Phys.: Conde. Matt. 2002;14(43):R1099. DOI: 10.1088/0953-8984/14/43/202.

[43] Fauster T., Steinmann W. Two-photon photoemission spectroscopy of image states. In: Halevi, P., editor. Electromagnetic Waves: Recent Developments in Research. 2 ed. Photonic Probes of Surfaces, North Holland: Elsevier; 1995. p. 347–411. ISBN: 9780444821980.

[44] Zhu X. Y. Electron transfer at molecule-metal interfaces: a two-photon photoemission study. Ann. Rev. of Phys. Chem. 2002;53(1):221–247. DOI: 10.1146/annurev.physchem.53.082801.093725.

[45] Chang C.-W., Sud D., Mycek M.-A. Fluorescence lifetime imaging microscopy. In: Sluder G., Nordberg J. J., editors. Methods in Cell Biology. 3rd ed., Vol. 81. Amsterdam: Elsevier; 2007. p. 495–524. DOI: 10.1016/S0091-679X(06)81024-1

[46] Krichevsky O., Bonnet G. Fluorescence correlation spectroscopy: the technique and its applications. Rep. Prog. Phys. 2002;65(2):251–297. DOI: 10.1088/0034-4885/65/2/203.

[47] Mayboroda O. A., van Remoortere A., Tanke H. J., Hokke C. H., Deelder A. M. A new approach for fluorescence correlation spectroscopy (FCS) based immunoassays. J. Biotechnol. 2003;107(2):185–192. DOI: 10.1016/j.jbiotec.2003.10.007.

[48] Roiter Y., Minko S. AFM single molecule experiments at the solid-liquid interface: in situ conformation of adsorbed flexible polyelectrolyte chains. J. Am. Chem. Soc. 2005;127(45):15688–15689 . DOI: 10.1021/ja0558239.

[49] Huang B., Bates M., Zhuang X. Super resolution fluorescence microscopy. Annu Rev Biochem. 2009;78(1):993–1016. DOI: 10.1146/annurev.biochem.77.061906.092014.

[50] Svoboda K., Yasuda R. Principles of two-photon excitation microscopy and its applications to neuroscience. Neuron. 2006;50(6):823–839. DOI: 10.1016/j.neuron. 2006.05.019.

[51] Dürig U., Pohl D. W., Rohner F. Near-field optical scanning microscopy. J. Appl. Phys. 1986;59(10):3318. DOI: 10.1063/1.336848.

[52] Nasdala L., Götze J., Hanchar M. Luminescence spectroscopy and imaging: analytical advances and perspectives in the Earth sciences and related disciplines. Miner Petrol. 2013;107:349–351. DOI: 10.1007/s00710-013-0292-4.

[53] Götze J. Potential of cathodoluminescence (CL) microscopy and spectroscopy for the analysis of minerals and materials. Anal. Bioanal. Chem. 2002;374:703–708. DOI 10.1007/s00216-002-1461-1.

[54] Nevin A., Cesaratto A., Belle S., D'andrea C., Toniolo L., Valentini G., Comelli D. Time-resolved photoluminescence spectroscopy and imaging: new approaches to the analysis of cultural heritage and its degradation. Sensors. 2014;14:6338–6355. DOI: 10.3390/s140406338.

[55] Dalipi R., Borgese L., Casaroli A., Boniard M., Fittschen U., Tsuji K., Depero L. E. Study of metal release from stainless steels in simulated food contact by means of total reflection X-ray fluorescence. J. Food Engineering. 2016;173:85–91. DOI: 10.1016/ j.jfoodeng.2015.10.045.

Chapter 2

The Dynamics of Luminescence

Luyanda L. Noto, Hendrik C. Swart,
Bakang M. Mothudi, Pontsho S. Mbule and
Mokhotjwa S. Dhlamini

Additional information is available at the end of the chapter

http://dx.doi.org/10.5772/65050

Abstract

Domestic light providing devices have always been an important component of life and continue to provide us light beyond sunset. These devices continue to be improved frequently to allow ease of use and to enhance their efficiency. The tungsten electric light bulbs are widely used, which are based on incandescence of a continuously heated tungsten element. However, their use will soon be short lived because of the increased usage of fluorescent tubes and light-emitting diode (LED) devices, which are based on luminescence emission. These emission materials that display luminescence are called phosphors, and their emission is based on electron transitions. In the following chapter, we shall look into photoluminescence from both intrinsic and extrinsic defects, covering both down- and upconversion (UP). We will look into the concept of energy transfer and persistent luminescence and lastly provide related applications of luminescence in the modern days.

Keywords: electron traps, fluorescence, phosphorescence, kinetics, defects, persistent luminescence

1. Intrinsic and extrinsic luminescence

Luminescence occurs when a material absorbs radiation that brings about the transition of electrons from the valence band to the conduction band [1]. This is followed by de-excitation of the electrons back to the valence band via a luminescence center, which converts their energy to electromagnetic waves [2]. The luminescence center can either be from intrinsic defects [3] or from extrinsic defects [4]. **Figure 1a** shows a schematic diagram illustrating both intrinsic and extrinsic defects. Intrinsic defects occur as a result of an atom in a matrix being displaced from its original position to an interstitial position, where it generates a point defect, leaving behind a vacancy defect [5]. Positive defects occupy the donor level (D in **Figure 1a**), and the negative defects occupy the acceptor level (A in **Figure 1a**) within the band gap of a material [6].

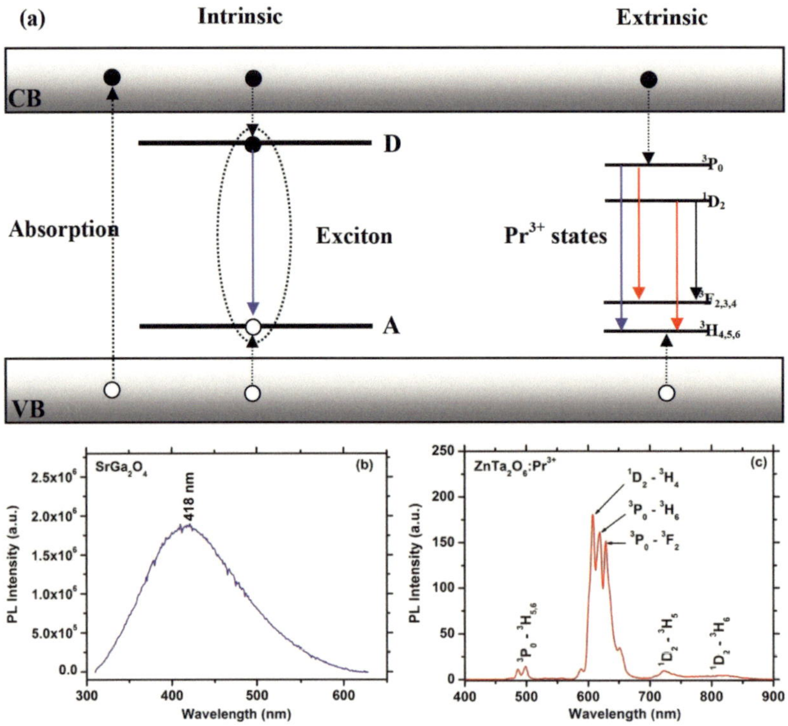

Figure 1. (a) A schematic diagram showing an emission from both intrinsic and extrinsic defects [2], (b) an emission spectrum of $SrGa_2O_4$ displaying luminescence from intrinsic defects, (c) shows an emission of $ZnTa_2O_6:Pr^{3+}$ from an extrinsic luminescent center [8].

When a material with intrinsic defects absorbs radiation, the electrons are excited from the valence band to the conduction band, leaving behind positively charged holes in the valence band [4, 5]. The positive charge of the donor level exerts a coulombic force unto the excited electrons within the conduction band and attracts them. Similarly, the negative charge of the acceptor level attracts the free holes in the valence band [7]. The two will exist in a temporary bound state, and eventually the electrons will de-excite to recombine with the holes. When electrons de-excite, they lose energy, which is converted to electromagnetic waves [4, 5]. An emission spectrum showing an intrinsic defect emission of $SrGa_2O_4$ (unpublished data) is presented in **Figure 1b**.

Extrinsic defects are intentionally incorporated dopants into a host matrix, in order to generate a luminescent center (**Figure 1a**). When the excited electrons de-excite to the luminescent center (Pr^{3+} is used as an example of the luminescent center), further de-excite to lower state of the

http://dx.doi.org/10.5772/65050

center radiatively, giving electromagnetic waves with different wavelengths, depending on the ion adopted for the luminescent center [2]. It is also worth noting that direct excitation to the luminescent center occurs simultaneously with the excitation to the conduction band [5]. **Figure 1c** shows photoluminescence emission spectrum of $ZnTa_2O_6$:Pr^{3+} phosphor, which displays blue and red emission lines from different metastates of Pr^{3+} as electrons de-excite further to their ground state. The blue emission lines are attributed to ${}^3P_0 \rightarrow {}^3H_4$ transitions at 447–449 nm, and the red emission lines are attributed to ${}^1D_2 \rightarrow {}^3H_4$, ${}^3P_0 \rightarrow {}^3H_6$ and ${}^3P_0 \rightarrow {}^3F_2$ transitions at 608, 119, and 639 nm, respectively [8].

2. Fluorescence and phosphorescence

The luminescence emission is differentiated by the length of its lifetime, which can either be fluorescent, phosphorescence, or persistent [5]. On one hand, with an emission lifetime, fluorescence is an emission lasting up to 10 ns [9]. On the other hand, phosphorescence is an emission that is longer than 10 ns, and it lasts up to 10 s [9]. Additionally, there is also an emission that lasts for a couple of minutes up to several hours after the excitation source has been removed, and it is referred to as persistent luminescence [10].

The difference between the two is explained using a simplified Jablonski diagram (**Figure 2**), which only shows transitions between vibrational states of a dopant ion. Fluorescence (F—in **Figure 2**) occurs when the energy of the incoming radiation excites an electron (A—in **Figure 2**) residing in the ground state (S_0) to higher singlet energy states (S_1, S_2...S_n). From where the electron will de-excited to the lowest excited state (S_1), then radiatively de-excite to the ground state (S_0) within 10 ns. In the case of phosphorescence emission (P in **Figure 2**), there are triplet energy states between metastates, from which the electron stabilizes then de-excites the one below and so from T_2 to T_1 (**Figure 2**). This delayed transition of an electron results in an emission that may be delayed up to 10 s from the time of radiation absorption [11–13].

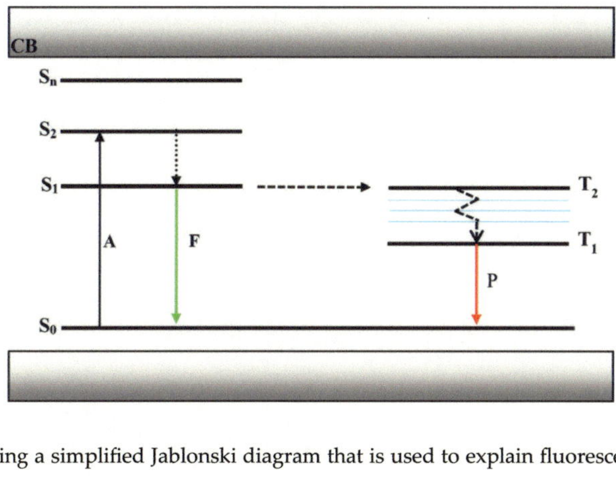

Figure 2. Schematic showing a simplified Jablonski diagram that is used to explain fluorescence and phosphorescence emission [2].

3. Persistent luminescence

Persistent luminescence differs from both fluorescence and phosphorescence, because it does not depend on the dopant ion, but the electron-trapping centers, which are generated by intrinsic defects [10]. In this case, when a sample is irradiated, the excited electrons are trapped by electron traps, and as a result of thermal energy, they gradually migrate to the luminescent center (Pr^{3+} in $GdTaO_4:Pr^{3+}$) [10, 14].

This type of luminescence may last up to several minutes or hours [10]. The decay curve (**Figure 3**) of $GdTaO_4$:Pr shows the luminescence emission intensity change over 1200 s. It is divided into two components, the fast and the slow components, attributed to shallower and deeper electron-trapping centers [2], with time parameters that can be extracted by fitting the curve with a second-order exponential equation (Eq. (1)): where $i(t)$ is the luminescence intensity, A and B are constants, and t is the measurement time. The first and second terms describe the decay of the first and the second components; τ_1 and τ_2 represent the lifetime of the two components [15].

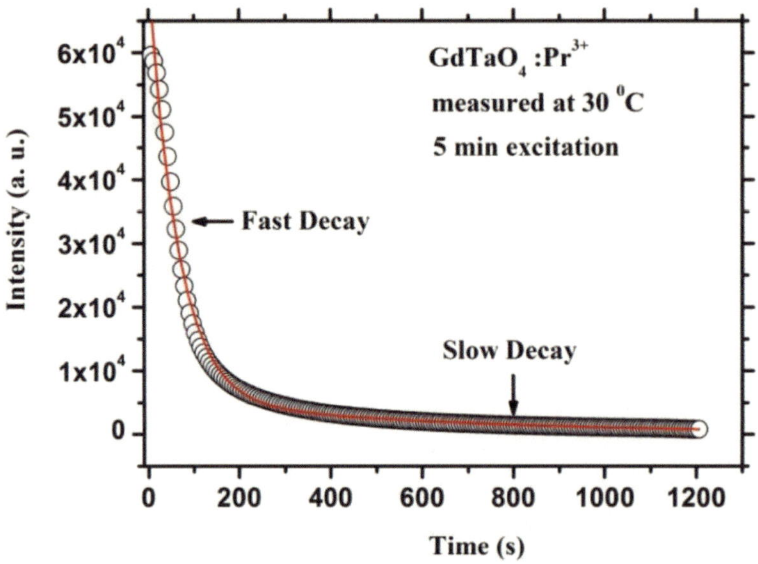

Figure 3. Decay curve of $GdTaO_4:Pr^{3+}$ persistent luminescence [2].

$$I(t) = Ae^{-t/\tau_1} + Be^{-t/\tau_2} \quad (1)$$

Persistent luminescence relies on the electron-trapping (**Figure 4**) [14] centers, which may be more than one in a particular system, and among them are positively charged oxygen vacancies ($V_O^{\cdot}$ and $V_O^{\cdot\cdot}$) [14–18]. These defects (oxygen vacancies) have an energy level that overlaps with energy of the conduction band, and the positively charged defects occupy the donor level, as mentioned earlier. They also occupy different energy levels relative to the conduction band [2].

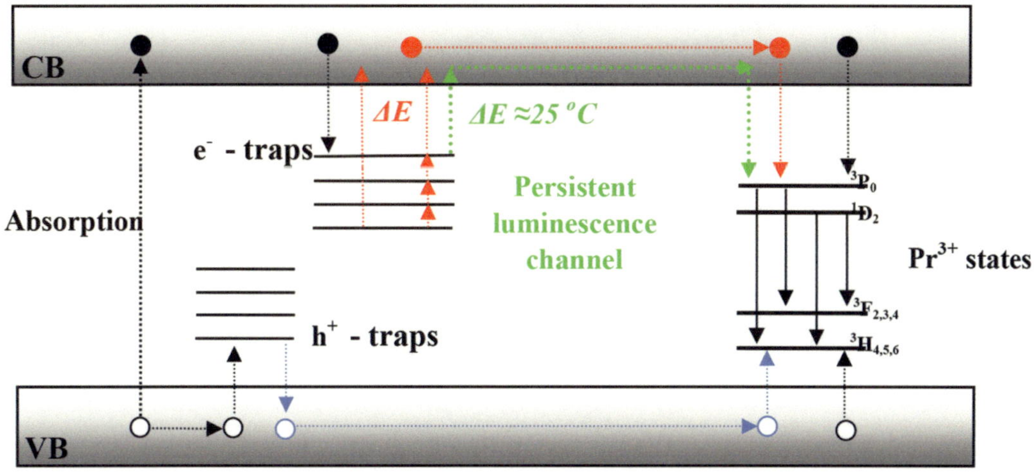

Figure 4. Schematic showing persistent luminescence mechanism for $GdTaO_4:Pr^{3+}$ [2].

The process begins by exciting the electrons to the conduction band, from where they are trapped by the electron-trapping centers [10, 14, 15], by exerting a coulombic force [7] to them. Those trapped by centers with energy levels that are equivalent to thermal energy corresponding to room temperature (shallow traps), will be detrapped and migrated to the luminescent center, via the conduction band [19]. The detrapping and migration process may last up to several minutes or hours, resulting in a phosphor displaying persistent luminescence [14–18]. The decay curve (**Figure 3**) is a measure of how the luminescence intensity resulting from the migrating electrons, changes in time. As observed, in time the luminescence intensity decreases, and this is attributed to electrons being depleted from the shallow electron traps [19].

4. Thermal-stimulated luminescence

The electrons trapped (**Figure 4**) within deeper electron-trapping centers may be thermally stimulated back to the conduction band by temperatures higher than room temperature [20]. Such luminescence is presented as a function of temperature (**Figure 5**) [21]. The resulting glow curve (**Figure 5**) is then used to approximate depth of the electron-trapping centers. Different methods may be used to calculate the depth of the electron, ranging from initial rise, Chen's peak geometry, isothermal analysis, variable heating rate, and computerized glow curve deconvolution (CGCD), among many methods [20].

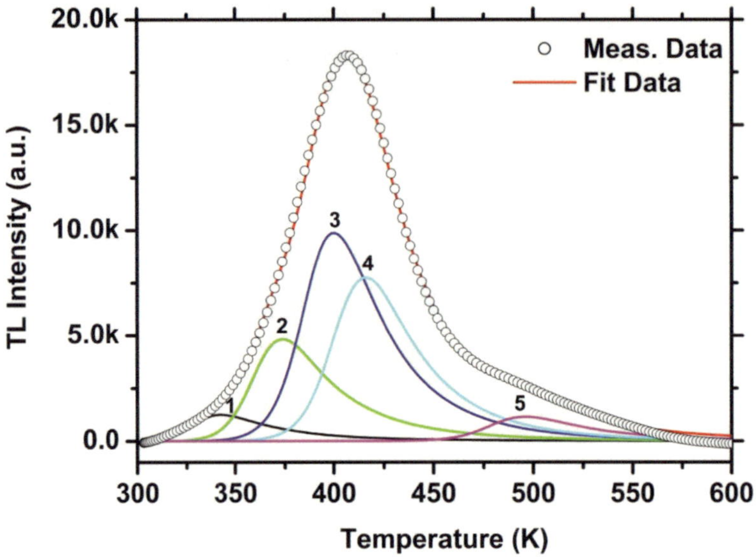

Figure 5. A deconvoluted glow-curve of $ZnTa_2O_6:Pr^{3+}$ [19].

The glow curve of $ZnTa_2O_6:Pr^{3+}$ (**Figure 5**) was deconvoluted into five thermal peaks using CGCD method, guided by the T_m–T_{stop} measurements (not shown). Thermal peaks correspond to prominent electron-trapping centers with very close energy distribution within the forbidden region [22, 23]. The deconvolution was performed by the CGCD from a software package (TLAnal) developed by Chung et al. [24]. The general-order kinetics-related functions were used to compute for the activation energy (Eq. (2)), frequency factor (Eq. (3)) and the concentration of the electrons trapped within electron-trapping centers (Eq. (4)). where I_M and T_M are the TL intensity and temperature (K) at the glow peak maximum, respectively, E is the activation energy (eV), k is the Boltzmann constant, β is the heating rate, n_o is the concentration of the trapped electrons, and b is the kinetic parameter. With the above model, the depth of the electron-trapping centers can be determined along with the quantity of electrons that were trapped [20–23].

$$I(T) = I_M b \frac{b}{b-1} exp(\frac{E}{kT} \times \frac{T - T_M}{T_M})[1 + (b-1)\frac{2kT_M}{E} + (b-1)(1 - \frac{2kT_M}{E}) \times (\frac{T^2}{T_M^2}) exp(\frac{T - T_M}{T_M}\frac{E}{kT})]^{\frac{-b}{b-1}} \quad (2)$$

$$s = \frac{\beta E}{kT_M^2(1 + \frac{2kT_M(b-1)}{E})} exp(\frac{E}{kT_M}) \quad (3)$$

http://dx.doi.org/10.5772/65050

$$I(T) = sn_o exp\left(-\frac{E}{kT}\right)\left[1 + \frac{s(b-1)}{\beta}\int_{T_0}^{T} exp\left(-\frac{E}{kT}\right) dT\right]^{-\frac{b}{b-1}} \quad (4)$$

5. Energy transfer upconversion

In order to optimize the luminescence efficiency of a phosphor, several methods are adopted, which include energy transfer [2, 25], charge compensation [2], to mention a few. The earlier involves the energy transfer between two dopants: the activator (accepter) and the sensitizer (donor). For the process to begin, an interaction between the two ions is required, which can either be exchange interaction, radiation reabsorption (resonant non-radiative energy transfer) or magnetic-multipole interaction [25]. There are several types of multipolar interactions involved in the energy transfer, such as dipole-dipole (d-d), dipole-quadrupole (d-q), and quadrupole-quadrupole (q-q) interactions. The process where the energy transfer occurs as radiation reabsorption is shown in **Figure 6**, where the excited donor, emits at the region where the acceptor is excited, resulting in an enhanced luminescence emission of the acceptor [26].

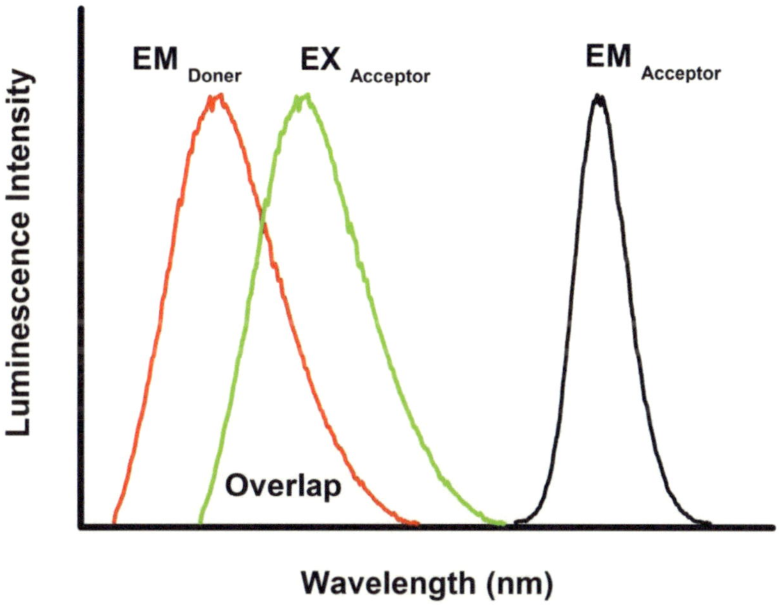

Figure 6. Schematic showing the overlap between the donor emission and the acceptor excitation.

Figure 7 shows the spectra of SiO_2:Er^{3+} and SiO_2:Er^{3+},Yb^{3+}. The emission of SiO_2:Er^{3+} is attributed to radiative electronic transitions of Er^{3+}, which has luminescent emission peaks at 432 from $^4F_{5/2}$ to $^4I_{15/2}$, 486 from $^4F_{5/2}$ to $^4I_{15/2}$, 542 from $^4F_{5/2}$ to $^4I_{15/2}$, 611 and 708 nm from different meta-states of $^4F_{5/2}$ to $^4I_{15/2}$, as shown in **Figure 8** [27].

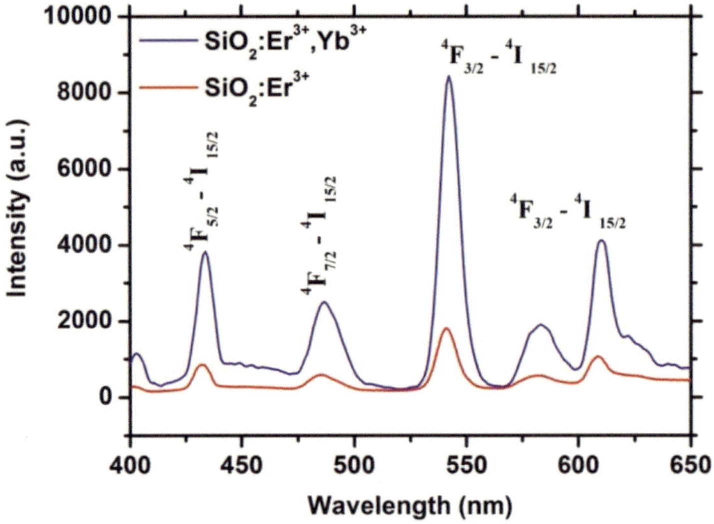

Figure 7. Luminescence from different levels of Er^{3+} energy states in SiO_2:Er^{3+} and SiO_2:Er^{3+},Yb^{3+} that were excited at 980 nm to achieve upconversion.

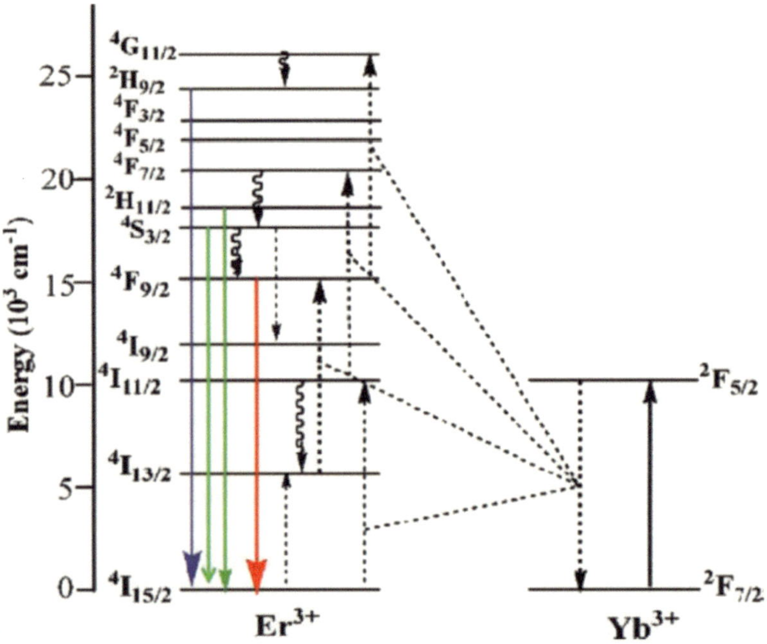

To

Figure 8. Jablonski energy diagram illustrating energy transfer from Er^{3+} to Yb^{3+} [28].

The luminescence of SiO_2:Er^{3+} was achieved upon exciting the sample with a 980 nm LED laser, which excited the electrons from $^4I_{15/2}$ to the nearest $^4I_{11/2}$ metastate or other, metastates, as illustrated in **Figure 8**. Eventually all the small photon energies are combined to excite the electrons in a stair like process until the electrons are excited up-to the highest metastate of Er^{3+} [27]. The electrons positioned at the higher energies may de-excite directly to $^4I_{15/2}$, which is the ground state, and emit photons with energies > that of the exciting source [28]. This process is referred to as upconversion (UC) luminescence [27]. Which is different from the case, where a sample is excited with high-energy photons (ultraviolet light) and yields lower-energy photons (visible, near infrared or infrared emission), as we have discussed earlier.

$SiO_2:Er^{3+}$ was co-doped with Yb^{3+} that acts as a sensitizer, in order to enhance the luminescence emission intensity of the phosphor. When the 980 nm laser is used to pump energy to $SiO_2:Er^{3+},Yb^{3+}$, the electrons are excited from $^4I_{15/2}$ to $^4I_{11/2}$ metastates of Er^{3+}, and from $^2F_{7/2}$ to $^2F_{5/2}$ metastates of Yb^{3+} (**Figure 8**). The spectral overlap between the two ions results in the emission of Yb^{3+} that is faster than that of Er^{3+}, being absorbed, therefore transferring energy to Er^{3+} via the resonant non-radiative energy transfer channel [27]. This results in the emission of Yb^{3+} from $^2F^{5/2}$ to $^2F_{7/2}$ transition being reabsorbed to enhance the electron excitation from $^4I_{15/2}$ to $^4I_{11/2}$, $^4I_{13/2}$ to $^4F_{9/2}$ [28] and the rest of the transitions as shown in **Figure 8**.

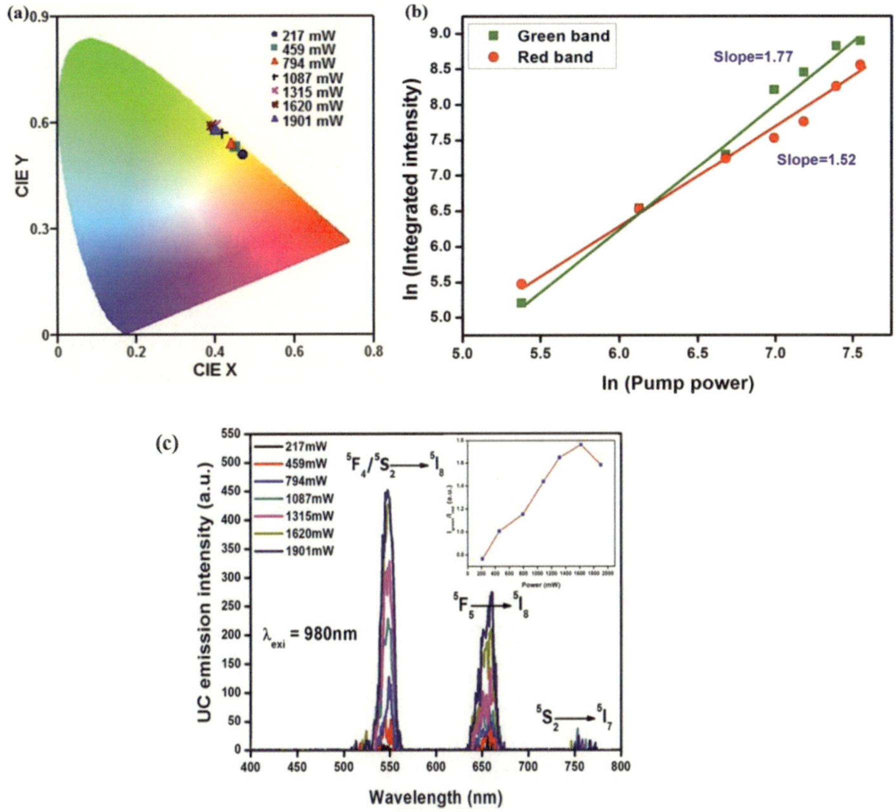

Figure 9. (a) UC emission spectra of 1 mol% Ho^{3+}-doped TZO glass on increasing excitation power (inset-green to red emission intensity ratio plot), (b) logarithmic dependence of the pump power versus integrated UC intensity, of 1 mol % Ho^{3+}-doped TZO glass, and (c) emission spectra of TZO glass doped with Ho^{3+} [30].

It is possible to determine the number of pump photons (n) involved in the UC emission process. For this, it is necessary to measure the variation of the pump power (P) as a function of the UC emission intensity (I) which is related by the following equation [29]

$$I \propto P^n \tag{5}$$

The UC emission spectra of a Ho^{3+}-doped TeO_2–ZnO (TZO) glass recorded upon a 980 nm excitation wavelength in the 400–800 nm range at different excitation powers are shown in **Figure 9a**. Three UC emission bands were observed around 547, 660, and 760 nm in the green, red, and NIR regions, assigned to $^5F_4/^5S_2 \rightarrow {}^5I_8$, $^5F_5 \rightarrow {}^5I_8$, and $^5S_2 \rightarrow {}^5I_7$ optical transitions of Ho^{3+} ion. These transitions were aided by multiphoton excitation, and the increased intensity came about as a result of increasing the excitation power [30].

The relative intensity ratio of the green to the red emission bands changed corresponding to the excitation power as shown in the inset of **Figure 9a**. This variation in intensity was attributed to the change in their excited energy level population. **Figure 9b** shows the ln-ln plot of power versus UC emission intensity for the green and red emission bands of the Ho^{3+} activated TZO glass. Linear fittings of the experimental data resulted into slopes with values of 1.77 and 1.52 for the emission bands observed through the $^5F_4/^5S_2 \rightarrow {}^5I_8$ and $^5F_5 \rightarrow {}^5I_8$ transitions, respectively (**Figure 9c**). Thus, a two pump photon process is responsible for the UC emission from the mentioned glass system.

6. Cathodoluminescence

So far, we have mainly focused on emission that results from photon excitation and also from thermal stimulation of pre-excited electrons. In true sense, luminescence can be generated by exciting electrons via photoluminescence and cathodoluminescence and may be stimulated mechanically or thermally. Cathodoluminescence results from exciting a phosphor by an electron beam and photoluminescence by a photon beam [25]. Several properties of a phosphor may be explored by exciting it with an electron beam. However, in this chapter, we will focus on mapping as method of determining the homogeneity of the luminescence center [31], and luminescence degradation as a method of determining the chemical stability of a phosphor [2].

A commercial phosphor ($CaS:Eu^{2+}$) was exposed to the electron beam irradiation, which was accelerated using an energy of 20 keV, in 30 Pa vacuum pressure. A CL emission spectrum was obtained (**Figure 10a**), showing a broad red emission peak that is positioned at 650 nm. The peak corresponds to the radiative relaxation of the electrons from the $4f^6\,5d^1$ (t_{2g}) to the $4f^7$ ($^8S_{7/2}$) of Eu^{2+}. A CL map (**Figure 10b**) was obtained by collecting several spectra from a 44 × 32 μm field of view, at the same time as the backscattered electron (BSE) image (**Figure 11**) was obtained. The CL map comes with a scale bar indicating the intensity changes of the luminescence emission at different points along the scanned area. More luminescence intensity came from the smaller and spherical-shaped particles, than from the bigger and octahedral-shaped particles. This is an indication that the dopant (Eu^{2+}) is better incorporated in the smaller

http://dx.doi.org/10.5772/65050

particles and not homogenously incorporated throughout the particles of the phosphor powder. Along with the BSE image (**Figure 11**), the X-ray maps (**Figure 11**) were obtained, confirming the presence Ca and S from CaS, and O attributed to a secondary phase ($CaSO_4$) [31].

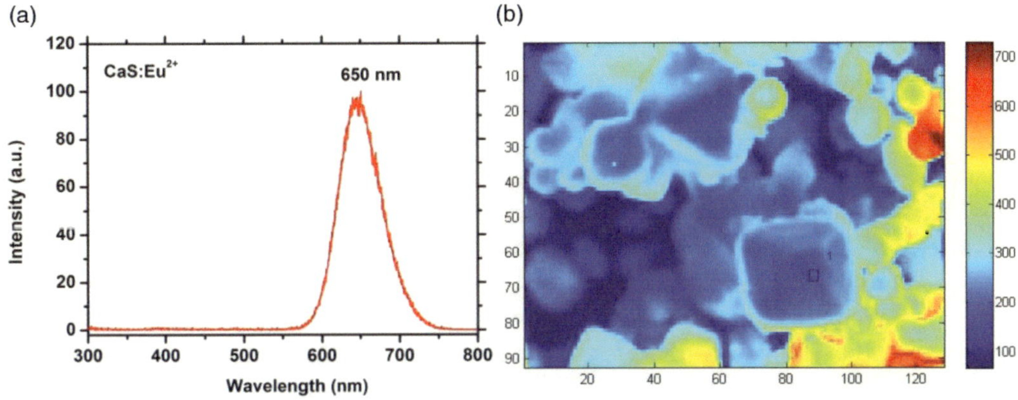

Figure 10. (a) CL spectrum and (b) map of CaS:Eu^{2+} [31].

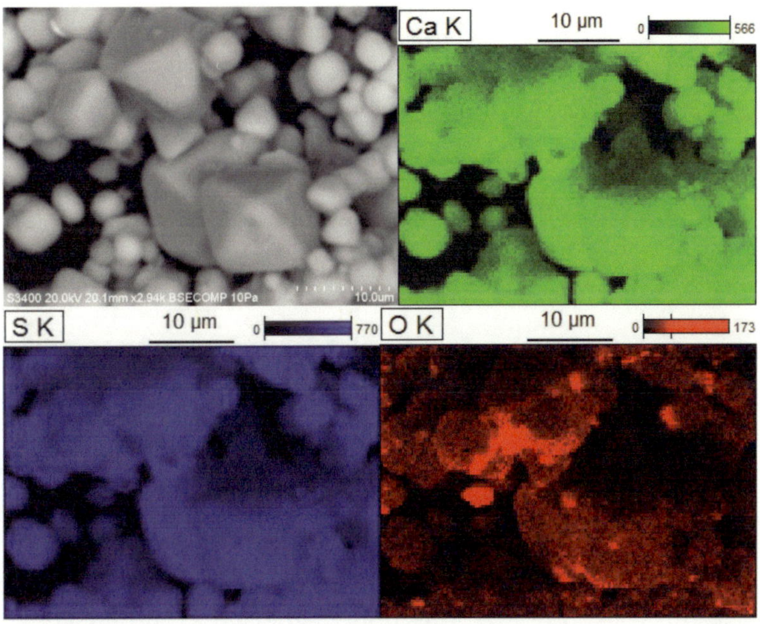

Figure 11. The BSE image and X-ray maps of CaS:Eu^{2+} commercial phosphor [31].

The surface chemical stability is an important parameter for phosphors that are projected for industrial purposes, such as manufacturing of the television and mobile phone display screens, to mention a few. The adopted laboratory procedure to investigate the chemical stability of a particular compound is to expose its surface to harsh conditions and then monitor CL degradation. Such is achieved by accelerating a prolonged electron beam on the surface of a sample. $CaTiO_3:Pr^{3+}$ was subjected to a prolonged electron beam irradiation *in-situ*, using an Auger electron spectroscopy at 1×10^{-6} Torr O_2, which was oxygen backfilled [2].

The resulting effects may lead to a completely degraded CL intensity, if the sample surface is not chemically stable [2]. The CL spectra of $CaTiO_3:Pr^{3+}$ (**Figure 12a**) before and after degradation are presented, which show that the luminescence intensity degraded by approximately 50%.

During the degradation process, the variation of the chemical species on the surface was monitored and plotted as the Auger peak to peak height (APPH) profile (**Figure 12b**). The introduction of O_2 in the system led to a fast oxidation of the surface of the sample, which is observed from 0 to 500 C cm^{-2} electron doses (**Figure 12b**). This is attributed to the reactive O^- species that attacked the surface of the phosphor, which was generated by the electron beam interaction with either the O_2 or H_2O or both inside the chamber, according to the ESSCR mechanism [2, 32, 33]. As a result of the electron-stimulated surface chemical reactions at the interface of the new forming surface and the real surface, a defect grows, which is responsible for the quenched luminescence intensity [2].

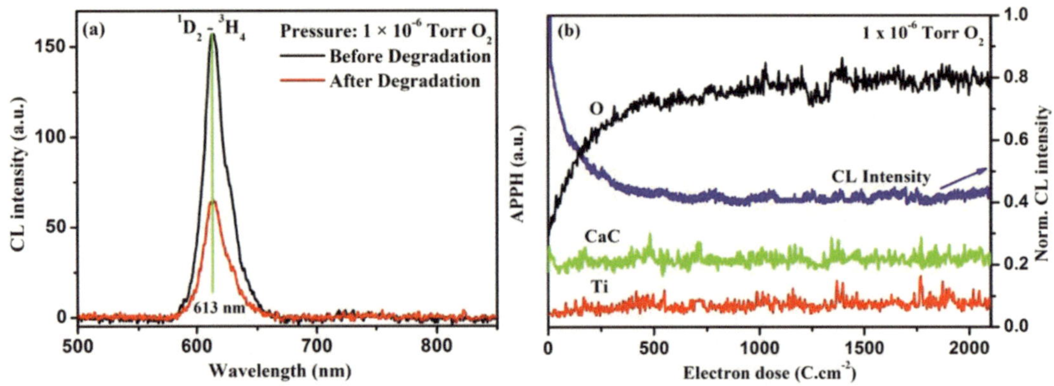

Figure 12. (a) CL spectra and (b) APPH profile of $CaTiO_3:Pr^{3+}$ irradiated with a prolonged electron beam [2].

After having probed the sample with an electron beam, X-ray photoelectron spectroscopy (XPS) was used to investigate the chemical changes that took place during the process. **Figure 13**, shows fitted XPS spectra, indicating the chemical state of the surface of the sample. The oxygen peak O 1s consists of five peaks prior degradation, contributed by peaks at 529.2 eV binding energy (BE) from the matrix ($CaTiO_3$), 530.4 eV BE from TiO_2 and Ti_2O_3, 531.6 eV BE from the OH^- group bonded species. The last two peaks at 532.7 and 534.2 eV correspond to the Si-O-Si species from the silica crucible that was used to prepare the material, and from the chemisorbed species, respectively. The O 1s XPS peak of the sample that was subjected to electron beam degradation, was deconvoluted into seven peaks, with the additional peaks at

528.4 from the CaO formed on the surface as a result of surface oxidation, and at 529.8 eV BE as a result of CaO_x sub-oxide formation [2].

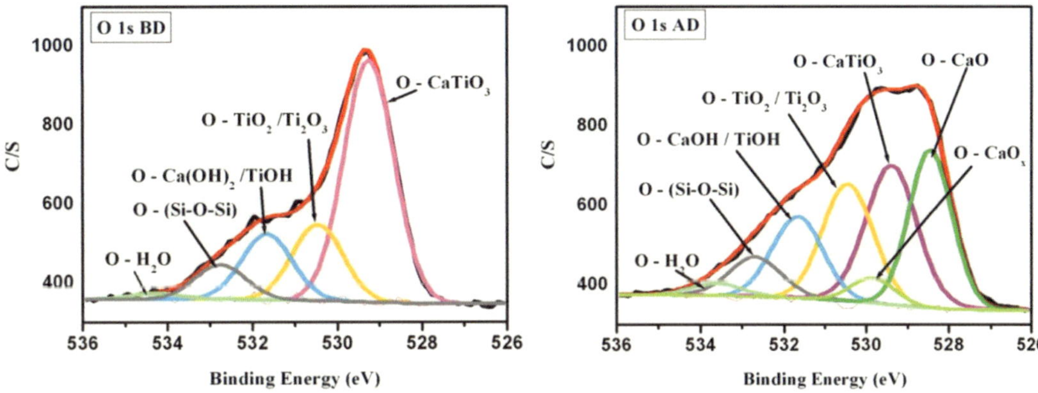

Figure 13. Spectra of O 1s before (O 1s BD) and after (O 1s AD) degradation [2].

7. Applications

There are several applications of luminescent materials; however, we will focus only on a few, which are of modern technological innovation. Such as the persistent luminescence for home lighting, luminescence for biological imaging, temperature sensing, white phosphor converted light-emitting diodes, and phosphors for a television display.

7.1. Home lighting

Persistence luminescence offers an alternative lighting that is cost-effective and energy conservative. It offers the possibility of having a light bulb that gives out light without any

Figure 14. Light bulbs made of luminescent materials [34].

electrical connection (**Figure 14**). Such a bulb will be made of a persistent luminescent material, can be placed outside during the day to absorb the sunlight, and then placed back inside the house, where it will continue to glow in the absence of the excitation source [19].

7.2. Biological imaging

Photoluminescence mapping enables tracking of drug delivery to assess the effectiveness of the drug release. The photoluminescence functionalized drugs are employed to detect and quantify a particular disease [35, 36]. A similar approach was adopted using zinc gallate doped with chromium ions, to map the path of the luminous drugs through the gastrointestinal tract, after an oral consumption [35]. An LED was used to excite the luminous drug that was fed to a mouse (**Figure 15**) [35], which glows in the spots where the luminous drug was situated.

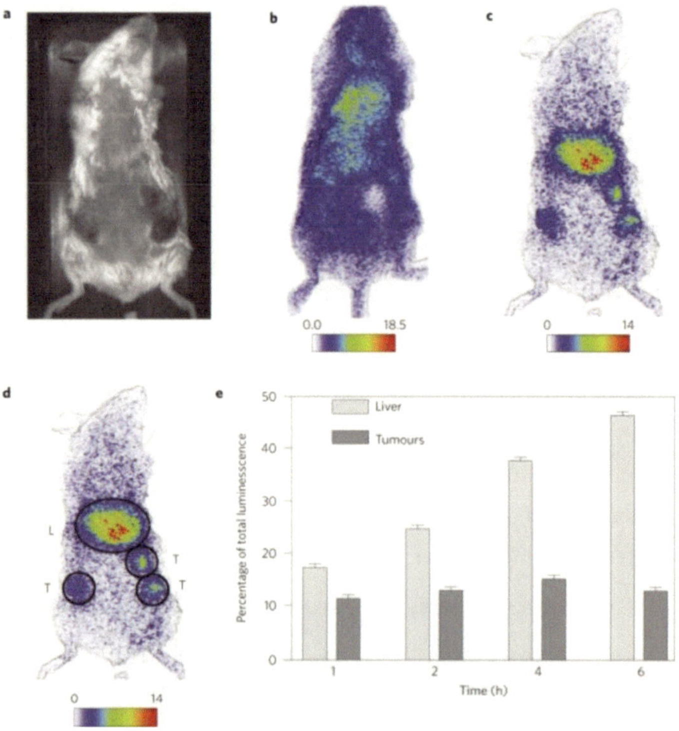

Figure 15. Schematic showing mapped luminescence of zinc gallate and poly-ethylene glycol composite [35].

7.3. Temperature sensing

Distant objects can best be measured using optical temperature sensors. Upconversion sensors modified from inorganic materials doped with rare-earth ions are proving to be better candidates in this regards [36]. Upconversion-modified sensors make use of the luminescence changes from two close metastates of a luminescence center to derive the optical temperature changes [37]. A total of 2000 cm^{-1} is a maximum required energy difference in between these metastates, for the sensor to be effective [38]. The fluorescent intensity ratio (FIR) is then used to approximate the temperature that corresponds to the fluorescence of the thermally coupled levels [36, 39]. The fluorescence intensity ratio of thermally coupled transitions can be presented as (Eq. (6)): where I_{525} and I_{547} are the integrated intensities corresponding to the $^2H_{11/2} \rightarrow {}^4I_{15/2}$ and $^4S_{3/2} \rightarrow {}^4I_{15/2}$ transitions, respectively, B is the pre-exponential constant, ΔE is the energy difference between the $^2H_{11/2}$ and $^4S_{3/2}$ levels, k is Boltzmann's constant, and T is absolute temperature [40, 41].

$$FIR = \frac{I_{525}}{I_{547}} = B \exp\left(\frac{-\Delta E}{kT}\right) \tag{6}$$

The optical temperature sensing behavior of Er^{3+}–Yb^{3+} co-doped $SrWO_4$ phosphor upon 980 nm excitation in the 510–570 nm range on increasing the temperature up-to 518 K have been recorded keeping the power constant as shown in **Figure 16** [42]. The integrated intensity of the UC emission bands around 525 and 547 nm (assigned through the $^2H_{11/2} \rightarrow {}^4I_{15/2}$ and $^4S_{3/2} \rightarrow {}^4I_{15/2}$ transitions of the Er^{3+} ion, respectively) varied when increasing the temperature of the sample. At room temperature (300 K), the intensity of both the transitions was nearly equal, whereas at 518 K, the intensity corresponding to the $^4S_{3/2} \rightarrow {}^4I_{15/2}$ transition was more reduced than that of the $^2H_{11/2} \rightarrow {}^4I_{15/2}$ transition. The observed change of the two transitions is plotted as the intensity ratio (I_{525nm}/I_{547nm}), which is a function of temperature (inset of **Figure 16**) [31].

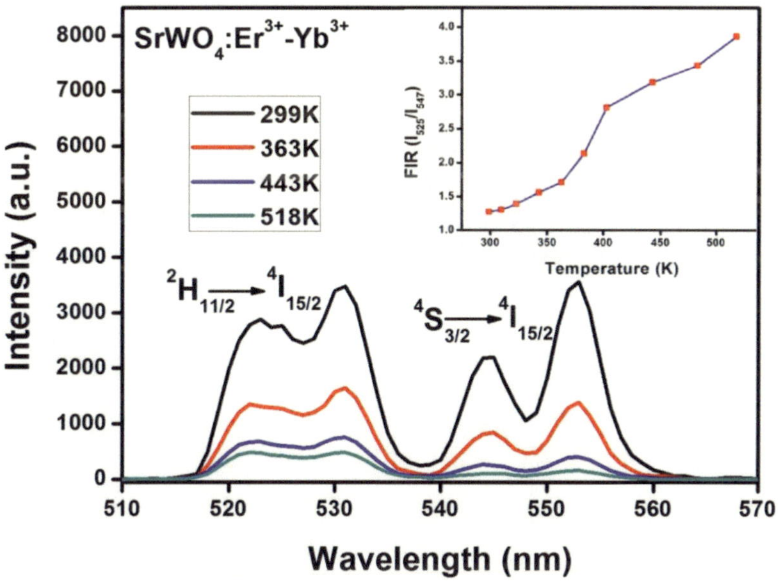

Figure 16. Green UC emission spectra of $SrWO_4$:Er^{3+}–Yb^{3+} phosphor on increasing temperatures and the variation of FIR as function of absolute temperature (inset) [42].

In accordance with Eq. (6), the ln(I_{525}/I_{547}) is plotted against the inverse absolute temperature as shown in **Figure 17**. The experimental data were linearly fitted and gave a slope equal to 866. This resulted into the energy difference value ΔE of the two thermally coupled levels of about 600 cm^{-1}.

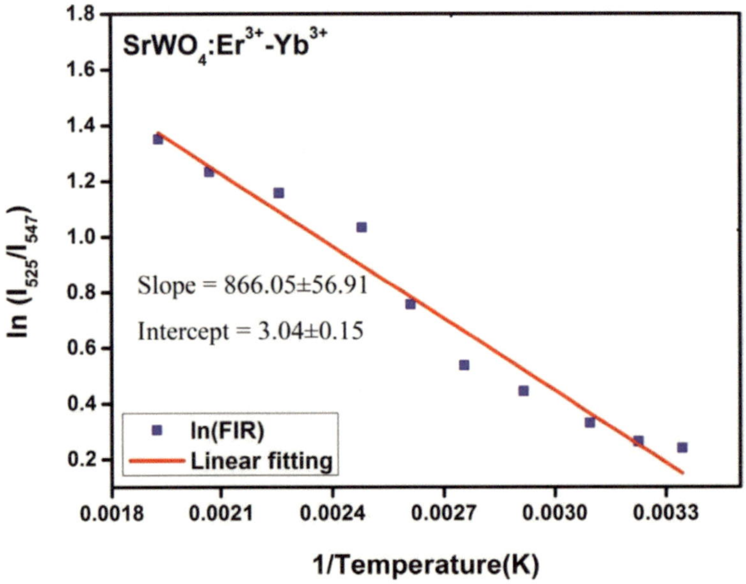

Figure 17. Plot of logarithmic of FIR versus the inverse absolute temperature [42].

7.4. Wp-LEDs

LEDs are a form of solid-state lighting technology that relies on the inorganic compounds to convert electricity to light. Recently, a blue-emitting light diode was achieved using InGaN by a team of scientists led by Isamu Akasaki, Hiroshi Amano and Shuji Nakamura [37]. An LED displaying an efficiently white emission compared to the conventional incandescent lighting

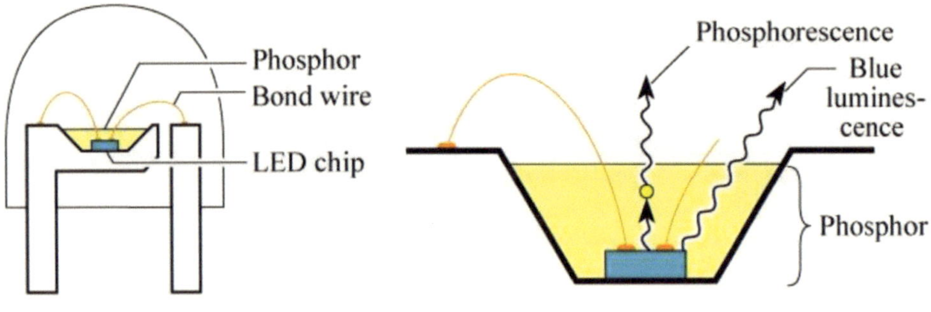

Figure 18. Schematic of phosphor converted white LED [38].

was later achieved by converting the blue LED to white LED by coupling the diode with a yellow light-emitting phosphor (**Figure 18** [38]), $Y_3Al_5O_{12}:Ce^{3+}$ [39].

7.5. Television display

One of the important applications of phosphors is in the display technology. We rely on television to watch live news and events, in full color. We use mobile phones to capture moments in color and for live video calling. All these are made possible by phosphors for display [40]. There are several technologies that are used to achieve different types of television displays, ranging from cathode ray tubes [41], liquid crystal display [42], field emission display [43], to plasma display panels [44], to mention a few. For an example, the plasma display panels (**Figure 19**) [45] have pixels that consist of small gas-discharge cells [37]. The gas is made of Xe-Ne plasma, which excites the phosphor with a vacuum ultra-violet source of 147 and 172 nm [44, 46].

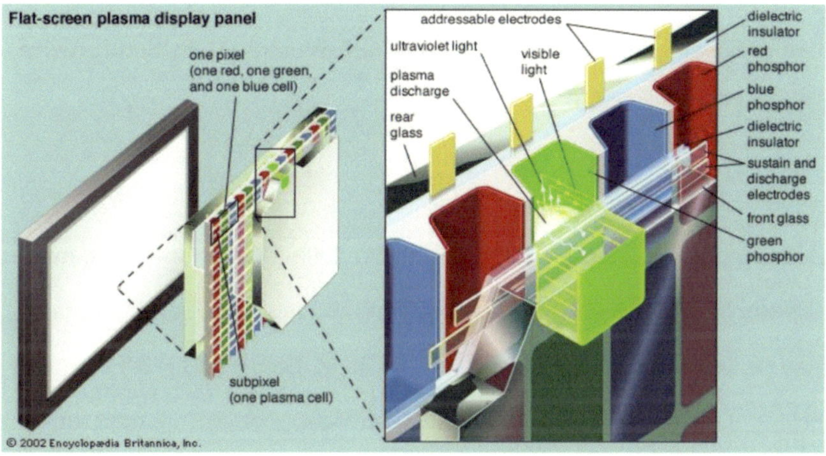

Figure 19. A schematic showing pixels of a plasma display panel [45].

8. Conclusions

The concept of luminescence, though not fully understood in the light of persistent luminescence, it has advanced the human life. The mentioned applications are of great significance in the way we live today and are only a handful of the broader applications in the technological applications. Some of which are not listed. The luminescence expected to contribute vastly in different fields, such as home lighting using electricity free light bulbs and in improving the efficiency of the solar cells.

Acknowledgements

We sincerely appreciate the funding from the South African National Research Foundation and the Erasmus Mundus EU-SATURN. Many thanks to Raphael Nyenge, who allowed me to do experiments on his sample, and the Lumilab in Belgium for opening doors for me to carry out research work. We extend gratitude to the University of the Free State, where most of the experiments were performed.

Author details

Luyanda L. Noto[1], Hendrik C. Swart[2], Bakang M. Mothudi[1], Pontsho S. Mbule[1] and Mokhotjwa S. Dhlamini[1*]

*Address all correspondence to: dhlamms@unisa.ac.za

1 Department of Physics, University of South Africa, Pretoria, South Africa

2 Department of Physics, University of the Free State, Bloemfontein, South Africa

References

[1] D.D. Busch, Nikon D200 Digital Field Guide, 2006, Wiley Publishers, Indianapolis, IN.

[2] L.L. Noto, M.Sc. Thesis, 2011, University of the Free State, South Africa.

[3] F. Stavale, N. Nilius, H.J. Freund, J. Phys. Chem. Lett. 4 (2013) 3972.

[4] I. Pelant, J. Valenta, Luminescence Spectroscopy of Semiconductor, 2012, Oxford University Press, Oxford, UK.

[5] R.C. Ropp, Luminescence and the Solid State, 2012, Elsevier Science Publishers B.V., Amsterdam, Netherlands.

[6] J. Breithaupt, New Understanding Physics for Advanced Level, 2000, 4th ed., Nelson Thornes Ltd., London, UK.

[7] C.T. Sah, Fundamentals of Solid-state Electronics: Solution Manual, 1996, World Scientific Publishing Co. Pte Ltd., London.

[8] L.L. Noto, M.L. Chithambo, O.M. Ntwaeaborwa, H.C. Swart, Powder Technol. 247 (2013) 147.

[9] J.R. Lakowicz, Principles of Fluorescence Spectroscopy, 2006, 3rd ed., Springer Publishers, New York, NY.

[10] J. Holsa, The Electrochemical Society Interface meeting, Winter 2009, http://www.electrochem.org/dl/interface/wtr/wtr09/wtr09_p042-045.pdf [09 March 2016].

[11] T.G. Chasteen, S. Houston, Relaxation mechanism for excited state molecules http://www.shsu.edu/chm_tgc/chemilumdir/JABLONSKI.html [09 March 2016].

[12] A.J. Pesce, C. Rosen, T.L. Pasby, Fluorescence Spectroscopy—an Introduction for Biology and Medicine, 1971, Marcel Dekker Inc, New York, NY, p 34.

[13] P.W. Hawkes, Advances in Electronics and Electron Physics, 1990, vol. 79, Academic Press Inc, Cambridge, UK, p285.

[14] L.L. Noto, S.S. Pitale, M.A. Gusowki, O.M. Ntwaeaborwa, J.J. Terblans, H.C. Swart, J. Lumin. 145 (2014) 907.

[15] P.T. Diallo, P. Boutinaud, R. Mahiou, J.C. Cousseins, Phys. Stat. Sol. (a) 160 (1997), 255.

[16] B.J. Nyman, M.E. Bjorketun, G. Wahnstrom, J. Sol. Stat. Ionics. 189 (2011), 19.

[17] P.J. Deren, R. Pazik, W. Strek, P. Bautinaud, R. Mahiou, J. Alloy Compd. 451 (2008), 595.

[18] E.J. Popovici, M. Nazarov, L. Muresan, D.Y. Noh, E. Bica, M. Morar, I. Arellano, E. Indrea, Phys. Proc. 2 (2009), 185.

[19] L.L. Noto, PhD Thesis, 2014, University of the Free State, South Africa.

[20] V. Pagonis, G. Kitis, C. Furetta, Numerical and Practical Exercises in Thermoluminescence, 2006, Springer and Business Media Inc., New York, NY.

[21] L.L. Noto, O.M. Ntwaeaborwa, M.Y.A. Yagoub, H.C. Swart, Mater. Res. Bull. 70 (2015) 545.

[22] C. Furetta, Handbook of Thermoluminescence, 2003, World Scientific Publishing, Singapore.

[23] S.W.S McKeever, Thermoluminescence of Solids, 1985, Cambridge University Press, New York, NY.

[24] K.S. Chung, H.S. Choe, J.I. Lee, J.L. Kim, S.Y. Chang, Radiat. Prot. Dosim. 115 (2005) 1.

[25] M.S. Dhlamini, PhD Thesis, 2008, University of the Free State, South Africa.

[26] J. Chen, J.X. Zhao, Sensors 12 (2012) 2414. http://krauthammerlab.med.yale.edu/imagefinder/ImageDownloadService.svc?articleid=3438444&file=fendo-03-00100-g001&size=SMALL [18 May 2016].

[27] F. Artizzu, F. Quochi, L. Marchiò, E. Sessini, M. Saba, A. Serpe, A. Mura, M.L. Mercuri, G. Bongiovanni, P. Deplano, J. Chem. Lett. 4 (2013) 3062.

[28] A. Pandey, V. K. Rai, V. Kumar, V. Kumar, H. C. Swart, Sens. Actuators B, 209 (2015) 352.

[29] P. Anurag, C. Hendrik. Swart, J. Lumin., 169 (2016) 93.

[30] R.L. Nyenge, PhD thesis, 2015, University of the Free State, South Africa.

[31] H.C. Swart, J.S. Sebastian, T.A. Trottier, S.L. Jones, P.H. Holloway, J. Vac. Sci. Technol. 14(3) (1996) 1697.

[32] P. Boutinaud, R. Mahiou, E. Cavalli, M. Bettinelli, Chem. Phys. Lett. 418 (2006) 185.

[33] http://blocksandbricks.co.za/blocks-and-bricks/meet-the-glow-in-the-dark-brick/ [28 April 2014].

[34] T. Maldiney, A. Bessiere, J. Seguin, E. Teston, S.K. Sharma, B. Viana, A.J.J. Bos, P. Dorenbos, M. Bessodes, D. Gourier, D. Scherman, C. Richard, Nat. Mater. 13 (2014) 418.

[35] V. K. Rai, Appl. Phys. B, 88 (2007) 297–303.

[36] Y. Shen, X. Wang, H. He, Y. Lin, C.-W. Nan, Compos. Sci. Technol., 72 (2012) 1008–1011.

[37] S.F. Leon-Luis, U.R. Rodriguez-Mendoza, E. Lalla, V. Lavin, Sens. Actuators B, 158 (2011) 208.

[38] S. A. Wade, S. F. Collins, G. W. Baxter, J. Appl. Phys., 94 (2003) 4743.

[39] S. K. Singh, K. Kumar, S. B. Rai, Sens. Actuators A, 149 (2009) 16.

[40] V. K. Rai, A. Pandey, R. Dey, J. Appl. Phys., 113 (2013) 083104.

[41] Anurag Pandey, Vineet Kumar Rai, Vijay Kumar, Vinod Kumar, H. C. Swart, Sens. Actuators B: Chem., 209 (2015) 352–358.

[42] Z.J. Li, Y.J. Zhang, H.W Zhang, H.X. Fu, J. Micro. Meso. Mater., 176 (2013) 48.

[43] NobelPrize, http://www.nobelprize.org/nobel_prizes/physics/laureates/2014/press.html (20 March 2016).

[44] https://www.ecse.rpi.edu/~schubert/Light-Emitting-Diodes-dot-org/chap21/F21-07%20Nichia%20wh%20LED%20structu.jpg (20 March 2016).

[45] X. Zhang, W. Liu, G.Z. Wei, D. Banerjee, Z. Hu, J. Li, J. Am. Chem. Soc., 136 (40) (2014) 14230.

[46] W.M. Yen, S. Shionoya, H. Yamamoto, Phosphor Hand Book, 2nd ed., 2007, CRC Press, Boca Raton, FL.

Chapter 3

Luminescence in Rare Earth Ion-Doped Oxide Compounds

Carlos Ruvalcaba Cornejo

Additional information is available at the end of the chapter

http://dx.doi.org/10.5772/65185

Abstract

This chapter focuses on the study on luminescent materials, which consist of oxide compounds (host material) and rare earth ions (as the activator) in the valence state, mostly 3+. The first part begins with a background study about the luminescence phenomenon, its stages, and the configurational coordinate diagram. Then, we review the notation often used for rare earth ions, such as the term symbols associated with the energy levels of the ground state. Lastly on the first part, we establish a relationship between the configurational coordinate model and the electronic transitions of such ions. The second part shows the photoluminescence results in some oxide compound materials doped with rare earth ions that have been reported in research articles with potential applications.

Keywords: rare earth, luminescence, oxides, glasses, crystals

1. Introduction

1.1. Background

Luminescence can be defined as the capability of a body to emit light when exposed to electromagnetic radiation or other means such as energy from an electron, a chemical reaction, and so on. Examples are as follows: when the body is excited by low-energy photons, often ultraviolet radiation (*photoluminescence*), by cathode rays (*cathodoluminescence*), by an electric field strength (*electroluminescence*), by X-rays (*X-rays luminescence*), and so on. The light emitted by a body (a luminescent material) appears in the visible part of the electromagnetic spectrum, but can be in the infrared (IR) or ultraviolet (UV) regions.

A first application of a luminescent material is that the photoluminescence occurs in a fluorescent lamp. Such lamp consists of a glass tube in which a low-pressure mercury discharge generates ultraviolet radiation. This material converts the UV radiation in white light. Visible light emission processes due to thermal radiation are termed *incandescence*, and not luminescence. This distinction explains why the efficiency of conversion of electricity to light is in a fluorescent lamp considerably higher than in an incandescent lamp [1]. **Figure 1** shows an example of this situation.

In the example of **Figure 1(a)**, if a photon of 3.35 eV (≈370 nm of wavelength in ultraviolet part of the electromagnetic spectrum) is absorbed by the phosphor, then it might emit a photon of say 2.48 eV (≈500-nm green light), which is common in Stokes' process [1].

The luminescence of *inorganic* materials is composed of the following processes: (1) absorption and excitation, (2) energy transfer, and (3) emission. Most luminescent materials consist of a so-called *host* material to which certain dopant ions, also termed *activators*, are added. In such cases, the host lattice plays two distinct roles: as a *passive matrix* to define the spatial locations of the activator ions; and as an *active participant* in the luminescence process, exerting its own specific influence on the spectroscopic behavior of the activator. In the latter, it helps to shape the structure of the energy levels of the activator and also introduces vibrations of various energies, the so-called *phonons*, which influence the kinetics of the luminescence phenomena.

Consider the luminescence phenomenon in the simple case as follows. **Figure 2(a)** shows that the exciting radiation is absorbed by the activator ion (A) into a host material. A system of two electronic states, the ground energy state (*g*) and a higher-lying "excited" level (*e*), is schematized in **Figure 2(b)**. The process of excitation, which supplies energy to the host matrix, raises

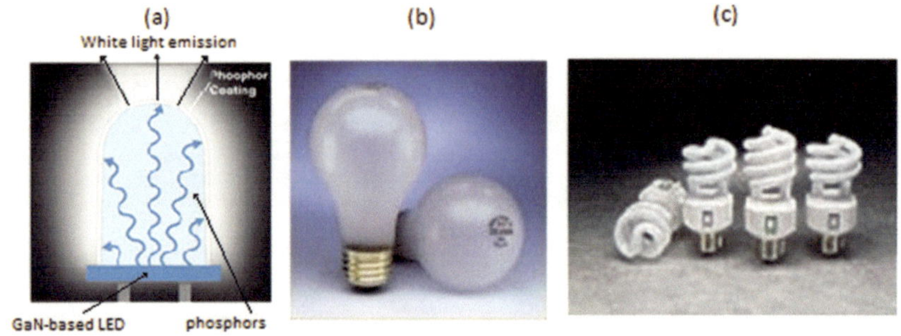

Figure 1. (a) Solid-state lighting: based on GaN blue- (400 nm) or UV- (370–400-nm) light-emitting diodes (LEDs). (b) and (c) white-light lamps: traditional designs.

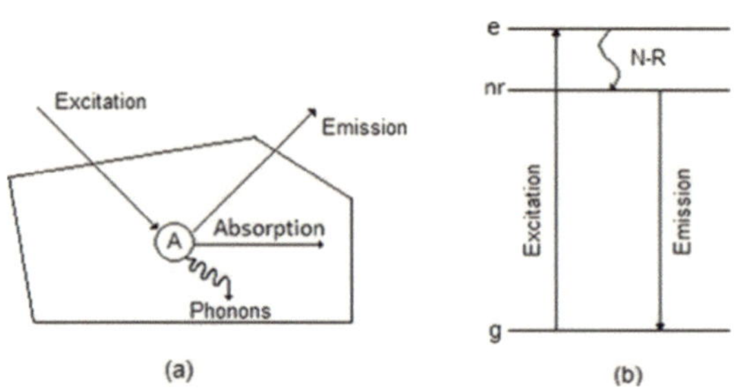

Figure 2. (a) An activator ion A in a passive host lattice and (b) schematic depiction of excitation (absorption) and emission processes in an activator A.

an electron from the g state to the e level. Most of the excitation energy moves to a different lower-lying state, with the remainder being dissipated without emission of light, through the so-called *non-radiative* (N-R) *relaxation*, $e \rightarrow (nr)$. This configuration then relaxes to the ground state by sending off a photon of light, a *radiative emission* $(nr) \rightarrow g$. This completes the process, allowing the system to reach a state of minimum energy.

On the other hand, if energy absorbed by an activator in a crystal (a passive host) is transferred to a second activator of a different kind with the result that luminescence occurs in the second activator, the process is called *sensitized luminescence*. The activator that is responsible for the absorption of energy is called the *sensitizer*, and the activator that luminesces is the *emission center*. For this situation, the next stages occur: the absorption $g^{S} \rightarrow e^{S}$ transition is due to the sensitizer, an *energy transfer* process occurs from the sensitizer S to the center A in its higher-lying "excited" level. The level e^{A} populated by the energy transfer decays non-radiatively to the slightly lower (nr), and finally the radiative emission, $(nr) \rightarrow g^{A}$, occurs.

Sometimes, the host lattice can function as the absorber in a luminescent system, that is, as an active matrix, in which case we refer to *"lattice-sensitized" luminescence*.

1.2. Configurational coordinate diagram (CCD)

For electronic transitions within a solid, it is very common to find bands rather than isolated absorption and emission lines. The configurational coordinate model is a model to explain the *width* of bands in solids, which should correspond to discrete levels such as is depicted next.

Figure 3 shows the configurational coordinate diagram (CCD) for a *single* metal ion (a luminescent center). The ordinate is the energy E of the activator-ligand system, whereas R (the configurational coordinate) is an interaction coordinate in the drawing and represents interatomic distance in a general way that defines the configuration of the ligands.

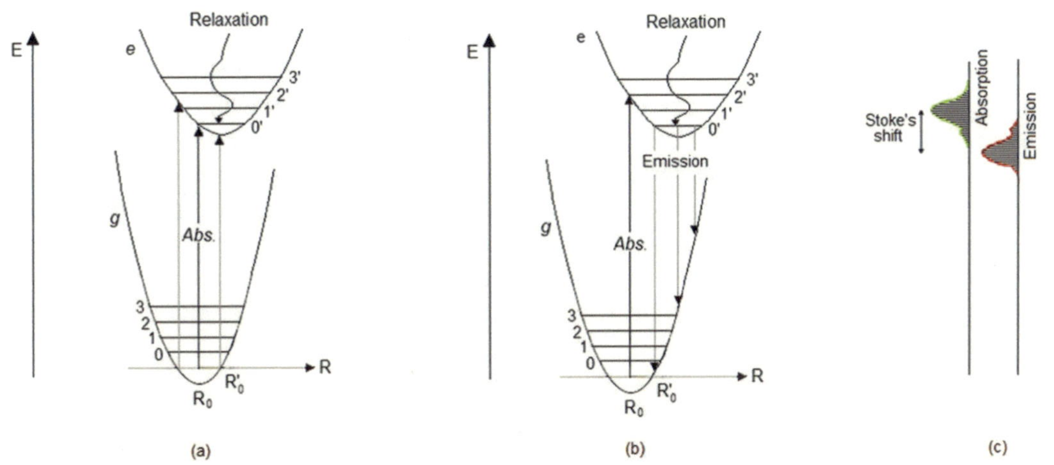

Figure 3. Configurational coordinate diagram showing (a) three vertical rows (absorption transitions) and (b) three vertical rows (emission transitions). (c) Stokes' shift between absorption and emission bands.

As a first approximation, for small displacements the center ion behaves as a harmonic oscillator. Their two electronic states are represented as parabolas. The equilibrium distance between the center ion and the ligand will be represented by a quantity R_0 for the g state, and to the first excited state (e) will be $R_{0'}$. The vibrational levels are shown by horizontal lines for both states (denoted by 0, 1, 2, 3 and 0′, 1′, 2′, 3′ the levels of g and e states, respectively). The absorption and non-radiative relaxation processes are indicated in **Figure 3(a)**. The vertical arrows represent the absorption of a photon (or energy from some other source: electric field, X-rays, neutrons, etc.) which excite a *single* center from the g state to the e state, and is usually referred as an optical absorption transition. Emission takes place as the electronic transition of the center from the lowest vibrational levels of the excited state to the ground state, **Figure 3(b)**.

However, the absorption, relaxation, and emission processes cover the dynamic behavior of all centers (metal ions) present in the solid, which is considered as a collective process. As the vibrations of the host lattice are random, they will affect the position of each center, so they will make different radiative transitions, depending on the value that the coordinate R takes.

If the luminescent material is cooled to 0°K, will only present electronic transitions between vibrational level lower of the ground state and the lowest vibrational level of the excited state, but at higher temperatures the vibrational coupling originates that electronic transitions made by all centers forming *bands* of excitation and emission into the solid in a continuous range of values of the energy [2]. The energy difference between the maximum of the absorption band and that of the emission band is called *Stokes' shift*; see **Figure 3(c)**.

This scheme of CCD assumes that there is an offset between the parabolas of the g and e states. *What does this mean*? In the next section, we focus the study on rare earth (RE) ions. Lastly on this section, the CCD will resume in the case of a crystal doped with a specific type of rare earth ions to show how the formation of bands in the crystal is due to this type of activators.

1.3. Luminescent properties of rare earth ions into a crystalline material

1.3.1. Chemistry of the rare earths

The rare earths (RE) elements are La, Ce, Pr, Nd, Pm, Sm, Eu, Gd, Tb, Dy, Ho, Er, Tm, Yb, and Lu, and can possess up to 14 identical electrons involving the *4f-shell*. All the REs exhibit the +3 valence state. The +4 and +2 valence states are stable mainly for RE ions with completely empty, half-full, or completely filled *f* shells, that is, $[Xe]4f^0$, $[Xe]4f^7$, and $[Xe]4f^{14}$ electron configurations. Examples are as follows: cerium has a stable 4+ valence state with a $4f^0$ electron configuration, and Eu^{2+} is the most stable divalent ion and has a $4f^7$ electron configuration (see **Table 1**). The RE^{3+} ions have the maximum possible numbers of unpaired *f electrons* (up to 7 for Gd^{3+}) and exhibit complicated magnetic behavior due, in general, that the electrons of the unfilled shells provide a net magnetic moment that may be oriented by the use of an *external magnetic field B* [3, 4].

1.3.2. Term symbols

For many-electron systems such as the transition metal, rare earth and actinide ions are considered, is convenient to represent the electron states with *term symbols* [5]. A *term symbol* has the general form $^{2S+1}L_J$, where 2S + 1 is the multiplicity of the term, S is the quantum

http://dx.doi.org/10.5772/65185

Z Element	Neutral atom configuration	$(RE)^{3+}$ configuration	Term $(RE)^{3+}$	$(RE)^{2+}$ configuration	Term $(RE)^{2+}$
57 La	$4f^06s^25d$	$4f^0$	$^2F_{5/2}$		
58 Ce	$4f^16s^25d$	$4f^1$	$^2F_{5/2}$		
59 Pr	$4f^36s^2$	$4f^2$	$^3H_4$	$4f^3$	$^4I_{9/2}$
60 Nd	$4f^46s^2$	$4f^3$	$^4I_{9/2}$		
61 Pm	$4f^56s^2$	$4f^4$	$^5I_4$		
62 Sm	$4f^66s^2$	$4f^5$	$^6H_{5/2}$	$4f^6$	$^7F_0$
63 Eu^{3+}	$4f^76s^2$	$4f^6$	$^7F_0$	$4f^7$	$^8S_{7/2}$
64 Gd	$4f^86s^25d$	$4f^7$	$^8S_{7/2}$		
65 Tb^{3+}	$4f^96s^2$	$4f^8$	$^7F_6$		
66 Dy	$4f^{10}6s^2$	$4f^9$	$^6H_{15/2}$	$4f^{10}$	$^5I_8$
67 Ho	$4f^{11}6s^2$	$4f^{10}$	$^5I_8$	$4f^{11}$	$^4I_{15/2}$
68 Er	$4f^{12}6s^2$	$4f^{11}$	$^4I_{15/2}$	$4f^{12}$	$^3H_6$
69 Tm	$4f^{13}6s^2$	$4f^{12}$	$^3H_6$	$4f^{13}$	$^2F_{7/2}$
70 Yb	$4f^{14}6s^2$	$4f^1$		$4f^{14}$	$^1S_0$
71 Lu	$4f^{14}6s^25d$	$4f^{13}$	$^2F_{7/2}$	$4f^{14}6s^1$	$^2S_{1/2}$

Table 1. Electron configurations of RE ions in the ground state.

number of the total spin angular momentum, *L* represents the quantum number of the total orbital angular moment and is denoted by capital letters as follows:

- *L*: 0 1 2 3 4 5 6,...
- capital letter: *S P D F G H I*....

(It is an extension of the notation for a one-electron atom or ion: $s = 0$, $p = 1$, $d = 2$, etc.). *J* is the quantum number of the total angular momentum with allowed values being $|L–S|$, $L–S+1$, …, $L+S–1$, $L+S$ [6].

The term symbols can be obtained using *Hund's rule*, which are of great help in evaluating the ground state of atomic or ionic systems [7]. They are stated as follows:

1. Terms allowed by the Pauli principle are ordered according to the quantum number *S*. The term of lowest energy level will be one of maximum value of *S* (highest multiplicity).
2. For two or more terms with the same maximum multiplicity that with greatest value of *L* will be the lowest energy level.
3. For configurations consisting of electrons in *a less than half-filled shell*, the ground *multiplet* (a term symbol) has the minimum *J* value, whereas for electron configurations with *more than half-filled shell* the multiplet has the maximum *J* value. That is, the lowest value of *J* is $|L–S|$ and its maximum value is $L + S$.

The term symbols are exemplified next. The Ce^{3+} ion ($4f^1$) contains a single electron into *4f-shell*. Therefore, the quantum numbers: $S = ½ = s$ for the spin of an electron, $L = 3 = l$ (for one *f* electron) and the values $J = j = l–s$ (=5/2), $l + s$ (=7/2). Thus, the term symbols to Ce^{3+} are $^2F_{5/2}$ and $^2F_{7/2}$, and they have the same multiplicity: 2. In agreement with third Hund's rule, $^2F_{5/2}$ term corresponds to the lowest energy level of the ground state.

Other case is to Eu^{2+} ion ($4f^7$), it has seven f electrons, first half-filled *4f-shell*, and the lowest value for total orbital angular momentum is $L = 0$ and the highest value for total spin angular momentum is $S = 7/2$ and hence $J = |L \pm S| = 7/2$. Thus, the maximum multiplicity is 8, and $^8S_{7/2}$ is a single-term symbol corresponding to the ground state.

A third example: Yb^{3+} has the configuration consisting of 13 electrons. Using the *principle of equivalence of electrons and holes* [7], we obtain the term symbols $^2F_{5/2}$ and $^2F_{7/2}$. However, as the electrons are occupying more than half-filled *4f-shell*, now, the lowest energy level will be $^2F_{7/2}$ (see the ground multiplet for Ce^{3+}, Eu^{2+}, and Yb^{3+} ions in **Table 1**).

A point to keep in mind: *Hund's rule is not to be applied to excited states.* Appendix 1, particularly, describes the terms that correspond to the electron configuration $4f^n$, being $n = 1, 2, 3, 4, \ldots, 14$ for the rare earths.

The former two Hund's rule show their evidence: *the same largest* $(2S + 1)$ *occurs in several terms.* An example is Pr^{3+} ($4f^2$ electron configuration). Their terms are ^1S, ^1D, ^1G, ^1I, ^3P, ^3F, ^3H (see Appendix 1). For this case, the largest multiplicity is 3, but the greatest value of L is 5(= H). Thus, the term multiplet is $^3H_{J\,=\,4,5,6}$.

1.3.3. *Individual rare earth ions*

To understand the energy levels of RE ions in a crystal, it is necessary to determine them in detail in the free atom. Since the lines originating are *intra-4f^n configuration transitions* only, the energy levels of the *f-shell* have to be obtained [1, 8].

The Hamiltonian of a many-electron ion (assuming the nucleus fixed) in a magnetic field B is given by [4, 6]

$$H = \sum_i \frac{p_i^2}{2m} - \sum_i \frac{Ze^2}{r_i} + \frac{1}{2}\sum_{i>j} \frac{e^2}{r_{ij}} + \lambda \mathbf{L} \cdot \mathbf{S} \qquad (1)$$

In Eq. (1), the first term represents the kinetic energy of the electrons, with $i = 1, 2, \ldots, n$ electrons into *d-shell*, *f-shell*, and so on. The second is their Coulomb energy in the field of the nucleus. The third represents the energy associated with the mutual Coulomb repulsion between the electrons (electrostatic interactions), where the factor $^1/_2$ is justified by the Pauli principle [6]. The fourth is the spin-orbit interaction due to all the electrons, where λ is known as the *spin-orbit parameter*, and the $\boldsymbol{L}$ and $\boldsymbol{S}$ vectors are the total orbital and spin angular momenta, respectively. This interaction is denoted as $\boldsymbol{H}_{so}$.

For the magnetic ions of much interest such as rare earth and transition metal ions, the next assumption is appropriate: the electrostatic interactions are certainly larger than the magnetic interactions [4]. Particularly, in rare earth ions, spin-orbit interaction is much smaller than electrostatic interaction, and, however, this must be taken into account due to coupling between $\boldsymbol{L}$ and $\boldsymbol{S}$.

The energy levels of the free ion, such as an RE or a transition metal ion, in the absence of spin-orbit interaction, are characterized by different terms ^{2S+1}L. *What does this mean*?

Example. Consider us again the Ce^{3+} ion. When the spin-orbit interaction is taken into account over the $4f^1$ level, it splits in other two levels, $^2F_{5/2}$ and $^2F_{7/2}$.

The total angular momentum vector is $\mathbf{J} = \mathbf{L} + \mathbf{S}$. The eigenvalues of the $\boldsymbol{J}{\cdot}\boldsymbol{J}$ operator are $J(J+1)$ and in similar way $\boldsymbol{L}{\cdot}\boldsymbol{L}$ have eigenvalues $L(L+1)$, and $S(S+1)$ to the $\boldsymbol{S}{\cdot}\boldsymbol{S}$ operator [6]. Using the relation

$$2\boldsymbol{L}\cdot\boldsymbol{S} = (\boldsymbol{L}+\boldsymbol{S})\cdot(\boldsymbol{L}+\boldsymbol{S}) - \boldsymbol{L}\cdot\boldsymbol{L} - \boldsymbol{S}\cdot\boldsymbol{S} = J(J+1) - L(L+1) - S(S+1) \tag{2}$$

It can be shown that, for example, with λ positive, for the single electron in *4f-shell* of Ce^{3+}, the energy level to $J = 7/2$ is higher than that to $J = 5/2$, by $7\lambda/2$ due to the perturbation $\lambda\, \boldsymbol{L}\cdot\boldsymbol{S}$.

1.3.4. Crystal field theory

Crystal field theory is based on the hypothesis that a magnetic ion (an RE, a transition metal, or an actinide ion) in a crystal site feels the influence of its neighbors, the ligand ions, as an electric field which has the symmetry of the site. RE ions in crystals present spectra which are *sharp f-f transition* similar to free atoms [1, 3]. This is a consequence of the shielding of the *4f-shell* from the surroundings by the filled *5s* and *5p* shells.

When RE ions are embedded in a host, crystal field $\boldsymbol{H}_{\mathrm{cryst}}$ effects are small (typically 100 cm^{-1}) because f electrons are deeper in the ion [1, 8]. Hence, the spin-orbit interactions would be stronger due to the electrons that are nearer to nucleus: $\boldsymbol{H}_{\mathrm{cryst}} < \boldsymbol{H}_{\mathrm{so}}$.

The effects of crystal field strength and spin-orbit interaction are important, according to the order of magnitude [1, 3]. Thus, $\boldsymbol{H}_{\mathrm{crys}}$ is included in the Hamiltonian as follows:

$$H = \sum_i \frac{p_i^2}{2m} - \sum_i \frac{Ze^2}{r_i} + \frac{1}{2}\sum_{i>j} \frac{e^2}{r_{ij}} + \lambda \boldsymbol{L}\cdot\boldsymbol{S} + \boldsymbol{H}_{\mathbf{cryst}} \tag{3}$$

For instance, consider the CaS:Eu luminescent compound. The CaS host lattice has the NaCl structure; therefore, each one of Eu^{2+} ion prefers to occupy the Ca^{2+} site, because the ionic radii of Eu^{2+} (1.12 Å) is compatible with those of Ca^{2+} lattice site and it has an octahedral symmetry as shown in **Figure 4**. In this situation, the $^8S_{7/2}$ level (orbitally nondegenerate, because $L = 0$) of the ground state is very stable and cannot be *split* by the crystal field.

1.4. Absorption and emission spectra of RE ions in crystals

1.4.1. Intraconfigurational-$4f^n$ transitions: narrow absorption and emission lines

Figure 5 shows an energy-level diagram for some RE^{3+} ions. The effect of crystal field is very small, but produces a *splitting* on these energy levels to each one of RE^{3+} (Gd^{3+} is an exception; it has seven electrons in *4f-shell*, and thus the $^8S_{7/2}$ level of the ground state is very stable). The electronic transitions within the *4f*-shell are only weakly influenced by crystal fields and

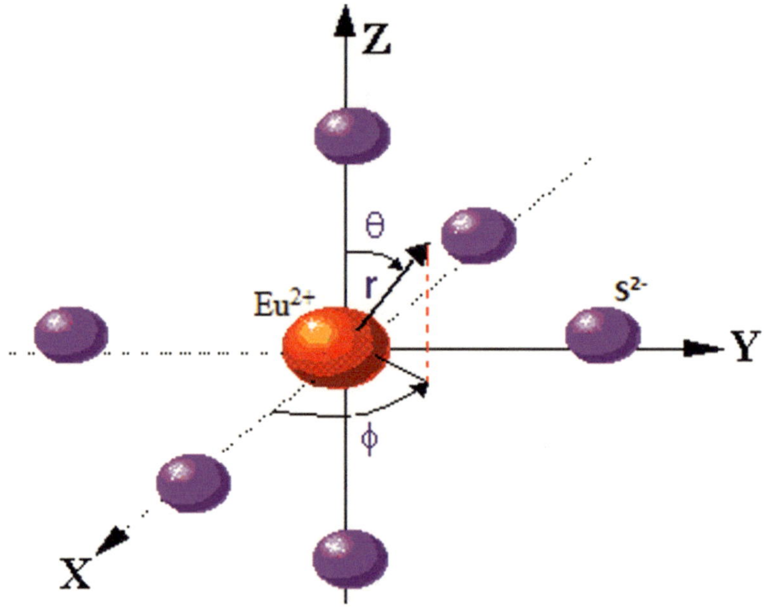

Figure 4. Eu^{2+} (central ion) substitute for Ca^{2+} lattice site has octahedral symmetry around six S^{2-} ligands.

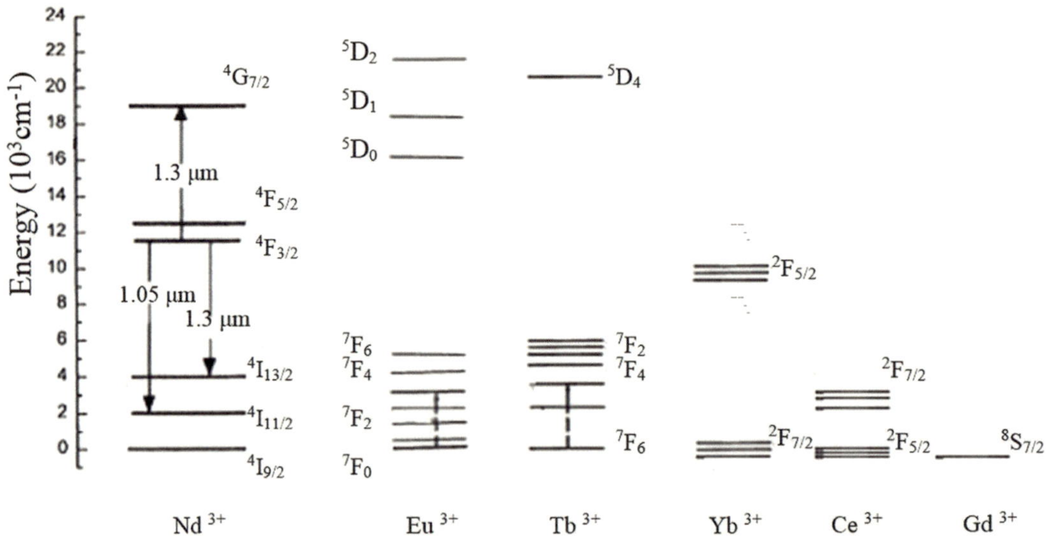

Figure 5. Energy-level diagram of rare earth ions: Nd^{3+}, Eu^{3+}, Tb^{3+}, Yb^{3+}, Ce^{3+} and Gd^{3+}.

covalency effects [1, 4]. Actually, $\Delta R = R(e){-}R(g){\sim}0$, the change in configurational coordinate between ground and excited states is very small or zero, so that narrow absorption and emission *lines* will be observed, *rather* than broad bands as is shown on the left-hand side of **Figure 6(a)** [2].

http://dx.doi.org/10.5772/65185

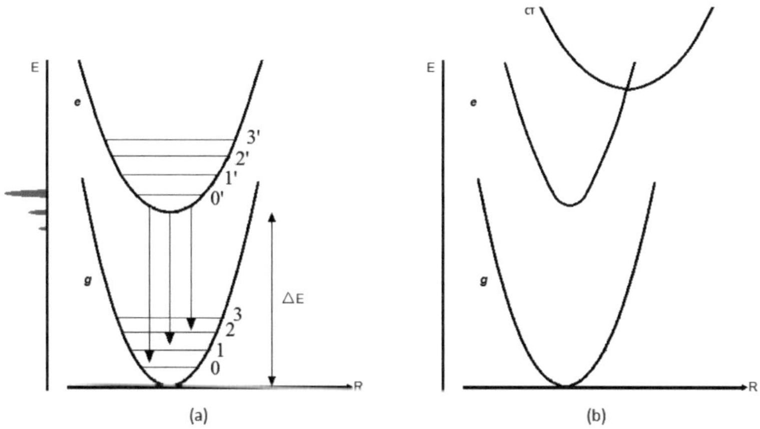

Figure 6. (a) In a CCD, these levels appear as parallel parabolas. On the left-hand side of **Figure 6(a)**, narrow lines appear due to *intra-4f* ^{n} *configuration transitions*. (b) The charge-transfer state has a larger offset (highest parabola).

1.4.2. Interconfigurational transitions: absorption and emission bands

The allowed optical absorption transitions of the rare earth ions are *interconfigurational* and consist of two different types: $4f^n \rightarrow 4f^{n-1}5d$ transitions and charge-transfer transitions ($4f^n \rightarrow 4f^{n+1}L^{-1}$, where L denotes an anion such as Cl^-, S^{2-}, O^{2-}, etc.) [1, 2].

1.4.2.1. $4f^n \rightarrow 4f^{n-1}5d$ transitions

The $4f^n \rightarrow 4f^{n-1}5d$ transitions are strongest in rare earth that have a tendency to became divalent (from RE^{3+} to RE^{2+}), such as Eu^{2+}, Sm^{2+}, and Yb^{2+}, where broad and intense absorption bands are observed in the UV of electromagnetic spectrum.

Example 1: Eu^{2+} ions into a host lattice. Its interconfigurational transition $4f^7 \rightarrow 4f^6 5d$ corresponds to a *broad absorption band*, and it shows an *5d →4f emission band* which can vary from UV-yellow range of the electromagnetic spectrum.

Example 2: Ce^{3+} ions into a crystal. Remember the next: spin-orbit coupling causes $^2F_{5/2}$ and $^2F_{7/2}$ levels of the ground-state configuration ($4f^1$) of Ce^{3+} ion. The crystalline field yields a splitting of the configuration ($5d^1$) in several levels. Thus, due to *4f →5d* transitions, absorption bands appear in the UV region of electromagnetic spectrum because Ce^{3+} tends to become tetravalent. Hence, the luminescent spectrum contains two *sharp emission bands* due to next transition: from the lowest *5d* crystal field level to the ground state ($4f^1$).

1.4.2.2. Charge-transfer transitions

Figures 3 and **6(a)** have been referred to two different schemes on configurational coordinate diagram: the former for two non-parallel parabolas and the second one to parallel parabolas. Now, we consider a suitable concentration *centers* inside host lattice, schematized through three parabolas as shown in **Figure 6(b)**. *What does the highest parabola mean?*

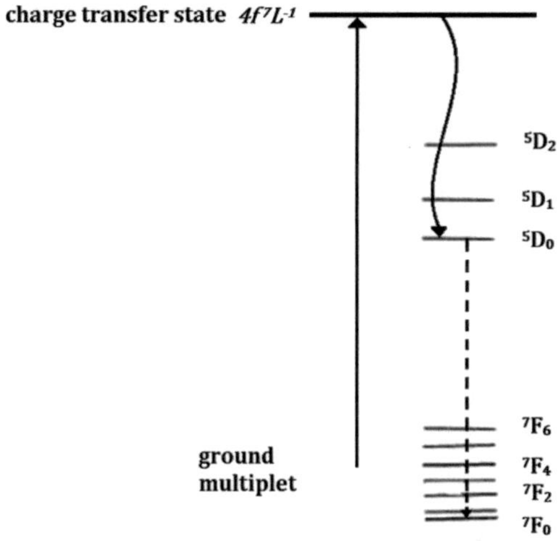

Figure 7. The role of charge-transfer state. It populates, at least partly, the ground-state levels $^7F_{J=0,1,2,3,4,5,6}$ of the electron configuration of the Eu^{3+}.

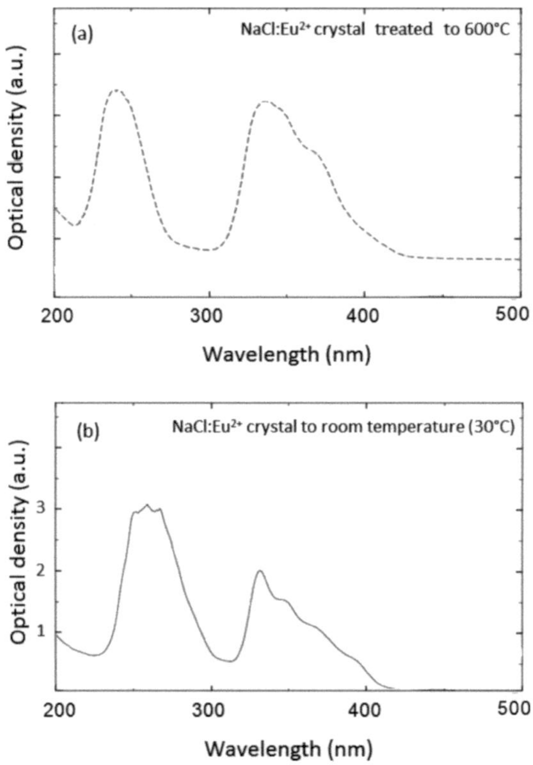

Figure 8. Both absorption spectra consist of two bands: (a) NaCl:Eu^{2+}crystal to 600°C and (b) NaCl:Eu^{2+} without treatment thermal.

Intraconfigurational-$4f^{\,n}$ transitions (the two parallel parabolas) yield very weak emission lines. In this scheme, the non-parallel parabola means a *charge-transfer state*, which originates from a different configuration and is connected to the $4f^{\,n}$ configuration by an allowed transition as follows. *Excitation* occurs now from the ground state to the charge-transfer state (hence the yield of the *absorption band* is a charge-transfer transition in the RE^{3+}-L^{-} bond, L^{-}being an anion such as Cl^{-}, S^{2-}, O^{2-}, etc.). From here, the system relaxes to the relaxed excited state of the second parabola (the *non-radiative transition* between the two upper parabolas is possible). *Emission* occurs from the second parabola (*line emission*) to the lower parabola.

For example, consider an Eu^{3+} ions-doped host lattice. Levels of the ground state ($4f^{6}$) of Eu^{3+}, according to Appendix 1, are ^{7}F, ^{5}D, and so on. Luminescent processes are $^{7}F_J \rightarrow$ charge-transfer state *excitation*, charge-transfer state $\rightarrow {}^{5}D_0$ *relaxation*, and $^{5}D_0 \rightarrow {}^{7}F_0$ *emission*. **Figure 7** shows some energy level of Eu^{3+} in a certain semiconductor crystal.

Luminescence process can occur, mainly, because the host contains an *appropriate concentration of ions* which luminesce. Also, bands involved in an absorption spectrum maintains a *dependence with the temperature* [7]. Consider the next example.

A previous example to the results, which is presented in Section 2.1, *is referred to an NaCl crystal doped with* Eu^{2+}. This crystal has dimensions of 8 × 8 × 1 mm^3, and grows by the Czochraski method. The concentration of Eu^{2+} aggregated to the powder of NaCl was of 0.1 wt%. This sample was heated in the seven next stages: 100, 240, 340, 390, 440, 540, and 600°C. All absorption spectra consist of two bands in the range of 200–440 nm, but in **Figure 8(a)** only the spectrum referred to 600°C is shown. With respect to the spectrum of continuous curve shown in **Figure 8(b)**, the high-energy band lost its structure in relation to the relative maxima observed in the continuous curve and the low-energy band does not considerably change its structure with such treatments [9, 10]. *Remark: the high-energy band has shown a dependence with the treatment temperature.*

1.5. Types of luminescence spectra and its relationship with photoacoustic technique

In this section, we review some techniques related with the luminescent spectra for measuring PL decay lifetimes, time-resolved PL, and photoacoustic (PA) spectra.

1.5.1. Technique of optical absorption

1.5.1.1. Theoretical model

For a homogeneous material, the relative fraction of the light intensity absorbed in traversing a thickness dx depends on absorption coefficient α (a constant characteristic of the material) as follows:

$$\frac{dI}{I} = -\alpha dx \tag{4}$$

Thus, if the incident photons intensity at the sample surface is $I_0 = I(x = 0)$ and I is the transmitted intensity, we obtain Eq. (6):

$$I(x) = I_0 e^{-\alpha x} \tag{5}$$

Transmittance is defined as $(\frac{I}{I_0})100\%$, but also the optical density (OD) is often used:

$$\mathrm{OD} = \ln(\frac{I_0}{I})/2.303 \tag{6}$$

where the reflection at the sample surface is neglected for simplicity (so, e.g., 1% transmittance corresponds to an OD = 2). Of Eqs. (6) and (7), the absorption coefficient can be expressed as

$$\alpha = \left(\ln(\frac{I_0}{I})\right)/x, \tag{7}$$

with α in cm^{-1}. For example, if the thickness of an amorphous film is 1 μm and $\alpha = 100\ cm^{-1}$, thus, the transmittance is 99%.

1.5.1.2. Experimental equipment

The absorption spectrophotometer equipment often measures the quantity OD.

1.5.2. Technique of photoluminescence

Consider again a two-level system: the ground state (g) to refer to a level in the ground state, and (e) for another level but in the excited state. Consider N_i to be the number of centers that exist in the host lattice. At the time of excitation of the sample with light coming from a power source, a fraction f of the total centers will be taken from g to e state, and will occur after a *stimulated emission* from e to the g state. This process will continue until the condition of an equilibrium is established:

$$fNi = (C_g + C_e) \tag{8}$$

That is, the total number of centers that succeeded to pass through level g to level e. C_e is a number of centers in the state e and a number C_g in the ground state that made it through the level g [10].

The energy gap between the levels e and g is the energy of emission of centers and is given by the equation

$$E_{\mathrm{photon}} = hc/\lambda \tag{9}$$

λ being the wavelength of the emitted light and hc is a constant depending on the system of units used (if E_{photon} is expressed in cm^{-1} and λ is in nm then the constant hc has the value 10^7, and when E_{photon} is measured in eV, the constant hc has the value 1239.8 eV-nm].

On the other hand, the intensity of emission I due to n photons emitted every interval of time t is

$$I = \frac{n}{t} \tag{10}$$

In these circumstances to say, the intensity of emission I reaches the value of *steady state* [2, 10].

The process of excitation emission to measure the photoluminescence spectra is schematized in **Figure 9**.

1.5.3. Techniques of time-resolved photoluminescence

1.5.3.1. Theoretical aspect

For the two-level system, e and g, the population of the excited state decreases according to [2, 10]

$$\frac{dn}{dt} = -P_{eg}n \tag{11}$$

In Eq. (5), the value of n gives the number of luminescent centers in the excited state after an excitation pulse, t is the time, and P_{eg} the probability for *spontaneous* (or radiative) *emission* from the excited to the ground state. Integration yields

$$n(t) = n(0)e^{\frac{-t}{\tau}} \tag{12}$$

$\tau = {}^{1}/_{P_{eg}}$ being the *radiative decay time*. The emission intensity is the rate of decay of the population of the excited state, that is,

$$I = -\frac{dn}{dt} \tag{13}$$

$$I = \frac{n(0)}{\tau}e^{\frac{-t}{\tau}} \tag{14}$$

1.5.3.2. Experimental

The time-resolved PL spectroscopy is very productive for obtaining the recombination rates (or lifetimes) of various transitions. Thus, determining the dynamical processes including

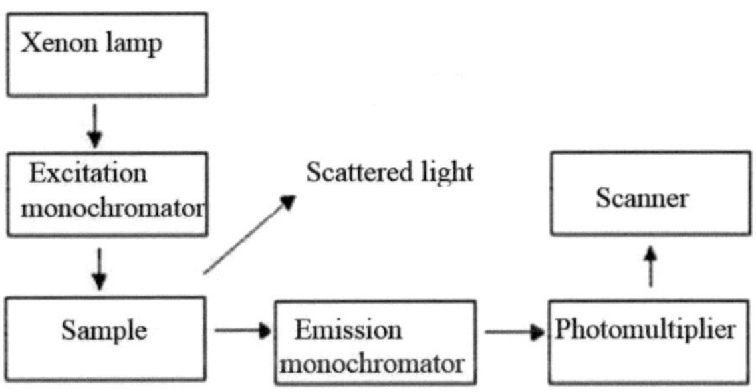

Figure 9. Schematic diagram of a spectrofluorimeter brand SPEX Fluorolog to obtain the excitation-emission spectra of the phosphor materials.

emission line energies and the associated recombination rates as well as quantum efficiencies is one of the basic purposes of time-resolved PL studies [11].

Time-resolved PL spectroscopy (or *pulsed laser-excited time-resolved luminescence*) is a relatively new technique usually paired with pulse laser excitation. If the laser pulse is brief enough (typically a few nanoseconds), the excitation is considerably shorter than most excited state lifetimes, offering the possibility of separating emission from different electronic states. In practice, this technique is applied by the synchronous use of laser pulses and gated detectors. One-timed signal will activate the laser, while the next, after a predetermined wait, will turn on the detector. Another signal will turn off the detector after a set dwell period. In this way, over many thousands of pulses, a complete emission spectrum can be acquired from electronic states with certain lifetimes. This is a particular value when trying to sort out the emission of activators whose emission spectra heavily overlap but differ in lifetime. **Figure 10** shows an arrangement for obtaining the time-resolved PL spectra.

1.5.4. Photoacoustic spectroscopy

1.5.4.1. General description

The photoacoustic spectroscopy (PAS) of solid materials was revived around the year 1970, and is widely used in the research of amorphous films to investigate the phenomena physical and chemical in a number of fields including biology and medicine. More frequent use photoacoustic detection mode includes a microphone of gas cell, and is unique because it makes a direct monitoring of non-radiative relaxation and hence channel that complements the spectroscopic techniques of optical absorption and photoluminescence. This complementarity is because the PA spectrum is a spectrum of excitation which is related to the amount of heat generated by many non-radiative processes [12, 13].

1.5.4.2. Experimental

This technique is used to obtain optical absorption spectra more defined near the *edge of absorption*, from which E_g, the bandwidth of optical energy [12], can be calculated.

The PA spectrometer consists of a stabilized 1000-W Xe lamp and a (1/8)-m-grating monochromator (Oriel, Model 77250). The monochromatic output beam was intensity modulated at 17

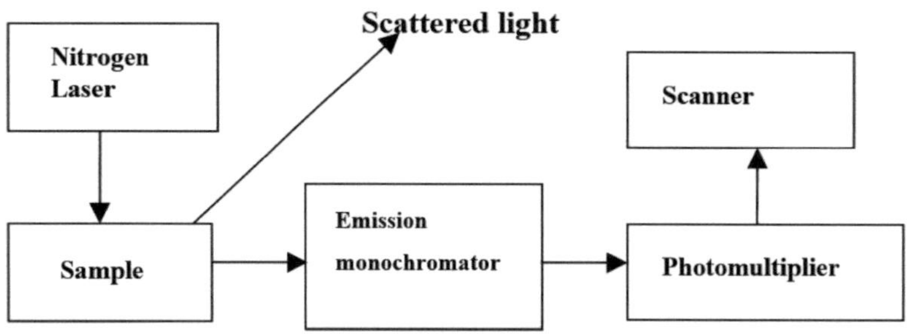

Figure 10. Arrangement for measuring PL decay and PL emission intensity at different delay times.

http://dx.doi.org/10.5772/65185

Hz with a mechanical chopper. The PA signal is recorded by an electret microphone and a lock-in amplifier. More details of this technique will be reviewed in Section 2.2.3.

1.6. Luminescent glasses

The preceding sections have dealt exclusively on solids in their crystalline form. However, also many glassy solids are of practical use today rather than crystalline.

When a liquid solidifies without crystallizing, it is currently said to form a *glass*, that is, to *vitrify*, or to pass to a *vitreous state*. This definition was initially reserved for inorganic solids, but such definition is too restrictive, due to that a large number of organic liquids also form glasses [14, 15].

Most of the inorganic glasses encountered in the laboratory or in daily life are composed of mixed oxides of several elements. The oxides are usually described as the anion or polyanion of a nonmetallic element of high electronegativity (silica, borate, silicate, phosphate, germinate, and tellurite glasses). Examples of oxide glasses are SiO_2, B_2O_3, P_2O_5, SiO_2-Na_2O, TeO_2-V_2O_5, B_2O_3-Al_2O_3-CaO, TeO_2-GeO_2-ZnO, and PbO-P_2O_5-TeO_2 [14, 15].

Like crystals, glasses consist of an extended three-dimensional *network* but the diffuse character of the X-ray diffraction (XRD) spectra shows that the network is not symmetric and periodic as in crystals (i.e., there is no long-range order).

The glass network may be compared to a unique molecule or a system with a giant unit cell. Its structure can be analyzed in terms of *coordination polyhedra* of cations surrounded by a variable number of anions. In crystalline solids, polyhedra can have common corners, edges, or faces. For example, in NaCl crystals, each *octahedron* consists of six Cl^- ions surrounding Na^+ cation.

Zachariasen sought the manner in which the polyhedra could be joined to build a disordered network related to that of a crystal [15]. For instance, SiO_2 in their different *crystalline forms* (quartz, cristobalite, tridymite, etc.), its network is built with SiO_4 *tetrahedra* (four oxygen ligands to a single *Si* form a single tetrahedron) joined at their corners. The same is true for *vitreous* SiO_2 but due to its non-periodicity, the mutual orientation of the consecutive tetrahedra is variable.

According to Zachariasen, the term *network former* has been adopted for an oxide which belongs to vitreous network, and *network modifier* for an oxide which does not participate directly in the network [15]. Certain oxides can function as glass formers or as modifiers depending upon the glass compositions involved. They are called *intermediate oxides*. **Figure 11** classifies some oxide compounds as glass formers, modifiers, and intermediates [14].

In particular, oxide compounds such as SiO_2, B_2O_3, P_2O_5, and GeO_2 alone have the ability to form glass, and when mixed with other oxides such as those of alkaline metals, alkaline earth metals, and transition metal form glasses [15]. These are the *classic or conventional glasses* that are referred in the literature as *silicate, borate, phosphate,* and *germinate* glasses, respectively. On the other hand, oxides TeO_2, V_2O_3, SeO_2, MoO_3, and Bi_2O_3 are known as *conditional network formers* not able to vitrify themselves when a *conventional melt-quenching method* of

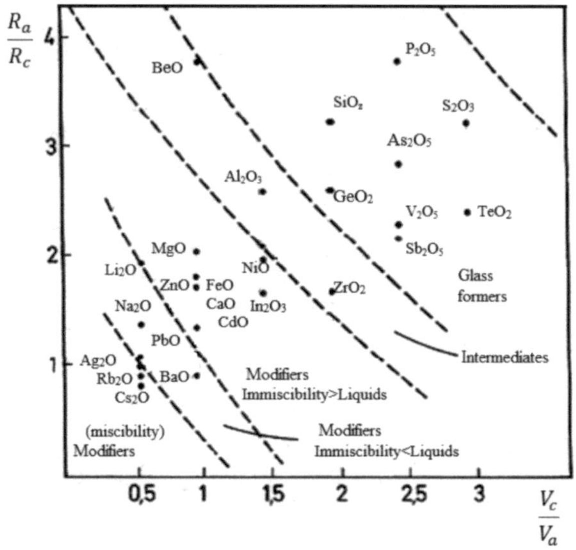

Figure 11. Classification of some oxide compounds in accordance with the role played in the glass formation.

vitrification is used, but when mixed with other oxides they are so-called *non-conventional* glasses [16].

Consider now TeO_2, a conditional network former. The structure of TeO_2 *crystalline* consists of trigonal di-pyramids (tdp) TeO_4. **Figure 12** shows a tdp: the equatorial plane is occupied by the alone pair of electrons (*5s-electrons*) of the tellurium atom and two atoms of oxygen O_{eq}-Te-O_{eq}, and in the axial positions are occupied by two atoms of oxygen O_{ax}-Te-O_{ax} [17]. X-ray and neutron diffraction structural studies have shown that the atomic arrangement in glassy TeO_2 may be well described as a three-dimensional network of TeO_4 *tdp* [18, 19].

Actually, a *tellurite glass* is constituted by TeO_2 in high percentage and other network nonformer oxides as second component, third component, and so on. The structural units of *tellurite glasses* are assumed to be TeO_4 and trigonal pyramids TeO_3, the latter with three bond Te-O shorter distances concerning the TeO_4, where it has been observed that a change of TeO_4 in TeO_3 happens when the *modifier oxide* (a second component) content increases at the expense of that high content TeO_2 decreases [20]. **Figure 13** shows that the polyhedra could be joined to build a disordered network in tellurite glasses. The relative abundance of these units depends on the composition [21]. Section 2.2 presents the photoluminescence and photoacoustic spectra (PAS) measured of a type of tellurite glasses.

Some glass-forming oxides, for examples, TeO_2, B_2O_3, S_2O_3, in themselves *do not luminesce*, but the presence of suitable activators leads to numerous interesting phenomena. Contrary to the case of crystalline solids, the lack of periodicity in atomic arrangements in glasses leads to a *broadening* of the luminescence spectra. Glasses are an intriguing material with which to study luminescence phenomena; however, their power of dissolving most additives makes systematic studies of the effect of composition and, hence, of structure on luminescence possible [14, 22].

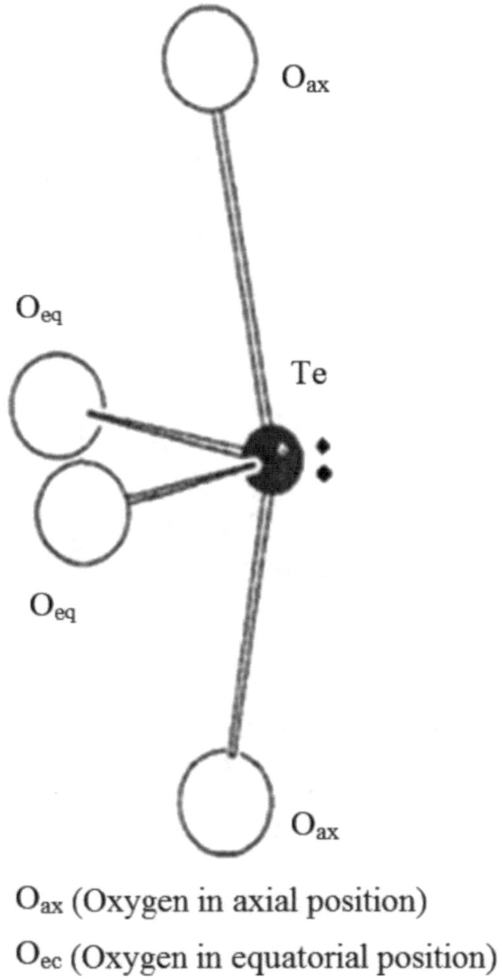

Figure 12. Structural model for TeO_4 units in TeO_2 crystalline: the Te presents coordination four with the first neighbors (the four oxygen), which are occupying four of the vertices of the tdp. These units form a three-dimensional-ordered network, which is not shown here.

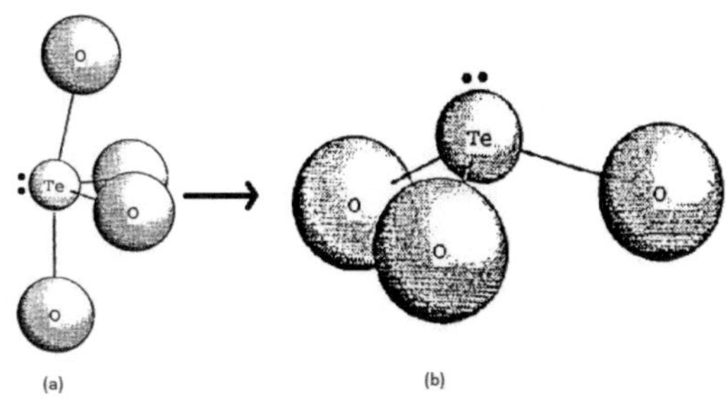

Figure 13. In the structure of glasses with containing high amount of TeO_2, the building units of these materials are (a) TeO_4 and (b) TeO_3 groups. TeO_4 represents the partial transformation of TeO_3 units due to the incorporation of a *modifier* of the network in a tellurite glass.

Luminescent centers in a host glassy consist of *(a) "energetically isolated" activators*, such as metal atoms, which may be considered to exist as a metal vapor in the glass, *(b)* molecules or groups that possess such a highly covalent nature as to be considered molecular groups, and *(c)* ions that participate directly in the glass structure and are, thus, more greatly affected by the host glass composition. In the latter case, the *luminescence of rare earth ions* is influenced by changes in glass structure less than that of other ions because of the shielding of the inner *4f* electrons in which the transitions leading to emission occur. Rare earth ions can be easily incorporated into many glasses, so these materials represent in principle an interesting area of research. In a host glassy, the local environment of a rare earth ion is roughly the same as that in a crystal, giving rise to a spectrum consisting of distinct lines.

Luminescent glasses serve as a useful function in products such as lasers, dosimeters, scintillation counters, artificial teeth, and electroluminescent devices. Furthermore, luminescent glasses are useful in research for studying such problems as the constitution of glasses and the formation of a vitreous phase in a crystalline substance.

2. Experimental results

This section covers the results on the study in luminescent materials doped with a rare earth compound (with an appropriate concentration of RE ions). In order to start using some techniques of photoluminescence, a first result of my research presents the time-resolved photoluminescence spectra of an alkali halide crystal doped with Eu^{2+} ions; the next results on photoluminescence and photoacoustic were obtained for oxide glassy compounds containing a concentration of a one type of rare earth ions.

2.1. Results of time-resolved PL spectroscopy: NaCl: Eu^{2+} crystal

Consider the NaCl:Eu^{2+} luminescent compound as an example. *$EuCl_2$* was added to the powder of NaCl. The Eu^{2+} ions enter the NaCl network and substitute some Na^+ ions during the crystal growth [9, 10]. The concentration of Eu ions into the NaCl powder was of 0.1 wt%.

In this section, the results of luminescent decay and spectral measures with temporal resolution (pulsed luminescence) are presented, in order to corroborate the fact that the *Eu* centers have a different interaction with the environment (Cl^- ligands and cation vacancies). To measure the decay of luminescence, the crystal was exposed to 500-ps *and* 337-nm laser pulses sent to the crystal with a frequency approximate of 10 pulses/s. By every pulse of excitation that makes a measure of intensity, this means that the intensity of the emitted light could vary according to the intensity of the pulse. For the analysis of the measures, the intensity of the excitation pulse has been considered constant. In this way, already every measure of intensity corresponding to a single pulse, intensity can be seen in relative and in this way we obtain the set of measures that represent the decay of the broadcast of the crystal at a fixed wavelength. Taking into account the luminescence results reported in the interval (380, 540) nm, the emission monochromator was set at corresponding wavelength of luminescence spectral region [9]. In **Figure 14**, a particular case of luminescent decay occurs on a semilogarithmic plot. As a way of presenting, it is to be noted that the decay of the intensity is very approximately exponential, in agreement with Eq. (15). These measures correspond to a freshly annealed crystal of 600°C to room temperature and to 427-nm emission wavelength. The slope of the graph obtained by least squares adjustment turned out to be 1.10 μs, which coincides within the experimental error with that reported as in [23]. Measures of decay emissions taken between 400 and 500 nm showed times of life around 0.85 μs to the crystal without treatment thermal.

Using the technique of temporal resolution, the *emission spectra* in the region of 390–490 nm were obtained. Two typical examples are presented in **Figures 15** and **16**. Curves in **Figure 15** correspond to the newly annealed crystal and those in **Figure 16** to the crystal without heat treatment. Both curves in **Figures 15** and **16** were taken with 100- and 1300-ns delay times. We see a similar behavior in both cases. In the case of the newly annealed crystal, it dominates the

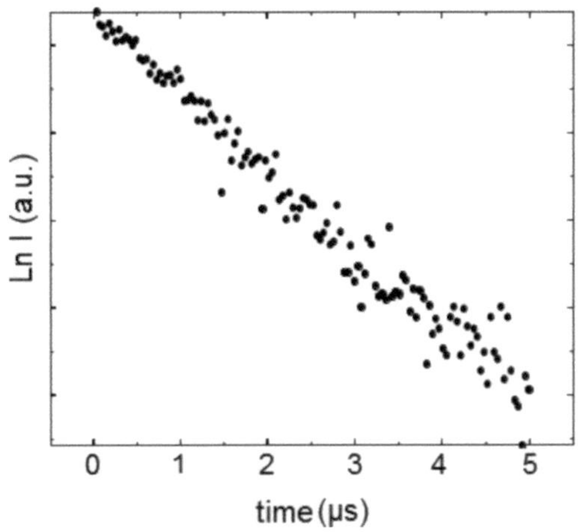

Figure 14. Temporal response of PL emission from a NaCl:Eu^{2+} freshly annealed crystal to 600°C (to a light emitted in 427 nm).

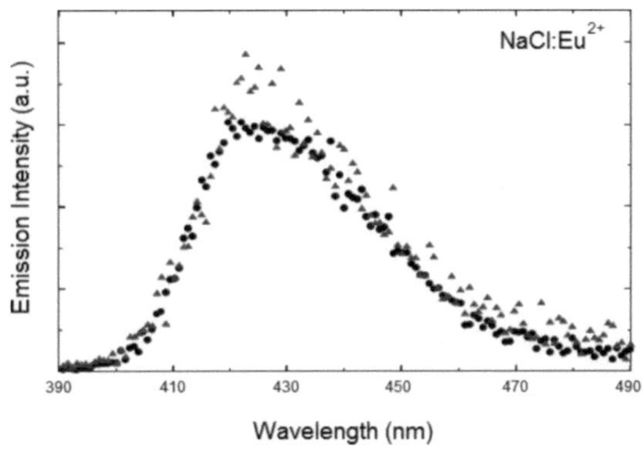

Figure 15. Emission spectra of a *NaCl:Eu crystal* freshly annealed crystal to 600°C with a temporal resolution of 100 (circles) and 1300 *ns* (triangles).

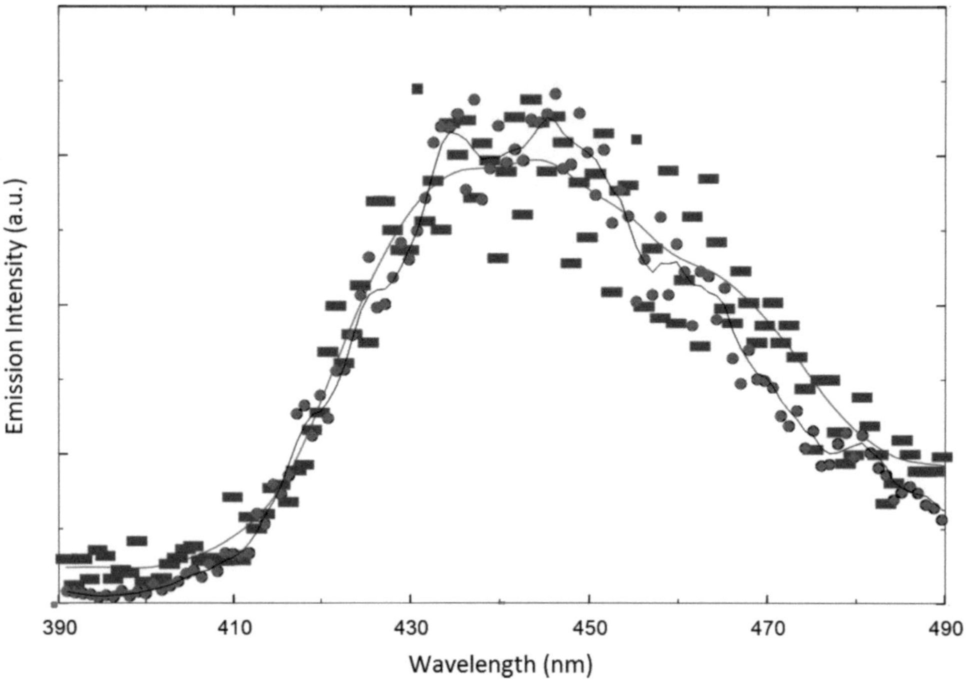

Figure 16. Emission spectra of a *NaCl:Eu crystal* without prior heat treatment, with a temporal resolution of 100 (circles) and 1300 *ns* (rectangles).

emission of free dipoles, while in the case of unheated crystal, it dominates the aggregate emissions [10].

2.2. Results of the photoacoustic and photoluminescence spectra of rare earth ions-doped cadmium-tellurite glasses

Research on tellurite glasses is currently being performed because several of its properties can be used in different types of modern devices [24–26]. These glasses show wide transmission in the 0.4–5.0-μm range, high linear and nonlinear refractive index, good corrosion resistance, thermal and chemical stability, and they are capable of incorporating large concentrations of rare earth ions into the matrix [27, 28]. Tellurite glasses represent a compromise between the desire for a low phonon energy host (800 cm^{-1}) coupled with the need to retain mechanical strength and low-processing temperatures [29]. When tellurite glasses are doped with rare earths, high-intensity, narrow-peak emissions can be obtained. This last property makes these materials good candidates for laser applications [30].

RE^{3+}-doped zinc-tellurite glasses have been studied for RE = Nd [31].

The following results are part of a wide research about the study of a ZnO-CdO-TeO_2 ternary system doped with rare earth ions. Four papers have been published early in the which the photoluminescence properties, structural studies, optical, and thermal analysis about this

http://dx.doi.org/10.5772/65185

Sample	V1	V2	V3
ZnO	9.52	9.52	9.52
CdO	9.52	19.04	28.58
TeO_2	76.20	66.68	57.14
REc	4.76	4.76	4.76

REc (rare earth compound): Tb_4O_7,$YbBr_3$, $NdCl_3$, $EuCl_3$. Their concentrations are (2.5, 2.9) range mol%.

Table 2. Bath original compositions (wt%) of ZnO-CdO-TeO_2:REc glasses.

matrix containing ytterbium and terbium [32, 33], neodymium [34], and europium ions [35] are reported.

Particularly, *tellurite glasses in the glass-formation region* (near the corner rich in TeO_2 in a composition triangle), for the ZnO-CdO-TeO_2 system, have been studied [36].

2.2.1. *Conventional melt-quenching method*

Three batches of ZnO-CdO-TeO_2 system were prepared, as can be seen in **Table 2**, by mixing appropriate amounts of the oxides: TeO_2 99.999% purity (Sigma Aldrich), CdO 99.999% purity (Sigma Aldrich, St. Louis, MO), ZnO 99.5% purity (Merck, Dakota, MN), and a rare earth (Yb, Tb, Nd, or Eu) compound 99.99% purity (Alfa Aesar, Ward Hill, MA). The glasses were fabricated by using a conventional melt-quenching method in a platinum crucible in the 1000–1200°C range.

Remark: V1, V2, and V3 glasses containing Tb^{3+} ions are referred as Tb118, Tb127, and Tb136, respectively; meanwhile, those doped with Yb^{3+} ions are Yb118, Yb127, and Yb136. However, VI, V2, and V3 denote the other two types, glasses containing Nd^{3+} or Eu^{3+}. All the measurements were recorded at room temperature.

2.2.2. *XRD results on ZnO-CdO-TeO_2:REc glasses*

These results are summarized as follows:

a. Yb118, Yb127, and Yb136 glasses display the pattern of an amorphous structure, and small particles of $CdTeO_3$ crystals are observed in the sample Yb118, see Refs. [32, 33].

b. Tb118, Tb127, and Tb136 glasses also show a crystalline phase $CdTeO_3$, see Refs. [32, 33].

c. Of the V1, V2, and V3 glasses containing Nd^{3+}, all exhibit a vitreous structure, which can correspond to short ordering of ZnO_4 tetrahedra embedded in the glassy structure, see Ref. [34].

d. Of the V1, V2, and V3 glasses containing Eu^{3+} ions, that with the highest content of CdO modifier helps in the formation of a crystalline phase of $CdTe_2O_5$ compound, see Ref. [35].

2.2.3. *Photoacoustic spectra*

The results of optical absorption, in the majority of cases, have pointed out that these glasses are transparent in the visible and near IR regions of the electromagnetic spectrum [37]. PA spectroscopy was employed in order to obtain more defined absorption spectra around the *edge absorption*.

PAS is the most direct method of obtaining the spectral dependence of the optical absorption coefficient in *edge absorption*. The absorption coefficient α is proportional to PA intensity, that is,

$$\alpha(hv)\propto \mathrm{PA} \qquad (15)$$

The PA spectrometer equipment often measures the quantity PA signal intensity versus α.

$CdTeO_3$ is a semiconductor of *direct band*, and the same is observed to the ZnO, CdO, and TeO_2 constituents of these glasses. Hence, near *edge absorption* spectral behavior follows [32]:

$$\alpha(hv)\propto(hv-Eg)^{q} \qquad (16)$$

In this relation, α is the absorption coefficient, h is Planck's constant, v is the frequency of the incident light, E_g is the optical energy measures in eV, and q = ½ for a *direct band*. Thus, of Eqs. (16) and (17), a measurement of $|\mathrm{PA}|^2$ versus hv typical plot for the E_g calculation [12].

In **Figure 17**, the PA spectra for the samples Yb127 and Yb136 are displayed, but Yb118 is not shown here. The E_g is estimated, its value was 3.47 eV for Yb136 and 3.6 eV for Yb127.

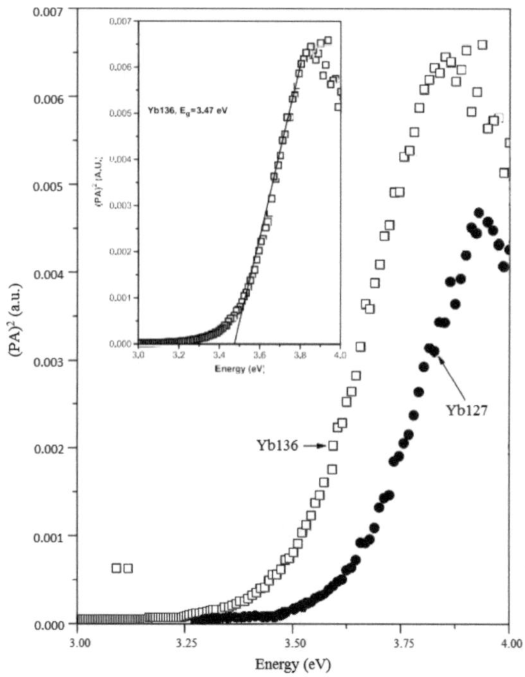

Figure 17. PA spectra of Yb127 and Yb136 glasses. In the inset, a typical plot used to determine E_g is illustrated.

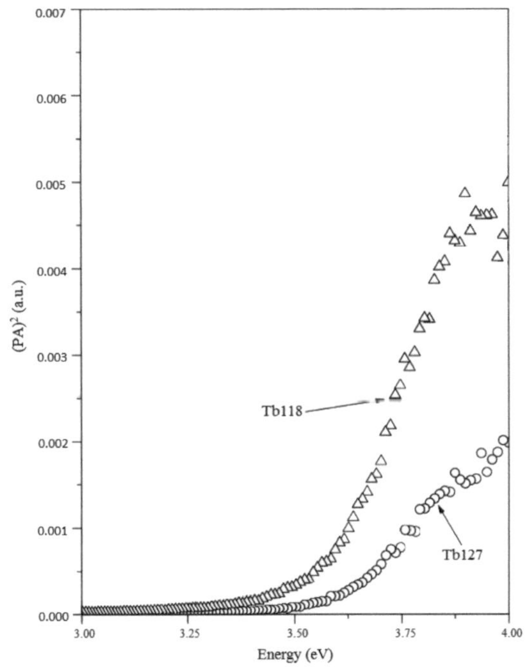

Figure 18. PA spectra of Tb118 and Tb127 glasses.

Figure 18 shows PA spectra for Tb118 and Tb127 glasses, but Tb136 is not displayed here. The E_g is estimated, its value was 3.51 eV for Tb118 and 3.6 eV for Tb127. In both cases, **Figures 17** and **18**, the value E_g is limited in the interval (3.47, 3.60) eV, which is very near of E_g = 3.9 eV, the bandgap of $CdTeO_3$ [32].

Figure 19 shows the PA spectra of the glasses doped with Nd^{3+}: VI, V2, and V3. By considering that one of the compounds in the tellurite matrix is ZnO, a material of direct bandgap (E_g) thus, a same value of E_g = 3.45 eV was calculated, for the three samples (see the inset of **Figure 19**, a plot to determine E_g). This value of E_g indicates a wide bandgap material, and was an expected result for this type of glass, but this value is lower than those reported for other glasses of the same matrix doped with other ions [34].

Now, the PA spectra for the V1, V2, and V3 glasses containing Eu^{3+} are exhibited in **Figure 20**. These spectra were acquired with steps of 1 nm/s, taking into account an average thickness of 1 mm. Obviously, this case is different from (a) to (c) cases because the crystalline phase of $CdTe_2O_5$ corresponds to a semiconductor more complex.

In the region between ~ 10^1 and ~ 10^4 cm^{-1}, α obeys a simple relation,

$$\alpha \propto e^{(h\nu/E_0)} \tag{17}$$

which is termed the *Urbach tail*, being E_0 hardly depends on temperature [12]. In amorphous semiconductors, it is reported that E_0 varies *linearly* in the range of 0.05–0.1 eV. Using the *Urbach relation and* $\alpha(h\nu)\propto$PA, we obtain that

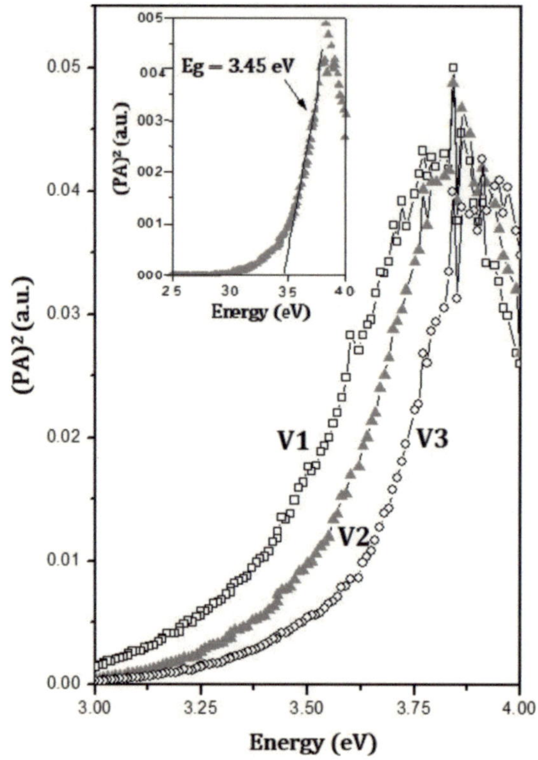

Figure 19. PA spectra of three glasses doped with Nd^{3+}. In the inset, a typical plot used to determine E_g is illustrated, for the V2 glass.

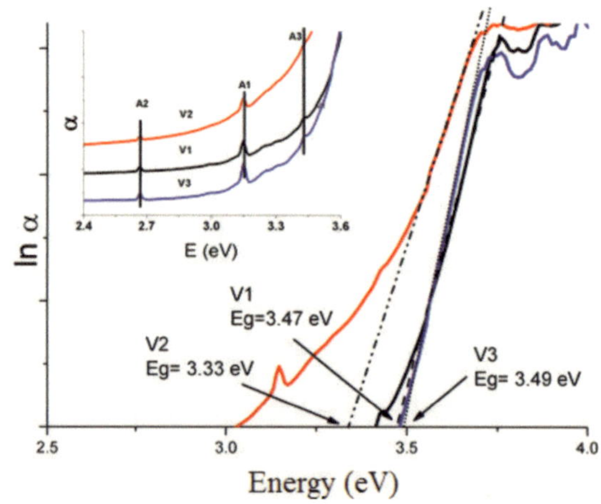

Figure 20. PA spectra of three glasses doped with Eu^{3+}. In the inset, a typical plot used to determine E_g is illustrated.

$$\ln(\alpha) \propto h\upsilon \tag{18}$$

By applying Eq. (19), $\ln(\alpha)$ versus $h\upsilon$, for the V1–V3 glasses doped with europium, the energy of the optical absorption edge (E_g) was determined (see **Figure 20**). The obtained E_g values were 3.47, 3.33, and 3.49 eV for the V1, V2, and V3, respectively. The E_g for the starting oxides (ZnO, CdO, and TeO_2) is 3.3, 2.5, and 3.3 eV, respectively [35].

2.2.4. Photoluminescence spectra

Figure 21 shows the PL intensity of the samples Yb118, Yb127, and Yb136. Only the $^2F_{5/2}\rightarrow{}^2F_{7/2}$ transition was detected, as expected because this is the only transition reported until now for Yb-doped materials. The gradual diminution of the PL signal in the samples from Yb118 (the higher signal) until Yb136 (the lower signal) can be due to surface effects. This fact can be due to the difficulty to polish this kind of samples, in this way, some of them can present more roughness than others. The PL emissions of Yb-doped samples are accompanied by the second harmonic of the laser line (976 nm). The effects of the luminescence of the ion Yb^{3+} have already been observed in other hosts of tellurite glasses, showing a typical transition [38].

In **Figure 22**, the PL spectra for the samples Tb118, Tb127, and Tb136 are displayed. The signal of luminescence is very intense, with the emissions being identified as transitions between levels: $^5D_4\rightarrow{}^7F_5$, $^5D_4\rightarrow{}^7F_4$, $^5D_4\rightarrow{}^7F_3$, of the ion Tb^{3+}, respectively. It is important to stand out that some of the luminescent emissions were intense enough that could easily be seen with naked eye. As a matter of fact, the signal of three of the six samples at 548 nm (transition $^5D_4\rightarrow{}^7F_5$) saturated our detector. On the other hand, it can be observed that it does not matter if the type of the matrix is crystalline or amorphous, since the position of the peaks does not change. PL in crystalline and amorphous matrices of RE-doped materials has already been reported in the literature.

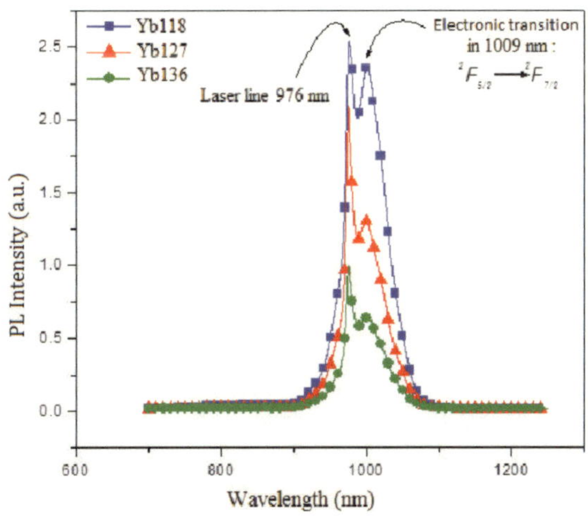

Figure 21. The photoluminescence []spectra of the Yb118, Yb127, and Yb136 samples are displayed.

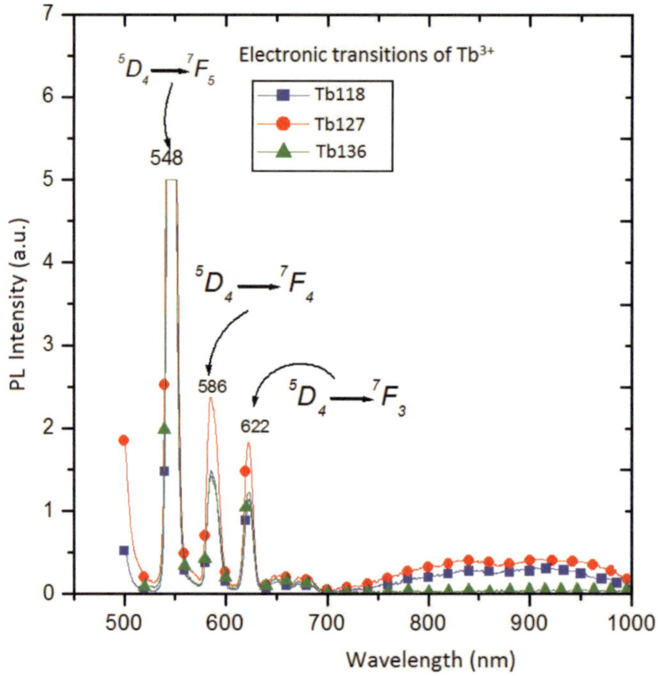

Figure 22. The photoluminescence spectra of the Tb118, Tb127, and Tb136 samples are displayed.

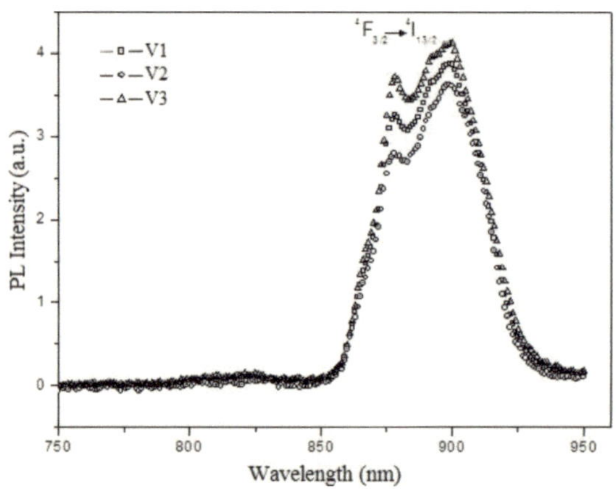

Figure 23. Photoluminescence spectra of V1, V2, and V3 glasses containing Nd^{3+}.

PL spectra of glasses are shown in **Figure 23**, where only one transition $^4F_{3/2}{\rightarrow}^4I_{9/2}$ appears due to the emission of the Nd^{3+} ions. Such a transition is typical in Nd^{3+}-doped glasses and has also been reported by other authors [31].

PL spectra of glasses are shown in **Figure 24**, indicating the characteristic transitions due to the Eu^{3+} ion, which were identified as follows: $^5D_0 \rightarrow {}^7F_1$, $^5D_0 \rightarrow {}^7F_2$, $^5D_0 \rightarrow {}^7F_3$, and $^5D_0 \rightarrow {}^7F_4$, which appear around 590, 620, 650, and 700 nm, respectively. Such transitions are typical in

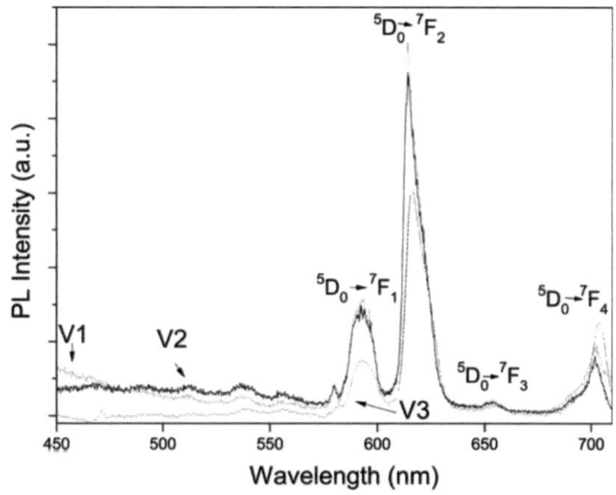

Figure 24. Photoluminescence spectra of V1, V2, and V3 glasses containing Eu^{3+}.

Eu-doped-glasses and have also been reported previously in glasses containing this dopant [39, 40]. The emission spectra for the three samples are dominated by the $^5D_0 \rightarrow {}^7F_2$ transition, the highest intensity, indicating that Eu^{3+} occupies a low symmetry site. The intensity ratio of $R = I(^5D_0 \rightarrow {}^7F_2)/I(^5D_0 \rightarrow {}^7F_1)$, which is a measure of Eu^{3+} ion site symmetry, is calculated to be upper to the unit. This result is an indication that Eu^{3+} ions occupy mainly the lattice site of noninversion symmetry [1].

3. Conclusions

1. For NaCl crystals containing a considerable concentration of Eu^{2+} ions (≈200 ppm or 0.1 wt%), the following results were confirmed:

   1.1 The decay curve of the luminescent intensity due to Eu^{2+} ions had an approximately exponential behavior.

   1.2 Luminescent emission spectra consist of a single broadband centered approximately between 430 and 450 nm, which is typical of Eu^2 + ions in crystalline matrices mainly alkali halides.

2. Glasses of the ZnO-CdO-TeO_2 system were doped with Yb^{3+}, Tb^{3+}, Nd^{3+}, or Eu^{3+} ions:

   2.1 Optical absorption spectra showed that these glasses are transparent in much of the optical region of the electromagnetic spectrum.

   2.2 The photoluminescent emissions of these glasses investigated were assigned to typical transitions of those rare earth ions that already have been identified in other systems vitreous.

   2.3 The results seen in the photoacoustic spectra, and their correlation with XRD patterns, have allowed to obtain the bandwidth of optical energy (optical bandgap) of crystalline phases that are embedded in the vitreous structure characteristic of the material under study.

2.4 PA spectroscopy allows to determine the optical energy gap values that were found with those reported in other systems already studied by this technique.

2.5 Tellurite glasses containing Yb^{3+} are potential candidates for parent, a single line of excitation laser $^2F_{7/2} \rightarrow {}^2F_{5/2}$.

2.6 Of the glasses containing $CdTeO_3$, crystals can be used in the technology of devices semiconductor, and also may be applied to nanotechnology.

Acknowledgements

The authors thank Geologist Rojas Santiago, Department of Geology, University of Sonora, for his technical help in this chapter. The authors also thank Mechatronic Engineer Tadeo R. Marielvis J., Department of Industrial Engineering, University of Sonora, for her technical help in this chapter.

f^1, f_{13}	^2F			
f^2, f_{12}	$^1S, ^1D, ^1G, ^1I$	$^3P, ^3F, ^3H$		
f^3, f_{11}	$^2P, ^2D, ^2F, ^2G, ^2H, ^2I, ^2K, ^2L$	$^4S, ^4D, ^4F, ^4G, ^4I$		
f^4, f_{10}	$^1S, ^1D, ^1F, ^1G, ^1H, ^1I, ^1K, ^1L, ^1N$	$^3P, ^3D, ^3F, ^3G, ^3H, ^3I, ^3K, ^3L, ^3M$	$^5S, ^5D, ^5F, ^5G, ^5I$	
f^5, f^9	$^2P, ^2D, ^2F, ^2G, ^2H, ^2K, ^2L, ^2M, ^2N, ^2O$	$^4S, ^4P, ^4D, ^4F, ^4G, ^4H, ^4I, ^4K, ^4L, ^4M$	$^6P, ^6F, ^6H$	
f^6, f^8	$^1S, ^1P, ^1D, ^1F, ^1G, ^1H, ^1I, ^1K, ^1L, ^1M, ^1N, ^1Q$	$^3P, ^3D, ^3F, ^3G, ^3H, ^3I, ^3K, ^3L, ^3M, ^3N, ^3O$	$^5S, ^5P, ^5D, ^5F, ^5G, ^5H, ^5I, ^5K, ^5L$	^7F
f^7	$^2S, ^2P, ^2D, ^2F, ^2G, ^2H, ^2I, ^2K, ^2L, ^2M, ^2N, ^2O, ^2Q$	$^4S, ^4P, ^4D, ^4F, ^4G, ^4H, ^4I, ^4K, ^4L, ^4M, ^4N$	$^6P, ^6D, ^6F, ^6G, ^6H, ^6I$	^8S

Appendix 1. Table of terms corresponding to the 4f n electron configurations.

Author details

Carlos Ruvalcaba Cornejo

Address all correspondence to: carlruval@yahoo.com.mx

University of Sonora, Hermosillo, Sonora, Mexico

References

[1] Blasse G. and Grabmaier B.C. Luminescent Materials. Berlin: Springer-Verlag; 1994. 232 p.

[2] Ropp R.C. Luminescence and the Solid State. 2nd ed. Amsterdam: Elsevier; 1991. 711 p.

[3] Jorgensen C.K., Reisfeld, R. Chemistry and spectroscopy of the rare earths. Topics in Current Chemistry. 1982; **100**:127–167.

[4] Stevens K.W.H. Magnetic Ions in Crystals. NJ: Princeton University Press; 1997. 249 p.

[5] Sobelman I.I. Atomic Spectra and Radiative Transitions. Berlin Heidelberg: Springer-Verlag; 1994. DOI: 10.1007/978-3-662-05905-0

[6] Cohen-Gannoudji C., Diu B., Laloe F. Quantum Mechanics. New York, NY: Wiley-VCH; 1991. 640 p.

[7] Di Bartolo B. Optical Interactions in Solids. Cambridge, MA: John Wiley & Sons; 1968. 541 p.

[8] Hüfner S. Optical Spectra of Transparent Rare Earth Compounds. New York, NY: Academic Press; 1978. 237 p.

[9] López F.J., Murrieta S.H., Hernández A.J., Rubio O.J. Optical absorption and luminescence investigations of the precipitated phases of Eu in NaCl and KCl single crystals. Physical Review B. 1980; **22**:6428. DOI: http;//dx.doi.org/10.1103/PhysRevB.22.6428

[10] Ruvalcaba-Cornejo C. Master's Thesis. 2002. Pages 30–32. Effect of the symmetry of crystalline field on Eu2+ ions in NaCl crystals [Master's thesis]. Hermosillo, Sonora: University of Sonora; 2002. 46 p.

[11] Manasresh M.O., Jiang H.X. III-Nitride Semiconductors: Optical Properties I. Great Britain: Taylor & Francis Books; 2002. 417 p.

[12] Mandelis A. Photoacoustic and Thermal Wave Phenomena in Semiconductors. 1st ed. Toronto, Canada: Elsevier Science Publishing; 1987. 480 p.

[13] Rosencwaig A. Photoacoustic and Photoacoustic Spectroscopy. New York, NY: John Wiley & Sons; 1987.

[14] Zarzycki J. Glasses and the Vitreous State. Great Britain: Cambridge University Press; 1991. 505 p.

[15] Fernandez J.M. The Glass (University texts). Madrid: High Council for Scientific Research; 1991.

[16] D'Amore F., Di Giulio M., Pietralunga S.M., Zappettini A., Nasi L., Rigato V., Martinelli M. Sputtered stoichiometric TeO_2 glass films: Dispersion of linear and nonlinear optical properties. Journal of Applied Physics. 2003;**94**(3):1654.

[17] Mirgorodsky A.P., Merle-Méjean T., Champarnaud J.C., Thomas P., Frit B. Dynamics and structure of TeO_2 polymorphs: model treatment of paratellurite and tellurite; Raman

scattering evidence for new γ- and δ-phases. Journal of Physics and Chemistry of Solids. 2000;**61**(4):501–509. DOI: 10.1016/S0022-3697(99)00263-2

[18] Dimitriev Y., Dimitrov V., Gattef E., Kashchieva E and Petkov H. Effect of the mode formation on the structure of tellurite glasses. Journal of Non-Crystalline Solids. 1987;**95**:937—944. DOI: 10.1016/S0022-3093(87)80701-9

[19] Petkov V., Stachs O., Gerber Th., Ilieva D. and Dimitriev Y. Atomic ordering in $xGa_2O_3\cdot(100-x)TeO_2$ glasses (x=10, 17.5, 25) by X-ray diffraction. Journal of Non-Crystalline Solids. 1999;**248**(1):65–74. DOI: 10.1016/S0022-3093(99)00091-5

[20] Zwanziger J.W. Structure and chemical modification in oxide glasses. International Reviews in Physical Chemistry. 1998; **17**(1):65–90.

[21] Hoppe U., Yousef E., Rüssel C., Neuefeind J. and Hannon A. C. Structure of zinc and niobium tellurite glasses by neutron and x-ray diffraction. Journal of Physics: Condensed Matter. 2004; **16**(9):1645–1663. DOI: 10.1088/0953-8984/16/9/013

[22] Doremus R.H. Glass Science. 2nd ed. New York, NY: John Wiley & Sons; 1994. 339 p.

[23] Muñoz J.E. Extinction of the Luminescence of Eu^{2+} on Alkaline Halides Induced by Precipitation [PH thesis]. Madrid: Autonomous University of Madrid; 1991.

[24] Wang J.S., Vogel E.M., Snitzer E., Jackel J.L., Da Silva V.L., Silberberg Y., et al. 1.3 μm emission of neodymium and praseodymium in tellurite-based glasses. Journal of Non-Crystalline Solids. 1994;**178**:109–113

[25] Miyajima T., Tojyo T., Asano T., Yanashima K., Kijima S., Hino T., et al. GaN-based blue laser diodes. Journal of Physics: Condensed Matter. 2001;**13**(32):7099.

[26] Bey H. J. Light-Emitting Diodes-Photolithography Patterns Plastic LEDs. Laser Focus World. 1997; **33**:36.

[27] Lenth W., Macfarlane R.M. Upconversion lasers. Optics & Photonics News. 1992;**3**(3):8.

[28] Ozen G., Demirata B., Ovecoglu M.L., Genc A. Thermal and optical properties of Tm^{3+} doped tellurite glasses. Spectrochimica Acta Part A: Molecular and Biomolecular Spectroscopy. 2001;**57**(2):273.

[29] Kim S.H., Yoko T. Nonlinear optical properties of TeO2- based glasses: $MOx-TeO_2$ (M=Sc, Ti, V, Nb, Mo, Ta, and W) binary glasses. Journal of the American Ceramic Society. 1995;**78**(4):1061. DOI: 10.1111/j.1151-2916.1995.tb08437.x

[30] Rakov N., Maciel G.S., Sundheimer M.L., Menezes L. de S., Gomes A.S.L., Messaddeq Y., et al. Blue upconversion enhancement by a factor of 200 in Tm^{3+}-doped tellurite glass by codoping with Nd^{3+} ions. Journal of Applied Physics. 2002;**92**(10):6337.

[31] D.L. Sidebottom, M.A. Hruschka, B.G. Potter, R.K. Brow. Increased radiative lifetime of rare earth-doped zinc oxyhalide tellurite glasses. Applied Physics Letters. 1997;**71**(14):1963–1965.

[32] Ruvalcaba-Cornejo C., Flores-Acosta M., Zayas Ma. E., Lozada-Morales R., Palomino-Merino R., Espinosa J.E., et al. Photoluminescence properties of the ZnO-CdO-TeO_2 system doped with the Tb^{3+} and Yb^{3+} ions. Journal of Luminescence. 2008;**128**:213. DOI: 10.1016/j.jlumin.2007.07.004

[33] Rubalcava-Cornejo C., Zayas Ma. E., Castillo S.J., Palafox J.J., Lozada-Morales R., and Rincón J. Ma. Structural properties of rare earth doped ZnO-CdO-TeO_2 system. Journal of Glass Science Technology. B. 2009;**50**(1):19–22.

[34] Ruvalcaba-Cornejo C., Zayas Ma. E., Castillo S.J., Lozada-Morales R., Pérez-Tello M., Díaz C.G., et al. Optical and thermal analysis of ZnO-CdO-TeO_2 glasses doped with Nd^{3+}. Optical Materials. 2011;**33**:823–826. DOI: 10.1016/j.optmat.2011.01.001.

[35] Ruvalcaba-Cornejo C., Zayas Ma. E., Lozada-Morales R., Rincón J. Ma. and De L. Flores A. Effect of the Eu^{3+} addition on photoluminescence and microstructure of ZnO-CdO-TeO_2 glasses. Journal of the American Ceramic Society. 2013;**96**:3084–3088. DOI:10.1111/jace.12542

[36] Zayas M.E., Arizpe H., Castillo S.J., Medrano F., Diaz G.C., Rincón J.M., et al.. Glass formation area and structure of glassy materials obtained from the ZnO-CdO-TeO_2 ternary system. Physics and Chemistry of Glasses. 2005;**46**(1):46–57.

[37] Ruvalcaba C.C. Optical and Structural Study of Rare Earth Ions Doped Glasses of the ZnO-CdO-TeO_2 System [PH thesis]. University of Sonora: Department of Research in Polymers and Materials; 2009. 90 p.

[38] M. Dejneka, E. Snitzer, R.E. Riman. Blue, green and red fluorescence and energy transfer of Eu^{3+} in fluoride glasses. Journal of Luminescence. 1995;**65**(5):227–245. DOI: 10.1016/0022-2313(95)00073-9

[39] J. Ozdanova, H. Ticha, L. Tichy, Optical gap and Raman spectra in some $(Bi_2O_3)x(WO_3)y(TeO_2)100xy$ and $(PbO)x(WO_3)y(TeO_2)100xy$ glasses. Journal of Non-Crystalline Solids. 2009;**355**:2318–2322.

[40] Hongpeng Y., Masayuki N. Three-photon-excited fluorescence of Al_2O_3-SiO_2 glass containing Eu^{3+} ions by femtosecond laser irradiation. Applied Physics Letters. 2004;**84.**

Chapter 4

Crystal Symmetry and Polarized Luminescence on Nonpolar ZnO

Hiroaki Matsui and Hitoshi Tabata

Additional information is available at the end of the chapter

http://dx.doi.org/10.5772/64724

Abstract

We introduce excitonic polarized photoluminescence (PL) of nonpolar ZnO layers and related quantum well (QW) structures in terms of crystal symmetries and lattice distortions. Polarized PL characters are attributed to in-plane anisotropic strains in the host, which are fully demonstrated on *A*-plane ZnO. Theoretical evaluations propose that in-plane compressive strains induced in ZnO layers play an important role in obtaining highly polarized optical properties. We experimentally achieve polarized PL responses in strain-controlled *A*-plane ZnO layers. Furthermore, we find interesting relationship between polarization degree of PL and in-plane anisotropic strains. Finally, highly polarized PL at room temperature is obtained by controlling well width in $Cd_{0.06}ZnO_{0.94}O$/ZnO QWs as a consequence of change in crystal symmetry from C_{6v} to C_{2v} at interfaces between $Cd_{0.06}Zn_{0.94}O$ well and ZnO barrier layers in the QW samples.

Keywords: ZnO, nonpolar, luminescence, anisotropy, crystal symmetry

1. Introduction

Zinc oxide (ZnO) has been one of the candidates of important materials towards the fabrications of optoelectronic platforms such as transistors, light-emitting diodes, transparent electrodes, and magnetism. ZnO has large exciton energy of 64 meV [1], which has received much attention for the possibility of utilizing excitonic-based optical applications at room temperature. In addition, MgO–ZnO and CdO–ZnO alloys are attracting great deal of interests since these alloys possess higher and lower band gaps than that of ZnO [2–6], which have been applied to $Mg_xZn_{1-x}O$/ZnO and $Cd_xZn_{1-x}O$/ZnO multiple and single quantum wells (QWs) [7–10]. These low-dimensional heterostructures can produce interesting quantum phenomena such as an increased excitonic binding energy and two-dimensional (2D) electron transport. These physical properties have been received much attention by many researchers for science and practical applications.

ZnO has a non-centrosymmetric structure with no center of inversion. This crystal symmetry builds spontaneous and piezoelectric polarizations in the host, which provides interesting excitonic luminescence. When the growth direction is chosen along the polar [0001] axis, these polarizations result in potentially detrimental effects such as the lowering of the probability of radiative recombination of active layers in QWs due to spatial separation of electron and hole carriers because of the quantum-confinement Stark effect (QCSE) [11, 12]. To suppress the QCSE, it is needed to deposit ZnO films and quantum structures on substrates other than *Zn*-(0001) and *O*-(000-1) planes. The most promising are the nonpolar planes such as *A*-plane (11–20) and *M*-plane (10–10), which are no spontaneous polarization fields since the polar *c*-axis lies in the growth planes. Thus far, nonpolar ZnO layer growth has been performed using various oxide substrates, such as $LaAlO_3$, $LiGaO_2$, $SrTiO_3$ and (La, Sr)(Al, Ta)O_3 [13–16]. In particular, nonpolar ZnO films can be easily grown using *r*-plane sapphire substrates by pulsed laser ablation (PLD), molecular beam epitaxy (MBE), and chemical vapor deposition (CVD) [17–19]. Layer growths of nonpolar ZnO have provided some important tasks for scientific studies [20–23], which have played key factors in applying for photodetector and transistor devices [24–26]. In addition, artificial control of carrier polarity (p-type) and magnetism has also been reported [27, 28]. Accordingly, research involving ZnO has recently been focused on the growth of nonpolar axes rather than that of polar axes, which have been extended to nonpolar (*A*- and *M*-plane) $Mg_xZn_{1-x}O$/ZnO and $Cd_xZn_{1-x}O$/ZnO QWs with no internal polarization fields along the growth conditions [29, 30].

To date, some studies have reported on nonpolar ZnO-based QWs in terms of preventing the QCSE caused by *c*-polar QWs [31, 32]. The merit of nonpolar ZnO is not only the avoidance of QCSE but also the generation of polarized optical phenomena. These properties are expected to be promising for attractive optical devices such as polarized photodiodes, light-emitting diodes and solar applications [33–36]. The polarized light emissions of nonpolar ZnO are associated with the conduction (CB) and valence band (VB) levels [37]. When using nonpolar ZnO, the polarized photoluminescence (PL) of $Cd_xZn_{1-x}O$/ZnO QWs was higher than that of $Mg_xZn_{1-x}O$/ZnO QWs because the crystal symmetry in a $Cd_xZn_{1-x}O$ well layer was changed from C_{6v} to C_{2v} by introducing an anisotropic compressive strain on the growing surface [38, 39]. This lattice distortion causes a strain-induced modification of the VB, which results in polarized PL. Therefore, the polarized PL is highlighted as the consequence of anisotropic lattice distortions introduced in nonpolar ZnO layers and their related QWs.

This chapter is organized as follows. Theoretical evaluations of optical anisotropies are described in Section 2 in order to investigate polarized PL in nonpolar *A*-plane ZnO. In particular, we discuss on relationship between optical transitions at the excitonic edge and in-plane strains. In Section 3, we state film growths and polarized PL properties in nonpolar ZnO layers grown using homoepitaxial schemes. We experimentally clear correlation between polarized PL properties and in-plane strains. Furthermore, on the basis of above preliminary results, we fabricate nonpolar *A*-pane $Cd_xZn_{1-x}O$/ZnO QWs in order to show highly polarized

http://dx.doi.org/10.5772/64724

PL at room temperature by introducing in-plane compressive strains in the QWs, which is reported in Section 4. We report that anisotropic lattice distortions play an important role in determining the degree of polarized PL. Finally, some concluding remarks and future research directions are given in Section 5.

2. Optical polarization and electronic band structure

2.1. Electronic band calculations

Polarization control is theoretically demonstrated on the basis of excitonic selection rules at the band edge in ZnO, relating to the electronic band structures (EBSs) of the CB and VB. The transition energies of anisotropic strained *A*-plane ZnO were calculated using the $\boldsymbol{k} \cdot \boldsymbol{p}$ approximation with the 6 × 6 Bir-Pikus Hamiltonian for the VB. When treating the VBs together with the CB, we make use of the 8 × 8 Hamiltonian for the VB in combination with the 2 × 2

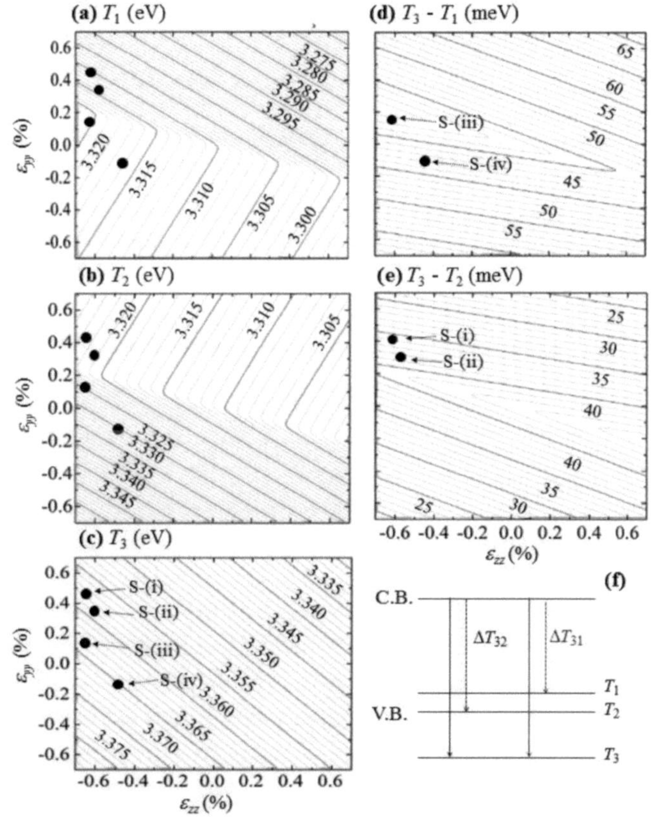

Figure 1. E_1 (a), E_2 (b), and E_3 (c) transition energies as a function of in-plane strain ε_{yy} and ε_{zz} for an *A*-plane ZnO layer at 300 K. Energy difference T_3–T_2: ΔT_{32} (d) and T_3–T_1: ΔT_{32} as a function of in-plane strain ε_{yy} and ε_{zz} (e). Schematic diagram of the transition energies of ΔT_{32} and ΔT_{31}. Black dots indicate experimental results in ZnO layers with different in-plane strains (Figure 1 of [51]). Copyright 2014 by the American Institute of Physics.

matrix for the CB. A high polarization is obtained when the energy shift (ΔE) between two inter-band transitions with $E//c$ and $E\perp c$ is large [40]. The $E//c$ and $E\perp c$ indicate electric fields parallel and perpendicular to the *c*-axis. We evaluated the in-plane anisotropic strains in order to obtain a large ΔE at the excitonic edge of the Γ point. For strain-free *A*-plane ZnO, the VBs consist of $|X \pm iY$ symmetry for the two topmost VBs (*A*- and *B*-excitons) and $|Z$ for the third VB (*C*-exciton). The VBs are mixed because of in-plane anisotropic strains, which provide polarized optical properties. Three transitions involving $|X$-like, $|Y$-like, and $|Z$-like VB require polarized lights along the *x*, *y*, and *z* directions, named T_1, T_2, and T_3 in order to increasing transition energies, respectively. The residual lattice strain $\varepsilon_{\beta\beta}$ ($\beta = x, y, z$) in the films was defined by three orthogonal directions: the *x*-[11-20] direction along the out-of-plane components, and the *y*-[1-100] direction and *z*-[0001] direction along the in-plane component. For calculations at 300 K, the deformation potentials D_j (j = 1–6) for the VB were given by Wrzesinski et al. [41] and Lambrecht et al. [42]. The crystal-splitting (Δ_{so}) and spin orbit splitting (Δ_{so}) were 43 and 16 meV, respectively [37]. The excitonic binding energy was taken to be 60 meV [43], and the band gap was 3.370 eV [44]. The *a*- and *c*-axis lengths were 0.3250 and 0.5204 nm, respectively [45]. The relationship between in-plane and out-of-plane strains can be expressed by the following equation [46].

$$\varepsilon_{xx} = -\frac{C_{12}}{C_{11}}\varepsilon_{yy} - \frac{C_{13}}{C_{11}}\varepsilon_{zz}. \tag{1}$$

where C_{ij} indicates the elastic stiffness constants [47]. **Figure 1a–c** shows the counter plots for variation of the transition energies T_1, T_2, and T_3 as a function of the in-plane strains. The calculations were extended to an arbitrary in-plane strain for the range $|\varepsilon_{yy}|$ and $|\varepsilon_{yy}| \leqq 0.7\%$. The energy and polarization properties of T_1, T_2, and T_3 at a zero strain are matched with those of the *A*-, *B*-, and *C*-excitons, respectively.

In **Figure 1**, the energy of the T_3 transition exhibits a linear correlation between ε_{yy} and ε_{zz}. In contrast, the transition energies of T_1 and T_2 showed larger energy shifts on ε_{zz} compared to ε_{yy}. The dependence of the energy shift on ε_{yy} changed in the vicinity of the boundary between tensile ($\varepsilon_{yy} > 0$) and compressive ($\varepsilon_{yy} < 0$) regions. This relation is well reflected by the oscillation strengths of T_1, T_2, and T_3 transitions (**Figure 2**). This provided a clue for an understanding of the dependence of the energy shift in each transition. However, the dependences of energy shifts of T_1, T_2, and T_3 transitions are determined by both material parameters and the relationship between ε_{xx}, ε_{yy}, and ε_{zz}. **Figure 1(d)** shows the dependence of the difference (ΔT_{31}) in transition energy $T_3 - T_1$ on the in-plane strain. An increase in ΔT_{31} was found in regions of in-plane biaxial tensile ($\varepsilon_{yy} > 0$ and $\varepsilon_{zz} > 0$) and in-plane biaxial compressive ($\varepsilon_{yy} < 0$ and $\varepsilon_{zz} < 0$) strains. On the other hand, the energy difference (ΔT_{32}) in the transition energy $T_3 - T_2$ gradually decreased in both regions of in-plane tensile and compressive strains. This result had an opposite tendency for the difference in ΔT_{31}.

The polarization anisotropy is dependent on the three polarization components of the oscillation strength for the transitions T_1, T_2, and T_3, which are obtained from momentum matrix elements of the type $\varphi^c | p_l | \varphi^v$ with $l = x, y, z$ [48, 49]. The orbital part of the CB can be described as $\varphi^c | = S |$, while that of the VB can be expressed by the following equation:

$$| \varphi^v = \frac{1}{\sqrt{2}}(a_{1n} + a_{2n} + a_{5n} + a_{6n}) | X + \frac{i}{\sqrt{2}}(a_{1n} + a_{2n} - a_{5n} - a_{6n}) | Y + (a_{3n} + a_{4n}) | Z \tag{2}$$

where the coefficients of a_{jn} (j = 1–6) were obtained from the eigenvectors of the Hamiltonian. The relative values of the oscillator strengths $S | p_x | X^2$, $S | p_{ly} | X^2$ and $S | p_z | Z^2$ are normalized to the same value [50]. The oscillator strengths for the transitions T_1, T_2, and T_3 are shown in **Figure 2**. The x polarization (out-of-plane) is not accessible with a normally incident beam when using *A*-plane ZnO. The oscillation strength of the T_3 transition was polarized for z polarization (*E*//*c*) in all strain range. However, the oscillation strengths of the T_1 and T_2 transitions changed substantially. For the in-plane biaxial compressive strains, the oscillation strength of the T_1 transition has a significant component for y polarization ($E \perp c$), providing the ΔT_{31} transition at the band edge. In contrast, the T_2 transition was polarized along the y direction for the in-plane biaxial tensile strains. Optical properties at the band edge for $E \perp c$

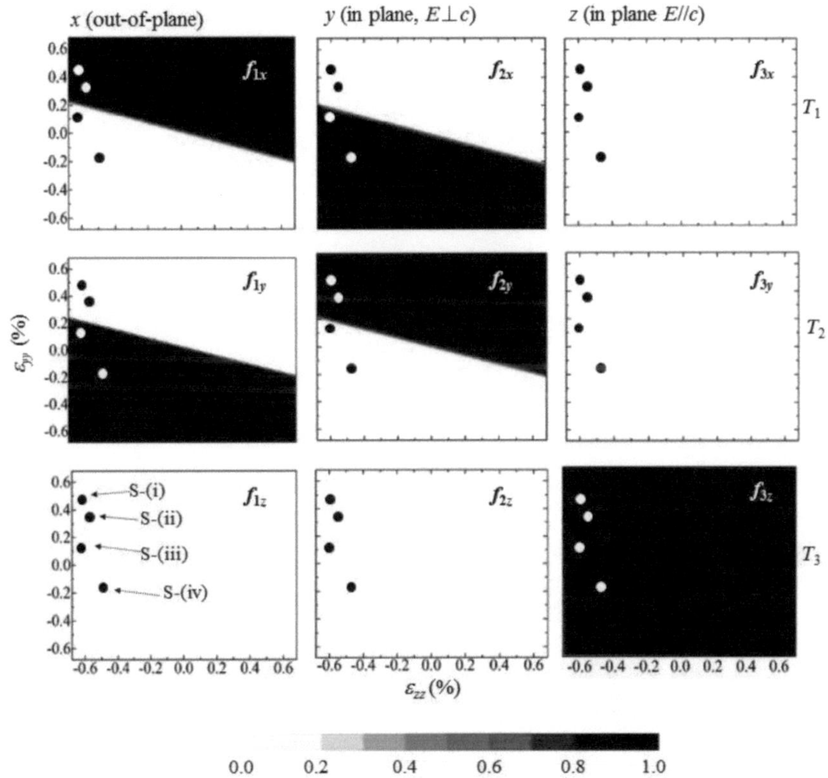

Figure 2. Relative x, y, and z components of the oscillator strength of T_1, T_2, and T_3 transitions as a function of the in-plane strain ε_{yy} and ε_{zz} of an A-plane ZnO layer. Black dots indicate experimental results in ZnO layers with different in-plane strains (Figure 2 of [51]). Copyright 2014 by the American Institute of Physics.

and $E//c$ consisted of a single inter-band transition. A large ΔE was observed within regions of in-plane biaxial compressive strains in A-plane ZnO.

2.2. In-plane strains and crystal symmetries

A-plane ZnO layers were deposited on sapphire (10-12) substrates using PLD. ArF excimer laser pulses (193 nm, 1 J/cm^2 and 3 Hz) were focused on ZnO targets (99.999%: 5 N) located 4 cm from the substrates in an oxygen flow of 1.8×10^{-2} Pa. The substrate temperatures were changed within the range of 450–730°C. ZnO layers grown at 450, 550, 650, and 730°C were named to S-(i), S-(ii), S-(iii), and S-(iv), respectively [51].

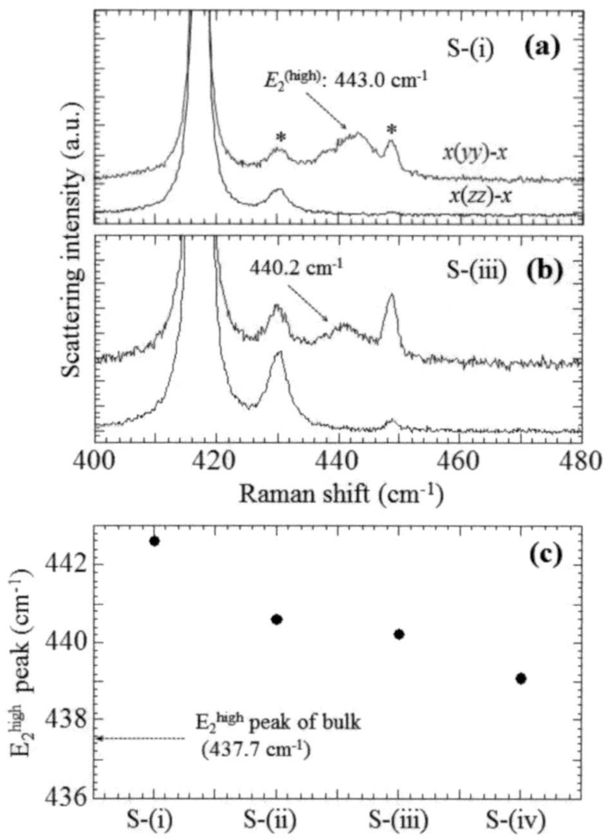

Figure 3. Polarized Raman spectra for ZnO layers at T_g = 450 (a) and 650°C (b) measured in two distinct backscattering geometries. Raman peaks of the sapphire substrate are indicated by*. (c) Peak frequency of E_2^{high} mode in ZnO layers at different in-plane strains (Figure 3 of [51]). Copyright 2014 by the American Institute of Physics.

We investigated crystallographic polarizations of the layers using micro-Raman scattering. The polarized Raman scattering was measured in two distinct backscattering geometries under a 514.5 nm laser at 300 K. We select the E_2^{high} peak based on the A_{1g} mode located at around 442

cm^{-1}. The E_2^{high} peak was found in the $x(yy)$-x geometry (**Figure 3a** and **b**). In contrast, we could not confirm the E_2^{high} peak in the $x(zz)$-x geometry due to crystal symmetry of nonpolar ZnO. In addition, the E_2^{high} peak is very sensitive to the in-plane strains, and their positions of all layers shifted to high-frequency region as compared to that of strain-free *A*-plane ZnO (**Figure 3c**). The frequency shift ($\Delta\omega$) of the E_2^{high} peak can be described as follows [52].

$$\Delta\omega = a(\varepsilon_{xx} + \varepsilon_{yy}) + b\varepsilon_{zz} \pm c(\varepsilon_{xx} + \varepsilon_{yy}) \tag{3}$$

where the coefficients of *a*, *b*, and *c* represent the corresponding deformation potentials per unit strain. In Eq. (3), the red-shifted E_2^{high} peak cleared presence of a compressive strain in the layer.

The lattice parameters of the layers were determined by high-resolution x-ray diffraction (HR-XRD). ZnO layers deposited at different T_g produced various in-plane strains. Large compressive strains along the *c*-axis direction were observed in all layers, which were also supported from the Raman data. All layers had orthorhombic distortion of C_{2v} symmetry. The relationship between the in-plane and out-of-plane strains was satisfied with Eq. (1) in all layers, which makes it possible to compare theoretical and experimental results. The layers grown at below T_g = 650°C showed compressive and tensile strains along the *z*- and *y*-directions. The values of (ε_{yy} and ε_{zz}) of the layers at T_g = 450, 550, and 650°C were (+0.42% and −0.61%), (+0.27% and −0.56%), and (+0.15% and −0.65%), respectively. On the other hand, the layer at T_g = 730°C realized in-plane biaxial compressive strains (ε_{yy} = −0.11% and ε_{zz} = −0.46%). Details were reported to Ref. [32].

2.3. In-plane strains and optical absorptions

Polarized absorption spectra were measured for the $E\perp c$ ($\alpha_\perp$) and $E//c$ ($\alpha_{//}$) geometries. **Figure 3** shows significant polarization anisotropy at the band edge. A plot of α^2 as a function of energy *E* for the ZnO layers at T_g = 450°C [S-(i)] showed that the absorption edge shifted to higher energy for $E//c$ relative to $E\perp c$ by 18 meV. The lines can be fitted by the following equation for direct interband transition:

$$\alpha^2(E) = \beta^2[E - E_{exp}] \tag{4}$$

where the values of β for both polarizations are related to oscillation strengths. We obtained the extrapolated absorption edge (E_{exp}) of 3.320 and 3.3378 eV for $E\perp c$ and $E//c$, respectively (**Figure 4a**). The energy separation (ΔE) at the absorption edge for $E\perp c$ and $E//c$ indicates the difference in the band gap. Values of E_{exp} for a layer at T_g = 650°C [S-(iii)] were 3.321 and 3.353 eV for $E\perp c$ and $E//c$, respectively (**Figure 4b**), resulting in an increased ΔE of 32 meV. The highest ΔE of 42 meV was obtained in the layer at T_g = 730°C [S-(iv), **Figure 4c**]. Herein, the transition energies in layers at T_g = 450°C [S-(i)] and 550°C [S-(ii)] were defined by T_2 and T_3 transitions, which dominated the optical transitions for $E\perp c$ and $E//c$, respectively. In contrast,

E_{exp} values for $E \perp c$ and $E//c$ in layers at T_g = 650°C [S-(iii)] and 730°C [S-(iv)] were equivalent to the T_1 and T_3 transitions, respectively. Therefore, a decrease in ε_{yy} resulted in the change of the inter-band transitions from ΔT_{32} to ΔT_{31}. The large ΔE at the band edge was caused by the energy difference between T_1 and T_3 transitions (**Figure 4c**).

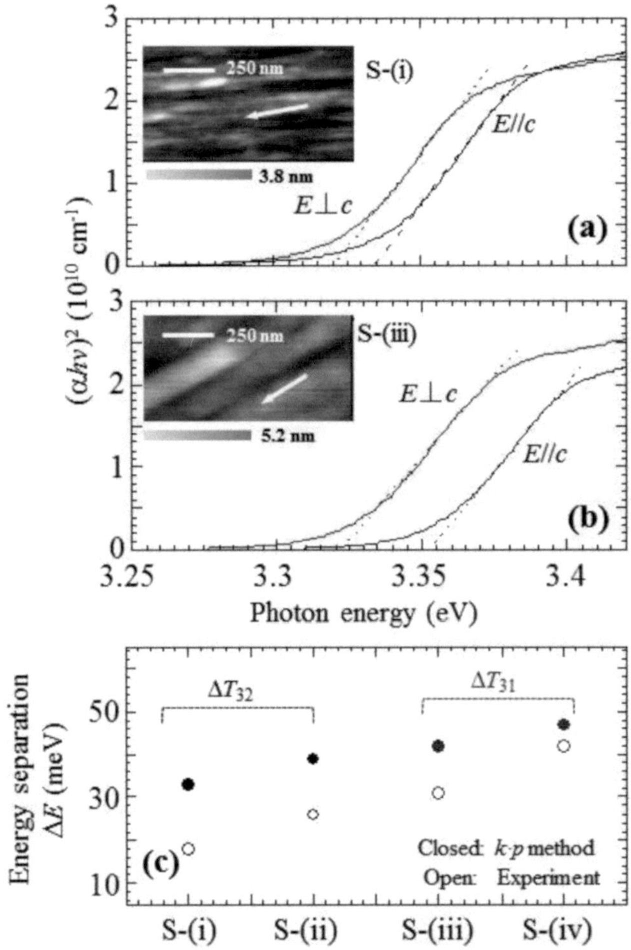

Figure 4. α^2E^2 as a function of photon energy E and AFM image (inset) in a ZnO layer at T_g = 450°C (a) and T_g = 650°C (b). (c) Correlation between experimental ΔE and theoretical ΔT between the two polarization directions for ZnO layers at different in-plane strains (Figure 4 of [51]). Copyright 2014 by the American Institute of Physics.

In this section, we theoretically investigated relationship between energy separations and in-plane anisotropic strains on A-plane ZnO, which were sufficiently verified from the experimental results. The large ΔE at the band edge was obtained by introducing the in-plane compressive strains in the layers, providing the large difference in absorption measured in the $E//c$ and $E \perp c$ geometries. It is expected that the introduction of in-plane compressive strains into A-plane ZnO can produce highly polarized PL.

http://dx.doi.org/10.5772/64724

3. In-plane lattice strains and polarized luminescence

3.1. *A*-plane ZnO homoepitaxial layer growth

We report homoepitaxial growth for strain-free and strained *A*-plane ZnO layers from viewpoints of morphologies, structural and optical properties. In particular, we present relationship between polarized PL and in-plane lattice strains because any in-plane anisotropic strain changes the crystal symmetry in the layers in terms of significant modifications of the EBS of ZnO.

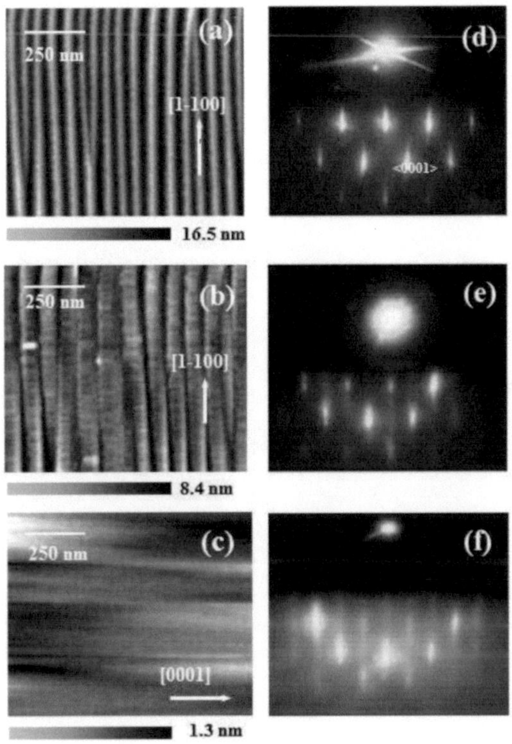

Figure 5. AFM images of strain-free ZnO layers with thicknesses of (a) 10 nm, (b) 28 nm, and (c) 140 nm. RHEED patterns with the [0001] azimuth of strain-free ZnO layers with thicknesses of (d) 10 nm, (e) 28 nm, (f) 140 nm (Figure 1 of [53]). Copyright 2012 by the American Institute of Physics.

The hydrothermally synthesized substrates were supplied by Crystec GmbH (Germany) and Goodwill (Russia). Both substrates were annealed at 1100°C for 1 h prior to PLD growth. *A*-plane ZnO layers were homoepitaxially grown on both substrates at 550°C in an oxygen flow of 10^{-4} mbar. ArF excimer laser pulses were focused on a ZnO target (5 N) 4.0 cm from the substrate [53].

Growth processes of the homoepitaxial layers on Crystec ZnO substrates were monitored using reflection high electron energy diffraction (RHEED) with the [0001] azimuth. Atomic force microscopy (AFM) was used to observe surface morphologies. At the beginning of layer growth, up to 10 nm in thickness, the V-groove was seen with a RHEED pattern of three-dimensional (3D) spot (**Figure 5a** and **d**). However, continued layer growth of ZnO up to 28 nm, changed to a slightly smooth surface by the filling of the V-groove structure (**Figure 5b**). As a consequence, the 3D spot of the RHEED pattern were weakened (**Figure 5c**). Finally, the layer morphology at a thickness of 140 nm showed a very flat surface. The RHEED pattern showed a sharp blight stripe with a high contrast to the background, which resulted from a clean and smooth surface (**Figure 5c** and **f**).

A cross-sectional transmittance electron microscopy (X-TEM) image with the [0001] zone axis showed that layer surface had nano-facet structures (**Figure 6a** and **b**). The nano-facets were consisted of the *M*-plane from the angle between the nano-facets and (11-20) plane due to higher surface energy of *A*-plane ZnO than *M*-plane ZnO (**Figure 6d**) [54]. The XRD showed that the 2θ pattern of the (11-20) plane was accompanied by Pendellosung fringes, which is related to interference between the layer surface and the layer/substrate interface. The XRD profile cleared coherent growth of *A*-plane ZnO.

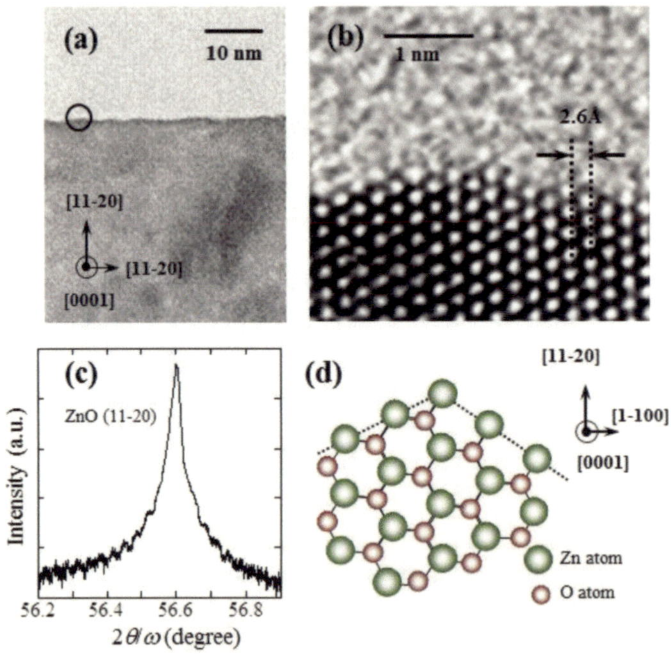

Figure 6. (a) Low-magnification and (b) high-magnification X-TEM images of a strain-free ZnO layer with a thickness of 140 nm. (c) Schematic cross-section of the atomic lattice of ZnO viewed along the [0001] direction (Figure 2 of [53]). Copyright 2012 by the American Institute of Physics.

Strained *A*-plane ZnO layers could be obtained using the goodwill ZnO substrates. The 2θ pattern of the (11-20) plane was separated from that of the ZnO substrate. The out-of-plane strain (ε_{xx}) expanded 0.47% with x being parallel to the [11-20] direction (**Figure 7a**). In addition, the layer showed a broad ω-rocking curve with a line-width of 0.14° because of a lattice relaxation at the heterointerface (**Figure 7b**). As a result, the in-plane lattice strain (ε_{yy} and ε_{zz}) showed anisotropic compressions of 0.04 and 0.78%, with y and z, respectively, being parallel to the [10-10] and [0001] direction. These values were determined from reciprocal space mapping of the (0002) and (10-10) planes (**Figure 7c** and **d**). For the local strain in A-plane wurtzite, ε_{xx} can be expressed using Eq. (1). The calculated ε_{xx} was estimated as +0.42%, which was close to the experimental ε_{xx}. The lattice expansion along the [11-20] direction resulted from the anisotropic compressive strains along the in-plane directions.

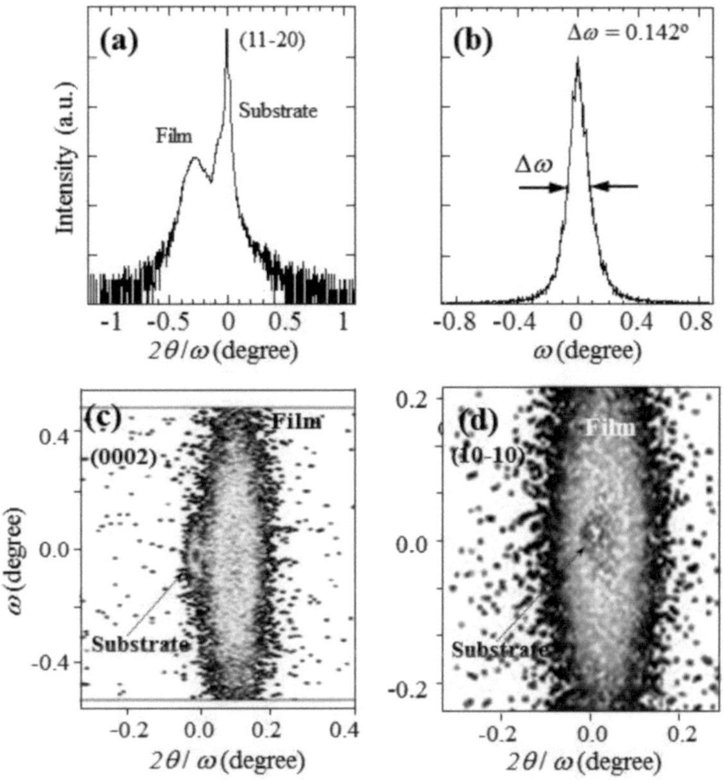

Figure 7. (a) $2\theta/\omega$ and (b) ω-rocking profiles of the (11–20) plane for strained ZnO layers. Reciprocal space mapping of (c) (0002) and (d) (10–10) for the strained ZnO layer (Figure 3 of [53]). Copyright 2012 by the American Institute of Physics.

3.2. Polarized PL from anisotropic strained ZnO layers

The in-plane lattice strains were highlighted to polarized PL. An oxygen pressure [$p(O_2)$] during the PLD growth could successfully produce the different lattice strains. The value of ε_{xx} gradually increased with increasing $p(O_2)$. In contrast, the values of ε_{yy} and ε_{zz} decreased. The relationship between ε_{xx}, ε_{yy}, and ε_{zz} was consistent with the theoretical calculation [Eq. (1)]. The increased ε_{xx} along the x direction was attributed to the anisotropic compressive strains along the in-plane ($y - z$) directions. The VBs of ZnO consist of p-like orbitals with wave functions of $|X \pm iY$ states for Γ_9 and Γ_{7upper} bands and a $|Z$ state for the Γ_{7low} band. The lowest A-exciton (Γ_9) occurs because of $E \perp c$, while the C-exciton (Γ_{7low}) occurs at the higher energy under $E//c$ [37].

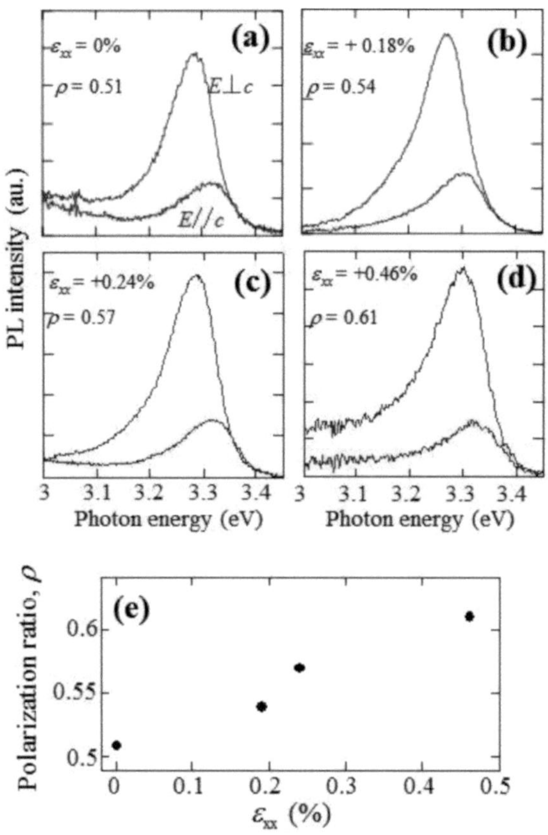

Figure 8. (a) Polarized PL spectra of a strain-free ZnO layers. Polarized PL spectra of strained layers with ε_{xx} = +0.19% (b), +0.24% (c), and +0.46% (d). (e) Polarized ratio (ρ) as a function of lattice strain (ε_{xx}) (Figure 5 of [53]). Copyright 2012 by the American Institute of Physics.

The strain-free ZnO layer showed polarization to $E \perp c$ with a polarization ratio (ρ) of 0.51 (**Figure 8a**). A value of ρ can be expressed by the following relation:

$$\rho = \frac{I_{\perp} - I_{//}}{I_{\perp} - I_{//}} \quad (5)$$

where $I_{\perp}$ and $I_{//}$ indicate the peak intensities for $E \perp c$ and $E//c$, respectively. Peak energies of PL were determined as 3.286 and 3.327 eV at 300 K for $E \perp c$ and $E//c$, respectively, which resulted in an energy separation (ΔE) of 41 meV. This value was close to the theoretical value between *A*- and *C*-excitonic transitions. The PL intensities of all strained ZnO layers were polarized to $E\perp c$ (**Figure 8b–d**). In-plane anisotropic strains change the EBS of ZnO. The original $|X \pm iY$ states are separated into $|X$-like and $|Y$-like states. The $|Y$-like state is raised, while that of the $|X$-like state is lowered. Therefore, a large polarization ratio of PL would be measured by the large energy separation between $|Y$-like state and $|Z$-like states for $E\perp c$ and $E//c$, respectively. The strained *A*-plane ZnO layers showed the blueshift of the PL peak energy when changing from $E\perp c$ to $E//c$. ρ gradually increased by up to 0.62 with increasing out-of-plane strain (ε_{xx}) (**Figure 8e**) because photo-excited holes occupy $|Y$-like and $|Z$-like states of the VBs by the Boltzmann-like distribution. The increase in ρ reflects a large difference in transition energies for $E\perp c$ to $E//c$, which originated from a lift of the degeneracy of VB with the anisotropic compressive strains. The degree of PL polarization was well correlated with the in-plane anisotropic strains.

In this section, the in-plane anisotropic strains were introduced in *A*-plane ZnO layers by changing substrate-type and growth condition. Strain-free layers were obtained when using Crystec ZnO substrates, while the use of Goodwill ZnO substrates provided strained ZnO layers. The magnitude of lattice strain was systematically changed by controlling the oxygen pressure during the PLD growth. The correlation between PL polarization and in-plane anisotropic strain was evidenced experimentally.

4. Control of polarized luminescence by quantum well geometries

4.1. $Cd_xZn_{1-x}O$/ZnO quantum wells

The anisotropic lattice distortion causes strain-induced modification of the EBS in ZnO, leading to polarization modulation of PL. That is, polarization ratio of PL is strongly highlighted by the anisotropic lattice distortion introduced in nonpolar ZnO. In this section, we report on quantum size effects of polarized PL on *A*-plane $Cd_xZn_{1-x}O$/ZnO QWs. A change in a quantum well width of a QW has a remarkable influence on a quantum size effect. There is a strong quantum confinement following a narrowing of well width (L_W). In our work, we observe highly polarized PL from *A*-plane $Cd_{0.06}Zn_{0.94}O$/ZnO QWs when using a narrow well width [30]. This finding suggests that a valence sub-band and a corresponding thermal carrier distribution are dependent on well width. The $Cd_{0.06}Zn_{0.94}O$ well layer is very sensitive to an epitaxial lattice strain owing to lattice mismatch between $Cd_{0.06}Zn_{0.94}O$ well and ZnO layers. The contribution of well width to the optical properties on *A*-plane $Cd_{0.06}Zn_{0.94}O$/ZnO QWs becomes very important to achieve highly polarized PL at room temperature.

$Cd_{0.06}Zn_{0.94}O$/ZnO QWs with different L_W were grown on *A*-face ZnO substrates by PLD. A $Cd_{0.06}Zn_{0.94}O$ well layer was embedded between a 150 nm-thick ZnO buffer and 5 nm-thick ZnO capping layers. The ZnO buffer was deposited at T_g = 650°C, while the well and capping layers were both grown at T_g = 260°C. An AFM image of an *A*-plane QW with L_W = 4.8 nm showed that the morphology of ZnO capping of the QW showed a small surface undulation elongated along the [0001] direction due to the anisotropic migrations of adatoms on the growing surface (**Figure 9a**) [55–57]. The height scale in the image was 1.4 nm with a roughness of a few angstroms, which was also confirmed using RHEED with a streaky pattern (**Figure 9b**). The whole specimens gave the same surface states, being independent of well width. The scanning electron microscopy (STEM) image with the [0001] zone axis revealed that an interface between the well and barrier layers was very smooth (**Figure 9c**). Therefore, we could keep the structural disorder of the interface between the barrier and the well layers to a minimum, which can reduce a partial loss of polarization deriving from the breaking of the selection rule owing to the structural disorder at the interface.

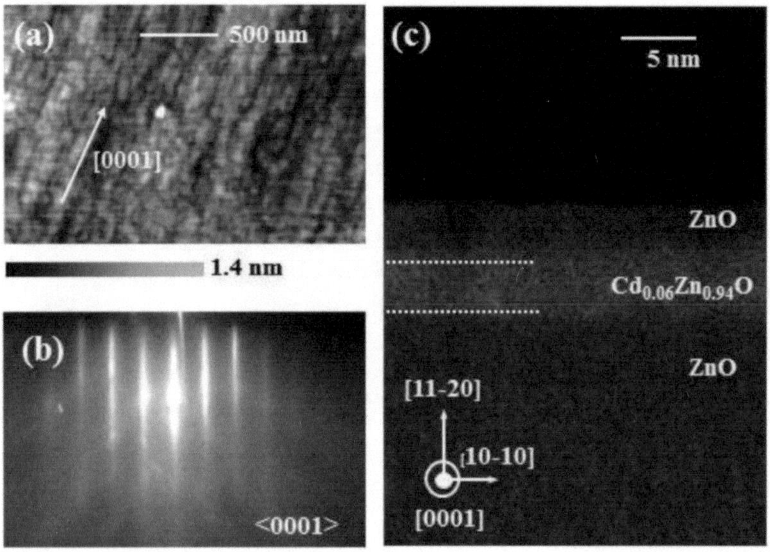

Figure 9. (a) AFM image and (b) RHEED pattern of a QW with L_W = 4.8 nm. (c) Cross-sectional STEM image taken along the [0001] zone axis of a QW with L_W = 4.8 nm (Figure 1 of [30]). Copyright 2012 by the American Institute of Physics.

4.2. Polarized PL properties in *A*-plane $Cd_{0.06}Zn_{0.94}O$/ZnO QWs

Figure 10a and **b** shows polarized PL spectra of a QW with L_W = 4.8 and 2.0 nm. The PL intensity for $E \perp c$ was higher than that for $E // c$. Energy separations (ΔE) between PL peak energies of $E \perp c$ and $E // c$ were 41 and 47 meV for QWs with L_W = 4.8 and 2.0 nm, respectively. These values were larger than that for the strain-free ZnO layer (ΔE: 35 meV). The large ΔE values are attributed to the strain-induced modification of the EBS in ZnO. In particular, the VB band is separated to three levels: |*Y*-like, |*Z*-like, and |*X*-like states in order to decrease energy by introducing anisotropic compressive strains on the growing surfaces of well layers. The y- ($E \perp c$) and z-polarized ($E // c$) light components are dominated by the |*Y*-like and |*Z*-like states, respectively.

The polarization ratio (ρ) of the PL intensity for $E \perp c$ and $E // c$ was dependent on temperature (T) (**Figure 10c** and **d**). Excited carriers (electrons and holes) at 300 K fully distributed in the $|Z$-like state, providing ρvalues of 0.63 and 0.71 for QWs with an L_W of 4.8 and 2.0 nm, respectively. The values of ρ gradually increased with decreasing T because the distributions of holes in the $|Y$-like state was higher than that in the $|Z$-like state. The high ρ of 0.82 and 0.88 was observed at 200 K by a negligible carrier distribution in the $|Z$-like state. The driving force to elevate carriers up to the $|Z$-like state is related to k_BT. When two different VB bands with the same effective mass are separated by ΔE, and carriers obey a Boltzmann-like distribution. ρ is described by the following expression [58]:

$$\rho = \frac{1-\exp(-\frac{\Delta E}{k_B T})}{1-\exp(-\frac{\Delta E}{k_B T})} \tag{6}$$

The calculated ΔE was 41 and 45 meV for QWs with L_W = 4.8 and 2.0 nm, respectively. These values were close to the experimental ΔE obtained from polarized PL spectra at 300 K (**Figure 10c** and **d**). Optical anisotropy for a temperature in the range 200–300 K obeys the

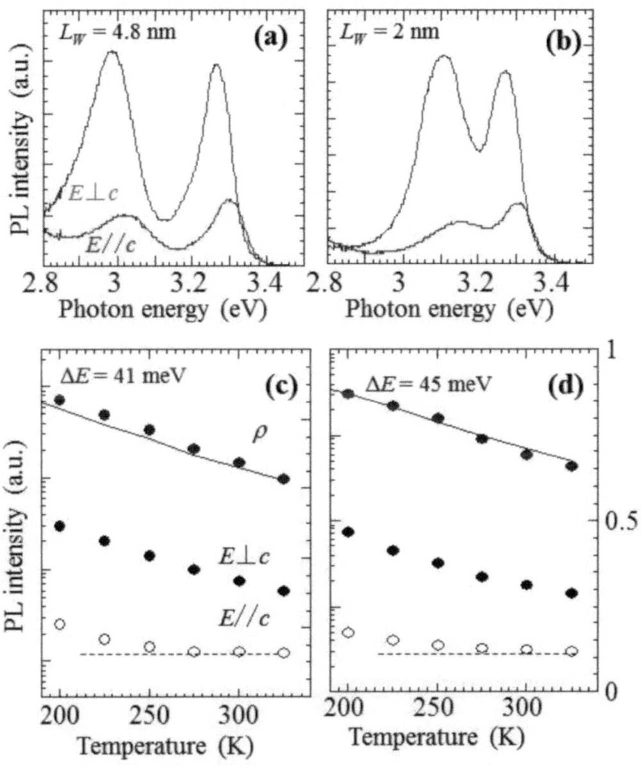

Figure 10. Polarized PL spectra at 300 K of a QW with LW = 4.8 nm (a) and 2.0 nm (b) for $E\perp c$ and $E//c$. Temperature dependence of PL peak energy for $E\perp c$ and $E//c$ and polarization (ρ) on QWs with L_W = 4.8 nm (c) and 2.0 nm (d) (Figure 2 of [30]). Copyright 2012 by the American Institute of Physics.

polarization selection rule based on a Boltzmann-like distribution. On the other hand, this selection rule breaks at low temperatures below 200 K because of the weak mixing between the different VB states, which is related to excitonic localization in the well layers [59].

4.3. Control of polarized PL in the QWs

Figure 11a shows the PL peak energies of QWs at 300 K as a function of well width. The PL peak energies systematically shifted to the higher energy with decreasing L_W because of quantum confinement. In our QW samples, no shift was observed when varying the excitation power from 0.1 to 10 W/cm^2, indicating no carrier screening effect owing to the absence of QCSE along the nonpolar growing direction. These trends were similar to those reported earlier for homoepitaxial ZnO/$Mg_xZn_{1-x}O$ QWs on nonpolar ZnO substrates. Moreover, PL peak energies at 300 K for QWs with L_W = 2.8 and 6 nm are also applicable to Eq. (6) because

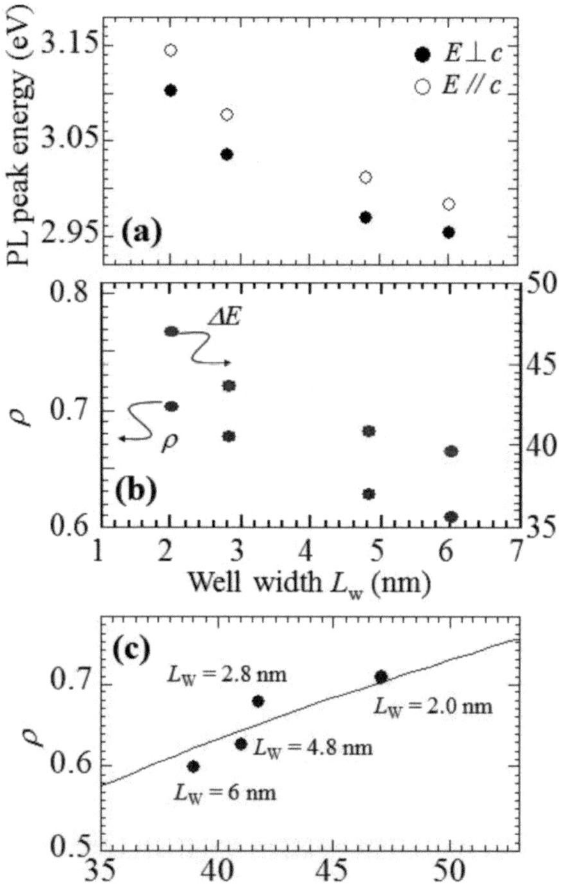

Figure 11. (a) Dependence of PL peak energy at 300 K for $E \perp cs$ and $E//c$ on L_W. (b) Dependence of ΔE and ρ on L_W at 300 K. (c) Correlation between ΔE and ρ in A-plane $C_{d0.06}Zn_{0.94}O$/ZnO QWs with different L_W. The solid line is described using Eq. (6) (Figure 3 of [30]). Copyright 2012 by the American Institute of Physics.

localization of excitons was 10.2 and 9.8 meV in QWs with L_W = 2.8 and 6.0 nm, respectively. Tuning of L_W in a QW influenced ΔE and ρ of the polarized PL at 300 K. **Figure 11b** shows the values of ρ and ΔE between the topmost VB levels as a function of L_w. At 300 K, ρ and ΔE gradually decreased with a widening of L_W and acquired values as low as 0.61 and 39 meV, respectively, when L_w was set to 6.0 nm. The values of ρ and ΔE were strongly dependent on a L_W. In order to investigate the cause of this dependence, we plotted the correlation between ρ and ΔE using Eq. (6). The calculated fit the experimental data quite well (**Figure 11c**). This indicated that the PL anisotropy of all QWs was derived from the selection rule based on a Boltzmann-like distribution. The energy separation of the topmost VB levels determined the polarization ratio of QWs.

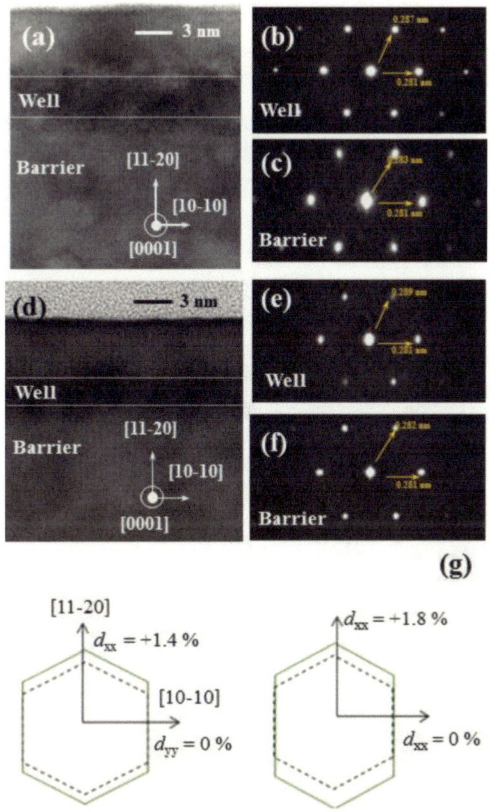

Figure 12. (a) Cross-sectional TEM image of a QW with L_W = 4.8 nm, and RSD patterns taken in the well (b) and barrier layer (c) regions. (d) Cross-sectional TEM image of a QW with L_W = 2.0 nm and RSD patterns taken in the well and barrier layers. (g) Schematic cross-section of a ZnO host viewed along the [0001] direction. The black dot and green solid lines represent ZnO layers and lattice disordered $Cd_{0.06}Zn_{0.94}O$ layers in QWs with L_W = 4.8 and 2.0 nm, respectively (Figure 4 of [30]). Copyright 2012 by the American Institute of Physics.

TEM images of QWs with L_W = 4.8 and 2.0 nm are shown in **Figure 12a** and **d**, respectively. The location of well layers in the QWs was identified by z-contrast STEM images. A fast Fourier transform (FFT) analysis was performed to examine a plane interval on the basis of a reciprocal space diffractogram (RSD) pattern. The plane intervals (d_{xx}) along the x-axis in the well layers were 1.65 and 1.66 Å for QWs with L_W = 4.8 and 2.0 nm, respectively. For a QW with L_W = 4.8 nm, the value of d_{xx} in the well layer expanded by 1.4% compared with that in the barrier layer (**Figure 12b** and **c**). Expansion of d_{xx} to the order of 1.8% was obtained for a QW with L_W = 2.0 nm (**Figure 12e** and **f**). In contrast, the in-plane intervals (d_{yy}) of the well layer were consistent with those of barrier layers for both QWs. As a consequence, the plane intervals along the x-axis only increased with a narrowing of L_W, which resulted in anisotropic lattice distortions of the basal hexagon of the $Cd_{0.06}Zn_{0.94}O$ unit cell in the well layers as shown schematically in **Figure 12g**.

Makino et al. reported lattice parameters of relaxed $Cd_xZn_{1-x}O$ (up to x = 0.073) epitaxial layers on $ScMgAlO_4$ substrates [60]. From their estimations, lattice parameters of a- and c-axis lengths at x = 0.06 are calculated as 3.260 and 5.241 Å, respectively, which indicates the a-axis length of $Cd_{0.06}Zn_{0.94}O$ slightly expanded by 0.3% compared to that of ZnO. Thus, the lattice distortions of well layers are larger than those of relaxed $Cd_xZn_{1-x}O$ layers, which lead to a change of structural symmetry from C_{6v} to C_{2v}. It is thought that the anisotropic lattice distortions along the growing direction are related to anisotropic compressive strains on the in-plane growing surfaces. Mata et al. reported that the highly polarized PL of a-plane GaN/AlN QWs exhibited a narrowing of L_W because of the increased anisotropic lattice distortions in the GaN well layers [61]. Moreover, the in-plane optical anisotropy was found even in (001) zinc-blende GaAs/AlGaAs QWs when its D_{2d} symmetry is reduced to C_{2v} in the GaAs well layers [62]. Accordingly, the highly polarized PL resulting from quantum confinement in a-plane $Cd_{0.06}Zn_{0.94}O$/ZnO QWs is associated with the anisotropic lattice distortions of well layers giving the change of crystal symmetry. This phenomenon is derived from an increase in the energy separation between the two topmost VB levels in the well layers of the QWs.

In this section, the dependence of polarized PL on well width was studied in A-plane $Cd_{0.06}Zn_{0.94}O$/ZnO QWs. The polarization ratio of PL gradually enhanced with a narrowing of L_W, resulting from the energy separation between the two topmost VB states. These effects were a result of the anisotropic compressive distortions in the well layers. The lattice distortions in the well layers played an important role in determining the degree of polarized PL, which could be controlled by well width.

5. Conclusion remarks

This chapter was reported polarized PL of nonpolar ZnO layers and their QW structures in terms to the crystal symmetry and the in-plane lattice strain during PLD growth. Anisotropic optical properties were closely related to in-plane anisotropic strains introduced into the layers, which were demonstrated on A-plane ZnO. Overall, the in-plane compressive strains could obtain highly anisotropic optical properties from consistency between experimental and theoretical viewpoints. We actually observed the polarized PL spectra at 300 K in the strain-controlled A-plane ZnO homoepitaxial layers. Furthermore, we showed systematic correlation between the polarization degree of PL and the in-plane lattice strain. Finally, we achieved highly polarized PL at room temperature by controlling a well width in the QW structures as a consequence of change in crystal symmetry from C_{6c} to C_{2v} at a hetero interface between the well and barrier layers.

Acknowledgements

This work was supported in part by a grant-in-aid from the JSPS Core-to-Core Program, A. Advanced Research Network, a grand from the Japan Science and Technology Agency (JST: A-Step) and a grant-in-aid from Exploratory Research and Scientific Research (B).

Author details

Hiroaki Matsui[1*] and Hitoshi Tabata[2]

*Address all correspondence to: hiroaki@ee.t.u-tokyo.ac.jp

1 Department of Bioengineering, The University of Tokyo, Bunkyō, Japan

2 Department of Electric Engineering and Information Systems, The University of Tokyo, Bunkyō, Japan

References

[1] D. G. Thomas. The exciton spectrum of zinc oxide. Journal of Physical and Chemical Solids. 1960;15:86–89. doi:10.1016/0022-3697(60)90104–9

[2] A. Ohtomo, M. Kawasaki, T. Koida, K. Masubuchi, H. Koinuma, Y. Sakurai, Y. Yoshida, T. Yasuda and Y. Segawa. $Mg_xZn_{1-x}O$ as a II–VI widegap semiconductor alloy. Applied Physics Letters. 1998;72:2466–2468. doi:10.1063/1.121384

[3] T. Gruber, C. Kircher, R. Kling, F. Reuss, A. Waag, F. Bertram, D. Forster, J. Christen and M. Schreck. Optical and structural analysis of ZnCdO layers grown by metalorganic vapor-phase epitaxy. Applied Physics Letters. 2003;83:3290–3292. doi: 10.1063/1/1620674

[4] S. Sadofev, P. Schäfer, Y. H. Fan, S. Blumstengel, F. Henneberger, D. Schulz and D. Klimm. Radical-source molecular beam epitaxy of ZnMgO and ZnCdO alloys on ZnO substrates. Applied Physics Letters. 2007;91:201923(1)–(3). doi:10.1063/1.2815662

[5] F. Bertram, S. Giemsch, D. Forster, J. Christen, R. Kling, C. Kirchner and A. Waag. Direct imaging of phase separation in ZnCdO layers. Applied Physics Letters. 2006;88:061915(1)–(3). doi:10.1063/1.2172146

[6] M. D. Neumann, C. Cobet, N. Esser, B. Laumer, T. A. Wassner, M. Eickhoff, M. Feneberg and R. Goldhan. Optical properties of MgZnO alloys: Excitons and exciton-phonon complexes. Journal of Applied Physics. 2011;110:013520(1)–(8). doi:10.1063/1.3606414

[7] T. Makino, C. H. Chia, N. T. Tuan, H. D. Sun, Y. Segawa, M. Kawasaki, A. Ohtomo, K. Tamura and H. Koinuma. Room-temperature luminescence of excitons in ZnO/(Mg, Zn)O multiple quantum wells on lattice-matched substrates. Applied Physics Letters. 2000;77:975–977. doi:10.1063/1.1289066

[8] A. Y. Azariv, T. C. Zhang, B. G. Svensson and A. Y. Kuznetsov. Cd diffusion and thermal stability of CdZnO/ZnO heterointerfaces. Applied Physics Letters. 2011;99:11903(1)–(3). doi:10.1063/1.3639129

[9] T. Makino, A. Ohtomo, C. H. Chia, Y. Segawa, H. Koinuma and M. Kawasaki. Internal electric field effect on luminescence properties of ZnO/(Mg, Zn)O quantum wells. Physica E. 2004;21:671–675. doi:10.1063/1.3481078

[10] S. H. Park and D. Ahn. Internal field engineering in CdZnO/MgZnO quantum well structures. Applied Physics Letters. 2009;v:083507(1)–(3). doi:10.1063/1.390490.

[11] H. Teisseyre, A. Kaminska, S. Birner, T. D. Young, A. Suchocki and A. Kozanecki. Influence of hydrostatic pressure on the built-in electric field in ZnO/MgZnO quantum wells. Journal of Applied Physics. 2016;119:215702(1)–(8). doi:10.1063/1.4953251

[12] K. Sato, T. Abe, R. Fujimura, K. Yasuda, T. Yamaguchi, H. Kasada and K. Ando. Stark effects of ZnO thin films and ZnO/MgZnO quantum wells. Physica Status Solidi (c). 2012;209:1801–1804. doi:10.1002/201100592

[13] W. L. Wang, C. Y. Peng, Y. T. Ho and L. Chang. Microstructure of *a*-plane ZnO grown on $LaAlO_3$ (001). Thin Solid Films. 2010;518:2967–2970. doi:10.1016/j.tsf.2009.09.186

[14] T. Huang, S. Zhou, H. Teng, H. Lin, J. Wang, P. Han and R. Zhang. Growth and characterization of ZnO films on (001), (100) and (010) $LiGaO_2$ substrates. Journal of Crystal Growths. 2008;310:3144–3148. doi:10.1016/j.jcrsgro.2008.03.037

[15] X. H. Wei, Y. R. Li, J. Zhu, W. Huang, Y. Zhang, W. B. Luo and H. Ji. Epitaxial properties of ZnO thin films on $SrTiO_3$ substrates grown by laser molecular beam epitaxy. Applied Physics Letters. 2007;90:151918(1)–(3). doi:10.1063/1.2719026

[16] M. M. C. Chou, D. R. Hang, S. C. Wang, C. Chen, S. C. Wang and C. Y. Lee. Nonpolar *a*-plane ZnO growth and nucleation mechanism on (100) (La, Sr) (Al, Ta)O_3 substrate. Materials Chemistry and Physics. 2011;125:791–795. doi:10.1016/j.atchemphys.2010.09.057

[17] C. Y. Peng, J. S. Tian, W. L. Wang, Y. T. Ho and L. Chang. Morphology evolution of *a*-plane ZnO films on *r*-plane sapphire with growth by pulsed laser deposition. Applied Surface Science. 2013;265:553–557. doi:10.1016/j.apsusc.2012.11.044

[18] S. K. Han, S. K. Hong, J. W. Lee, J. Y. Lee, J. H. Song, Y. S. Nam, S. K. Chang, T. Minegishi and T. Yao. Structural and optical properties of nonpolar *A*-plane ZnO films grown on *R*-plane sapphire substrates by plasma-assisted molecular-beam epitaxy. Journal of Crystal Growth. 2007;309:121–127. doi:10.1016/j.jcrysgro.2007.09.025

[19] T. Moriyama and S. Fujita. Growth behaviour of nonpolar ZnO on M-plane and R-plane sapphire by metalorganic vapor phase epitaxy. Japanese Journal of Applied Physics. 2005;44:7919–7291. doi:10.1143/JJAP.44.7919

[20] M. D. Neumann, N. Esse, J. M. Chauveau, R. Goldhahn and M. Feneberg. Inversion of absorption anisotropy and bowing of crystal field splitting in wurtzite MgZnO. Applied Physics Letters. 2016;108:221105(1)–(4). doi:10.1063/1.4953159

[21] H. Lin, H. J. Su, C. H. Lu, C. P. Chang, W. R. Liu and W. F. Hsieh. Pump polarization dependent ultrafast carrier dynamics and two-photon absorption in an *a*-plane ZnO epitaxial film. Applied Physics Letters. 2015;107:142107(1)–(4). doi:10.1063/1.4933038

[22] C. M. Lai, Y. E. Huang, K. Y. Lou, C. H. Chen, L. W. Tu and S. W. Feng. Experimental and theoretical study of polarized photoluminescence caused by anisotropic strain relaxation in nonpolar a-plane textured ZnO grown by a low-pressure chemical vapour deposition. Applied Physics Letters. 2015;107:022110(1)–(4). doi:10.1063/1.4926978

[23] S. Yang, H. C. Hsu, W. R. Liu, B. H. Lin, C. C. Kuo, C. H. Hsu, M. O. Eriksson, P. O. Holtz and W. F. Hsieh. Recombination dynamics of a localized exciton bound at basal stacking faults within the *m*-plane ZnO film. Applied Physics Letters. 2014;105:011106(1)–(5). doi:10.1063/1.4887280

[24] C. H. Chao, W. J. Weng and D. H. Wei. Enhanced UV Photodetector response and recovery times using a nonpolar ZnO sensing layer. Journal of Vacuum Science and Technology A. 2016;34:02D106(1)–(7). doi:10.1116/4939751

[25] F. J. Klüpfel, H. V. Wenckstern and M. Grundmann. Low frequency noise of ZnO based metal-semiconductor field-effect transistors. Applied Physics Letters. 2015;106:033502(1)–(4). doi:10.163/1.4906292

[26] G. Tabares, A. Hierro, M. Lopez-Ponce, E. Munoz, B. Vinter and J. M. Chauveau. Light polarization sensitive photodetectors with *m*- and *r*-plane homoepitaxial ZnO/MgZnO quantum wells. Applied Physics Letters. 2015;106:061114(1)–(3). doi:10.1063/1.4908183

[27] A. Shao, P. S. Ku, X. L. Wang, W. F. Cheng, C. W. Leung and A. Ruotolo. Growth and characterization of nonpolar, heavily Mn-substituted ZnO films. Journal of Applied Physics. 2014;115:17D703(1)–(3). doi:10.1063/1.4862846

[28] H. H. Zhang, X. H. Pan, Y. Li, Z. Z. Ye, B. Lu, W. Chen, J. Y. Huang, P. Ding, S. S. Chen, H. P. He, J. G. Lu, L. X. Chen and C. L. Ye. The role of band alignment in p-type

conductivity of Na-doped ZnMgO: Polar versus nonpolar. Applied Physics Letters. 2014;104:112106(1)–(4). doi:10.1063/1.4869481

[29] J.-M. Chauveau, M. Teisseire, H. Kim-Chauveau, C. Deparis, C. Morgain and B. Vinter. Benefits of homoepitaxy on the properties of nonpolar (Zn, Mg)O/ZnO quantum wells on a-plane ZnO substrates. Applied Physics Letters. 2010;97:081903(1)–(3). doi: 10.1063/1.3481078

[30] H. Matsui and H. Tabata. The contribution of quantum confinement to optical anisotropy of *a*-plane $Cd_{0.0}6Zn_{0.94}O/ZnO$ quantum wells. Applied Physics Letters. 2012;100:171910(1)–(4). doi:10.1063/1.4707384

[31] H. Matsui, N. Hasuike, H. Harima and H. Tabata. Growth evolution of surface nanowires and large anisotropy of conductivity on MgZnO/ZnO quantum wells based on M-nonpolar (10–10) ZnO. Journal of Applied Physics. 2008;104:094309(1)–(6). doi: 10.1063/1.3009959

[32] T. S. Ko, T. C. Lu, L. F. Zhuo, W. L. Wang, M. H. Liang, H. C. Kuo, S. C. Wang, L. Chang and D. Y. Lin. Optical characteristics of a-plane $Zn/Zn_{0.8}Mg_{0.2}O$ multiple quantum wells grown by pulse laser deposition. Journal of Applied Physics. 2010;108:073504(1)–(3). doi:10.1063/1.3488898

[33] G. Tabares, A. Hierro, B. Vinter and J.-M. Chauveau. Polarization-sensitive schottky photodiodes based on *a*-plane ZnO/ZnMgO multiple quantum wells. Applied Physics Letters. 2011;99:071108(1)–(3). doi:10.1063/1.3624924

[34] Z. Zhang, H. V. Wenckstern, J. Lenzner and M. Grundmann. Wavelength-selective ultraviolet (Mg, Zn)O photodiodes: Tuning of parallel composition gradients with oxygen pressure. Applied Physics Letters. 2016;108:243503(1)–(4). doi: 10.1063/1.4954375

[35] K. H. Baik, H. Kim, J. Kim, S. Jung and S. Jang. Nonpolar light emitting diode with sharp near-ultraviolet emissions using hydrothermally grown ZnO on p-GaN. Applied Physics Letters. 2013;103:091107(1)–(4). doi:10.1063/1.4819847

[36] V. Awasthi, S. K. Pandey, V. Garg, B. S. Brajendra, S. Sengar and P. Sharma. Plasmon generation in sputtered Ga-doped MgZnO thin films for solar cell applications. Journal of Applied Physics. 2016;119:233101(1)–(12). doi:10.1063/1.4953877

[37] D. C. Reynolds, D. C. Look, B. Jogai, C. W. Litton, G. Cantwell and W. C. Harsch. Valence-band ordering in ZnO. Physical Review B. 1999;60:2340–2344. doi:10.1103/Phys.RevB.60.2340

[38] H. Matsui and H. Tabata. In-plane anisotropy of polarized photoluminescence in *M*-plane (10–10) ZnO and MgZnO/ZnO multiple quantum wells. Applied Physics Letters. 2009;94:161907(1)–(3). doi:10.1063/1.3124243

[39] H. Matsui and H. Tabata. In-plane light polarization in nonpolar *m*-plane $Cd_xZn_{1-x}O/ZnO$ quantum wells. Applied Physics Letters. 2011;98:261902(1)–(3). doi: 10.1063/1.3603931

[40] P. Misra, Y. J. Sun, O. Brandt, H. T. Grahn. In-plane polarization anisotropy and polarization rotation for M-plane GaN films on $LiAlO_2$. Applied Physics Letters. 2003;83:4327–4329. doi:10.1063/1.1630168

[41] J. Wraesinski and D. Frohlich. Two-photon and three-photon spectroscopy of ZnO under uniaxial stress. Physical Review B. 1997;56:13087–13093. doi:10.1103/PhysRevB.56.13087

[42] W. R. L. Lambrecht, A. V. Rodina, S. Limpikumnong, B. Segall and B. K. Meyer. Valence-band ordering and magneto-optic exciton fine structure in ZnO. Physical Review B. 2002;65:075207(1)–(12). doi:10.1103/PhysRevB.65.075207

[43] S. F. Chichibu, T. Sota, G. Cantwell, D. B. Wason and C. W. Litton. Polarized photoreflectance spectra of excitonic polaritons in a ZnO single crystal. Journal of Applied Physics. 2003;93:756–758. doi:10.1063/1/1527707

[44] A. Ohtomo, M. Kawasaki, I. Ohkubo, H. Koinuma, T. Yasuda and Y. Segawa. Structural and optical properties of $ZnO/Mg_{0.2}Zn_{0.8}O$ superlattices. Applied Physics Letters. 1999;75:980–982. doi:10.1063/1/124573

[45] H. Karzel, W. Potzel, M. Kofferlein, W. Schiessl, M. Steiner, U. Hiller, G. M. Kalvius, D. W. Mitchell, T. P. Das, P. Blaha, K. Schwarz and M. P. Pasternak. Lattice dynamics and hyperfine interactions in ZnO and ZnSe at high external pressures. Physical Review B. 1996;53:11425–11438. doi:10.1103/PhysRevB.53.11425

[46] H. H. Huang and Y. R. Wu. Study of polarization properties of light emitted from a-plane InGaN/GaN quantum-well-based light emitting diodes. Journal of Applied Physics. 2009;106:023106(1)–(6). doi:10.1063/1.3176964

[47] T. B. Bateman. Elastic moduli of single-crystal zinc oxide. Journal of Applied Physics. 1962;33:3309–3312. doi:10.1063/1.1931160

[48] S. Ghosh, P. Waltereit, O. Brandt, H. T. Grahn and K. H. Ploog. Electronic band structure of wurtzite GaN under biaxial strain in the M-plane investigated with photoreflectance spectroscopy. Physical Review B. 2002;65:075202(1)–(7). doi:10.1103/PhysRevB.65.75202

[49] V. A. Fonoberov and A. A. Balandin. Excitonic properties of strained wurtzite and zinc-blende $GaN/Al_xGa_{1-x}N$ quantum dots. Journal of Applied Physics. 2003;94:7178–7183. doi:10.1063/1.1623330

[50] M. Suzuki and T. Uenoyama. Theoretical study of momentum matrix elements of GaN. Japanese Journal of Applied Physics. 1996;35:543–545.

[51] H. Matsui and H. Tabata. Engineering of optical polarization based on electronic band structures of A-plane ZnO layers under biaxial strains. Journal of Applied Physics. 2014;116:113505(1)–(7). doi:10.1063/1.4895842

[52] V. Darakchieva, T. Paskova, M. Schubert, H. Arwin, P. P. Paskov, B. Monemar, D. Hommel, M. Heuken, J. Off, F. Scholz, B. A. Haskell, P. T. Fini, J. S. Speck and S. Nakamura. Anisotropic strain and phonon deformation potentials in GaN. Physical Review B. 2007;75:195217(1)–(11). doi:10.1103/PhysRevB. 75.195217[26]

[53] H. Matsui and H. Tabata. Lattice strains and polarized luminescence in homoepitaxial growth of *a*-plane ZnO. Applied Physics Letters. 2012;101:231901(1)–(4). doi: 10.1063/1.4769036

[54] O. Dulub, L. Boatner and U. Diebold. STM study of the geometric and electronic structure of ZnO (0001)-Zn, (000–1)-, (10–10) and (11–20) surfaces. Surface Science. 2002;519:201–217.

[55] A. Ishii, Y. Oda and K. Fujiwara. DFT study for the anisotropic epitaxial growth of a-face ZnO (11–20). Physcia Status Solidi (c). 2008;5:2726–2728. doi: 10.1002/pssc200779288DFT (study for the anisotropic epitaxial growth of a-face ZnO(1120)).

[56] H. Matsui and H. Tabata. Self-organized nanostripe arrays on ZnO (10–10) surfaces formed during laser molecular-beam-epitaxy growth. Applied Physics Letters. 2005;87:143109(1)–(3). doi:10.1063/1.2081133

[57] H. Matsui and H. Tabata. Correlation of self-organized surface nanostructures and anisotropic electron transport in nonpolar ZnO (10–10) homoepitaxy. Journal of Applied Physics. 2006;99:124307(1)–(8). doi:10.1063/1.2207551

[58] K. Kojima, H. Kamon, M. Funato and Y. Kawakami. Theoretical investigations on anisotropic optical properties in semipolar and nonpolar InGaN quantum wells. Physcia Status Solidi (c). 2008;5:3038–3041. doi:10.1002/pssc200779277

[59] H. Matsui, T. Osone and H. Tabata. Band alignment and excitonic localization of ZnO/$Cd_{0.8}Zn_{0.92}O$ quantum wells. Journal of Applied Physics. 2010;107:093523(1)–(6). doi:10.1063/1.3359720

[60] T. Makino, Y. Segawa, M. Kawasaki, A. Ohtomo, R. Shiroki, K. Tamura, T. Yasuda and H. Koinuma. Band gap engineering based on $Mg_xZn_{1-x}O$ and $Cd_yZn_{1-y}O$ ternary alloy films. Applied Physics Letters. 2001;78:1237–1239. doi:10.1063/1.1350632

[61] R. Mata, A. Cros, J. A. Budagosky, A. Molina-Sanchez, N. Garro, A. Garcia-Cristobal, J. Renard, S. Founta, B. Gayral, E. Bellet-Amalric, C. Bougerol and B. Daudin. Reversed polarized emission in highly

strained a-plane GaN/AlN multiple quantum wells. Physical Review B. 2010;82:125405(1)–(9). doi:10.1103/PhysRevB.82.125405

[62] C. G. Tang, Y. H. Chen, B. Xu, X. L. Ye and Z. G. Wang. Well-width dependence of in-plane optical anisotropy in (001) GaAs/AlGaAs quantum wells induced by in-plane uniaxial strain and interface asymmetry. Journal of Applied Physics. 2009;105:103108(1)–(6). doi:10.1063/1.3132089

Chapter 5

Excitation-Intensity (EI) Effect on Photoluminescence of ZnO Materials with Various Morphologies

Prasada Rao Talakonda

Additional information is available at the end of the chapter

http://dx.doi.org/10.5772/64937

Abstract

The chapter discusses about excitation-intensity effects on photoluminescence emission peaks of zinc oxide (ZnO) material. ZnO is an ideal material for optoelectronic devices due to its wide band gap of 3.37 eV and some exciting optical properties. The performance of optoelectronic devices is greatly affected by the vibrational properties of the material, which are influenced by the interaction of phonons with free and bound electron-hole pairs. The photoluminescence (PL) spectroscopy is used to understand the extrinsic and intrinsic defects in ZnO materials. Understanding PL of ZnO nanostructures/thin films may lead to development of more efficient ZnO-based optoelectronic devices.

Keywords: ZnO, Defects, excitation intensity, photoluminescence

1. Introduction

Luminescence is one of the fastest growing and most useful analytical techniques in science and technology. The applications of luminescence can be found in various subject areas as materials science, microelectronics, clinical chemistry, biology, physics, environmental science, chemistry, toxicology, biochemistry, pharmaceuticals, and medicine. The term luminescence includes a wide variety of light-emitting processes, which originate their names from the varied sources of excitation energy that power them.

Photoluminescence (PL) spectroscopy is a sensitive nondestructive technique, suitable for identifying point defects including extrinsic and intrinsic defects in materials [1]. It is a useful tool to evaluate the quality and to study the physical parameters of semiconductor materials in the form of thin films and nanomaterials. Zinc oxide (ZnO) is one of the promising semiconductor materials of II–VI group, because of its wider band gap (3.37 eV) and large excitation binding energy (60 meV) at room temperature [2]. ZnO thin films and nanostructures are increasingly being used in light-emitting diodes, solar cells, chemical sensors, photocatalysis, and antibacterial materials, etc., because of their unique physical properties [2, 3].

In general, defects and impurities are playing an important role in the semiconductor industry to develop devices. In the case of ZnO material, it is very difficult to understand and study the role of defects or impurities in order to develop optoelectronic devices. Native or intrinsic defects are existing in the ZnO material such as oxygen vacancies (V_O), zinc interstitials (Zn_i), oxygen interstitial (O_i), zinc vacancies (V_{Zn}), zinc antisites, and oxygen antisites. Understanding the incorporation and behavior of intrinsic defects in ZnO material is essential to its successful application in optoelectronic devices. These native defects have long been believed to play an important role in ZnO-based devices. The band-to-band excitation of ZnO promotes electrons from the valence band to the conduction band, leaving holes in the valence band. The holes migrate from the valence band to deep levels and recombination occurs between electrons from either the conduction band or shallow donor levels and trapped holes on deep levels [4–6]. Basically, the PL of ZnO is related to the presence of holes in the valence band [4]. The PL spectroscopy can help to understand the extrinsic and intrinsic defects in ZnO materials. Understanding of PL on ZnO nanostructure/thin films may lead to development of more efficient ZnO-based optoelectronic devices.

The PL properties of ZnO materials have been widely studied by many research groups and several research groups noticed that excitation intensity (EI) also has important influence on the PL spectra of ZnO material [7, 8]. The current chapter is providing more details about the EI effects on ZnO materials with different morphologies. Three different sets of ZnO samples were deposited on various substrates with three kinds of morphologies. PL and Raman spectra were recorded using LABRAM-HR spectrometer excited with a 325 nm He-Cd laser at room temperature. The excitation intensity-dependent PL and Raman spectra were obtained at room temperature by changing the laser intensity with the change of optical filters. The laser intensity was fixed as P_0 for without filter (no filter). The EI for different filters denoted as $D_{0.03} = P_0/2$, $D_{0.06} = P_0/4$, $D_1 = P_0/10$, $D_2 = P_0/100$, $D_3 = P_0/1000$, $D_4 = P_0/10{,}000$. Low-temperature PL and Raman spectra were obtained using WITec instrument. The crystal structure and morphology of the samples were characterized by X-ray diffraction (XRD) and scanning electron microscopy (SEM).

2. Results

2.1. ZnO thin films

ZnO thin films with thickness of 450 nm were deposited on glass substrates using spray pyrolysis (SP) technique. The precursor solution for spray pyrolysis was prepared by dissolving appropriate amounts of zinc acetate dihydrate in a mixture of deionized water and ethanol at room temperature. In this mixture, ethanol concentration was 10 ml in 100 ml solution. A solution of 0.3 M Zn $(CH_3CO_2)_2$ was used as a precursor, prepared by the dissolving in deionized water and ethanol [3]. A few drops of acetic acid were added to aqueous solutions to prevent the formation of hydroxides. ZnO thin films were deposited on glass substrates at a substrate temperature of 623 K. The glass substrates were cleaned thoroughly with acetone, isopropanol, deionized water, and finally cleaned with the help of an ultrasonic bath for 30 min and dried [3, 9]. The spray nozzle was at a distance of 20 cm from the substrate during deposition [3, 9]. The solution flow rate was held constant at 3 ml/min. Air was used as the carrier gas at the pressure of 2 bar [3, 9]. The structural and morphological properties of ZnO thin films are shown in **Figure 1**.

Figure 1. (a) XRD, (b) SEM, and (c) AFM.

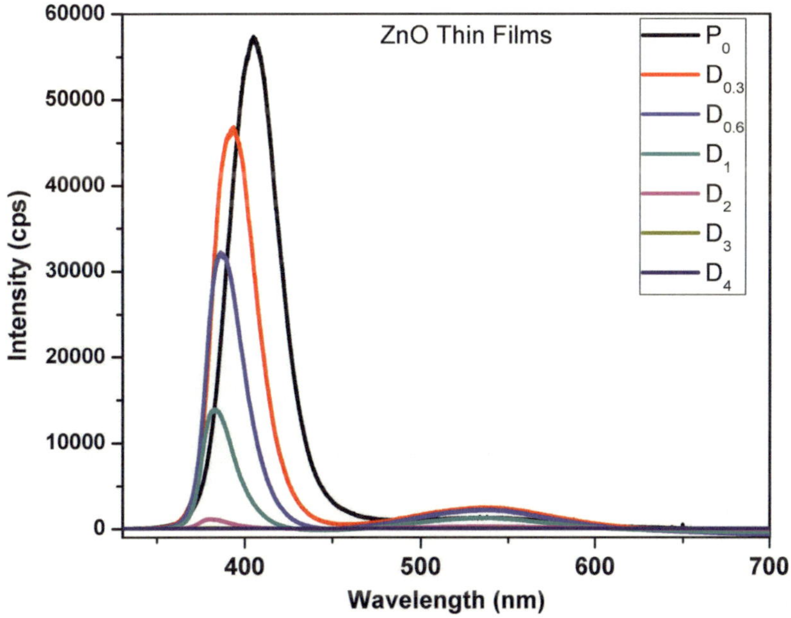

Figure 2. PL emission peak intensity variation with EI intensity.

The XRD studies have shown that all the diffraction peaks belong to ZnO wurtzite structure and the (0 0 2) diffraction peak related to the *c*-axes is predominant. No other phase was observed. The ZnO thin films are having uniform grains on the surface. These grains are uniformly distributed on the surface of the films without any voids. This can be understood from the SEM and AFM studies (**Figure 1**). The room temperature PL studies are shown in **Figure 2**.

The PL spectrum of ZnO thin films consists of two major emission peaks: a sharp peak located at ultraviolet region, and a weak broad peak in the visible region. According to the previous reports, the appearance of UV emission and visible emission indicates that the existence of native defects in ZnO. The intensity of UV peak of ZnO thin films is decreases with the decreasing EI. The UV emission peak position moved toward lower wavelength side (406–380 nm) with the decreasing of EI. The UV emission peak variation with EI could not explained completely by previous reports. Some previous reports were explained this kind of effect using the concept of laser-induced heating effect. The EI variation studies have extend for different ZnO samples with different morphology (nanorods, tetrapods, etc.) to understand EI effect on ZnO material.

2.2. ZnO nanorods

Nanorods structures like ZnO material were deposited on ITO substrate by electrochemical deposition (EC) technique. Electrolytes were zinc nitrate hydrate ($Zn(NO_3)_2{\cdot}6H_2O$) and analytic-grade hexamethylenetetramine ($C_6H_{12}N_4$). **Figure 3** shows XRD and the FESEM images of rod-like ZnO nanostructures with different magnifications that have been grown on ITO substrate. The nanostructures are arbitrarily directed and a nonuniform distribution of dense, arbitrarily directed ZnO nanorods, and no branching is observed, which indicates that the ZnO nanorods are grown from spontaneous nucleation.

Figure 3. (a) XRD and (b) SEM.

Figure 4 shows that the PL spectrum of ZnO nanorods. The UV emission peak position exhibits a significant blue-shift from 417 to 377 nm with a range of about 40 nm with decrease of EI. The UV peak intensity also varied with the EI.

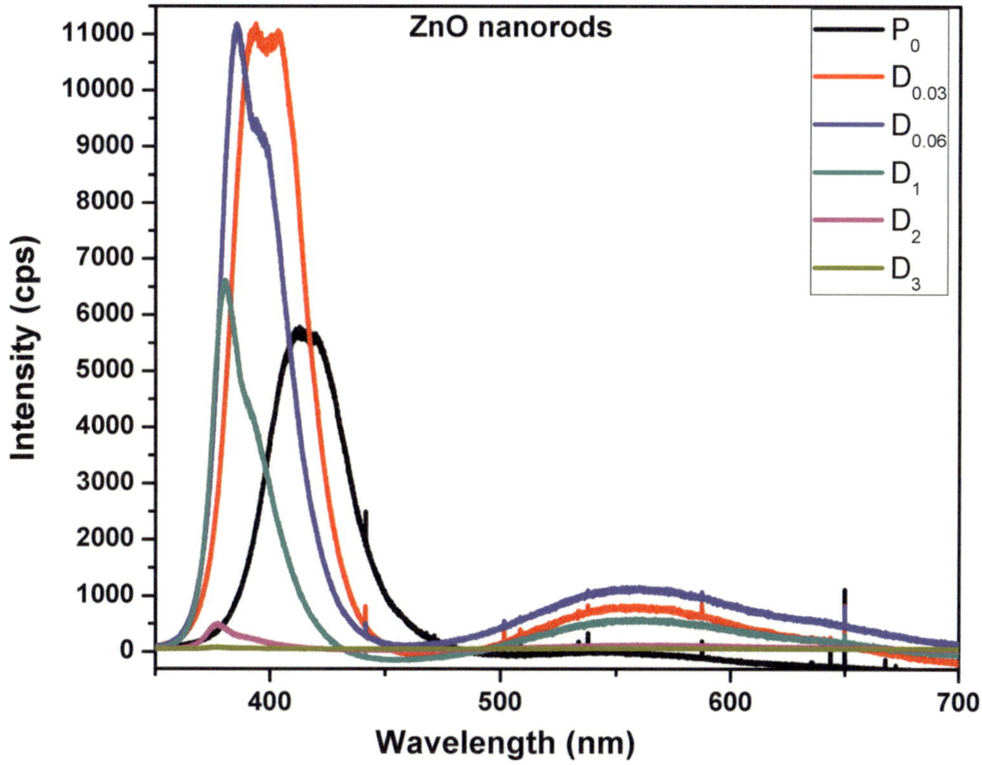

Figure 4. PL spectrum of ZnO nanorods.

2.3. ZnO tetrapods

ZnO tetrapods were synthesized on silicon substrate by vaporizing Zn granule (99.8% purity) in a quartz tube of 30 mm diameter and 100 cm in length at 1200°C in a horizontal tube furnace for 30 min duration at a heating rate of 10°C/min in ambient. **Figure 5** shows the XRD and SEM of ZnO tetrapods.

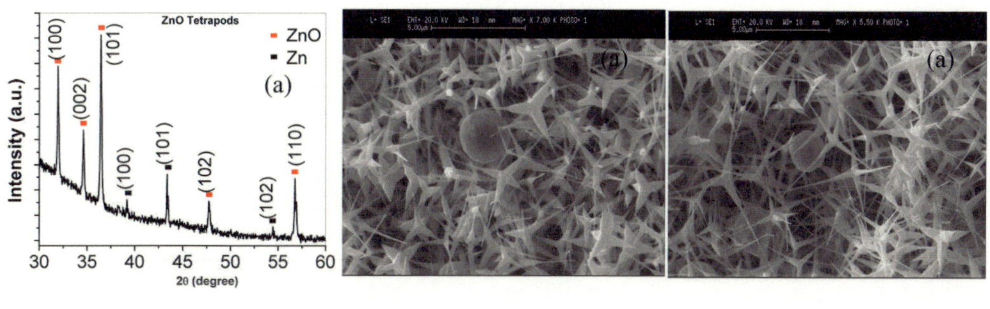

Figure 5. ZnO tetrapods (a) XRD and (b) SEM.

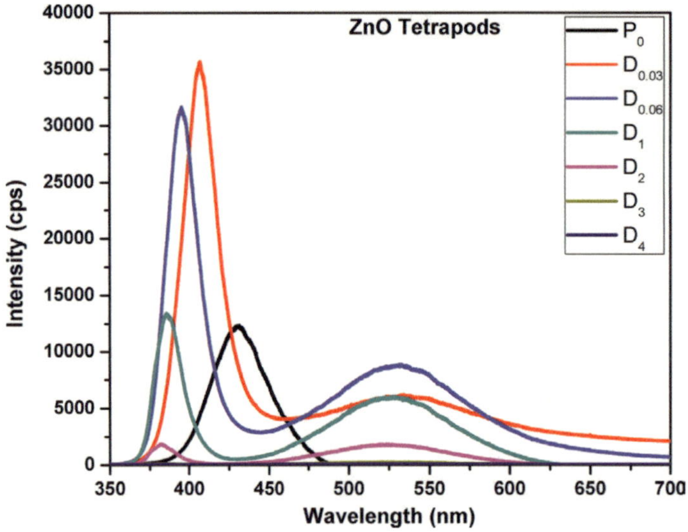

Figure 6. PL spectrum of ZnO tetrapods.

The PL spectrum of ZnO tetrapods is shown in **Figure 6**. The UV emission peak position of ZnO tetrapods moved significantly toward lower wavelength side (from 430 to 380 nm) with the range of about 50 nm when the EI intensity decreased from P_0 to D_4. The intensity of UV peak is increase with EI ranging from P_0 to $D_{0.06}$, and reaches highest at $D_{0.06}$. Then the UV peak intensity decreases and reaches lowest for the D_4.

2.4. ZnO bulk material

Bulk ZnO (ZnO powder) was taken from the Sigma Aldrich for the comparative studies to understand the EI effect on ZnO material. **Figure 7** shows the XRD and SEM of ZnO bulk powder. SEM showed micron-sized particles. The PL spectrum of ZnO bulk material is shown in **Figure 8**. The UV emission peak position and intensity variations are similar to the PL results of ZnO nanorods and ZnO tetrapods.

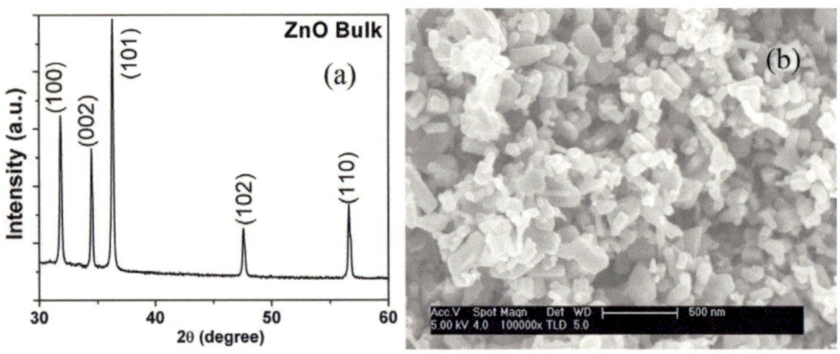

Figure 7. ZnO bulk powder (a) XRD and (b) SEM.

http://dx.doi.org/10.5772/64937

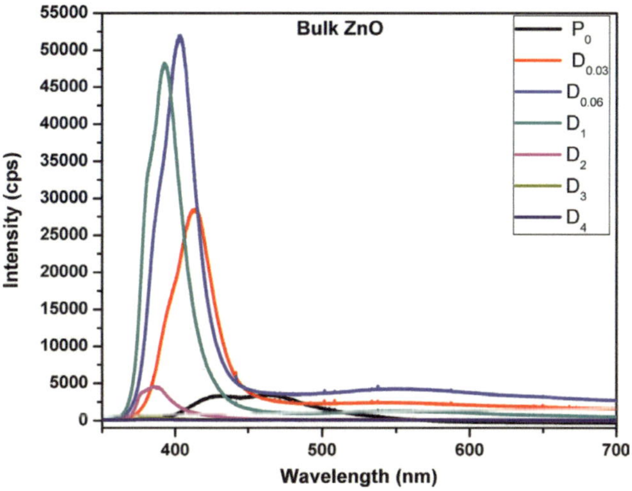

Figure 8. PL spectrum of ZnO bulk powder.

3. Discussion

PL studies show that the UV emission peak position is moved toward higher energy side (blue shift), i.e. lower wavelength side with EI decreases. The blue shift of UV emission peak is observed in all the samples irrespective of their morphology. In the PL spectra of ZnO, recently many research groups are observed that the UV emission peak intensity and peak positions varying with the excitation intensity of laser. In other words, the excitation intensity (EI) strongly influences the PL in ZnO nanostructures including nanocrystals, wires, rods, tetrapods, and ribbons [1, 7, 8, 10]. Hongwei et al. [11] studied the excitation-power dependence of the near-band-edge emission in ZnO inverse opals and nanocrystal films. Zhang et al. [8] studied the effects of excitation intensity (EI) on fluorescence spectra in various ZnO nanostructures, including tetrapod, rod, and sheet-like structures. The EI effects on PL spectra of ZnO nanostructures were explained by many authors using the UV-laser-induced heating effect/laser heating effect. Kurbanov et al. [1] explained UV-laser-induced heating effect using concept of thermal effusivity for ZnO nanocrystals grown on different substrates. In order to understand the local laser heating effects, we have assumed that the laser heating effects increase temperature in the semiconductor material (ZnO nanostructures/thin films). The rising temperature of semiconductor material due localized laser heat effects leads to the changes in optical and vibrational properties of the material [12, 13]. The phonon frequency shifts in ZnO nanostructures/thin films due to laser heating effects can be obtained using Raman spectroscopy. Raman spectroscopy is a nondestructive characterization technique, which is widely used to study the optical and vibrational properties of semiconductor materials including with thin films, hetrostructures, nanocrystals, nanowires, nanoribbons, nanorods, and tetrapods. ZnO is uniaxial crystals with wurtzite structure and it belongs to

$C^{4}_{(6\vartheta)}(P6_3mc)$ space group. Group theory predicts the existence of the following Raman modes [14]:

$$\Gamma_{opt} = A_1 + 2B_1 + E_1 + 2E_2$$

where Γ is point of the Brillouin zone and it involved in first-order Raman scattering for perfect ZnO crystals. Both A_1 and E_1 modes are polar. They split into transverse optical (A_1TO, E_1TO) and longitudinal optical (A_1LO, E_1LO) modes. E_2 is nonpolar mode consists of two modes of low- and high-frequency phonons (E_2 low, E_2 high). E_2 (high) is associated with oxygen atoms and E_2 (low) is associated with Zn sublattice [13]. These modes are Raman active according to Raman selection rule. B_1 branch is silent. The phonon frequency shifts in the Raman spectra of ZnO nanostructures are probably due to the optical phonon confinement, defects or impurities in the material, tensile strain, and laser-induced heating [13, 14]. The optical phonon confinement effect is neglected in this work because Bohr exciton radii (≈2.34 nm) are smaller than the nanocrystallite size of the grown ZnO nanostructure/thin films.

The PL spectra of ZnO nanostructures with the changes of EI are summarized as follows. In the case of bulk ZnO, the UV peak position moved significantly toward lower wavelength side (from 423.6 to 386.5 nm) with the range of about 37.1 nm when the EI intensity decreased from P_0 to D_4. The intensity of UV peak is increase with EI ranging from P_0 to $D_{0.06}$, and reaches highest at $D_{0.06}$. Then the UV peak intensity decreases and reaches lowest for the D_4. Similar behavior is observed in the case of ZnO nanorods and tetrapods. In the case of ZnO thin films, the intensity of UV peak decreases with the decreasing EI. The green emission peak position is not changed with the EI. The green band intensity is fluctuating with the EI. In other words, the changes in EI not affect by green band position, intensity, and width. Moreover, the appearance of green band in the spectrum is directly supporting the existence of oxygen vacancies or Zn interstitials in the ZnO nanostructure/film. **Figure 9** shows the Raman spectra of ZnO nanostructure/film sample excited using 514 nm laser. The spectrum of bulk ZnO exhibits one prominent peak at 439 cm^{-1} (E_2 high) in addition weak peak at 332 and 579 cm^{-1}. The nanorods also show similar spectrum. In the case of thin films, the prominent peak is observed at 438 cm^{-1} (E_2 high) and weak peaks are observed at 331, 493, and 577 cm^{-1}. The tetrapods of ZnO show prominent peak at 437 cm^{-1} (E_2 high) in addition to weak peaks at 330, 376, and 579 cm^{-1}. In all the cases, the intensity of all the Raman peaks decreases with decreasing EI. The observed values (Raman shift/peak position) are much closed to the previous reports [15, 16]. The Raman peak E_2 high (439/438/437 cm^{-1} representing bulk/rods/tetrapods, respectively) is corresponding to the wurtzite phase of ZnO. The EI variation of the laser (514 nm) does not produce any change in E_2 high peak and other additional peaks position (579/578/577 cm^{-1} representing bulk/rods/tetrapods, respectively). The most important peak is observed at 579/578/577 cm^{-1} in all the ZnO nanostructures. In general, the LO frequency modes are used to observe between 574 and 591 cm^{-1} for the relatively large size of ZnO nanocrystals. According to previous reports [17, 18], the 579/578/577 cm^{-1} mode is ascribed to the LO phonon of A_1 or E_1 symmetry. According to Zhang et al. [14], the peak observed at 577 cm^{-1} is a defect induced

http://dx.doi.org/10.5772/64937

mode in ZnO nanostructures. We believe that the 579/578/577 cm^{-1} corresponding to bulk/rods/tetrapods are defect-induced mode associated with the defects such as oxygen vacancies, zinc interstitials, and the free carrier lack [19]. In the most of the samples (see **Figure 9**), ZnO transversal modes and second order modes are absent in the spectra. This leads to understand poor crystallization quality of ZnO nanostructures.

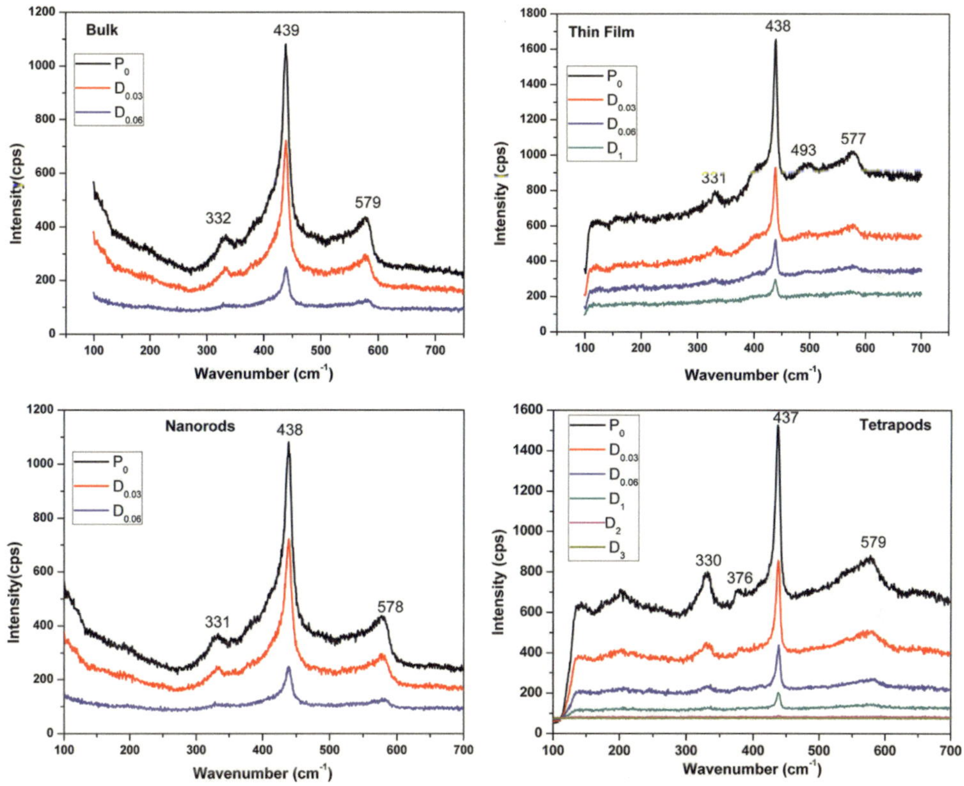

Figure 9. Raman spectrum of ZnO material for bulk, thin films, nanorods, and tetrapods.

Raman studies with 325 nm laser line are conducted on the ZnO nanostructures in order to cross check the laser heating effect because the same laser was used for the PL. The Raman spectra with 321 nm laser are shown in **Figure 10**. No E_2 (high) peak is observed. The LO mode is observed at 579/578/577 cm^{-1} for the different ZnO nanostructures. The intensity of LO is decreases with EI. Neither a blue nor a red shift is observed in LO position with a decrease in EI. According to previous report [13], the huge red shift of LO should be observed for the ZnO nanostructure when the intense local heating induced by UV laser. The observations from **Figure 10**, indirectly tells that the presence of the intrinsic impurity or defects in the sample. After careful analysis of Raman results, we understand that the huge shift of UV peak in PL may not be attributed to local heating induced by UV laser in the ZnO nanostructures. In order to verify this assumption, we recorded the PL and Raman spectra at low temperature for all the samples using WITec Alpha SNOM instrument with solid state lasers 355 and 532 nm, respectively. We change the instrument to avoid instrumental error. Here, the PL spectrum was recorded at the maximum and minimum of laser EI. The samples were tested in the temperature range of −190 to 30°C (83–303 K).

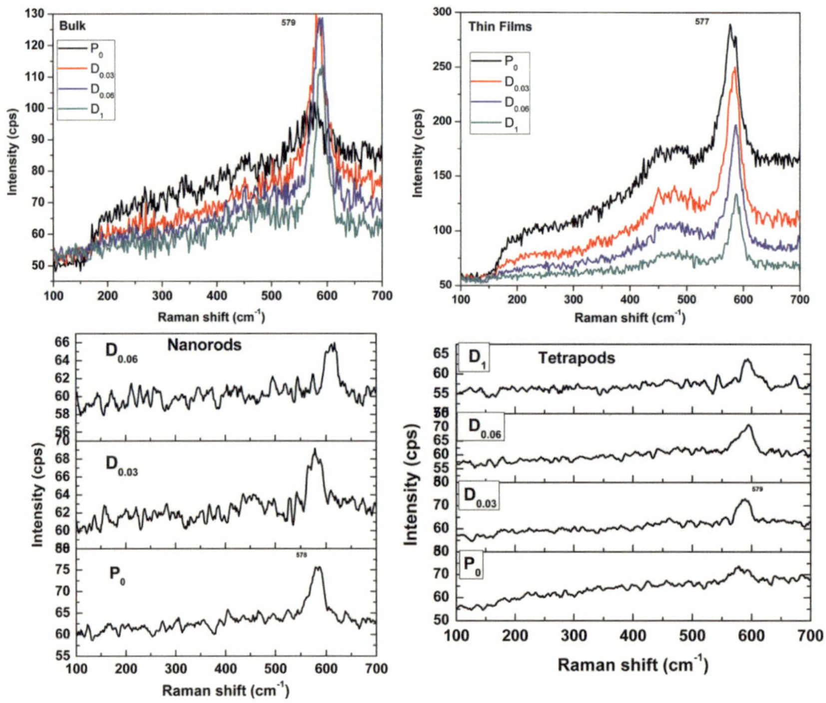

Figure 10. Raman spectrum of ZnO material with 325 nm excitation source for bulk, thin films, rods and tetrapods.

Figure 11 shows the low-temperature PL spectra. It is well known that the low-temperature PL from ZnO material with reasonably good crystalline quality is dominated by UV emission. From **Figure 11**, the origin of the UV emission peak shift explained based on the two major parameters one is laser EI the other one low-temperature variation.

From low-temperature PL, it can be argued that when the sample is in liquid nitrogen reservoir, then the reservoir temperature would be expected to compensate the local heating by laser. Hence, no-shift in the UV band (or the quenching of the UV emission) is expected. However, at liquid nitrogen temperature, the similar blue shift of UV emission has been observed for the samples (thin films, nanorods, and tetrapods) at *E*(max) and *E*(min). Moreover, we have tried to measure the surface temperature of the samples using thermocouple as well as IR thermometer during surface illumination by laser. No evidence of the increase in the surface temperature of the samples is observed. It is stated that such shift in UV emission is independent of the deposition parameter, morphology and structural parameters. The PL UV peak emission position blue shift of ZnO material can be understood from **Figure 12**.

http://dx.doi.org/10.5772/64937

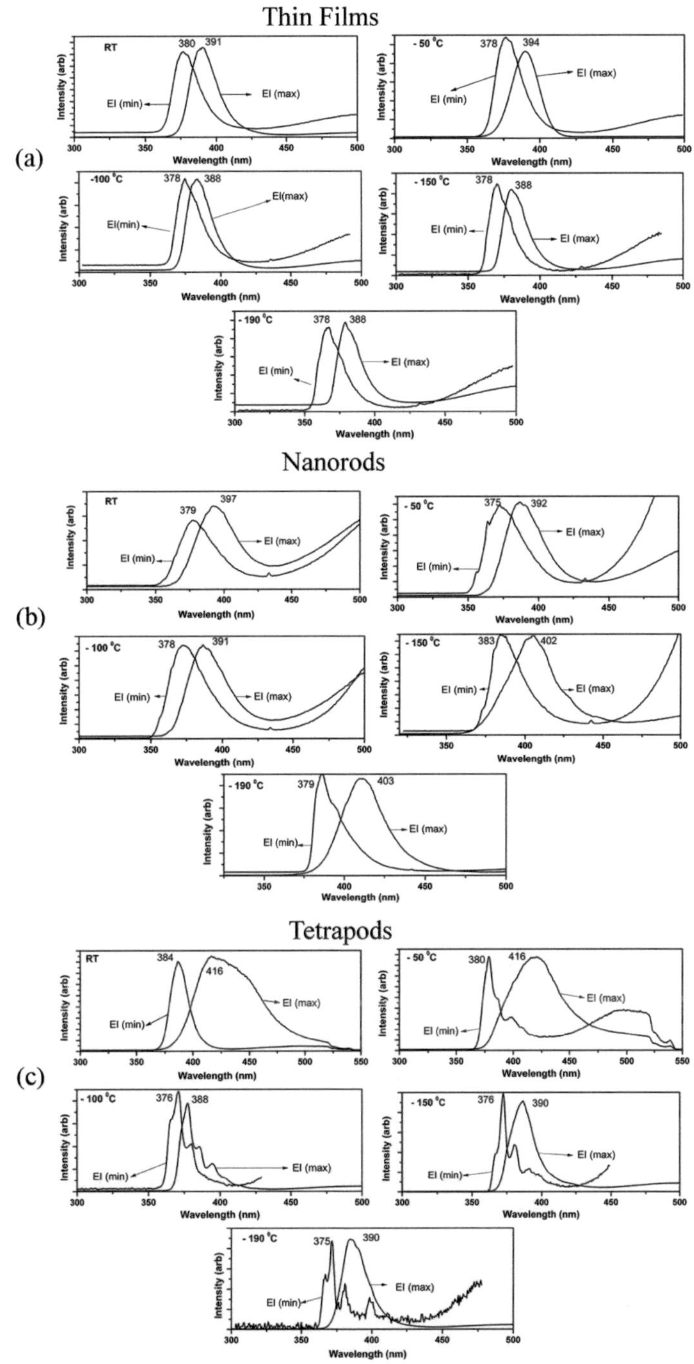

Figure 11. (a) Low-temperature PL of ZnO thin films. (b) Low-temperature PL of ZnO nanorod. (c) Low-temperature PL of ZnO tetrapods.

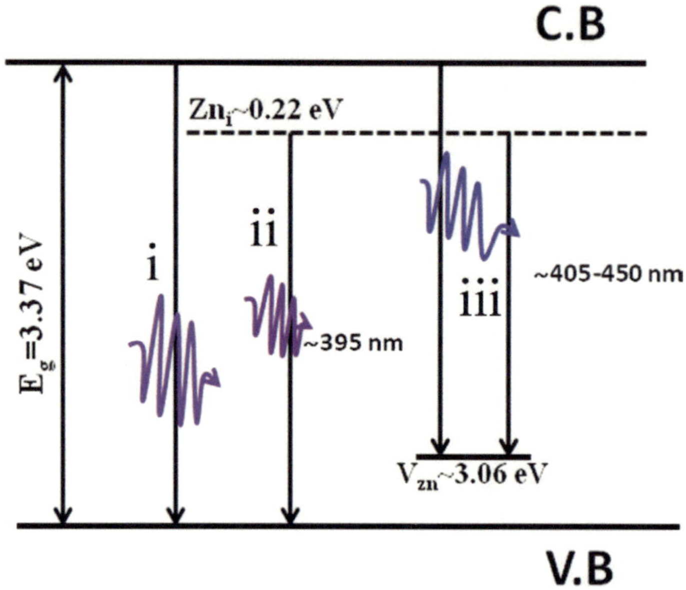

Figure 12. Energy band diagram of ZnO showing the probable transition processes [7].

After careful analysis of the RT and low-temperature PL studies, the shift in the UV emission can be due to intrinsic defect level emissions in ZnO as explained based on the energy band diagram shown in **Figure 12**. From **Figure 12**, the origin of UV emissions in ZnO are attributed to four factors. They are: (i) direct transition from conduction band (CB) to valence band (VB); (ii) transition of zinc interstitial (Zn_i) defect level to VB; (iii) transition from the CB to zinc vacancy (V_{Zn}) which is a result of blue-green emission of the material; and (iv) transition from Zn_i to V_{Zn} [7, 20]. It is believe that the Zn_i and V_{Zn} defects are playing crucial role in the blue shift of UV emission in ZnO. According to PL results, UV emission shifts largely from 377 to 415 nm with EIs for various samples. This UV emission can be due to the transitions (i), (ii), and (iii) involved between defects levels Zn_i, V_{Zn}, and CB due to which shift in the band-to-band transition is observed for different EIs. Hence, it is understood from the above discussion that defect levels present in the sample may be responsible for the shifts in the UV emission band with different EIs. In this regard, all ZnO samples exhibit the A_1LO Raman band (~580 cm^{-1}) which is ascribed to a defect induced mode [7] as evidenced from the Raman spectrum. Overall, all the samples possess considerable intrinsic defects which may be responsible for the shifts in UV emission band with EIs.

4. Conclusion

The excitation-intensity dependence of the UV emission for various ZnO samples has been studied in detail. The EI effects were studied systematically at room temperature and at liquid nitrogen temperature. It is observed that the UV emission band moves toward higher energy side (blue shift) when the EI is decreased and vice versa both room temperature as well as at liquid nitrogen temperature. The blue shift of UV emission is only observed for some samples whereas few samples do show a marginal shift with the EI variation. These results give an implication that the shift may not be the intrinsic effect of laser-induced heating as has been claimed by other research groups; rather other mechanism can also be equally responsible. For detailed understanding, Raman measurements are methodically recorded for different EIs. The marginal shift in A_1LO modes observed for different samples with different EIs and systematic analysis of all the results supporting the shift of UV emission. It may also be attributed to the native defects in ZnO nanostructures/thin films apart from laser-induced heating. Result analysis show that the PL properties of the ZnO nanostructures/thin films are a strong function of the excitation intensity. The Raman studies helped to identify the origin of the UV emission peak shift in ZnO nanostructures. These results strongly suggest that heating effect alone cannot explain the observed excitation intensity dependency of luminescence spectra. However, further study is underway for better understanding.

Acknowledgements

The author is thankful to Dr K.K. Nanda, MRC, IISc, Dr G.K. Goswami, MRC, IISc, and Dr M.C. Santhosh Kumar, Department of Physics, NIT Tiruchirappalli for their help and discussion.

Author details

Prasada Rao Talakonda

Address all correspondence to: prasadview@gmail.com

Department of Physics, K L University, Andhra Pradesh, India

References

[1] S.S. Kurbanov, Kh.T. Igamberdiev, T.W. Kang. The UV-laser induced heating effect on photoluminescence from ZnO nanocrystals deposited on different substrates. J. Phys. D: Appl. Phys. 2010;43:115401. DOI:10.1088/0022-3727/43/11/115401.

[2] M. Kashif, S.M. Usman Ali, M.E. Ali, H.I. Abdulgafour, U. Hashim, M. Willander, Z. Hassan. Morphological, optical, and Raman characteristics of ZnO nanoflakes prepared via a sol–gel method. Phys. Status Solidi A, 2012;209:143–147. doi: 10.1002/pssa.201127357.

[3] Prasada Rao Talakonda, ZnO Thin films for Optoelectronic Applications, LAMBERT Academic Publishing, Heinrich-Bocking-Str. 6–8, 66121 Saarbrucken, Deutschland/ Germany, 2013, ISBN: 978-3-659-37010-6.

[4] J.C. Ronfard-Haret. Electric and luminescent properties of ZnO-based ceramics containing small amounts of Er and Mn oxide. J. Luminesc. 2003;104:1–12. doi:10.1016/ S0022-2313(02)00574-4.

[5] Khedidja Bouzid, Abdelkader Djelloul, Noureddine Bouzid, Jamal Bougdira. Electrical resistivity and photoluminescence of zinc oxide films prepared by ultrasonic spray pyrolysis. Phys. Status Solidi A. 2009;206:106–115. doi: 10.1002/pssa.200824403.

[6] D.C. Reynolds, D.C. Look, B. Jogai, J.E. Van Nostrand, R. Jones, J. Jenny. Source of the yellow luminescence band in GaN grown by gas-source molecular beam epitaxy and the green luminescence band in single crystal ZnO. Solid State Commun. 1998;106:701–704. doi: 10.1016/S0038-1098(98)00048-9.

[7] T. Prasada Rao, G.K. Goswami, K.K. Nanda. Detailed understanding of the excitation-intensity dependent photoluminescence of ZnO materials: role of defects. J. Appl. Phys. 2014;115:213513. doi: 10.1063/1.4881779.

[8] Zhang YuGang, Zhang LiDe, Effect of excitation intensity on fluorescence spectra in ZnO nanostructures and its origin. Sci. China. Ser. G-Phys. Mech. Astron. 2009;52:4–12. doi: 10.1007/s11433-009-0008-2.

[9] T. Prasada Rao, M.C. Santhosh Kumar. Realization of stable p-type ZnO thin films using Li–N dual acceptors. J. Alloys Compd. 2011;509:8676–8682. doi:10.1016/j.jallcom. 2011.05.094.

[10] S. Dhara, P.K. Giri. Rapid thermal annealing induced enhanced band-edge emission from ZnO nanowires, nanorods and nanoribbons. Funct. Mater. Lett. 2011;4:25–29. doi: 10.1142/S1793604711001658.

[11] Hongwei Yan, Yingling Yang, Zhengping Fu, Beifang Yang, Jian Zuo, Shengquan Fu. Excitation-power dependence of the near-band-edge photoluminescence of ZnO inverse opals and nanocrystal films. J. Luminesc. 2008;128:245–249. doi: 10.1016/ j.jlumin.2007.08.002.

[12] E. Alarcon-Liado, J. Ibanez, R. Cusco, L. Artus, J.D. Prades, S. Estrade, J.R. Morante, Ultraviolet Raman scattering in ZnO nanowires:quasimodemixing and temperature effects. J. Raman Spectrosc. 2011;42:153–159. Doi: 10.1002/jrs.2664.

[13] K.A. Alim, V.A. Fonoberov, M. Shamsa, A.A. Balandin. Micro-Raman investigation of optical phonons in ZnO nanocrystals. J. Appl. Phys. 2005;97:124313. doi: 10.1063/1.1944222.

[14] R. Zhang, P.G. Yin, N. Wang, L. Guo. Photoluminescence and Raman scattering of ZnO nanorods. Solid State Sci. 2009;11:865–869. doi: 10.1016/j.solidstatesciences.2008.10.016.

[15] S.K. Panda, C. Jacob. Surface enhanced Raman scattering and photoluminescence properties of catalytic grown ZnO nanostructures. Appl. Phys A. 2009;96:805–811. doi: 10.1007/s00339-009-5309-9.

[16] I. Calizo, K.A. Alim, V.A. Fonoberov, S. Krishnakumar, M. Shamsa, A.A. Balandin, R. Kurtz. Micro-Raman spectroscopic characterization ZnO quantum dots, nanocrystals, and nanowires. Proc. SPIE. 2007;6481:64810N. doi: 10.1117/12.713648.

[17] K.A. Alim, V.A. Fonoberov, A.A. Balandin. Origin of the optical phonon frequency shifts in ZnO quantum dots. Appl. Phys. Lett. 2005;86:053103. doi: 10.1063/1.1861509.

[18] Y. Yang, H. Yan, Z. Fu, B. Yang, J. Zuo. Correlation between 577 cm^{-1} Raman scattering and green emission in ZnO ordered nanostructures. Appl. Phys. Lett. 2006;88:191909. doi: 10.1063/1.2202741.

[19] C. Xu, G. Xu, Y. Liu, G. Wang. A simple and novel route for the preparation of ZnO nanorods. Solid. State. Commun. 2000;122:175–179. doi:10.1016/S0038-1098(02)00114-X.

[20] A.N. Mallika, A. Ramachandra Reddy, K. Sowribabu, K. Venugopal Reddy. Structural and optical characterization of $Zn_{0.95-x}Mg_{0.05}Al_xO$ nanoparticles. Ceram. Int. 2015;41:9276–9284. doi:10.1016/j.ceramint.2015.03.096.

Section 2

Photo-Physical Properties and their Emerging Applications

Chapter 6

Photon-Upconverting Materials: Advances and Prospects for Various Emerging Applications

Manoj Kumar Mahata, Hans Christian Hofsäss and Ulrich Vetter

Additional information is available at the end of the chapter

http://dx.doi.org/10.5772/65118

Abstract

Rare-earth-doped upconversion materials, featuring exceptional photophysical properties including long lifetime, sharp emission lines, large anti-Stokes shift, low autofluorescence of the background, and low toxicity, are promising for many applications. These materials have been investigated extensively since the 1960s and employed in many optical devices. However, due to rapid development of synthesis strategies for nanomaterials, upconversion materials have been rehighlighted on the basis of nanotechnology. Herein, we discuss the recent advances in upconversion materials. We start by considering energy transfer processes involved in the basic study of upconversion emission phenomena, as well as synthesis strategies of these materials. Progress in different energy transfer processes, which play an important role in determining luminescence efficiency, is then discussed. Newer applications of these materials have been vastly reviewed.

Keywords: upconversion, luminescence, rare earth, 4f-4f transitions, nanomaterials, energy transfer, optical applications, nanophosphor

1. Introduction

Photon upconversion (UC) is a nonlinear optical process which yields high-energy photons through sequential absorption of two or more low-energy photons. The concept of UC was conceived by physicist N. Bloembergen [1] in 1959 to develop an infrared (IR) photon detector for counting infrared photons through the interaction of infrared (IR) photons with rare earth (RE) or transition-metal ions incorporated in crystalline materials. Though, due to the lack of coherent pumping sources, the prospect of achieving upconversion was not possible. With the advances in lasers and optical technology, the landmark experiment to observe IR to visible upconversion was reported first time by F. Auzel [2] in 1966 and explained energy transfer in Er^{3+}-Yb^{3+} and Tm^{3+}-Yb^{3+} systems. Since then, research on photon upconversion has inspired the design of near-infrared (NIR) to visible upconverter suitable for a wide range of applications. Later, technological advances in laser and optical tools compelled the application of upconversion lasers for converting low-energy laser radiation to high-energy laser radiation [3]. However, in early days, applications of UC materials in biomedical fields were not clear [4]. One of the major challenges was to control the size and shape of the materials that are suitable in biological environments.

Rare earth elements, including scandium and yttrium, are important in functional materials due to their similar electronic configurations ([Xe]$4f^{n-1}5d^{0-1}6s^2$). Because of this, they have similar physical and chemical properties in their trivalent ionic state. The energy levels of RE^{3+} from the 4f electronic configuration are abundant and thus allow for many intraconfigurational transitions (**Figure 1**). The intra-4f transitions are forbidden for free RE ions but these rules are partially broken when the REs are embedded in an inorganic lattice. Owing to many energy levels with intraconfigurational transitions, RE-ions act as promising luminescent centers. Moreover, the intrinsic character of RE ions due to shielding of partially filled 4f electrons by completely filled $5s^25p^6$ subshells makes them less sensitive to the environment of the host lattice. Apart from this, long life, sharp emission lines, large anti-Stokes shift, and photostability are the results of their unique 4f energy levels.

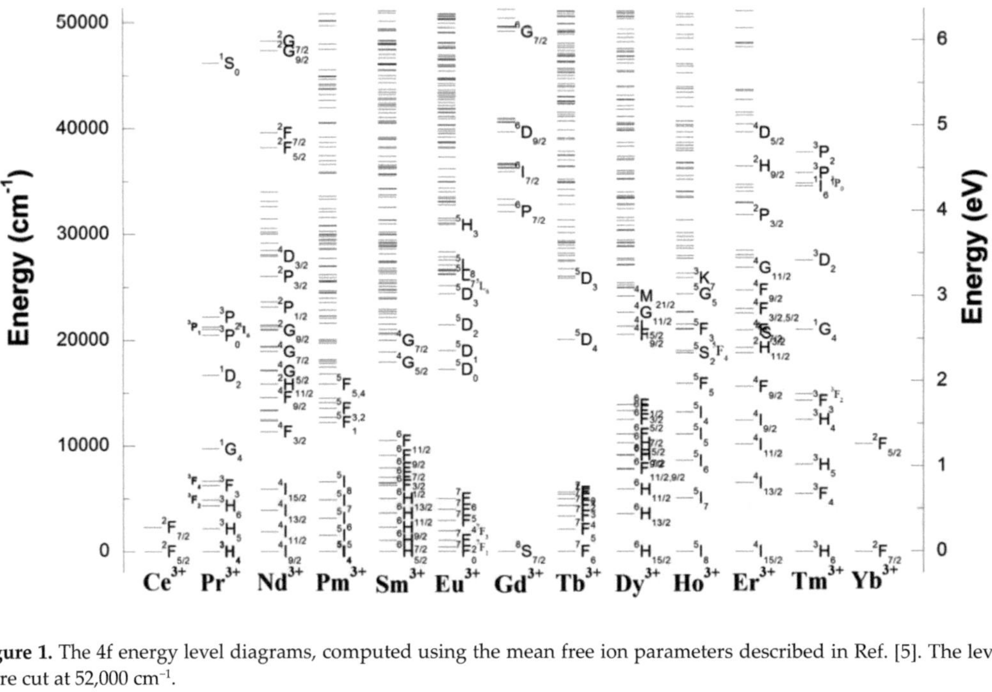

Figure 1. The 4f energy level diagrams, computed using the mean free ion parameters described in Ref. [5]. The levels were cut at 52,000 cm^{-1}.

http://dx.doi.org/10.5772/65118

In recent decades, UC materials have attracted much attention because of their special spectroscopic properties [6]. RE-doped luminescent materials have been employed in display devices and solid state lasers for long time. With the development of new synthesis strategies, RE-doped UC materials can be prepared at nanoscale for better prospects [3]. The emission bands of UC nanomaterials are less affected by the particle size unlike quantum dots, whose luminescence is dependent on particle size. Moreover, owing to their high resistance to photobleaching, photoblinking, specific NIR excitation, excellent penetration depth in biosystems without any autofluorescence from the background, UC particles have a wide range of applications in bioimaging and theranostics [7]. In spite of these features, several aspects of the UC nanomaterials are of great concern, such as enhancing UC efficiency, color outputs, and manipulation of energy transfer pathways to achieve desired UC emissions for specific applications.

Here, we review the recent advances in photon-upconverting materials. In this chapter, we discuss the composition of ideal UC materials, synthesis of UC particles, physics of UC followed by their applications and future prospects.

2. Composition of upconversion materials

Upconverting particles are generally composed of a host material doped with optically active sensitizer and activator ions [8]. Controlling the energy transfer processes among these three constituents is of great importance in selecting suitable dopant-host pairs. The host matrices differ in their coordination numbers, distances between dopants, and efficiencies of energy transfer. Thus proper selection of the host matrix is of significant importance. Host matrices must possess low-phonon energies to avoid degradation of efficiency through nonradiative relaxations and should be chemically and thermally stable. Moreover, host materials should have high tolerance for sensitizer and activator ions and high transparency for free migration of NIR photons in the lattice. Fluoride materials meet these criteria and are popular choices as host matrices for UC materials [8]. Among the reported host matrices, the $NaREF_4$ series has been proven to be highly effective due to their lower phonon energy and excellent chemical stability. In this family, the hexagonal-$NaREF_4$ series is more efficient than the cubic-$NaREF_4$ series due to its unique crystal structure. Oxide materials such as ZrO_2, Y_2O_3, vanadates (e.g., YVO_4, $GdVO_4$), phosphates (e.g., $LuPO_4$, YPO_4) are also examples of some common host materials because of their high chemical stability, even though they have larger phonon energies than the fluoride materials [6, 8].

The role of sensitizers is to sensitize the activators in the UC materials. An ideal sensitizer should have large absorption cross-section at the desired excitation wavelength and resonant energy levels to those of the activators, with suitable excited energy state located in the NIR region. Among all the RE ions, Yb^{3+} is the best choice as a sensitizer. The absorption cross-section of Yb^{3+} is 9.11×10^{-21} cm^{-2} at 980 nm, which is relatively large among RE ions [9]. Moreover, the energy level diagram of Yb^{3+} has only one excited state ($^2F_{5/2}$) that matches very well with the f-f transitions of many RE activators, e.g., Er^{3+}, Ho^{3+}, Tm^{3+}, etc. and therefore Yb^{3+}

is an excellent sensitizer to transfer energy to other RE ions. For example, the $^2F_{5/2}$ state of Yb^{3+} overlaps the $^4I_{11/2}$ state of Er^{3+}, allowing Yb to Er energy transfer (**Figure 1**).

UC emission increases with doping concentration of activator ions until a critical concentration. Beyond that concentration, quenching among activator ions occurs and decreases the upconversion emission [10]. To get rid of this problem, concentrations of activators are kept low while one or more types of RE ions with high concentration are incorporated as sensitizers. The activators should possess a large number of metastable energy levels. Once ground state electrons are excited to the metastable state, activators are likely to accept energy from nearby sensitizers to be excited at higher excited states. The energy levels of the activators should not be close enough to promote nonradiative relaxations. According to the energy level structures, Er, Tm, and Ho ions in their trivalent state are ideal to minimize nonradiative relaxations, whose transition rate decreases exponentially with increasing energy gap.

3. Synthesis of upconversion materials

In order to achieve high UC efficiency, synthesis of UC materials is very crucial. Recently, various methods have been developed for synthesizing RE-doped UC materials, including thermal decomposition, coprecipitation, hydrothermal, sol-gel, combustion, microwave, microemulsion, and so on [11–20]. The methods are to address specific requirements such as surface modification and monodispersity. The synthesis routes are not mutually exclusive, and often more than one method is combined to produce the suitable UC particles. Some important synthesis processes for preparing UC particles are discussed below.

3.1. Thermal decomposition

Thermal decomposition gives size-controlled well-shaped particles within short reaction time [11]. It usually involves surfactant-assisted decomposition of precursors in high-boiling organic solvents. Then the generated ions are combined into new nuclei at relatively high temperature. The trifluoroacetate and oleate compounds are commonly used organic precursors and polar capping groups such as oleic acid, oleylamine and octadecene are usually used as surfactants. It is believed that the solvents can control the growth of the particles via capping the surface of UC particles. A systematic investigation on the growth mechanism of UC nanocrystals was performed by Mai et al. [12] and inferred the passage of precursors in surfactant solutions through a slow nucleation pathway. They also showed that various sizes and shapes of Er^{3+}/Yb^{3+} doped $NaYF_4$ particles can be obtained by varying reaction time, reaction temperature, and concentration of reagents. Though this method has several important advantages, e.g., strong UC emission, high quality of products and pure phase crystals, it suffers from some disadvantages including operation difficulty, production of toxic materials, and presynthesis of $RE(CF_3COO)_3$ precursors.

http://dx.doi.org/10.5772/65118

3.2. Coprecipitation

The coprecipitation method is developed to overcome the limitations of the thermal decomposition method and considered as a significant method for synthesis of UC particles. It involves the precipitation of two substances simultaneously. In the coprecipitation route, UC nanocrystals are formed with organic surfactants and prevent agglomeration though their adsorption to the surface of the particles [6]. The benefits of the coprecipitation method include absence of toxic by-products, inexpensive equipment, and simple procedure. Er^{3+}/Yb^{3+}, Ho^{3+}/Yb^{3+}, and Tm^{3+}/Yb^{3+} doped LaF_3 nanoparticles [13] with particle size of 5 nm were prepared by Yi et al. In addition to that, RE-doped $LuPO_4$, $YbPO_4$, $NaYF_4$, $NaGdF_4$, and $Y_3Al_{15}O_{12}$ (YAG) nanoparticles were also produced using this method. Recently, our group has synthesized the Er^{3+}/Yb^{3+} doped $BaTiO_3$ [14] and YVO_4 [15] nanocrystals using the coprecipitation method. Despite the general advantages, this method suffers from the long-time operation of the experimental procedure, which sometimes takes more than 8 h. Moreover, to produce UC particles in industrial scales, this method is not suitable.

3.3. Hydrothermal

Thermal decomposition can only use organic solvents while hydrothermal synthesis which is mainly a solution-based method, can occur in a water-based system with low reaction temperature (around 200°C) in a relatively environment friendly condition [11]. Hydrothermal methods involve heating of the solvent at high pressure above its critical point. This process is convenient to produce controllable size and shape of the inorganic nanoparticles with diverse nanodimensional architectures. For example, yttrium orthovanadate (YVO_4) crystals have been prepared hydrothermally in both acidic and basic media with well-defined microcrystals with clear facets and dispersed nanograins and nanoflakes within 5–50 nm dimension [16]. Prism, disk, rod, and tube like $NaYF_4$ crystals were prepared by hydrothermal method [17]. An advantage of this route is that experimental parameters such as reaction temperature, reaction time, solvent type, and surfactant type can be varied to control the size and shape of the synthesized particles.

3.4. Sol-gel

The sol-gel method is generally used for the preparation of thin films, oxides, and fluoride nanocrystals [11]. The method is a typical wet-chemical process, starts with liquid solution of molecular precursors and it forms a new sol phase through hydrolysis and polycondensation reactions. With the addition of base, the sol is agglomerated into gel through a large macromolecular network, followed by annealing at high temperature for a few hours. Annealing increases crystallinity and removes the solvents from the gel. Patra et al. [18] prepared Er^{3+} doped TiO_2 and $BaTiO_3$ using basic acetate and titanium tetraisopropoxide. Patra et al. [19] developed the sol-gel method into sol emulsion-gel method to synthesize $ZrO_2:Er^{3+}$ nanoparticles. A thin film of $YVO_4:Eu^{3+}$ was fabricated by Cheng et al. [20] through the combination of pechini-type sol-gel method and inkjet printing. The key step in a sol-gel process is the annealing procedure. The quality of the synthesized materials is significantly dependent on the annealing temperature and time. The materials produced via sol-gel method yield high

UC intensity and the method can be applied for production on a large scale. However, there are some shortcomings in this process. For instance, the derived particles are of irregular shapes, sizes, and are not water-soluble.

3.5. Combustion

Compared to thermal decomposition, hydrothermal, and sol-gel methods, combustion synthesis is a high-throughput method of producing UC particles within a very short period of time [7]. A balanced mixture of reducing agent (fuel) and oxidizers (metal nitrates) is usually selected for the reaction to release maximum energy during the reaction time. In this process, a series of controlled explosions take place in the reaction materials in a self-sufficient condition without requiring any extra heat for the total reaction. Generally, the oxide and oxysulfide materials are prepared through this technique. Some reports on producing RE-doped Y_2O_3, La_2O_2S, Gd_2O_3, and $Gd_2Ga_5O_{12}$ can be found in the literature [6, 11]. One should keep in mind that the fuel and nitrates are chosen so that they do not produce toxic gases and the reaction should be relatively mild. However, combustion synthesis is a readily scalable, time and energy-saving, low cost, and efficient process that can be extended for commercial purpose.

4. Physics of upconversion

4.1. Upconversion mechanisms

Several mechanisms have been identified to be involved in UC process either alone or in combination [3, 4, 6, 10]. Though much work has been performed to know the energy transfer processes during UC event, only the basic possibilities, namely excited state absorption (ESA), energy transfer UC (ETU), photon avalanche (PA), cooperative UC (CUC), and energy migration-mediated UC (EMU) are given here. **Figure 2** depicts the basic UC processes.

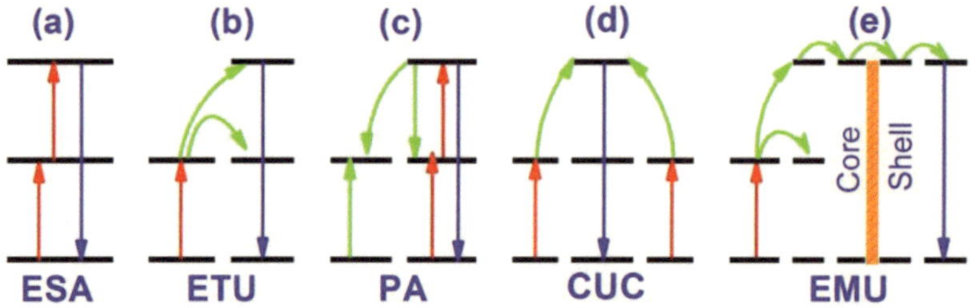

Figure 2. Simplified energy level diagrams describing upconversion processes: (a) ESA, (b) ETU, (c) PA, (d) CUC, and (e) EMU (red: excitation; blue: UC emission; green: energy transfer).

An ESA (**Figure 2a**) process refers to multistep excitation by sequential absorption of one or more low-energy photons from the ground state to intermediate state, and finally populates the intermediate state, from which upconversion emission occurs. The absorption cross-section of the excited ion should have the capability of absorbing the second pump photon. The ESA process generally occurs at low doping concentration (<1%) of the activator ions, as high doping concentration is likely to degrade the UC emission via nonradiative relaxation processes. Although the ESA process occurs in single RE ions, its efficiency is strongly suppressed due to weak absorption induced by parity-forbidden 4f-4f transitions within RE ions.

The ETU process (**Figure 2b**) is more efficient than ESA and it involves two types of luminescent centers—sensitizer and activator ions. Once the sensitizer is excited, it transfers energy to the nearest activator ion and UC emission is obtained from the activator when it drops back to the ground state or lower excited state. For sufficient energy transfer, the activator and sensitizer should possess resonant energy levels with closeness in spatial distance between them. Because of resonant energy absorption, the excitation lifetime is longer and probability of UC is higher which makes this process an efficient one.

The PA process (**Figure 2c**) is more complex than the ESA and ETU and only occurs after a critical level of pump density. The PA process was first proposed in 1979 using Pr^{3+} ion-based infrared quantum counter. If the pump density is sufficiently high, the intermediate reservoir level of many ions becomes populated initially by a nonresonant ground state absorption process, followed by resonant ESA or ETU from another excited ion to populate the UC emitting level. After this stage efficient cross relaxation takes place between excited and ground-state ions. As a result, population of the reservoir level and the UC emitting level increases and causes an "avalanche" effect of generating more excited ions through feedback looping, making this process the most efficient.

The CUC process (**Figure 2d**) is similar to ETU process and includes two types of luminescent centers—sensitizer and activator. In this process the UC emitting level of the activator is populated via cooperative energy transfer from two adjacent sensitizers. The basic difference between ETU and CUC is that in CUC the activator does not have adequate long-lived intermediate energy levels compatible to that of the sensitizers. CUC occurs mainly in Yb^{3+}/RE^{3+} doped UC materials, where Yb^{3+} acts as cooperative sensitizer. The efficiency of CUC is lower than that of the ETU.

Based on energy transfer within $NaGdF_4$:Yb^{3+},Tm^{3+}@$NaGdF_4$:Ln^{3+} (Ln = Tb, Eu, Dy, and Sm) core-shell nanostructures, the EMU process (**Figure 2e**) was first proposed by Liu and coworkers [21] in 2011. The EMU process involves four types of luminescent centers, namely sensitizer, accumulator, migrators, and activator. The sensitizer/accumulator and the activator are confined in separate layers of the core-shell and connected by migrators [10]. The sensitizer first excited by ground state absorption and transfers its energy to an accumulator, promoting it to higher excited state. The accumulator should possess energy levels with longer lifetimes to accept the energy from the sensitizer. Then, energy migration takes place from higher excited state of accumulator to migrator, followed by migration of the excitation energy through the migrators via core-shell interface [10]. Finally, the migrated energy is trapped by an activator in the shell and emits UC luminescence.

4.2. Excited-state dynamics

The excited-state dynamics of RE ions involved in UC processes has been extensively investigated in several bulk materials and most of these are also applicable to their equivalent nanoscale systems. Quantitatively, the UC process can be described by a set of rate equations [4, 10] by taking into account all the population and depopulation pathways involved in the concerned system.

$$\frac{dN_i}{dt} = \sum population\,rate - \sum depopulation\,rate$$

As an example of establishing rate equations, the rate equations for UC process upon 980 nm light excitation in Er^{3+}-Yb^{3+} system are discussed below [22].

As the absorption cross-section of Yb^{3+} is much larger than that of Er^{3+} at 980 nm excitation wavelength, it is reasonable to consider the energy transfer from Yb^{3+} to Er^{3+} [9]. The three possible energy ETU processes from Yb^{3+} to Er^{3+} are the following (refer to **Figure 3**):

(a) ETU-1: $^2F_{5/2}$ (Yb^{3+}) + $^4I_{15/2}$ (Er^{3+}) → $^2F_{7/2}$ (Yb^{3+}) + $^4I_{11/2}$ (Er^{3+})

(b) ETU-2: $^2F_{5/2}$ (Yb^{3+}) + $^4I_{11/2}$ (Er^{3+}) → $^2F_{7/2}$ (Yb^{3+}) + $^4F_{7/2}$ (Er^{3+})

(c) ETU-3: $^2F_{5/2}$ (Yb^{3+}) + $^4I_{13/2}$ (Er^{3+}) → $^2F_{7/2}$ (Yb^{3+}) + $^4F_{9/2}$ (Er^{3+})

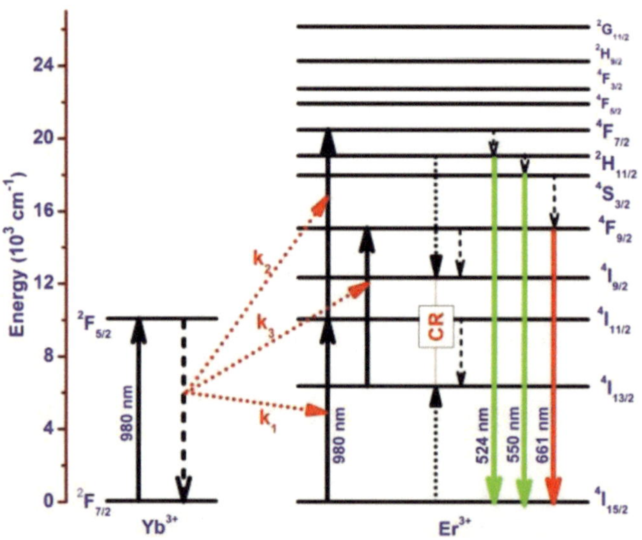

Figure 3. Energy transfer processes relevant for upconversion in Er^{3+}-Yb^{3+} system [22].

The Er^{3+} ions raised to the $^4F_{7/2}$ level by ETU-2 relaxes nonradiatively to the $^2H_{11/2}$/$^4S_{3/2}$ levels and subsequently, radiative transition to the ground state ($^4I_{15/2}$) yields green UC emission bands. Possible mechanisms for population of the $^4I_{13/2}$ level of Er^{3+} are (i) energy back-transfer:

$^2H_{11/2}/^4S_{3/2}$ (Er^{3+}) + $^2F_{7/2}$ (Yb^{3+}) → $^4I_{13/2}$ (Er^{3+}) + $^2F_{5/2}$ (Yb^{3+}) and (ii) cross-relaxation process of Er^{3+} ions: $^2H_{11/2}/^4S_{3/2}$ + $^4I_{15/2}$ → $^4{}_{19/2}$ + $^4I_{13/2}$.

The rate equations for Er^{3+}-Yb^{3+} codoped system are as follows:

$$\frac{dN_{Er,1}}{dt} = W_2 N_{Er,2} + C_{40} N_{Er,4} N_{Er,0} - k_3 N_{Er,1} N_{Yb,1} - W_1 N_{Er,1} - \rho_P \sigma_{13} N_{Er,1}$$

$$\frac{dN_{Er,2}}{dt} = k_1 N_{Er,0} N_{Yb,1} + \rho_P \sigma_{02} N_{Er,0} - k_2 N_{Er,2} N_{Yb,1} - W_2 N_{Er,2} - \rho_P \sigma_{24} N_{Er,2}$$

$$\frac{dN_{Er,3}}{dt} = k_3 N_{Er,1} N_{Yb,1} + \rho_P \sigma_{13} N_{Er,1} + W_4 N_{Er,4} - W'_3 N_{Er,3}$$

$$\frac{dN_{Er,4}}{dt} = k_2 N_{Er,2} N_{Yb,1} + \rho_P \sigma_{24} N_{Er,2} - W_4 N_{Er,4} - W'_4 N_{Er,4} - C_{40} N_{Er,4} N_{Er,0}$$

$$\frac{dN_{Yb,1}}{dt} = \rho_P \sigma_{Yb} N_{Yb,0} - \left(k_1 N_{Er,0} + k_2 N_{Er,2} + k_3 N_{Er,1}\right) N_{Yb,1} - W_{Yb} N_{Yb,1}$$

where $N_{Er,i}$ (i = 0, 1, 2, 3, 4) are the population densities of $^4I_{15/2}$, $^4I_{13/2}$, $^4I_{11/2}$, $^4F_{9/2}$ and $^2H_{11/2}/^4S_{3/2}$ levels of Er^{3+}. $N_{Yb,i}$ (i = 0, 1) are the population densities of the $^2F_{7/2}$ and $^2F_{5/2}$ levels of Yb^{3+}. W_1, W_2, and W_4 are the nonradiative decay rates of the $^4I_{13/2}$, $^4I_{11/2}$, and $^2H_{11/2}/^4S_{3/2}$ states, respectively; W'_3 and W'_4 are the radiative decay rates of the $^4F_{9/2}$ and $^2H_{11/2}/^4S_{3/2}$ states, respectively. The W_{Yb} is the radiative decay rate of the excited $^2F_{5/2}$ state of the Yb^{3+} ion. The k_1, k_2, and k_3 are the energy transfer rates of ETU-1, ETU-2, and ETU-3, respectively. The C_{40} is the cross-relaxation rate for $^2H_{11/2}/^4S_{3/2}$ + $^4I_{15/2}$ → $^4I_{9/2}$ + $^4I_{13/2}$; σ_{ij} is the absorption cross-section between levels i and j of Er^{3+}; σ_{Yb} is the absorption cross-section between levels $^2F_{5/2}$ and $^2F_{7/2}$ of Yb^{3+} and ρ_P is the pump constant which is proportional to the incident pump power, I_P. If the absorption cross-section of Er^{3+} is low compared to that of Yb^{3+} at 980 nm, the contribution of ground state absorption and ESA processes of Er^{3+} states can be neglected in comparison to the ETU (by Yb^{3+}) and relevant terms can be excluded from the rate equations. The energy transfer rates can be calculated using Judd-Ofelt theory [23, 24]. The multiphonon nonradiative relaxation rates can be derived using the modified energy gap law [4]. Once the transition rates are known, the characteristics of UC, namely, luminescence lifetime, spectral ratios, and quantum yields can be determined by solving these rate equations.

The rate equation model for UC process is generally made for theoretical analysis, but only very few reports are available on the use of this approach in UC nanomaterials mainly due to the complex nature of the involved UC process at the nanoscale and most of the spectroscopic characteristics of UC nanoparticles are determined experimentally. The investigation into the time-resolved photoluminescence behavior of UC emissions is often used to check the validity of the proposed model of UC processes, and the rate constants are extracted from the fittings [10]. The lifetime of the activator's excited state is shorter than those of the sensitizers. The ETU processes lengthen the lifetime of the activators due to energy transfer from the sensitizers. Thus, the apparent lifetime of the activators is dependent on the sensitizers. The measured lifetime in UC luminescence and downconversion luminescence (the excited states are directly excited) is different. However, in the nanomaterials several factors such as surface quenching effect, particle size, and surface structure affect the lifetime and deviate from the theoretical value. The UC process shows pump power-dependent UC luminescence due to its nonlinear character. The combination of pump power dependent of UC with the steady-state rate equations yields useful information on the UC kinetics. Theoretical models proposed by Pollnau et al. and Suyver et al. [25, 26] are used vastly to know the nature of UC, i.e., the number of photons absorbed in the UC processes. The UC emission intensity (I) is proportional to nth orders of the pump power density (P): $I \propto P^n$, where 'n' is the number of photons absorbed in the particular UC process. Thus, the slope of the plot of ln[I] vs. ln[P] reveals the nature of the UC process. The pump power dependence of the UC emission is valid in low power density and becomes complicated at high pump power due to the competition between linear decays and UC processes. In most of the cases, the slope value decreases with increasing excitation power and at very high pump power, the UC processes sometimes become pump independent due to saturation of some intermediate levels.

4.3. Photophysical processes

Basic understanding of the main photophysical processes of UC emission is necessary not only for design and optimization of performance of UC materials but also the proper interpretation of experimental results. Although the three major steps photon absorption, subsequent energy transfer, and emission of UC are key processes, various nonradiative processes that compete with the radiative processes play vital roles in determining UC efficiency. Luminescence intensities are connected to the quantum-mechanical transition rates that depend on the details of the energy states that take part for the particular UC emission. The details of the initial and final states involved in the UC emission can be determined by Judd-Ofelt theory [23, 24]. This theory is based on the approximation that all the Stark manifolds have equal population and the host matrix element is isotropic. This approach facilitates to estimate the radiative rates for RE^{3+} ions theoretically by using the experimental parameters of optical absorption or emission spectra. The radiative decay rate, W_{rad} (photons/s) for an electric dipole allowed transition between two manifolds of RE^{3+} ion can be described as [10]:

$$W_{rad} = \frac{4e^2\omega^3}{3\hbar c^3}\frac{1}{2J+1}n\left(\frac{n^2+2}{3}\right)^2 \sum_{\lambda=2,4,6} \Omega_\lambda \left|\langle SLJ||U^{(\lambda)}||S'L'J'\rangle\right|^2$$

where $|SLJ\rangle$ and $|S'L'J'\rangle$ are the initial and final states of the transition, respectively. e, ω, $\hbar$, c, and n are the electronic charge, average angular frequency of the optical transition, reduced Planck constant, speed of light, and refractive index, respectively. $[(n^2+2)/3]^2$ is the Lorentz local field

http://dx.doi.org/10.5772/65118

correction factor, and $(2J + 1)$ is the degeneracy manifold of the initial state. $|\langle SLJ||U^{(\lambda)}||S'L'J'\rangle|^2$ are the squared matrix elements of the electric dipole operator between the initial and final manifolds. The Ω_λ ($\lambda = 2, 4, 6$) are Judd–Ofelt parameters that can be obtained from the optical absorption characterization of the material and give the influence of the host matrix on the electric dipole transition probabilities. In order to understand the UC mechanism involving 4f electrons more precisely, the magnetic dipole transitions which are weakly affected by the crystal field of the host should be combined to the Judd-Ofelt rates. With magnetic dipole transitions factored in, the Hamiltonian of a free ion (RE^{3+}) can be treated in quantifying the magnetic part of the UC emission in RE^{3+}.

Within a sensitizer-activator pair, the excitation energy of the sensitizer is transferred resonantly to adjacent activator. The resonant energy transfer process may be radiative or nonradiative. The nonradiative energy transfer takes place via long-range dipole-dipole interactions while the former one requires emission and reabsorption of a photon [10]. According to Dexter's theory [27], the resonant energy transfer probability W_{SA} between the sensitizer and activator ions can be expressed using the spectral overlap of the emission of sensitizer and absorption of activator. The energy transfer probability is given by [3]:

$$W_{SA} = \frac{3h^4c^4}{64\pi^5 n^4 \tau_S} \frac{Q_{abs}}{r_{SA}^6} \int \frac{f_{em}^S(E) f_{em}^A(E)}{E^4} dE$$

where h, c, n, r_{SA}, and τ_S are Planck's constant, the speed of light, the refractive index, the sensitizer-activator distance, and the intrinsic lifetime of the sensitizer, respectively; Q_{abs} is the integrated absorption cross-section of the activator ion, $f_{em}^A(E)$ and $f_{em}^S(E)$ are the normalized spectral functions of the absorption band of the activator at intermediate levels ($\int f_{em}^A(E)dE = 1$) and the emission band of the sensitizer at excited states ($\int f_{em}^S(E)dE = 1$), respectively [3, 10]. Therefore, the sensitizer-activator separation (r_{SA}) and the spectral overlap (the integration part) are the two key parameters governing the energy transfer. Another requirement to get efficient energy transfer is the high absorbance (Q_{abs}) of the activator ion at the emission wavelength of the sensitizer ion.

If there is an energy mismatch between the energy levels of the sensitizer and activator ions, the resonant energy transfer probability, W_{SA} decreases exponentially with the magnitude of the mismatch and the energy transfer will be no more resonant. Nevertheless, energy transfer in such case is compensated by the nonradiative, phonon-assisted processes, known as phonon-assisted energy transfer [3, 10].

5. Applications

UC materials have been extensively used in solid-state-lasers and waveguide amplifiers for a long time [3]. Applications in rewritable optical storage and nondestructive optical memory devices are examples of some recent developments of exploiting UC materials in practical use [3]. Furthermore, the applications of these materials in biological field have been considered after the tremendous progress in preparation of UC nanomaterials with controllable particle size and shape. Recently, a diagnosis method for Ebola virus has been proposed by Tsang et al. using UC nanoparticles [28]. Therefore, advances of UC materials in newer applications will now be discussed in this section.

5.1. Lighting and displays

As a source of three primary colors—red, green and blue, UC emission offers many attractive features for display devices. An eminent advantage is the ability to operate displays with very high brightness and saturated colors without deterioration of the light emitting materials. Such displays are appropriate for high-brightness ambient lighting conditions.

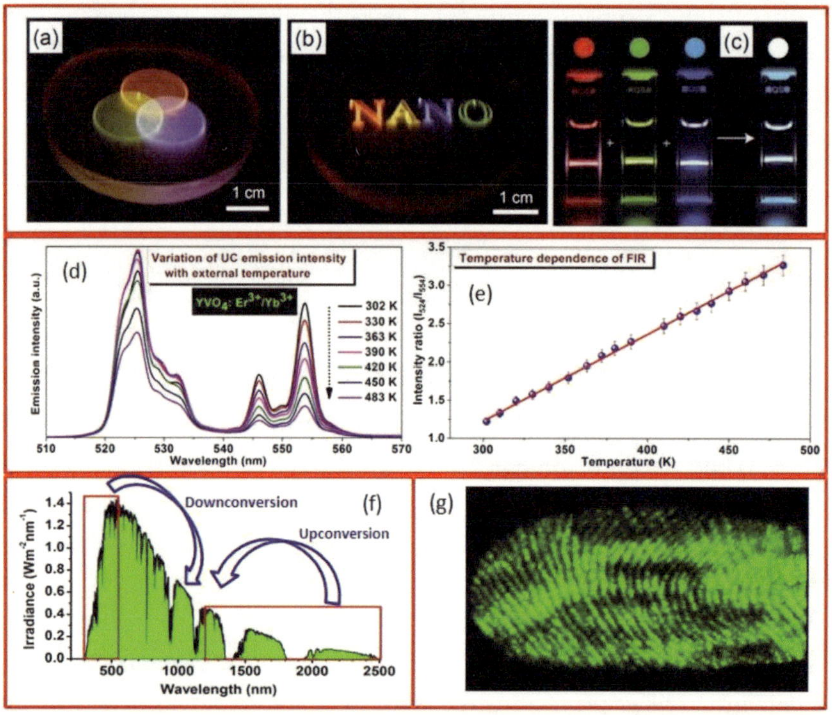

Figure 4. Luminescence color images generated in a RE-doped nanocrystal/polydimethylsiloxane composite monolith show the ability to display (a) additive colors [29], (b) three-dimensional objects [29], (c) volumetric three-dimensional display generated by combining three sets of conventional upconversion nanoparticles featuring monochromatic RGB emission (red, $KMnF_3$:Yb/Er; green, $NaYF_4$:Yb/Er; blue, $NaYF_4$:Yb/Tm) [29], (d) variation of UC emission intensity of green bands (524 nm and 554 nm) with external temperature of YVO_4:Er^{3+}/Yb^{3+} particles [36], (e) plot of fluorescence intensity ratio (FIR) as a function of external temperature [36], (f) AM1.5G solar irradiance spectrum (1000 W m^{-2}) and available for downconversion or downshifting (about 16% up to 500 nm) and upconversion (about 16% in the range of 1.2–2.5 μm), and (g) fresh fingerprint on glass by dry powdering with YVO_4:Er/Yb particles illuminated at 980 nm light [47]. Figures reproduced with permission from: (a–c) Ref. [29], Nature Publishing Group; (d and e) Ref. [36], Elsevier; and (g) Ref. [47], Optical Society of America.

The ability of dispersion of UC materials in polymers has further extended their applications in displays. For instance, $NaYF_4$:Er(Tm),Yb-PMMA nanocomposite producing white, blue, and green lights have been designed by Lin and his group [29] through *in situ* photopolymerization. Downing et al. [30] and Wang et al. [31] have shown that UC materials can be employed for three-dimensional displays. A very recent work by Deng et al. [32] shows the modulation of UC emission color of multilayer core-shell structure by controlling the pulse duration of the excitation laser. **Figure 4(a–c)** exhibits the volumetric three-dimensional display using pulse duration sensitive nanocrystals.

5.2. Temperature sensing

The luminescence from lanthanide-doped materials is an interesting solution to collect information about temperature in hazardous environment, microelectronic and photonic devices. Temperature sensing has been studied from the last decade using luminescent microcrystalline compounds at the tip of scanning thermal probe, at the tip of optical fiber or on the silica-on-silicon waveguides [33]. Although there were some drawbacks due to micron-sized particles; because the material may act as a thermal insulator and light scattering by rough surfaces of the bigger particles degrades the resolution of temperature sensing. The most important feature of UC nanoparticle-based thermometry is that these particles can be employed inside the cells. The temperature sensing of RE-doped UC materials is determined by two commonly used methods, i.e., decay time and fluorescence intensity ratio (FIR) methods. Besides these two major techniques, there are some other reports on temperature sensing based on temperature-dependent variation in emission bandwidth, variation of intensities of Stark sublevels, valley-to-peak intensity ratio, etc. [34, 35]. The most accepted method used for temperature sensing is based on the FIR technique which includes temperature-dependent variation of intensity ratio, emitted from two thermally coupled levels. The FIR method, in principle, uses low excitation power or short pulsed excitation in order to reduce the self-heating of the nanoparticles. The most studied RE element for temperature sensing application is Er^{3+}, which has well-known thermally coupled $^2H_{11/2}$ and $^4S_{3/2}$ levels. The sensor sensitivity is defined as the ratio in which the FIR changes with the temperature and is considered as the standard figure-of-merit. **Figure 4d** shows the upconversion spectra of an Er^{3+}, Yb^{3+} codoped YVO_4 powder recorded within 302–483 K [36]. The change in emission intensity from thermally coupled electronic multiplets is clearly observed. **Figure 4e** shows the plot of intensity ratio versus absolute temperature. As expected the system shows Boltzmann distribution of populations of the $^2H_{11/2}$ and $^4S_{3/2}$ levels. Among the luminescent thermometers proposed till now at submicron scale, only a few of them successfully illustrated the temperature sensing with a scanning microscope, e.g., thermal sensors using Er^{3+}/Yb^{3+} codoped fluoride glass [37] and $NaYF_4$:Er^{3+}/Yb^{3+} nanoparticles [38]. Temperature sensing performance up to 2000 K using metal-nanoparticle decorated, $(Gd,Yb,Er)_2O_3$ nanorods exploiting FIR technique and blackbody radiation has been reported recently by Carlos and his group [39].

5.3. Solar cell

Nonabundance of fossil fuel and public nonwillingness to accept nuclear power have encouraged researchers to search for alternative energy sources. Solar energy, a contemporary nonconventional source of energy is an obvious natural target due to its availability in large quantities. Crystalline silicon (c-Si) semiconductor was the key material for the earlier solar cells for converting solar energy into electricity. But the production of solar energy is still limited due to low efficiencies of power conversion. The multijunction semiconductor-based solar cells have maximum conversion efficiency 37.5% while the commercial silica-based solar cells have within 15–20% [40]. On the other hand, the dye-sensitized solar cells (DSSCs) yield conversion efficiencies below 12% [41]. The spectral mismatch between the incident photon energy of solar radiation and the band gap of c-Si semiconductor is one of the most important reasons for the low efficiency. Most of the solar energy is lost in several ways, including conversion to heat energy passing through the solar cell and thus it limits the efficiency [42, 43]. In order to enhance spectral response, wavelength converting layers are added to the solar cells. This layer is capable of absorbing the spectral range where photosensitive materials do not absorb well, although the addition of extra layers may modify the reflection, refraction and other optical responses in a negative or positive way. The photons of lower energy (UV/blue) can be converted into NIR through downconversion processes and NIR light can be upconverted by sequential absorption of two or more low-energy photons into visible or UV, which partially minimizes the optical mismatch in solar cells (**Figure 4f**). At the same time, the thermalization effect due to the carriers is greatly decreased and the conversion efficiency of the solar cell is enhanced. Nevertheless, examination of some available reports demonstrates that downshifting can add 1–2 absolute% to the conversion yield and upconversion 0.5–1% in silicon cells while in DSSC, the conversion yield is increased by 1–1.5% for downshifting and 0.2–0.5% for upconversion [44]. However, it is expected that these numbers will be increased in forthcoming years, especially if plasmonics are used to increase the UC efficiency.

5.4. Security

The unique properties of UC materials make them useful in security ink applications such as latent fingerprint and quick response code printing [45, 46]. Many UC particles are easily dispersible in well-known solvents, allowing printing or coating them onto various substrates. Moreover, multicolor emissions from UC layers upon IR excitation are useful for complex security patterns, giving another level of security or data protection. Earlier, we demonstrated the latent fingerprint detection (**Figure 4g**) using YVO_4:Er/Yb particles by a dry powdering method on a glass slide [47], which shows green fingerprint images upon 980 nm light illumination. The advantage of this technique is that it gives nearly no background and ridges can be detected clearly. The lifetimes of UC nanocrystals under a single-wavelength excitation vary from μs to ms and can be utilized for multiplexing in the time domain which is extendable to security and data storage technology including anticounterfeiting applications [48, 49].

5.5. Biomedical

With unique upconversion properties, UC nanoparticles yield high photostability, high sensitivity, and low optical background noise due to the absence of autofluorescence under NIR radiation which are suitable for bioimaging and biodetection [7, 28, 50, 51]. Moreover, NIR excitation enables deep penetration in tissue without damaging the biological specimens. Such properties of RE-doped UC nanomaterials are appropriate for a wide range of potential biological applications, including photodynamic therapy (PDT), drug delivery, biological imaging, and sensing. Most important point to be concerned for biomedical applications of UC nanomaterials is that the toxicity of the nanoparticles should be evaluated and must be nontoxic in nature for biological applications.

5.5.1. Biological imaging

The UC nanoparticle-based bioimaging has been widely used in cell, tissue, and small animal imaging due to their high contrast, absence of autofluorescence, and photostability against photobleaching and photoblinking [7, 50, 51]. The UC emissions are bright enough to be imaged at moderate excitation power of CW laser and can be used for single-molecule imaging. *In vitro* cellular imaging using RE-doped UC nanoparticles as targets has been demonstrated recently for colon cancer cells [52], ovarian cancer cells [53], HeLa cells [54], breast carcinoma cells [55], etc. In 2006, Lim and his coworkers [56] carried out pioneering work on live organism imaging on *Caenorhabditis elegans* worms using Y_2O_3:Er,Yb nanoparticles. The imaging of the digestive system of the worm under NIR excitation showed track of movement of the nanoparticles in intestines. Besides organism imaging, *in vivo* imaging of small mammals has been studied using $NaYF_4$:Er,Yb nanoparticles by Chatterjee et al. [52] and demonstrated that the luminescence from the nanoparticles can be clearly observed upon NIR excitation even when the nanoparticles are placed ~10 mm beneath the skin of the animal and the study opens up a new perspective for bioimaging.

5.5.2. Biological sensing and detection

In biological sensing and detection [28, 57], UC luminescence is turned on or off by chromophoric complexes, decorated with the UC nanoparticles and the techniques are based on mainly two processes: fluorescence resonance energy transfer (FRET) and non-FRET [7]. The energy transfer in FRET takes place between donor and acceptor at a very short distance, typically shorter than 10 nm. Wang et al. [58] reported detection of avidin using human biotin-functionalized $NaYF_4$:Er/Yb and gold nanoparticles. Recently, Liu and his coworkers [59] proposed a highly sensitive biosensing platform based on UC nanoparticles and graphene as donor-acceptor system. On the other hand, in non-FRET based biological sensing, RE-doped UC nanoparticles are used as reporters and the luminescence from the nanoparticles is observed directly. Van de Rijke et al. [60] used Y_2O_2S:Er,Yb nanoparticles in the detection of nucleic acids and observed four-fold increase in the detection limit compared to conventional reporter.

5.5.3. Photodynamic therapy

The UC nanoparticles have recently been developed for application in cancer photodynamic therapy (PDT) [61–63]. For photodynamic therapy, three primary elements—excitation light, photosensitizer, and reactive oxygen species (ROS) are required. The photosensitizers are excited to the higher energy states under the irradiation of suitable NIR (**Figure 5**). The photosensitizer then returns to lower energy state producing energy which is transferred to the nearby oxygen and creates ROS that damages the cancer cells in the vicinity. Zhang et al. [64] first time designed the UC nanoparticles, coated with layer of mesoporous silica shell where the photosensitizers were doped for photodynamic therapy for the treatment of bladder cancer cells. Yan and his group [65] constructed $NaGdF_4$:Yb,Er@CaF_2 core/shell UC nanoparticles loaded with hematoporphyrin and silicon phthalocyanine dihydroxide molecules, and showed excellent PDT efficiency in HeLa cancer cells upon NIR irradiation.

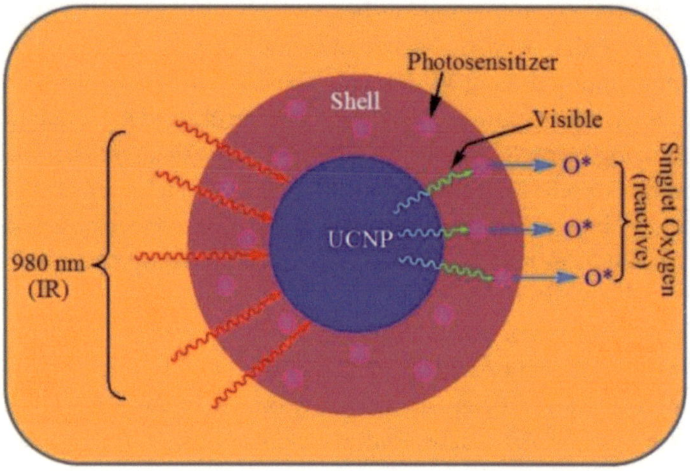

Figure 5. Schematic diagram of UC-based photodynamic therapy. The upconversion nanoparticles emit visible light which further excites the photosensitizers, resulting in the production of ROS and causes damage to the nearby cancer cells.

6. Summary and outlook

In summary, we presented an overview of the synthesis, properties, and recent development of RE-doped UC materials in various emerging applications. The UC materials show potential for a wide range of energy-related applications covering from lighting and display technologies to photovoltaic. At the same time, UC nanoparticles have proven to be effective in biological applications due to their unique optical and chemical properties such as absence of autofluorescence, low-toxicity, low photodamage to the biological cells, and high tissue-penetration depth of the excitation light. In spite of these benefits, there remain some limitations at the fundamental level associated with the practical use of UC materials. For example,

http://dx.doi.org/10.5772/65118

the efficiency of UC materials is restricted and to date, the maximum internal quantum yield has been obtained as 7.6% for $LiLuF_4$:Yb,Tm@$LiLuF_4$ [66] nanoparticles under NIR excitation. Another challenge for UC nanoparticles is surface engineering which is required for biomedical application and surface modification generally decreases the luminescence efficiency. Additionally, for biological applications, the UC nanoparticles should be the smallest (<10 nm) and the problem associated with smaller nanoparticles is that the upconversion efficiency decreases with decreasing particle size. However, manipulation of local environment around the RE ions in UC particles along with plasmons, quantum dots, dyes, or other dopants with a strong absorption might be effective ways to enhance the quantum yield. Therefore, more significant benefits of UC materials through constant and thriving research are yet to come in future for fundamental understanding as well as in emerging applications.

Acknowledgements

This work has been supported from the European Commission ("Interweave"—Erasmus Mundus project).

Author details

Manoj Kumar Mahata*, Hans Christian Hofsäss and Ulrich Vetter

*Address all correspondence to: mmahata@phys.uni-goettingen.de

Second Institute of Physics, University of Goettingen, Goettingen, Germany

References

[1] Bloembergen N. Solid state infrared quantum counters. Phys. Rev. Lett. 1959;2:84–85. DOI: 10.1103/PhysRevLett.2.84

[2] Auzel F. Upconversion and anti-Stokes processes with f and d ions in solids. Chem. Rev. 2004;104:139–173. DOI: 10.1021/cr020357g

[3] Zhou B., Shi B., Jin D., Liu X. Controlling upconversion nanocrystals for emerging applications. Nature Nanotech. 2015;10:924–936. DOI: 10.1038/NNANO.2015.251

[4] Zheng W., Huang P., Tu D., Ma E., Zhua H., Chen X. Lanthanide-doped upconversion nano-bioprobes: electronic structures, optical properties, and biodetection. Chem. Soc. Rev. 2015;44:1379–1415. DOI: 10.1039/c4cs00178h

[5] Walrand C. G., Binnemans K. Rationalization of crystal-field parametrization. In: Gscheidner Jr. K. A., Eyring L., editors. Handbook on the Physics and Chemistry of

Rare Earths. 23rd ed. Netherland: Elsevier Science B.V.; 1996. p. 121–283. DOI: 10.1016/S0168-1273(96)23006-5

[6] Sun L. D., Dong H., Zhang P. Z., Yan C. H. Upconversion of Rare Earth Nanomaterials. Annu. Rev. Phys. Chem. 2015;66:619–642. DOI: 10.1146/annurev-physchem-040214-121344

[7] Lin M., Zhao Y., Wang S. Q., Liu M., Duan Z. F., Chen Y. M. et al. Recent advances in synthesis and surface modification of lanthanide-doped upconversion nanoparticles for biomedical applications. Biotechnol. Adv. 2012;30(6):1551–1561. DOI: 10.1016/j.biotechadv.2012.04.009

[8] Haase M., Schäfer H. Upconverting nanoparticles. Angew. Chem. Int. Ed. 2011;50:5808–5829. DOI: 10.1002/anie.201005159

[9] Huang F., Liu X., Ma Y., Kang S., Hu L., Chen D. Origin of near to middle infrared luminescence and energy transfer process of Er^{3+}/Yb^{3+} co-doped fluorotellurite glasses under different excitations. Sci. Rep. 2015;5:8233(1–6). DOI: 10.1038/srep08233

[10] Nadort A., Zhao J., Goldys E. M. Lanthanide upconversion luminescence at the nanoscale: fundamentals and optical properties. Nanoscale. 2016;8:13099–13130. DOI: 10.1039/C5NR08477F

[11] Chang H., Xie J., Zhao B., Liu B., Xu S., Ren N. et al. Rare earth ion-doped upconversion nanocrystals: synthesis. Nanomaterials. 2015;5:1–25. DOI: 10.3390/nano5010001

[12] Mai H., Zhang, Y., Sun L., Yan C. Size- and phase-controlled synthesis of monodisperse $NaYF_4$:Yb,Er nanocrystals from a unique delayed nucleation pathway monitored with upconversion spectroscopy. J. Phys. Chem. C. 2007;111:13730–13739. DOI: 10.1021/jp073919e

[13] Yi G. S., Chow G. M. Colloidal LaF_3:Yb,Er, LaF_3:Yb,Ho and LaF_3:Yb,Tm nanocrystals with multicolor upconversion fluorescence. J. Mater. Chem. 2005;15:4460–4464. DOI: 10.1039/B508240D

[14] Mahata M. K., Kumar K., Rai V. K. Structural and optical properties of Er^{3+}/Yb^{3+} doped barium titanate phosphor prepared by co-precipitation method. Spectrochim. Acta A. 2014;124:285–291. DOI: 10.1016/j.saa.2014.01.014

[15] Mahata M. K., Koppe T., Hofsäss H., Kumar K., Vetter U. Host sensitized luminescence and time-resolved spectroscopy of $YVO_4:Ho^{3+}$ nanocrystals. Phys. Procedia. 2015;76:125–131. DOI: 10.1016/j.phpro.2015.10.023

[16] Wu H., Xu H., Su Q., Chen T., Wu M. Size- and shape-tailored hydrothermal synthesis of YVO_4 crystals in ultra-wide pH range condition. J. Mater. Chem. 2011;13:1223–1228. DOI: 10.1039/B210713

[17] Zeng J. H., Su J., Li Z. H., Yan R. X., Li, Y. D. Synthesis and upconversion luminescence of hexagonal-phase $NaYF_4:Yb^{3+},Er^{3+}$ phosphors of controlled size and morphology. Adv. Mater. 2005;17:2119–2123. DOI: 10.1002/adma.200402046

[18] Patra A., Friend C. S., Kapoor R., Prasad P. N. Fluorescence upconversion properties of Er^{3+}-doped TiO_2 and $BaTiO_3$ nanocrystallites. Chem. Mater. 2003;15:3650–3655. DOI: 10.1021/cm020897u

[19] Patra A., Friend C. S., Kapoor R., Prasad P. N. Upconversion in $Er^{3+}:ZrO_2$ nanocrystals. J. Phys. Chem. B. 2002;106:1909–1912. DOI: 10.1021/jp013576z

[20] Cheng Z. Y., Xing R. B., Hou Z. Y., Huang S. S., Lin J. Patterning of light-emitting $YVO_4:Eu^{3+}$ thin films via inkjet printing. J. Phys. Chem. C. 2010;114:9883–9888. DOI: 10.1021/jp101941y

[21] Wang F., Deng R. R., Wang J., Wang Q. X., Han Y., Zhu H., et al. Tuning upconversion through energy migration in core-shell nanoparticles. Nat. Mater. 2011;10:968–973. DOI: 10.1038/nmat3149

[22] Mahata M. K., Koppe T., Mondal T., Bruesewitz C., Kumar K., Rai V. K., et al. Incorporation of Zn^{2+} ions into $BaTiO_3:Er^{3+}/Yb^{3+}$ nanophosphor: an effective way to enhance upconversion, defect luminescence and temperature sensing. Phys. Chem. Chem. Phys. 2015;17:20741–20753. DOI: 10.1039/c5cp01874a

[23] Judd B. Optical absorption intensities of rare-earth ions. Phys. Rev. 1962;127(3):750–761. DOI: 10.1103/PhysRev.127.750

[24] Ofelt G. Intensities of crystal spectra of rare-earth ions. J. Chem. Phys. 1962;37 :511–520. DOI: 10.1063/1.1701366

[25] Pollnau M., Gamelin D., Lüthi S., Güdel H., Hehlen M. Power dependence of upconversion luminescence in lanthanide and transition-metal-ion systems. Phys. Rev. B: Condens. Matter. 2000;61:3337–3346. DOI: 10.1103/PhysRevB.61.3337

[26] Suyver J. F., Aebischer A., García-Revilla S., Gerner P., Güdel H. U. Anomalous power dependence of sensitized upconversion luminescence. Phys. Rev. B. 2005;71:125123. DOI: 10.1103/PhysRevB.71.125123

[27] Dexter D. L. A theory of sensitized luminescence in solids. J. Chem. Phys. 1953;21:836–850. DOI: 10.1063/1.1699044

[28] Tsang M. K., Ye W. W., Wang G., Li J., Yang M., Hao J. Ultrasensitive detection of Ebola virus oligonucleotide based on upconversion nanoprobe/nanoporous membrane system. ACS Nano. 2016;10(1):598–605. DOI: 10.1021/acsnano.5b05622

[29] Boyer J. C., Johnson N. J. J., van Veggel F. C. J. M. Upconverting lanthanide-doped NaYF4–PMMA polymer composites prepared by in situ polymerization. Chem. Mater. 2009;21(10):2010–2012. DOI: 10.1021/cm900756h

[30] Downing E., Hesselink L., Ralston J., Macfarlane R. A three-color, solid-state, three-dimensional display. Science. 1996;273(5279):1185–1189. DOI: 10.1126/science.273.5279.1185

[31] Wang F., Han Y., Lim C. S., Lu Y. H., Wang J., Xu J., et al. Simultaneous phase and size control of upconversion nanocrystals through lanthanide doping. Nature. 2010;463:1061–1065. DOI: 10.1038/nature08777

[32] Deng R., Qin F., Chen F., Huang W., Hong M., Liu X. Temporal full-colour tuning through non-steady-state upconversion. Nat. Nanotechnol. 2015;10:237–242. DOI: 10.1038/nnano.2014.317

[33] Henry D. M., Herringer J. H., Djeu N. Response of 1.6 μm $Er:Y_3Al_5O_{12}$ fiber-optic temperature sensor up to 1520. Appl. Phys. Lett. 1999;74;3447. DOI: 10.1063/1.12412

[34] Soni A. K., Dey R., Rai V. K. Stark sublevels in Tm^{3+}–Yb^{3+} codoped Na2Y2B2O7 nanophosphor for multifunctional application. RSC Adv. 2013;5:34999–35009. DOI: 10.1039/C4RA15891

[35] Li L., Xu W., Zheng L., Qin F., Zhou Y., Liang Z., et al. Valley-to-peak intensity ratio thermometry based on the red upconversion emission of Er^{3+}. Opt. Exp. 2016;24(12): 13244–13249. DOI: 10.1364/OE.24.013244

[36] Mahata M. K., Kumar K., Rai V. K. Er^{3+}–Yb^{3+} doped vanadate nanocrystals: a highly sensitive thermographic phosphor and its optical nanoheater behavior. Sensor Actuat. B-Chem. 2015;209:775–780. DOI: 10.1016/j.snb.2014.12.039

[37] Peng H. S., Stich M. I. J., Yu J. B., Sun L. N., Fischer L. H., Wolfbeis O. S. Luminescent europium(III) nanoparticles for sensing and imaging of temperature in the physiological range. Adv. Mater. 2010;22:716–719. DOI: 10.1002/adma.200901614

[38] Vetrone F., Naccache R., Zamarron A., Juarranz de la Fuente A., Sanz-Rodríguez F., Maestro L. M., et al. Temperature sensing using fluorescent nanothermometers. ACS Nano. 2010;4:3254–3258. DOI: 10.1021/nn100244a

[39] Debasu M. L., Ananias D., Pastoriza-Santos I., Liz-Marzan L. M., Rocha J., Carlos L. D. All-in-one optical heater-thermometer nanoplatform operative from 300 to 2000 K based on Er^{3+} emission and blackbody radiation. Adv. Mater. 2013;25:4868–4874. DOI: 10.1002/adma.201300892

[40] Green M. A., Emery K., Hishikawa Y., Warta W., Dunlop E. D. Progress in photovoltaics: research and applications. Photovoltaics. 2012;20:606–614. DOI: 10.1002/pip.2267

[41] Yella A., Lee H. W., Tsao H. N., Yi C. Y., Chandiran A. K., Nazeeruddin M. K., et al. Porphyrin-sensitized solar cells with cobalt (II/III)-based redox electrolyte exceed 12 percent efficiency. Science. 2011;334:629–634. DOI: 10.1126/science.1209688

[42] Hill S. P., Dilbeck T., Baduell E., Hanson K. Integrated photon upconversion solar cell via molecular self-assembled bilayers. ACS Energy Lett. 2016;1(1):3–8. DOI: 10.1021/acsenergylett.6b00001

[43] Rodríguez-Rodríguez H., Imanieh M. H., Lahoz F., Martín I. R. Analysis of the upconversion process in Tm3+ doped glasses for enhancement of the photocurrent in silicon solar cells. Sol. Energ. Mat. Sol. Cells. 2016;144:29–32. DOI: 10.1016/j.solmat.2015.08.017

[44] Bünzli J. C. G., Chauvin A. S. Lanthanides in Solar Energy Conversion. In: Bünzli J. C. G., Pecharsky V. K., editors. Handbook on the Physics and Chemistry of Rare Earths. 44th ed. Amsterdam: Elsevier Science, B.V.; 2014. p. 169–281. DOI: 10.1016/B978-0-444-62711-7.00261-9

[45] Wang J., Wei T., Li X., Zhang B., Wang J., Huang C., et al. Near-infrared-light-mediated imaging of latent fingerprints based on molecular recognition. Angew. Chem. Int. Ed. 2014;53:1616–1620. DOI: 10.1002/anie.201308843

[46] Meruga J. M., Baride A., Cross W., Kellar J. J., May P. S. Red-green-blue printing using luminescence-upconversion inks. J. Mater. Chem. C. 2014;2:2221–2227. DOI: 10.1039/C3TC32233E

[47] Mahata M. K., Tiwari S. P., Mukherjee S., Kumar K., Rai V. K. YVO_4:Er^{3+}/Yb^{3+} phosphor for multifunctional applications. J. Opt. Soc. Am. B. 2014;31(8):1814–1821. DOI: 10.1364/JOSAB.31.001814

[48] Gorris H. H., Wolfbeis O. S. Photon-upconverting nanoparticles for optical encoding and multiplexing of cells, biomolecules, and microspheres. Angew. Chem. Int. Ed. 2013;52:3584–3600. DOI: 10.1002/anie.201208196

[49] Kumar P., Singh S., Gupta B.K. . Future prospects of luminescent nanomaterial based security inks: from synthesis to anti-counterfeiting applications. Nanoscale. 2016;8:14297–14340. DOI: 10.1039/C5NR06965C

[50] Liu Y., Su Q., Zou X., Chen M., Feng W., Shi Y. et al. Near-infrared in vivo bioimaging using a molecular upconversion probe. Chem. Commun. 2016;52:7466–7469. DOI: 10.1039/C6CC03401B

[51] Tan G. R., Wang M., Hsu C.Y., Chen N., Zhang Y. Small upconverting fluorescent nanoparticles for biosensing and bioimaging. Adv. Opt. Mater. 2016;4(7):984–997. DOI: 10.1002/adom.201600141

[52] Chatterjee D. K., Rufaihah A. J., Zhang Y. Upconversion fluorescence imaging of cells and small animals using lanthanide doped nanocrystals. Biomaterials. 2008;29:937–943. DOI: 10.1016/j.biomaterials.2007.10.051

[53] Boyer J. C., Manseau M. P., Murray J. I., van Veggel F. C. J. M. Surface modification of upconverting NaYF4 nanoparticles doped with PEG-phosphate ligands for NIR (800 nm) biolabeling within the biological window. Langmuir. 2010;26:1157–1164. DOI: 10.1021/la902260j

[54] Cheng L., Yang K., Li Y. G., Chen J. H., Wang C. Y., Shao M. W., et al. Facile preparation of multifunctional upconversion nanoprobes for multimodal imaging and dual-targeted photothermal therapy. Angew. Chem. 2011;123:7523–7528. DOI: 10.1002/anie.201101447

[55] Yang Y. M., Shao Q., Deng R. R., Wang C., Teng X., Cheng K., et al. In vitro and in vivo uncaging and bioluminescence imaging by using photocaged upconversion nanoparticles. Angew. Chem. Int. Ed. 2012;51:3125–3129. DOI: 10.1002/anie.201107919

[56] Lim S. F., Riehn R., Ryu W. S., Khanarian N., Tung C. K., Tank D., et al. In vivo and scanning electron microscopy imaging of upconverting nanophosphors in *Caenorhabditis elegans*. Nano Lett. 2006;6:169–174. DOI: 10.1021/nl0519175

[57] Li H., Shi L., Sun D., Li P., Liu Z. Fluorescence resonance energy transfer biosensor between upconverting nanoparticles and palladium nanoparticles for ultrasensitive CEA detection. Biosens. Bioelectron. 2016;86:791–798. DOI: 10.1016/j.bios.2016.07.070

[58] Wang L. Y., Yan R. X., Huo Z. Y., Wang L., Zeng J. H., Bao J. et al. Fluorescence resonant energy transfer biosensor based on upconversion-luminescent nanoparticles. Angew. Chem. Int. Ed. 2005;44:6054–6057. DOI: 10.1002/anie.200501907

[59] Liu J., Liu Y., Liu Q., Li C., Sun L., Li F. Iridium(III) complex-coated nanosystem for ratiometric upconversion luminescence bioimaging of cyanide anions. J. Am. Chem. Soc. 2011;133:15276–15279. DOI: 10.1021/ja205907y

[60] van de Rijke F., Zijlmans H., Li S., Vail T., Raap A. K., Niedbala R. S. et al. Up-converting phosphor reporters for nucleic acid microarrays. Nature Biotechnol. 2001;19:273–276. DOI: doi: 10.1038/85734

[61] Dong H., Tang S., Hao Y., Yu H., Dai W., Zhao G. et al. Fluorescent MoS2 quantum dots: ultrasonic preparation, up-conversion and down-conversion bioimaging, and photodynamic therapy. ACS Appl. Mater. Interfaces. 2016;8(5):3107–3117. DOI: 10.1021/acsami.5b10459

[62] Lu F., Yang L., Ding Y., Zhu J. J. Highly emissive Nd^{3+}-sensitized multilayered upconversion nanoparticles for efficient 795 nm operated photodynamic therapy. Adv. Funct. Mater. 2016;26(26):4778–4785. DOI: 10.1002/adfm.201600464

[63] Chen J., Zhao J. X. Upconversion nanomaterials: synthesis, mechanism, and applications in sensing. Sensors. 2012;12:2414–2435. DOI: 10.3390/s120302414

[64] Zhang P., Rogelj S., Nguyen K., Wheeler D. Design of a highly sensitive and specific nucleotide sensor based on photon upconverting particles. J. Am. Chem. Soc. 2006;128:12410–12411. DOI: 10.1021/ja0644024

[65] Qiao X. F., Zhou J. C., Xiao J. W., Wang Y. F., Sun L. D., Yan C. H. Triple-functional core–shell structured upconversion luminescent nanoparticles covalently grafted with photosensitizer for luminescent, magnetic resonance imaging and photodynamic therapy in vitro. Nanoscale. 2012;4:4611–4623. DOI: 10.1039/C2NR30938F

[66] Huang P., Zheng W., Zhou, S., Tu D., Chen Z., Zhu H. et al. Lanthanide-doped $LiLuF_4$ upconversion nanoprobes for the detection of disease biomarkers. Angew. Chem. Int. Ed. 2015;53(5):1252–1257. DOI: 10.1002/anie.201309503

Chapter 7

Luminescent Glass for Lasers and Solar Concentrators

Meruva Seshadri, Virgilio de Carvalho dos Anjos and

Maria Jose Valenzuela Bell

Additional information is available at the end of the chapter

http://dx.doi.org/10.5772/65057

Abstract

Rare earth-doped glasses find applications in numerous photonic devices including color displays, infrared solid-state lasers, and indicators, among many. In this chapter, we will present and discuss several luminescent glasses doped with rare earth ions in their trivalent form (RE^{3+}) with general background and technological perspectives. Initially, we begin with a short introduction of RE^{3+} electronic energy-level structure in solids followed by the discussion of structural feature of glass lasers. While the lasing properties are mainly governed by the solubility of the ions and phonon interactions, the issue of ion interactions in solid hosts will be addressed since they hardly depend on the type of materials. Spectroscopic properties of Nd^{3+}-doped phosphate glasses are discussed in the framework of Judd-Ofelt theory. Rare earth-doped optical amplifiers are tackled from a technological point of view, as well as luminescent solar concentrators for enhancement of solar efficiency.

Keywords: glasses, rare earth ions, luminescence, optical amplifiers, solar concentrators

1. Introduction

A large number of ions from the lanthanide (rare earth) and actinide groups of periodic table exhibit laser action when doped into a large number of host crystals or glasses. The construction of high-power crystalline lasers requires substantial crystal sizes. In practice, crystals are grown by the small seed crystals while pulling the larger crystal from the melt. These methods are very difficult if larger crystals are needed. When glasses are used as host media, the environments of the ions vary much more than in a crystalline material because of the random structural character of the glass matrix. All glass lasers until present date have used trivalent lanthanides as the active ions. Glass plays many varied roles in rare earth laser systems, as it can be made with uniformly distributed rare earth concentrations and has great potential as a laser host medium. In addition, rare earth-doped fibers have received growing attention recently. They can be used as amplifiers in optical communication systems and as optical sources. Glass lasers are substantially inhomogeneously broadened, and usually present more broader line width than Nd:YAG ones. The laser transition line width and shape typically vary from one glass to another glass matrix. In this way, innovation of glass lasers remains a vibrant area in the development of science and technology. The luminescent glass materials are used to develop solid-state lasers operating in visible and NIR regions due to their potential applications in the fields of medical, eye safe lasers, atmosphere pollution monitoring, energy converters and telecommunications. Sharp fluorescent lines, strong absorption bands, and reasonable high quantum efficiency will determine the adequacy of laser material.

In this contribution, the introduction consists of brief review of electronic energy-level structure of RE^{3+} in solids, structural features, and general considerations in laser glasses. Laser glasses and amplifiers are studied to know the spectroscopic parameters using spectroscopic techniques. Especially, the current achievements on Nd^{3+}-doped phosphate glass lasers will be discussed and numerous data will be presented. Er^{3+}-doped glass fiber amplifiers subject will also be addressed. Finally, we will discuss rare earth-based luminescence concentrators for photovoltaic applications.

1.1. Rare earth ions as luminescent centers

Trivalent rare earth ions are well known for their special optical properties, which result from the fact that the electronic transitions within the 4f shell occur at optical frequencies. The 4f shell is shielded by the completely filled 5s and 5p shells, so the frequencies of the transitions are almost independent of the host. Generally, the electric dipole transitions are forbidden due to equal parity of the electronic levels within the 4f shell. Those transitions are possible in solids that have slightly mixed odd-parity wave functions. As a result, the absorption and emission cross sections are small (~10^{-20} or 10^{-21} cm^2), and the lifetime of the luminescent level is relatively long ranging from microseconds to several milliseconds. The influence of the electric field around the ions removes the degeneracy of the 4f levels, resulting in a Stark splitting of the levels. **Figure 1** shows the energy levels of the trivalent lanthanide ions (4f configurations). This figure provides useful information to predict and/or to make a proper assignment of the emission spectra corresponding to trivalent rare earth ions in crystals or glasses.

The energy levels of the 4f shell arise from spin-spin and spin-orbit interactions and are often denoted using Russell-Saunders notation $^{2S+1}L_J$ [1, 2], in which S is the total spin angular momentum, L is the total orbital angular momentum, and J is the total angular momentum. There exist 14 rare earth elements all having a different number of electrons in the incompletely filled 4f shell. However, due to the shielding by the outer lying shells, the magnitude of the splitting is small, resulting in relatively narrow lines.

Table 1 displays some important emission transitions of rare earth ions and their technological interest. More general views about the spectra and energy levels of rare earth ions can be found in literature [9–12]. The large number of excited states that are suitable for optical pumping and subsequent decay to metastable states having high quantum efficiencies and narrow emission lines is favorable for achieving laser action.

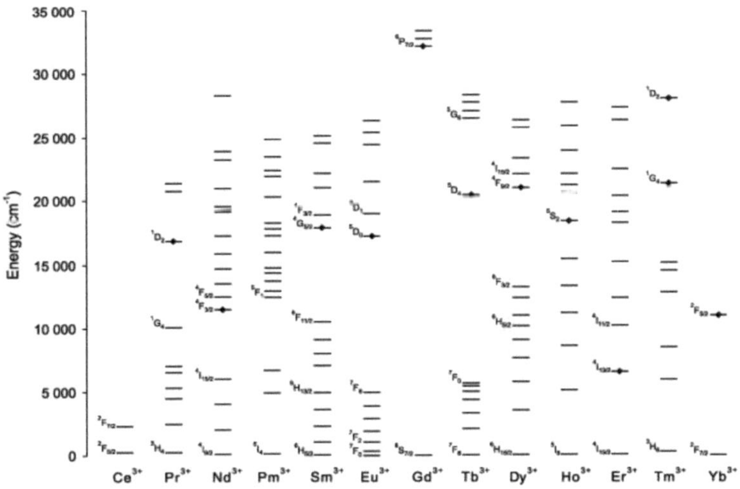

Figure 1. Schematic energy levels of rare earth ions.

RE^{3+} ions	Transitions	Wavelength (nm)	Application
Pr^{3+}	$^{1}G_{4}\rightarrow{}^{3}H_{5}$	1300	Optical amplifier [3]
Nd^{3+}	$^{4}F_{3/2}\rightarrow{}^{4}I_{11/2}$	1064	Solid-state lasers [4]
Eu^{3+}	$^{5}D_{0}\rightarrow{}^{7}F_{2}$	615	Displays, lighting [5]
Tb^{3+}	$^{5}D_{4}\rightarrow{}^{7}F_{5}$	545	Lighting [6]
Dy^{3+}	$^{6}F_{11/2}+{}^{6}H_{9/2}\rightarrow{}^{6}H_{15/2}$	1300	Optical amplifier [7]
Er^{3+}	$^{4}I_{13/2}\rightarrow{}^{4}I_{15/2}$	1530	Optical amplifier [8]
Tm^{3+}	$^{3}H_{4}\rightarrow{}^{3}F_{4}$	1480	Optical amplifier [7]
Yb^{3+}	$^{2}F5_{/2}\rightarrow{}^{2}F_{7/2}$	980	Sensitizer [6]

Table 1. Important emission lines of some lanthanide ions.

1.2. Glasses for lasers and amplifiers: structural features

Glasses are widely used nowadays and present several applications in different fields of life. There are several definitions for glassy materials. Grouping them one can define glass as an inorganic product of melting which has been cooled to a solid without crystallization [13]. It means that glass looks like an undercooled liquid. An undercooled liquid can crystallize at any moment but at room temperature it is not possible. Therefore, glass is an amorphous material as it does not exhibit long range order of atoms in a lattice. The transition of a glass melt to a crystallized state at the crystallization temperature does not take place if the cooling of the glass is fast. Below the crystallization temperature, glass behaves like a fluid down to the transformation temperature. Below this temperature, glass has the properties of a solid. Glass can also be defined as an amorphous solid lacking completely, in long range, periodic atomic structure and exhibiting a region of glass transformation behavior.

In general, glasses can be prepared either from high quality, chemically pure components or from a mixture of far less pure minerals. The batch materials can be divided in to five categories depending on the property of components used in the glass preparation: (1) *glass formers*, (2) *flux*, (3) *property modifier*, (4) *colorant*, and (5) *fining agent*. Every glass contains one or more components, which serve as the primary source of the structure. The same component may be classified into different categories when used for different purposes. For example, alumina serves as a glass former in aluminate glasses but in most silicate glasses, it works as property modifier [14]. Zachariasen [15] noted that those crystalline oxides that form open, continuous networks tended to form glasses and those glass-forming networks were associated with ions with particular coordination numbers (CN). The primary glass formers in commercial oxide glasses are silica (SiO_2), boric oxide (B_2O_3), and phosphoric oxide (P_2O_5), in addition to other compounds GeO_2, Bi_2O_3, As_2O_3, Sb_2O_3, TeO_2, Al_2O_3, Ga_2O_3, and V_2O_5 which act as glass formers under certain circumstances.

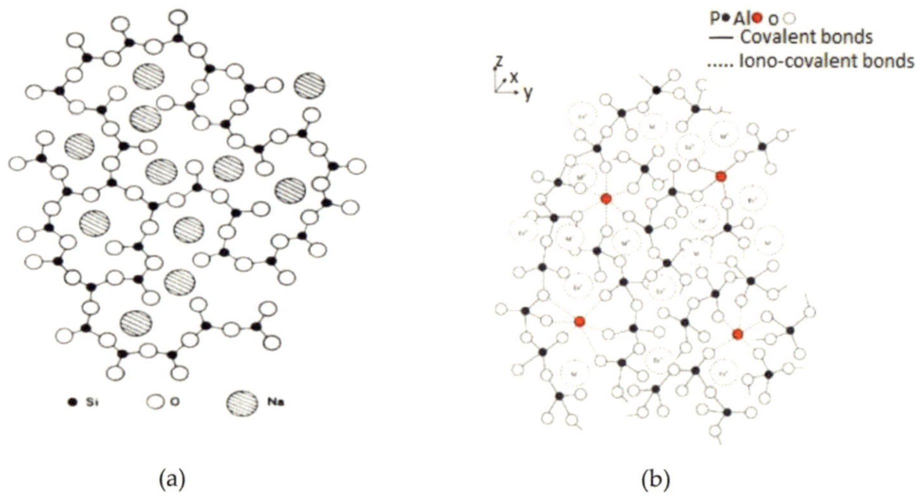

Figure 2. Schematic representation of (a) random network of alkali silicate and (b) glass structure of different atomic neighborhoods (3D).

These oxide glass formers play an important structural role in glasses when used as network or intermediate oxide modifiers. Thus, the oxide network modifiers will create strong bridging oxygen (BO) bonds in between the glass-forming polyhedra and weak nonbridging oxygen (NBO) bonds (see in **Figure 2a**). Moreover, the oxide modifiers will control many useful properties, lowering the melting temperature that is highly useful for developing technologically used glasses. The intermediate oxides also modify glass properties due to their coordination numbers and bond strengths in between the network formers and network modifiers. **Table 2** reports some of the oxide-based modifier glasses for realization of structure influence on various properties.

Glass composition	D (g/cm³)	Tg (°C)	n_d	τ (ms)
$60SiO_2$-$20Al_2O_3$-$20Li_2O$ [16]	2.40	693	1.531	2.65
$20Na_2O$ [16]	2.45	811	1.507	2.45
20MgO [16]	2.55	827	1.548	2.43
20CaO [16]	2.61	868	1.557	2.37
20ZnO [16]	2.84	742	1.573	2.51
$20La_2O_3$ [16]	4.08	863	1.709	2.01
$71SiO_2$-$14Al_2O_3$-15MgO [17]	2.44	831	1.520	2.48
10CaO [17]	2.61	829	1.549	–
10BaO [17]	2.86	841	1.549	2.39
15SrO [17]	2.73	834	1.549	–
TBZN−$0.05Ho_2O_3$ [18]	4.248	286	–	17.4
$0.1Ho_2O_3$	4.212	288	–	18.2
$0.5Ho_2O_3$	4.137	290	–	17.0
$1.0Ho_2O_3$	3.949	292	–	15.1
$1.5Ho_2O_3$	4.116	295	–	14.5

Table 2. Density (D), transition temperature (T_g), refractive index (n_d), and lifetime of respective ions ($\tau_{599\,nm}(Sm^{3+})$ [16, 17]; $\tau_{660\,nm}(Ho^{3+})$ [18]) in different oxide modifier-based glass compositions.

Owing to superior physical properties such as high thermal expansion coefficients, low melting and softening temperatures, and high ultraviolet transmission, phosphate glasses have several important attributes over conventional silicate and borate glasses. However, the poor chemical durability of these early optical glasses has temporarily discouraged for their further development. Interest in the amorphous alkali phosphates was stimulated in the 1950s by their use in a variety of industrial applications, including sequestering agents for hard water treatments and dispersants for clay processing and pigment manufacturing [19]. Studying such materials, Van Wazer [20] established the foundations for much of our current understanding of the nature of phosphate glasses. Kordes and co-workers [21, 22] re-examined the alkaline earth phosphate glasses, including UV-transmitting compositions and observed some "anomalous" trends in properties.

Phosphate glasses have unique characteristics that include high transparency, high thermal stability, low refractive index and dispersion, and high gain density due to high solubility for lanthanide ions and hence find growing field of applications [23–26]. An important step in laser glasses occurred in 1967 when phosphate-based compositions were first explored [27]. Several phosphate glasses find considerable applications in optical data transmission, detection, and sensing [28]. The phosphate glasses also find applications in fast ion conductors, optical filters, reference electrodes, stable storage medium for immobilizing high-level nuclear waste [29] and as additives in the manufacturing of insulating polymeric cables. Certain phosphate vitreous electrolytes are being used increasingly in electrochemical sensors, in prototype batteries, and in electrochemical devices. There have been many excellent reviews of structural studies on phosphate glasses including that of Van Wazer [20, 30, 31]. A schematic structure of different atomic neighborhoods (covalent and ionocovalent) in rare earth (Er^{3+})-doped phosphate glass is shown **Figure 2b**.

The large concentration of lasing atoms can be doped easily into different glasses, which can be made in different shapes and sizes depending on technological needs. The high-power laser threshold is possible from the ions-doped glasses compared to the same ions-doped crystal due to their large absorbing capability of incident energy and energy level broadening. The electronic transitions of rare earth ions have 4f-4f and 4f-5d configurations, which are weakly affected by the host material, since the 4f electrons are effectively screened by outer, filled electron shells. Thus, these transitions cause absorption and fluorescence patterns from the ultraviolet (UV) to infrared regions and they are narrow. Two important aspects of the optical behavior of rare earth ions are determined by the host material. One is the electric dipole transitions which occur between 4f states are strictly forbidden for an isolated ion, since the parity of the electronic configuration must change with an electric dipole transition. However, the strengths of these electric dipole transitions remains relatively weak due the perturbative nature of the admixing states, and as consequence, the radiative lifetimes of excited rare earth ions can be relatively long ($\sim 10^{-3}$ ms). Another aspect is related to the nonradiative rates originated from relaxation of the excited states of the rare earth ions, which is determined by the host material. Recently, Babu et al. [32] had studied different fluorophosphate glasses doped with 0.5 mol% of Er^{3+} ions through the Judd-Ofelt and McCumbers theories for potential broadband optical fiber lasers and amplifiers. Praseodymium (Pr^{3+})-doped high-aluminum phosphate (HAP) glasses with excellent chemical durability for thermal ion-exchanged optical waveguide have been investigated by Tian et al. [33]. In order to improve the solar cell efficiency, potential downconversion was studied in GeS_2-Ga_2S_3-CsCl glass for modifying solar spectrum and are found quantum yield to be below 1200 and 1650 nm are 51 and 76%, respectively [34].

2. Limiting factors in rare earth-doped glasses

2.1. RE^{3+} ions solubility in glasses

Higher concentration of RE^{3+} ions in glasses lead to clusters. Thus, the clusters of rare earth ion serve as luminescent quenchers, either by increasing ion-ion interactions between rare earth ions or by forming rare earth compounds that are not optically active. Therefore, much effort has been done into developing suitable host glass compositions for various rare earth doping levels. Some of the technological limitations for the development of new glasses are due to the rare earth ion density required for laser action. Currently, high-power lasers used Nd^{3+}-doped phosphate glasses and their nominal Nd doping levels in the range of about 3–4.2 × 10^{20} cm^{-3}.

2.2. Phonon interaction

Since the multiphonon decay process is essentially a competing process to the fluorescence, a low multiphonon decay rate can lead to an increased fluorescence efficiency for many important rare earth transitions. The efficiency of nonradiative transition rate depends on the energy gap between the ground and excited states, as well as the vibrational energy of the oscillators. The nonradiative transition decay rate can be expressed in terms of multiphonon relaxation, concentration quenching, energy transfer to another doping impurity such as transition metal ions or other rare earth ions, and energy transfer to hydroxyl groups OH^- [35]. The quenching efficiency is strongly dependent on the number of vibrational quanta that are needed to bridge the gap between the lowest emitting level and the highest nonemitting level of the lanthanide ion. Generally, hosts with low phonon energy tend to have a low multiphonon decay rate; thus, selection of lower phonon hosts such as fluoride or tellurite glasses can reduce the contribution of multiphonon relaxation and allow important radiative transitions.

2.3. Ion-ion interaction

Ion-ion interaction is due to multipolar interactions between neighbor rare earth ions. The analysis of ion-ion interaction and energy transfer provides essential information to the applications of laser glasses and display devices. In such process, the excitation energy transfers from an excited donor to a nearby unexcited activator (acceptor). Many theories have been put forward to give formulas for the rate of energy transfer by electric dipole-dipole interaction ($n = 6$), electric dipole-quadrupole interaction ($n = 8$), and quadrupole-quadrupole interactions ($n = 10$). These transfer mechanisms differ from one another in the dependence of the transfer rate on donor-acceptor distance, but common to all is the condition that an overlap between the donor emission spectrum and the acceptor absorption spectrum is essential for the transfer to occur. Such resonant transfer is analyzed most frequently through luminescence measurements: donor molecules are excited in the presence of acceptor ions, and the luminescence yield of donor and/or acceptor and the decay time of donor luminescence are measured as functions of the acceptor concentration. A more detailed description of ion-ion interactions is given by Inokuti-Hirayama theory [36], which explicitly deals with the dynamics of energy migration.

Among many energy transfer theories, Förster [37] and Dexter [38] theory on energy transfer is one of the most widely employed and the probability rate of energy transfer can be determined as [39] follows:

$$W_{DA}(R) = \frac{6cg_{low}^{D}}{(2\pi)^4 n^2 R^6 g_{up}^{D}} \sum_{m=0}^{\infty} e^{-(2\bar{n}+1)S_0} \frac{S_0^m}{m!} (\bar{n}+1)^m \int \sigma_{emis}(\lambda_m^+)\sigma_{abs}(\lambda)d\lambda = \frac{C_{DA}}{R^6} \quad (1)$$

where R is the distance of separation between donor and acceptor, C_{DA} is the energy transfer constant (cm^6/s), and $\bar{n}[= \ 1/e(\hbar\omega 0/KT) - 1]$ is the average occupation of phonon mode at temperature T. Then, the energy transfer constant is expressed as follows:

$$C_{DA} = \frac{6cg_{low}^{D}}{(2\pi)^4 n^2 g_{up}^{D}} \sum_{m=0}^{\infty} e^{-(2\bar{n}+1)S_0} \frac{S_0^m}{m!} (\bar{n}+1)^m \int \sigma_{emis}(\lambda_m^+)\sigma_{abs}(\lambda)d\lambda \qquad (2)$$

The critical radius of the interaction can be obtained by the product of energy transfer constant and intracenter lifetime of the donor excited level (i.e., $R_C^6 = C_{DA} \times \tau_D$).

In addition to the aforementioned mechanisms, in glasses containing only one RE^{3+} species (for example Nd^{3+}, Er^{3+}), a number of different ion-ion interactions occur such as energy migration, cross-relaxation among others. The most important are outlined below.

2.4. Energy migration

Energy migration is strongly dependent on rare earth ion concentration and due to the dipole-dipole Förster and Dexter interactions. An excited ion in the metastable state can interact with the nearby ground state ion then promoting it to the excited level (see **Figure 3**). The successive energy transfers between the ions increase the probability of nonradiative decay that lead to decrease the fluorescence efficiency [40, 41].

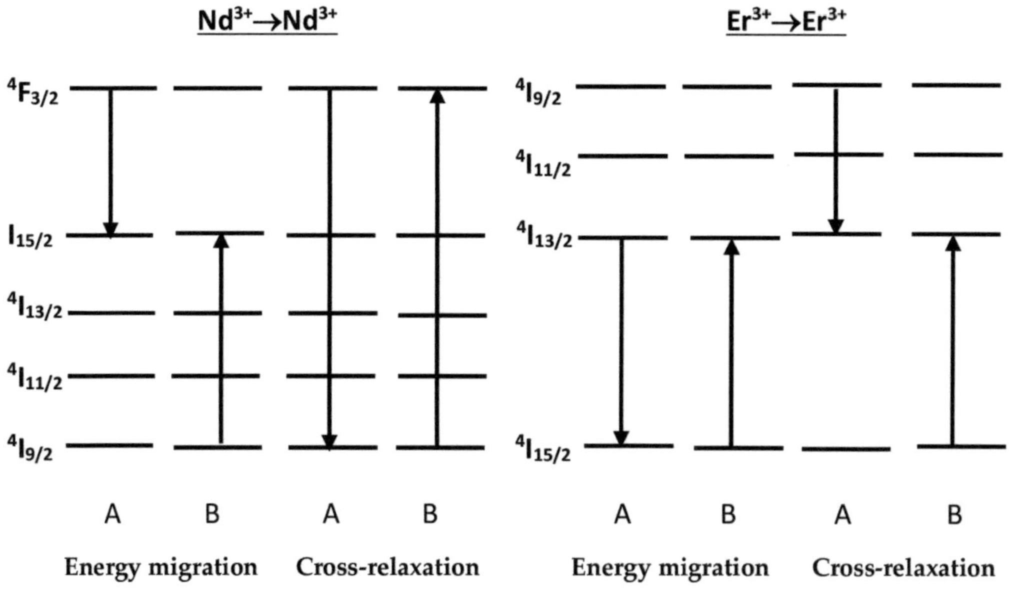

Figure 3. Schematic diagram of ion-ion interaction in the case of Nd^{3+} and Er^{3+} ions.

2.5. Cross relaxation

Cross relaxation involves the same RE ions and their schematic representation, which is shown in **Figure 3**. An excited state of ion A that gives half its energy to ion B is in its ground state. So that both ions end up in the intermediate energy level, from which they relax rapidly to the ground state through nonradiative relaxation. This relaxation is usually observed for Nd^{3+}-doped glasses that cause concentration quenching.

2.6. Upconversion mechanisms

By *Excited State Absorption* (*ESA*): The simplest representation of upconversion mechanism in trivalent Er^{3+} ion through the ground-state absorption (GSA)/excited state absorption (ESA) is shown in **Figure 4a**. The first excitation photon is absorbed by the ground-state N_0 and populates the intermediate-state N_1. Provided the lifetime of N_1 is long enough, a second incident photon can be absorbed, exciting further the ion from its intermediate-state N_1 to a higher-lying excited state N_2, from which upconversion luminescence arises.

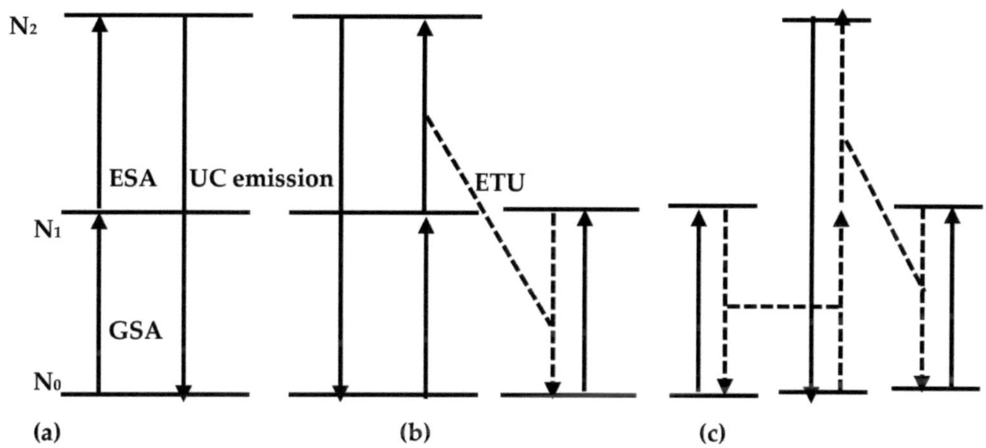

Figure 4. Schematic representation of upconversion mechanisms. (a) GSA/ESA process; (b) ETU process; (c) cooperative process.

By *sensitized energy transfer upconversion* (*ETU*): Sensitized energy transfer upconversion was first introduced by Auzel who called ETU from Addition of Photon for Energy Transfer (APTE) from (the French 'Addition de Photons par Transferts d 'Energie'). Upconversion involves nonradiative energy transfers between a sensitizer (e.g., Yb^{3+}) and an activator (e.g., Er^{3+}, Ho^{3+}, Tm^{3+}). Usually, sensitizer has a strong absorption cross section at the excitation wavelength. Once the sensitizer is excited (after absorption of an incident photon), it relaxes to a lower-energy state (the ground state in the case of Yb^{3+}) by transferring its energy to a neighboring activator, raising the latter to a higher-energy state (see **Figure 4b**). These energy transfer processes are generally based on electric dipole-dipole interactions.

By *cooperative luminescence*: **Figure 4c** shows the schematic representation upconversion by cooperative luminescence. Two excitation photons sequentially absorbed by two different ions (e.g., two Yb^{3+} ions, but the two ions do not need to be the same species), moving both of them into their excited state. Then, both ions decay simultaneously to their ground state, with the emission of one single photon that contains the combined energies of both ions. The cooperative emission occurs from a virtual level, and it explains why the emission probability is rather low.

3. Spectroscopic properties of Nd^{3+}-doped phosphate glasses

3.1. Judd-Ofelt theory and intensity parameters

Optical spectroscopy often used to measure optical absorption of trivalent rare earth ions in UV-vis-NIR regions, from which the effect of a host matrix on the local environment of a given rare earth cation with its first neighbor anions such as oxygen can be elucidated using the theory proposed by Judd [42] and Ofelt [43]. Each observed transition corresponds to a transition between two spin-orbit coupling levels. The Judd-Ofelt theory has been applied for the interpretation of these transitions by the three mechanisms: (1) magnetic dipole transitions, (2) induced electric dipole transitions, and (3) electric quadrupole transitions. The quantitative analysis of the intensities of these f-f transitions in the rare earth ions has been provided independently by Judd [42] and Ofelt [43]. The basic idea of Judd and Ofelt is that the intensity of f-f electric dipole transitions can arise from the admixture into the $4f^N$ configuration of opposite parity (e.g., $4f^{N-1}n^1d^1$ and $4f^{N-1}n^1g^1$). According to Judd-Ofelt theory, the intensity of magnetic and electric dipole transition is represented as follows:

$$\mathrm{f_{cal}} = f_{ed} + f_{md} \tag{3}$$

This means that experimentally measured oscillator strengths could be expressed to a good approximation in terms of absorption of light by electric dipole (f_{ed}) and magnetic dipole mechanisms. But, the order of magnitude of magnetic dipole oscillator strengths (f_{md}) is found to be very low for the observed intensities of rare earth ions and thus will not be considered further. The calculated oscillator strengths are therefore approximately equal to electric dipole oscillator strengths, that is

$$\mathrm{f_{cal}} = f_{ed} = \sum_{\lambda=2,4,6} T_\lambda \nu \left| \left\langle (S,L)J \left\| U^\lambda \right\| (S',L')J' \right\rangle \right|^2 \tag{4}$$

where υ is the mean energy of the transition $\psi J \rightarrow \psi' J'$, and $\left\| U^\lambda \right\|^2$ is the squared reduced matrix element of the unit tensor operator of the rank, λ (=2, 4 and 6). T_2, T_4, and T_6 are related to the radial part of $4f^N$ wave functions, the refractive index of the medium, and the ligand-field parameters that characterize the environment of the ion. These quantities are treated as parameters and are determined from the experimental oscillator strengths.

The intensity of an absorption band is expressed in terms of a quantity called the "oscillator strength." Experimentally, the oscillator strength (f) is a measure of the intensity of an absorption band and is proportional to the area under the absorption peak. The area under an absorption peak is a better measure of the intensity than the molar absorptivity at the peak maximum, because the area is the same for both the resolved and unresolved band. The oscillator strength (f) of each absorption band is expressed in terms of absorption coefficient $\alpha(\lambda)$ at a particular wavelength λ and is given by [44]

$$f_{exp} = \frac{mc^2\nu^2}{N\pi e^2}\int \varepsilon(\nu)d\nu \quad \text{or} \quad f_{exp} = 4.32\times10^{-9}\int \varepsilon(\nu)d\nu \tag{5}$$

where m and e are the mass and charge of the electron, c is the velocity of light, and N is the density of the absorbing ions obtained from the rare earth ion concentration and glass density values, respectively. $\int\varepsilon(\nu)d\nu$ represents the area under the absorption curve. The molar absorptivity $\varepsilon(\nu)$ of an absorption band at energy ν (cm^{-1}) is given by

$$\varepsilon(\nu) = (Cl)^{-1}\log\left(\frac{I_0}{I}\right) \tag{6}$$

where C concentration of rare earth ions per unit volume, l optical path length, and $\log\left(\frac{I_0}{I}\right)$ is the optical density.

As an example, optical absorption measurements were made at room temperature in the wavelength region 300–900 nm for an Nd^{3+}-doped phosphate glass using JASCO V-630, UV-vis spectrophotometer. **Figure 5** shows (a) absorption and (b) emission spectra of Nd^{3+}-doped phosphate glass. The validity of Judd-Ofelt theory is determined by the root-mean-square (δ_{rms}) deviation between the measured and calculated oscillator strengths by the relation [45].

$$\delta_{rms} = \left[\frac{\Sigma\left(f_{exp} - f_{cal}\right)^2}{P}\right]^{1/2} \tag{7}$$

Using the experimental oscillator strengths (f_{exp}), a best set of Judd-Ofelt intensity parameters Ω_λ ($\lambda = 2, 4$, and 6) for the Nd^{3+} ion-doped glass were determined using the procedure followed in Ref. [46]. The Judd-Ofelt intensity parameters represent the square of the charge displacement due to the induced electric dipole transition. The advantage of Ω_λ parameters is that a

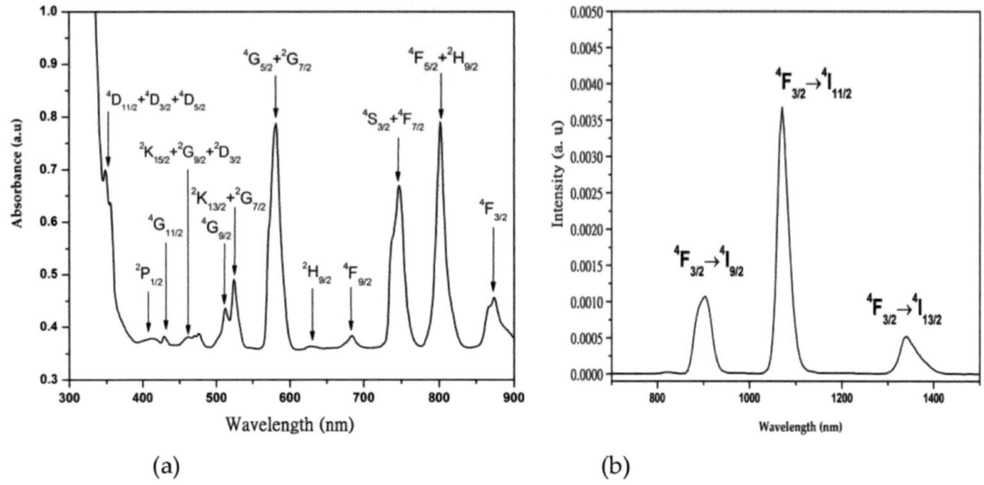

Figure 5. (a) Absorption and (b) emission spectra of Nd^{3+}-doped phosphate glass [33].

set of parameters is needed for describing both the absorption and emission processes. The Ω_λ parameters are important for the investigation of the local structure and bonding in the vicinity of rare earth ions. Reisfeld [47] indicated that Ω_2 parameter is sensitive to both asymmetry and covalency at rare earth sites. Oomen and van Dongen [48] pointed out that the rigidity or long range effects of glass hosts were responsible for changes in Ω_6. The Ω_4 parameters are affected by factors causing changes in both Ω_2 and Ω_6. **Table 3** shows Judd-Ofelt intensity parameters Ω_λ ($\lambda = 2, 4$, and 6) for the Nd^{3+}-ion doped various phosphate glasses.

Glass matrix	Ω_2	Ω_4	Ω_6	Ω_4/Ω_6
$40P_2O_5$-$20Al_2O_3$-$40Na_2O$ [49]	4.70	6.0	5.40	1.11
$41P_2O_5$-$17K_2O$-9.5CaO-$8Al_2O_3$-$24aF_2$ [50]	5.40	7.03	6.51	1.07
$58.5P_2O_5$-$17K_2O$-$14.5SrO_2$-$9Al_2O_3$ [51]	6.74	3.86	6.35	0.60
$75NaPO_3$-$24LiF_3$ [52]	3.44	4.14	6.28	0.65
($65P_2O_5$-$15Na_2O$)-$15Li_2O$ [45]	4.32	3.66	6.00	0.61
($65P_2O_5$-$15Na_2O$)-$15Na_2O$ [45]	5.42	4.93	8.06	0.61
($65P_2O_5$-$15Na_2O$)-$15K_2O$ [45]	7.68	8.96	11.71	0.76
($65P_2O_5$-$15Na_2O$)-$7.5Li_2O$-$7.5Na_2O$ [45]	4.01	3.69	5.92	0.62
($65P_2O_5$-$15Na_2O$)-$7.5Li_2O$-$7.5K_2O$ [45]	6.42	6.15	8.96	0.69
($65P_2O_5$-$15Na_2O$)-$7.5Na_2O$-$7.5K_2O$ [45]	4.90	3.88	6.18	0.63

Table 3. Judd-Ofelt intensity parameters (Ω_λ, $\lambda = 2$, 4, and 6×10^{-20} cm^2) in 1 mol% Nd_2O_3-doped phosphate glasses.

The covalency between the rare earth ion and the surrounding oxygen in the glass modifies the intensity of hypersensitive transitions (HST), as suggested by Reisfeld and co-worker [47,

48]. This can be observed more clearly in RE^{3+}-doped alkali- and mixed alkali-based glasses. For example, in Nd^{3+} ion, the transition ${}^4I_{9/2}\rightarrow{}^4G_{5/2}+{}^2G_{7/2}$ is found to be hypersensitive by selection rule, $\Delta J \leq 2$, $\Delta L \leq 2$, and $\Delta S = 0$. The observed oscillator strengths of the hypersensitive transition are higher when compared to other bands (see inset **Figure 5a**). **Figure 6** shows variation of J-O parameter, Ω_2 with (a) oscillator strength (b) energy (cm^{-1}) of HST transition of Nd^{3+} ion for alkali, mixed alkali phosphate glasses. From **Figure 6a**, among the alkali glass matrices, the oscillator strength of HST is found to be lower in lithium phosphate glass matrix and higher in potassium glass matrix. But the oscillator strength of HST is found to be higher in lithium-potassium phosphate glass matrix among the mixed alkali glass matrices. Thus, the results of oscillator strength show that potassium is playing active role in enhancing the oscillator strengths of Nd^{3+} ion in potassium phosphate glass matrices (K, Li-K, and Na-K) compared to other glass matrices. In lithium phosphate glass matrix, oscillator strength of HST indicates the lower crystal field symmetry at Nd^{3+} ion site.

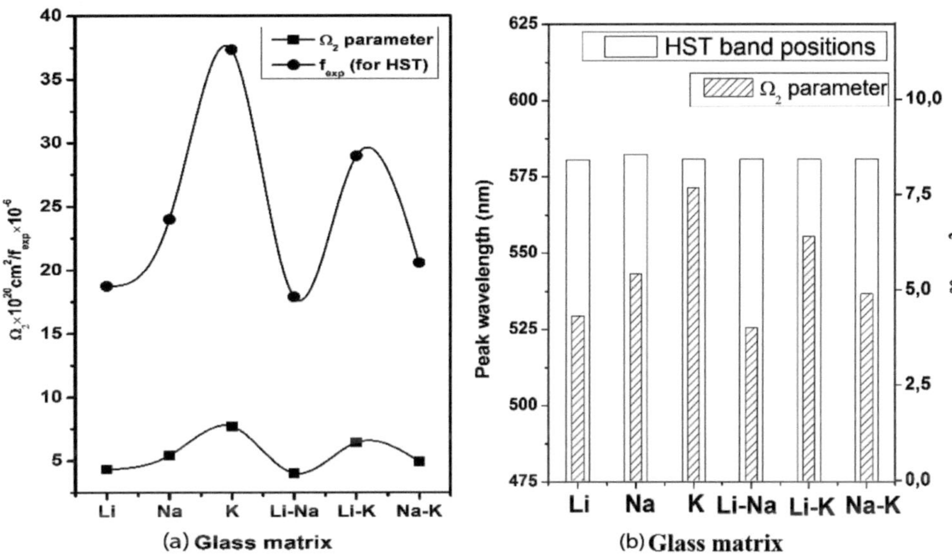

Figure 6. Variation of (a) Ω_2 with oscillator strength (f_{exp}) of HST and (b) Ω_2 with band position (E_{exp}) of HST in alkali and mixed alkali phosphate glasses [45].

The hypersensitive transition can also give the information regarding the covalency nature of rare earth-ligand interaction, which can be determined by the shift of hypersensitive band position to lower or higher wavelength due to nephelauxetic effect. From **Figure 6b**, it is observed that for lithium to sodium glass matrices, the peak wavelength of HST shifts toward higher wavelength side and the Ω_2 parameter also increases indicating that the structural changes are not influencing the covalence bond. For lithium to potassium and lithium-potassium glass matrices, the peak wavelength of HST does not change but the Ω_2 parameter increased indicating that the some structural changes are influencing the Nd-O bond. For sodium to potassium and sodium-potassium glass matrices, the peak wavelength of HST shifts toward lower wavelength side but Ω_2 parameter increased (for Na to K) and decreased (for Na to Na-K) indicating that the structural changes are influencing the Nd-O bond.

Due to the zero values of certain reduced matrix elements $\|U^{\lambda}\|$ of Nd^{3+} ion, certain lasing transitions can be uniquely characterized by the ratio of intensity parameters Ω_4 and Ω_6, which is known as spectroscopic quality factor (χ) and are shown in **Table 2**. If $\chi < 1$, the ${}^4F_{3/2} \rightarrow {}^4I_{11/2}$ transition shows stronger intensity than that of ${}^4F_{3/2} \rightarrow {}^4I_{9/2}$ transition. From the **Table 2** data, the obtained χ value is <1 for Nd^{3+}-doped phosphate glasses, indicating that the intensity of ${}^4F_{3/2} \rightarrow {}^4I_{11/2}$ transition at 1064 nm will be stronger for various phosphate glasses [53].

3.2. Fluorescence analysis and radiative properties

The emission spectra of Nd^{3+}-doped phosphate glass recorded at room temperature in the wavelength region 800–1500 nm under excitation wavelength, 514.5 nm of Ar^{3+} laser are shown in **Figure 5b**. Emission spectra show three emission peaks due to the transitions, ${}^4F_{3/2} \rightarrow {}^4I_{9/2}$, ${}^4F_{3/2} \rightarrow {}^4I_{11/2}$, and ${}^4F_{3/2} \rightarrow {}^4I_{13/2}$ nearly centered at 902, 1069, and 1340 nm, respectively. The stimulated emission cross section is an important parameter and its value is related to the rate of energy extraction from the optical material. The Judd-Ofelt theory can be applied to laser glasses and can successfully account for the induced emission cross sections that are observed.

The efficiency of a laser transition is evaluated by considering the stimulated emission cross section, and it is related to the radiative transition probability. It can be obtained from the emission spectra using Fuchtbauer-Ladenburg method [54]

$$\sigma_{\mathrm{p}} = \frac{\lambda_p^4}{8\pi c n^2 \Delta\lambda_{eff}} A_{rad}(J \rightarrow J') \tag{8}$$

where λ_p is peak wavelength, and $\Delta\lambda_{\mathrm{eff}}$ is the effective line width. The effective line width $\Delta\lambda_{\mathrm{eff}}$ is obtained from

$$\Delta\lambda_{\mathrm{eff}} = \int \frac{I(\lambda)}{I_{\mathrm{max}}} d\lambda \tag{9}$$

where $I(\lambda)$ is the integrated fluorescence intensity and I_{max} is the peak fluorescence intensity. The radiative transition probability $A_{\mathrm{rad}}(\psi J, \psi' J')$ for emission from an initial excited state ψJ to a final ground state $\psi' J'$ is

$$A_{rad}(J \rightarrow J') = \frac{64\pi^4 e^2}{3h(2J+1)\lambda^3}\left[\frac{n(n^2+n)^2}{9} S_{ed}\right] \tag{10}$$

http://dx.doi.org/10.5772/65057

Where the factor $\frac{n(n^2+2)^2}{9}$ represents the local field correction term for the ion in a medium, υ is the energy of transition, and n is the refractive index of the glass.

The radiative lifetime (τ_R) of an excited state ψ'J' is calculated from

$$\tau_R(J) = \frac{1}{\sum_{J'} A(J \rightarrow J')} \tag{11}$$

The fluorescence branching ratio, β_R, predicts the relative intensity of lines from a given excited states and characterizes the lasing potency of that particular transition. In order to choose suitable lasing transition, one has to select the transition having branching ratio >0.5 and the energy difference of about 3000 cm^{-1} between the emitting level and the next lower level. The fluorescence branching ratio (β_R) is given by

$$\beta_R(J \rightarrow J') = \frac{A(J \rightarrow J')}{\sum_{J'} A(J \rightarrow J')} \tag{12}$$

Table 4 shows critical parameters to the laser designer such as branching ratios and radiative life times, peak stimulated emission cross sections of 1062 nm laser line in various phosphate glasses.

Glasses	τ_{exp} (μs)	τ_{cal} (μs)	η (%)	β_{rad} (%)	Δλ (nm)	σ_p (cm^2)
$55P_2O_5$-$17K_2O$-11.5Bao-$6BaF_2$-$9Al_2O_3$ [55]	210	189	90	59	25.1	6.23
$58.5P_2O_5$-$17K_2O$-14.5MgO-$9Al_2O_3$ [56]	262	249	74	68	28.8	4.41
$57P_2O_5$-$14.5K_2O$-28.5BaO [57]	178	430	41	53	29.3	2.78
0.4MgO-$0.4BaF_2$-$0.1Al(PO_3)_3$-$0.1Ba(PO_3)_3$ (wt.%) [58]	185	308	60	49	32	2.97
APG1 [59]	385	361	106	–	27.8	3.4
APG2 [59]	464	456	101	–	31.5	2.4
HAP4 [60]	350	–	–	–	27.0	3.6
HAP3 [60]	380	372	102	–	27.9	3.2
Q89 [61]	350	–	–	–	21.2	3.8

τ_{exp}: radiative life time; τ_{rad}: radiative lifetime from J-O theory; η: quantum efficiency; β: branching ratio; σ_P: peak emission cross section (×10^{22} cm^2). (APG1, APG2, HAP4, HAP3, and Q89 are commercial phosphate glasses).

Table 4. Laser emission properties at 1060 nm of Nd^{3+}-doped phosphate glasses

4. Rare earth-doped optical amplifiers

Rare earth-doped fibers have received growing attention recently. They can be used as amplifiers in optical communication systems and as optical sources. Optical amplifiers amplify an incident weak light signal through the process of stimulated emission. The main ingredient of any optical amplifier is the optical gain realized when the amplifier is optically pumped to achieve population inversion. The optical gain, in general, depends mainly on the doping material, on the frequency (or wavelength) of the incident signal, and also on the local beam intensity at any point inside the amplifier. By a proper choice of the doping materials, the amplifier characteristics such as the operating wavelength and the gain bandwidth can be modified. Many different dopants such as erbium (Er^{3+}), holmium (Ho^{3+}), neodymium (Nd^{3+}), samarium (Sm^{3+}), thulium (Tm^{3+}), and ytterbium (Yb^{3+}) can be used to realize fiber amplifiers operating at different wavelengths covering a wide region extending over 0.5–3.5 μm and the most commonly used hosts are silicate, phosphate, fluoride, and tellurite glasses. **Figure 7** shows working of active fibers within a specific wavelength range determined by various rare earth ions. Specially, optical telecommunication transmission [wavelength division multiplexing (WDA)] systems work in the conventional C-band (1530–1565 nm) telecommunication window. This band can easily be observed in erbium (Er^{3+}) ions among the rare earths and is the most useful dopant for commercial optical amplifiers and erbium-doped fiber amplifiers (EDFAs) that are made from silicate and phosphate glass matrices. The schematic representation of Er^{3+}-doped fiber amplifiers (EDFA) is shown in **Figure 8**. An erbium-doped fiber is pumped optically by an infrared laser sources at 980 or 1480 nm and are compatible with InGaAs and InGaAsP laser diodes. The three-level pumping process in EDF is illustrated in **Figure 9**.

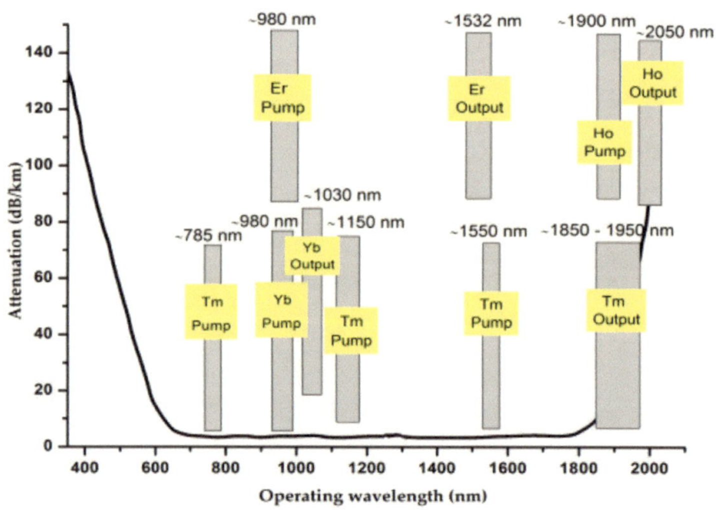

Figure 7. Rare earth-doped fibers working wavelength ranges.

http://dx.doi.org/10.5772/65057

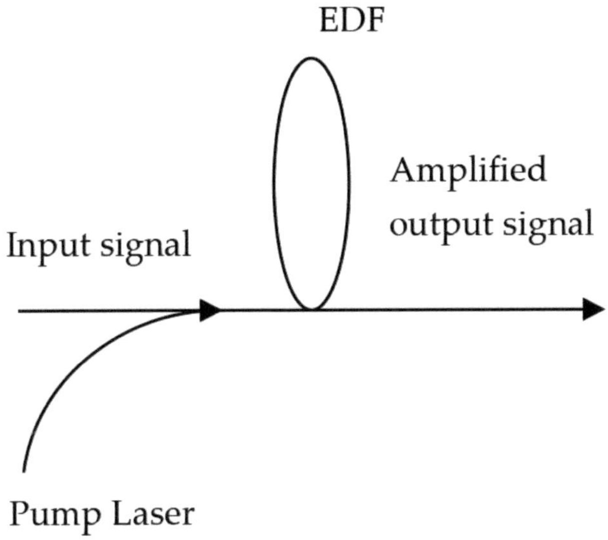

Figure 8. Erbium-doped fiber amplifier (EDFA).

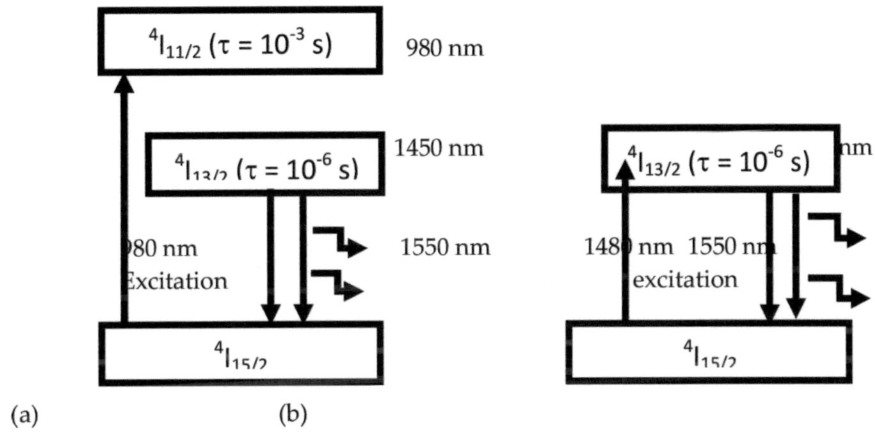

Figure 9. Simplified energy band diagram of Er^{3+}-doped silica fiber. (a) Three-level process. (b) Two-level process.

When a laser source tuned at 980 nm is used to pump the EDF, the Er^{3+} ions move from ground level ($^4I_{15/2}$) to excited level ($^4I_{15/2}$). The ions stay excited level only about 10^{-6} s and after that they decay into a metastable level through multirelaxation process. In this process, the energy loss is turned into mechanical vibrations in the fiber. Finally, ions stay in the order of 10^{-3} s in metastable level, which is longer than the ion lifetime in the excited level and decay to the ground level with emission of photons in the 1530 nm wavelength region. Therefore, under 980 nm pumping, almost all the ions will be accumulated in the metastable level, and the three-level system can be simplified into two levels for most of the practical applications, as shown in **Figure 9b**. Whereas in the case of 1480 nm wavelength pumping, ions excited from ground level to the metastable level directly. Therefore, 1480 nm pumping is more efficient than 980 nm pumping because it does not involve the nonradiative transition and is often used for high-power optical amplifiers. However, amplifiers with 1480 nm pump usually have higher noise than the ones pumped at 980 nm.

Optical pumping provides the necessary population inversion between the energy levels E_1 and E_2, which in turn provides the optical gain defined as $g = \sigma(N_2 - N_1)$, where σ is the transition cross section. Normally, in laser and amplifiers, the gain coefficient decreases as signal power increases and is called as gain saturation. The gain coefficient can be defined as [62] follows:

$$G = G_0 \left(\frac{P_s}{P_s - P} \right), \tag{13}$$

where G_0 is the small-signal gain coefficient at a given wavelength, P is the signal power, and P_s is the saturated signal power. The absorption and broadened gain spectra of EDFA (see **Figure 10**) are the advantages for Wavelength Division Multiplexing (WDM) applications.

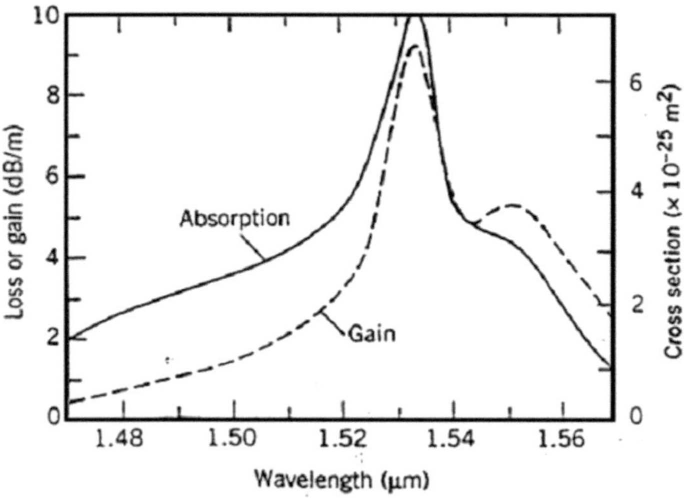

Figure 10. Absorption and gain spectra of EDFA.

Wave division multiplexing optical transmission system requires a flat gain spectrum of EDFA across the whole usable bandwidth. It is difficult to achieve the gain flatness in WDM system because EDFA has the narrow high gain in the C-hand wavelength region (1530–1570 nm) centered at 1550 nm. In recent decades, many glass hosts for Er^{3+} ions have been investigated to realize the optical amplifier. For the flattened gain performance of optical amplifier, Er^{3+}-doped fluoride [63], tellurite [64], and bismuth [65] based glasses are capable of realizing a flat gain over a broadband width of 1530–1560 nm.

http://dx.doi.org/10.5772/65057

5. Luminescent solar concentrators (LSC)

A schematic of luminescent solar concentrator (LSC) is shown in **Figure 11a**. LSC is working based on the absorption of sun light by a transparent material that has been doped with high quantum efficiency luminescent ions and subsequently re-emit by transparent material. The resulting luminescence is propagate by total internal reflection in transparent material and concentrated onto the solar cell, which attached to the edge of the transparent material to convert trapped emission light into electricity. The LSC was first proposed in late 1970s [66]. A typical design consists of a polymer plate doped with a luminescent material, such as a fluorescent organic dye, with solar cells optically matched to the plate edges. In recent decades, luminescent transparent material extends to availability of inorganic luminescence materials such as semiconductor quantum dots (QDs) [67], rare earth-doped materials [68] and semiconductor polymers [69]. Particularly, rare earth-doped glasses are still great interest due to their potential use as solid-state lasers and luminescence solar concentrators.

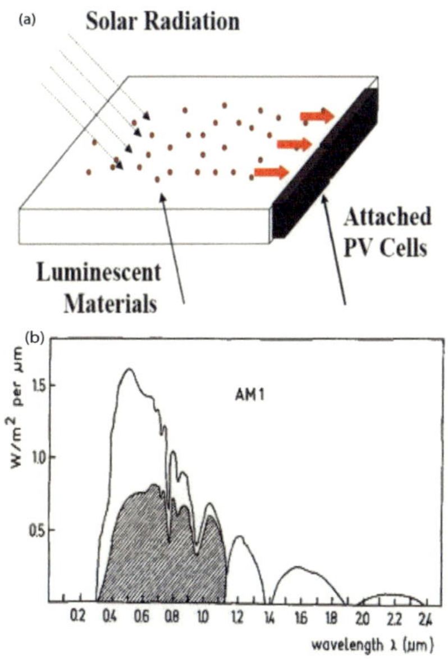

Figure 11. (a) Schematic diagram of luminescent solar concentrator. (b) Emission of solar radiation.

Figure 11b shows emission spectra of (AM1) solar radiation. It is known that the efficiency of solar cell is mainly limited by the loss of photons with much higher energy than the band gap of the photovoltaic solar cell. Normally, photovoltaic cells are made with silicon, which absorb photons only with energy greater than 1.1 eV and it is improving the performance of LSC. The luminescent material usually absorbs all wavelengths below 950 nm. Above 950 nm, the luminescence consists of a strong emission band in the range from 950 to 1000 nm [70]. This region is more reliable to increase the spectral response of the Si Photo-voltaic cell in LSC (see shaded region in figure). In addition the number of photons in the LSC is double while extending the absorption from visible (300–600 nm) out to the NIR (300–900 nm) range. The extending wavelength increased nonradiative recombination due to increasing molecular dimensions and decreasing probability of radiative transitions [71]. This leads to decrease in fluorescence quantum yield (FQY). In rare earth-doped materials, the FQY vary greatly depending on host materials and concentration, but values >90% have been reported in glass substrates [71]. The FQY is defined as follows:

$$\eta_{FQY} = \frac{\text{No. of emitted photons}}{\text{No. of absorbed photons}} \times 100\% \tag{14}$$

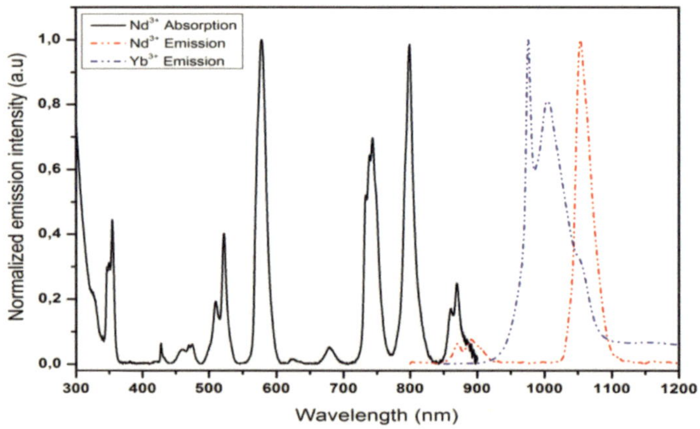

Figure 12. Absorption and emission spectra of Nd^{3+} and Yb^{3+} ions.

A high quantum yield (~95–100%) is essential for good LSC performance. Rare earth ions such as neodymium (Nd^{3+}) and ytterbium (Yb^{3+}) in glassy hosts exhibit high FQY (~90%) values due to range at 800–1300 nm (Nd^{3+}) and 1000 nm (Yb^{3+}) emission peaks. When Nd^{3+}/Yb^{3+} co-doped glasses are used as LCS, Nd^{3+} acts as sensitizer due to its numerous absorption bands in visible region, and Yb^{3+} acts as activator due to its single emission peak at ~1000 nm. Moreover, this material absorbs 20% of the solar spectrum in the range 440–980 nm range [72]. **Figure 12** shows absorption (for Nd^{3+}) and emission (for Nd^{3+}, Yb^{3+}) spectra for realizing the LCS. It is observed that the emission peak at ~1000 nm is matched perfectly to the maximum spectral response of a Si-Solar cell.

6. Conclusions

In summary, this chapter consists of a brief discussion on electronic energy-level structure of RE^{3+} in solids, structural features, and general considerations in lasers glasses such concentration quenching, phonon interaction, ion-ion interaction, and upconversion mechanisms. Judd-Ofelt intensity parameters (Ω_λ, $\lambda = 2, 4$, and 6) on Nd^{3+}-doped phosphate glasses and relevant theory were considered. We also included a study on the covalency between rare earth ion and the surrounding oxygen in the glass that increases the intensity of hypersensitive transitions for the case of alkali- and mixed alkali phosphate-based glasses. Laser parameters such as emission cross sections (σ_p), radiative lifetime (τ), branching ratios (β), peak bandwidth ($\Delta\lambda$), and quantum efficiency on Nd^{3+}-doped phosphate glasses are reported. Laser emission process in three-level and two-level-based silica-EDFA was discussed as well as significant features of rare earth-doped luminescent solar concentrators.

Acknowledgements

The authors would like to thank to Conselho Nacional de Desenvolvimento Científico e Tecnológico (CNPq) and Coordenação de Aperfeiçoamento de Pessoal de Nível Superior (CAPES), PNPD/CAPES, and FAPEMIG.

Author details

Meruva Seshadri*, Virgilio de Carvalho dos Anjos and Maria Jose Valenzuela Bell

*Address all correspondence to: seshumeruva@gmail.com

Departmento de Fisica – ICE, Laboratorio de Espectroscopia de Materiais, Universidade Federal de Juiz de Fora, Juiz de Fora, Minas Gerais, Brazil

References

[1] Moulton P.F. Laser Hand Book, 5th ed. M. Bass and M. L. Stitch. Netherland Publishers: Amsterdam; 1985. 203 p.

[2] Sutton D. Electron Spectra of Transition Metal Complexes. McGraw-Hill Publishing Company Ltd: New York; 1969.

[3] Belançon M.P., Marconi J.D., Ando M.F., Barbosa L.C. Near-IR emission in Pr^{3+} single doped and tunable near-IR emission in Pr^{3+}/Yb^{3+} codoped tellurite tungstate glasses for broadband optical amplifiers. Opt. Mater. 2014; 36: 1020–1026. doi:10.1016/j.optmat.2014.01.014

[4] Balakrishna A., Rajesh D., Ratnakaram Y.C. Structural and optical properties of Nd^{3+} in lithium fluoro-borate glass with relevant modifier oxides. Opt. Mater. 2013; 35: 2670–2676. doi:10.1016/j.optmat.2013.08.004

[5] Vishwakarma A.K., Jayasimhadri M. Significant enhancement in photoluminescent properties via flux assisted Eu^{3+} doped $BaNb_2O_6$ phosphor for white LEDs. J. Alloys Compound. 2016; 683: 379–386. doi:10.1016/j.jallcom.2016.05.052

[6] Gao Y., Hu Y., Ren P., Zhou D., Qiu J. Phase transformation and enhancement of luminescence in the Tb^{3+}-Yb^{3+} co-doped oxyfluoride glass ceramics containing $NaYF_4$ nanocrystals. J. Eur. Ceramic Soc. 2016; 36: 2825–2830. doi:10.1016/j.jeurceramsoc.2016.04.027

[7] Guo H., Liu L., Wang Y., Hou C., Li W., Lu M., Zou K., Peng B. Host dependence of spectroscopic properties of Dy3+ - doped and Dy3+, Tm3+-codoped Ge–Ga–$SCdI_2$ chalcohalide glasses. Opt. Express. 2009; 17: 15350–15358. doi:10.1364/OE.17.015350

[8] Seshadri M., Chillcce E.F., Marconi J.D., Sigoli F.A., Ratnakaram Y.C., Barbosa L.C. Optical characterization, infrared emission and visible up-conversion in Er^{3+} doped tellurite glasses. J. Non-Cryst. Solids. 2014; 402: 141–148. doi:10.1016/j.jnoncrysol.2014.05.024

[9] Diek G.H. Spectra and Energy Levels of Rera Earth Ions in Crystals. New York, Wiley: 1968.

[10] Peacock R.D. Rare earths. In: Structure and Bond, vol. 22. Springer, Berlin: 1975; p. 83.

[11] Reisfeld R. Radiative and nonradiative transition of rare earths in glasses. In: Structure and Bond, vol. 22. Springer, Berlin: 1975; p. 123.

[12] Wybourne B.G. Spectroscopy Properties of Rare earths. New York, Wiley: 1965.

[13] ASTM Standards on Glass and Glass Products, prepared by ASTM Committee C14, American Society for Testing Materials, April 1955.

[14] Shelby J.E. Introduction to Glass Science and Technology, 2nd ed. Royal Society of Chemistry; Cambridge: 2005; 291 p.

[15] Zachariasen W.H. The atomic arrangement in glasses. J. Am. Chem. Soc. 1932; 54: 3841–3851.

[16] Herrmann A., Kuhn S., Tiegel M., Russel C., Korner J., Klopfel D., Hein J., Kaluza M.C. Structure and fluorescence properties of ternary aluminosilicate glasses doped with samarium and europium. J. Mater. Chem. C. 2014; 2: 4328–4337. doi:10.1039/C4TC00036F

[17] Tiegel M., Herrmann A., Russel C., Korner J., Klopfel D., Hein J., Kaluza M.C. Magnesium aluminosilicate glasses as potential laser host material for ultrahigh power laser systems. J. Mater. Chem. C. 2013; 1: 5031–5039. doi:10.1039/C3TC30761A

[18] Seshadri M., Barbosa L.C., Radha M. Study on structural, optical and gain properties of 1.2 and 2.0 μm emission transitions in Ho^{3+} doped tellurite glasses. J. Non-Cryst. Solids. 2014; 406: 62–72. doi:10.1016/j.jnoncrysol.2014.09.042

[19] Van Wazer J.R. Phosphorous and Its Compounds, vol. 1. New York, Interscience; 1958.

[20] Kordes E., Vogel W., Feterowsky R. Physical chemistry studies About the properties and Feinban of phosphate glasses. Z. Elektrochem. 1953; 57: 282.

[21] Kordes E., Nieder R. Glasstech. 1968; 41: 41.

[22] Kordes E., Vogel W., Feterowsky R. Z. Elektrochem. 1953; 57: 282.

[23] Hench L.L. Bioceramics. J. Am. Ceram. Soc. 1991; 74: 1487.

[24] Volksch G., Hoche T., Szabo I., Holand W., Glasstech. Ber. Glass Sci. Technol. 1998; 71: 500.

[25] Pinheiro A.S., Freitas A.M., Silva G.H., Bell M.J.V, Anjos V., Carmo A.P., Dantas N.O. Laser performance parameters of Yb^{3+} doped UV-transparent phosphate glasses. Chem. Phys. Lett. 2014; 592: 164–169. doi:10.1016/j.cplett.2013.12.022

[26] Silva G.H., Anjos V., Bell M.J.V., Carmo A.P., Pinheiro A.S. Dantas N.O. Eu^{3+} emission in phosphate glasses with high UV transparency. J. Lumin. 2014; 154: 294–297. doi: 10.1016/j.jlumin.2014.04.043

[27] Deutschbein O., Pautrat C., Svirchevsky I.M. Phosphate glasses: New laser materials. Rev. Phys. Appl. 1967; 1: 29.

[28] Abd El-Ati M.I., Higazy A.A. Electrical conductivity and optical properties of gamma-irradiated niobiumphosphate glasses. J. Mater. Sci. 2000; 35: 6175. doi:10.1023/A:1026768925365

[29] Pinheiro A.S., da Costa Z.M., Bell M.J.V., Anjos V., Dantas N.O., Reis S.T. Thermal characterization of iron phosphate glasses for nuclear waste disposal. Opt. Mater. 2011; 33: 1975–1979. doi:10.1016/j.optmat.2011.03.050

[30] Martin S.W. J. Review of the Structures of Phosphate Glasses. Solid State Chem. 1991; 28: 163.

[31] Belliveau T.F., Simkin D.J. On the coordination environment of rare earth ions in oxide glasses: Calcium titanosilicate and sodium aluminosilicate glasses. J. Non-Cryst. Solids. 1989; 110: 127. doi:10.1016/0022-3093(89)90249-4

[32] Babu S., Seshadri M., Reddy Prasad V., Ratnakaram Y.C. Spectroscopic and laser properties of Er^{3+} doped fluoro-phosphate glasses as promising candidates for broad-band optical fiber lasers and amplifiers. Mater. Res. Bulletin. 2015; 70: 935–944. doi: 10.1016/j.materresbull.2015.06.033

[33] Tian Y.M., Shen L.F., Pun E.Y.B., Lin H. High-aluminum phosphate glasses for single-mode waveguide-typed red light source. J. Non-Cryst. Solids. 2015; 426: 25–31. doi: 10.1016/j.jnoncrysol.2015.06.015

[34] Fan B., Point C., Adam J.L., Zhang X., Fan X., Ma H. Near-infrared down-conversion in rare earth-doped chloro-sulfide glass GeS2-Ga2S3-CsCl: Er, Yb. J. Appl. Phys. 2011; 110: 13107. doi:10.1063/1.3665638

[35] Ehrmann P.R., Campbell J.H. Nonradiative energy losses and radiation trapping in neodymium-doped phosphate laser glasses. J. Am. Ceram. Soc. 2002; 85: 1061–1069. doi:10.1111/j.1151-2916.2002.tb00223.x

[36] Goncalves R.R., Carturan G., Montagna M., Ferrari M., Zampedri L., Pelli S., Righini G.C., Ribeiro S.J.L., Messaddeq Y. Erbium-activated HfO_2-based waveguides for photonics Opt. Mater. 2004; 25: 131–139. doi:10.1016/S0925-3467(03)00261-1

[37] Förster T. Zwischenmolekulare Energiewanderung und Fluoreszenz. Ann. Phys. 1948; 2: 55–75.

[38] Dexter D.L. A theory of sensitized luminescence in solids. J. Chem. Phys. 1953; 21: 836–850. doi:10.1063/1.1699044

[39] Tarelho T.V.G. Gomes L., Ranieri I.M. Determination of microscopic parameters for nonresonant energy-transfer processes in rare earth-doped crystals. Phys. Rev. B. 1997; 56: 14344. doi:10.1103/PhysRevB.56.14344

[40] de Sousa D. F., Batalioto F., Bell M.J.V., Oliveira S.L., Nunes L.A.O. Spectroscopy of Nd^{3+} and Yb^{3+} codoped fluoroindogallate glasses. J. Appl. Phys. 2001; 90: 3308–3313. 10.1063/1.1397289

[41] Batalioto F., de Sousa D.F., Bell M.J.V., Lebullenger R., Hernandes A.C., Nunes L.A.O. Optical measurements of Nd^{3+}/Yb^{3+} codoped fluoroindogallate glasses. J. Non-Cryst. Solids. 2000; 273: 233–238. doi:10.1016/S0022-3093(00)00132-0

[42] Judd B.R. Optical absorptionm intensities of rare earth ions. Phys. Rev. 1962; 127: 750. doi:10.1103/PhysRev.127.750

[43] Ofelt G.S. Intensities of crystal spectra o rare earth ions. J. Chem. Phys. 1962; 37: 511. doi:10.1063/1.1701366

[44] Broer L.F.J., Gorter C.J., Hoogschangen J. The intensities and multiple character in the spectra of rare earth ions. Physica 1945; XI: 231–250.

[45] Seshadri M., Venkata Rao K., Lakshmana Rao J., Koteswara Rao K.S.R., Ratnakaram Y.C. Spectroscopic investigations and luminescence spectra of Nd^{3+} and Dy^{3+} doped different phosphate glasses. J. Lumin. 2010; 130: 536–543. doi:10.1016/j.jlumin.2009.10.027

[46] Ratnakaram Y.C., Thirupathi Naidu D., Vijaya Kumar A., Rao J.L. Characterization of Tm^{3+} doped mixed alkali borate glasses - Spectroscopic investigations. J. Phys. Chem. Solids. 2003; 64 (12): 2487–2495. doi:10.1016/j.jpcs.2003.06.001

[47] Reisfeld R., Jorgensen C.K. Lasers and excited states of rare earths. Berlin, Springer-Verlag; 1977.

[48] Oomen E.W.J.L., Van Dongen A.M.A. Europium (III) in oxide glasses: dependence of the emission spectrum upon glass composition. J. Non-Cryst. Solids. 1989; 111: 205–213. doi:10.1016/0022-3093(89)90282-2

[49] Martins V.M., Messias D.N., Doualan J.L., Braud A., Camy P., Dantas N.O., Catunda T., Pilla V., Andrade A.A., Moncorgé R. Thermo-optical properties of Nd^{3+} doped phosphate glass determined by thermal lens and lifetime measurements. J. Lumin. 2016; 162: 104–107. doi:10.1016/j.jlumin.2015.02.013

[50] Linganna K., Dwaraka Viswanath C.S., Narro-Garcia R., Ju S., Han W.T., Jayasankar C.K., Venkatramu V. Thermal and optical properties of Nd^{3+} ions in K-Ca-Al fluorophosphate glasses. J. Lumin. 2015; 166: 328–334. doi:10.1016/j.jlumin.2015.05.024

[51] Upendra Kumar K., Babu P., Jang K.H., Jin Seo H., Jayasankar C.K., Joshi A.S. Spectroscopic and 1.06m laser properties of Nd^{3+}-doped K-Sr-Al phosphate and fluorophosphate glasses. J. Alloys and Compd. 2008; 458: 509–516. doi:10.1016/j.jallcom.2007.04.035

[52] Binnemans K., Van Deun R., Gorller-Walrand C., Adam J.L. Optical properties of Nd^{3+} -doped fluorophosphate glasses. J. Alloys and Compounds. 1998; 275–277: 455–460. doi:10.1016/S0925-8388(98)00367-3

[53] Tian C., Chen X., Shuibao Y. Concentration dependence of spectroscopic properties and energy transfer analysis in Nd^{3+} doped bismuth silicate glasses. Solid State Sci. 2015; 48: 171–176. doi:10.1016/j.solidstatesciences.2015.08.008

[54] Seshadri M., Barbosa L.C., Cordeiro C.M.B., Radha M., Sigoli F.A., Ratnakaram Y.C. Study of optical absorption, visible emission and NIR-Vis luminescence spectra of Tm^{3+}/Yb^{3+}, Ho^{3+}/Yb^{3+} and $Tm^{3+}/Ho^{3+}/Yb^{3+}$ doped tellurite glasses. J. Lumin. 2015; 166: 8–16. doi:10.1016/j.jlumin.2015.04.022

[55] Balakrishnaiah R., Babu P., Jayasankar C.K., Joshi A.S., Speghini A., Bettinelli M. Optical and luminescence properties of Nd^{3+} ions in K-Ba-Al phosphate and fluorophosphate glasses. J. Phys.: Condens. Matter. 2006; 18: 165–179. doi:10.1088/0953-8984/18/1/012

[56] Surendra babu S., Babu P., Jayasankar C.K., Joshi A.S., Speghini A., Bettinelli M. Luminescence and optical absorption properties of Nd^{3+} ions in K-Mg-Al phosphate and fluorophosphate glasses. J. Phys.: Condens. Matter. 2006; 18: 3975–3991. doi: 10.1088/0953-8984/18/16/007

[57] Ajroud M., Haouari M., Ben Ouada H., Maaref H., Brenier A., Garapon C. Investigation of the spectroscopic properties of Nd^{3+}-doped phosphate glasses. J. Phys.: Condens. Matter. 2000; 12: 3181–3193. doi:10.1088/0953-8984/12/13/324

[58] Choi J.H., Margaryan A., Margaryan A., Shi F.G. Judd-Ofelt analysis of spectroscopic properties of Nd^{3+}-doped novel fluorophosphate glass. J. Lumin. 2005; 114: 167–177. doi:10.1016/j.jlumin.2004.12.015

[59] Advanced Optics, SCHOTT North America Inc., Duryea, PA, USA.

[60] HOYA Corporation USA, Optics Division, 3285 Scott Blvd., Santa Clara, CA 95054 USA.

[61] KIGRE, Inc., 100 Marshland Road, Hilton Head, SC 29926, USA.

[62] Iizuka K. Lab instruction notes for "Erbium Doped Fiber Amplifiers". University of Toronto; Canada: 1998.

[63] Wang J.S., Vogel E.M., Snitzer E. Terullite glass: a new candidate for fiber devices. Opt. Mater. 1994; 3(3): 187–203. doi:10.1016/0925-3467(94)90004-3

[64] Mori A., Ohishi Y., Sudoi S. Erbium-doped tellurite-based fiber laser and amplifier. Electron. Lett. 1997; 33(10); 863–865.

[65] Sugimoto N., Ochiai K., Ohara S., Hayashi H., Fukasawa Y., Hirose T., Reyes M. Highly efficient and short length Lanthanum codoped Bh03-based EDF for extended L-band amplification. In: Proceedings of Optical Amplifiers and their applications. 2002, OAA'02, PD 5, Vancouver, 2002.

[66] Weber W.H., Lambe J. Luminescent greenhouse collector for solar radiation. Appl. Opt. 1976; 15(10): 2299–2300. doi:10.1364/AO.15.002299

[67] Barnham K., Marques J.L., Hassard J., Obrien P. Quantum-dot concentrator and thermodynamic model for the global redshift. Appl. Phys. Lett. 2000; 76(9): 1197–1199. doi:10.1063/1.125981

[68] Werts M.H.V., Hofstraat J.W., Geurts F.A.J., Verhoeven J.W. Fluorescein and eosin as sensitizing chromophores in near-infrared luminescent ytterbium(III), neodymium(III) and erbium(III) chelates. Chem. Phys. Lett. 1997; 276(3/4): 196–201. doi:10.1016/S0009-2614(97)00800-2

[69] Sholin V., Olson J.D., Carter S.A. Semiconducting polymers and quantum dots in luminescent solar concentrators for solar energy harvesting. J. Appl. Phys. 2007; 101: 1231141–1231149. doi:10.1063/1.2748350

[70] Brenda C.R., Wilson L.R., Richards B.S. Advanced material concepts for luminescent solar concentrators. IEEE J. Selected Topics in Quant. Electron. 14(5): 2008; 1321–1322.

[71] Goetzberger A., Greubel W. Solar energy conversion with fluorescent collectors. Appl. Phys. 1977; 14: 123–139. doi:10.1007/BF00883080

[72] Reisfeld R. Future technological applications of rare earth doped materials. J. Less Common Mater. 1983; 93: 243–251. doi:10.1016/0022-5088(83)90163-7

Chapter 8

Luminescent Devices Based on Silicon-Rich Dielectric Materials

Santiago A. Cabañas-Tay, Liliana Palacios-Huerta, Mariano Aceves-Mijares, Antonio Coyopol, Sergio A. Pérez-García, Liliana Licea-Jiménez, Carlos Domínguez and Alfredo Morales-Sánchez

Additional information is available at the end of the chapter

http://dx.doi.org/10.5772/64894

Abstract

Luminescent silicon-rich dielectric materials have been under intensive research due to their potential applications in optoelectronic devices. Silicon-rich nitride (SRN) and silicon-rich oxide (SRO) films have been mostly studied because of their high luminescence and compatibility with the silicon-based technology. In this chapter, the luminescent characteristics of SRN and SRO films deposited by low-pressure chemical vapor deposition are reviewed and discussed. SRN and SRO films, which exhibit the strongest photoluminescence (PL), were chosen to analyze their electrical and electroluminescent (EL) properties, including SRN/SRO bilayers. Light emitting capacitors (LECs) were fabricated with the SRN, SRO, and SRN/SRO films as the dielectric layer. SRN-LECs emit broad EL spectra where the maximum emission peak blueshifts when the polarity is changed. On the other hand, SRO-LECs with low silicon content (~39 at.%) exhibit a resistive switching (RS) behavior from a high conduction state to a low conduction state, which produce a long spectrum blueshift (~227 nm) between the EL and PL emission. When the silicon content increases, red emission is observed at both EL and PL spectra. The RS behavior is also observed in all SRN/SRO-LECs enhancing an intense ultraviolet EL. The carrier transport in all LECs is analyzed to understand their EL mechanism.

Keywords: silicon-rich dielectrics, photoluminescence, electroluminescence, conduction mechanisms

1. Introduction

The use of photonic signals instead of electrons to transmit information through an electronic circuit is an actual challenge. Unfortunately, it is well known that bulk silicon (Si) is an indirect bandgap semiconductor, making it an inefficient light emitter. Therefore, great efforts have been taken to obtain highly luminescent Si-based materials in order to get Si-based photonic devices, especially a light emitting device [1–3]. Such circumstances have led to explore new options for converting silicon into a luminescent material. Si nanoparticles (Si-nps) embedded in a dielectric material as silicon-rich oxide (SRO) or silicon-rich nitride (SRN) show a prominent photoluminescence (PL) emission in red and blue-green region, respectively [4–10]. Thus, SRN or SRO films have been considered as promising candidates for emissive materials due to their potential applications in Si-based optoelectronic devices, and their fully compatibility with the complementary metal-oxide-semiconductor (CMOS) processes [11–16].

Two main strategies have been explored: those that focus on the intrinsic emission from the matrix, either through emissions from defects or by the presence of Si-nps. The second one focus on extrinsic emission, which is produced by doping the material (usually introducing rare earth ions) [3, 17].

The most common strategy to obtain intrinsic emission is through silicon nanostructures, which significantly increase the emission due to the quantum confinement effect (QCE) [5, 18]. Furthermore, the dependence on the size of the Si-nps on the forbidden gap width allows that the emission can be adjusted in the visible and the red-near infrared region of the electromagnetic spectrum [19]. Several studies have reported both red and near infrared electroluminescence (EL) SRO which is mainly attributed to the recombination of excitons in Si-nps [20, 21]. On the other hand, the emission of green or blue light has been attributed to defects associated with oxygen [22, 23]. Some studies have reported that red (620 nm) EL emission could be attributed to non-bridging oxygen hole center (NBOHC) defects whose origin has been corroborated by the fact that the peak position does not change if the film is excited with different energies [24].

Another alternative to obtain intrinsic light emission is through an ordered structure of Si-nps by a superlattice, which is formed by the alternating of SRO and SiO_2 nano-films. Red or near infrared emission has been observed in these structures and has been related to both excitonic recombination taking place in confined states within Si-nps or relaxation of hot electrons [25, 26].

Intrinsic emission has been also observed in SRN films [27–31]. For example, an orange emission at 600 nm was observed at room temperature and has been related to the electron-hole pairs' recombination within Si-nps [27]. Also, green emission has been observed in nitrogen-rich silicon nitride, which was attributed to radiative recombination in localized states related to Si-O [28]. Some other authors have shown significant improvement of the green emission intensity using oxidized silicon-rich nitride [29]. Also, when a silicon nitride film is implanted with Si ions, violet and green-yellow emission bands are observed, giving rise to an intense white EL emission [30]. The violet band was related with the presence of defects states related to silicon dangling bonds (=Si^0 or centers K^0) located near the middle of the forbidden gap of silicon nitride and defect states related to the unit Si–Si= located near the edge of the valence band, while the green-yellow band was attributed to the transition from the =Si^0 state to nitrogen dangling bonds (=N–) in the tails of valence bands.

Red-near infrared EL (800 nm) has been reported in superlattices combining SRN and SiO_2 films and explained by the bipolar recombination of electron-hole pairs in Si-nps present within the SRN films [31]. A yellow EL emission has been reported when an SRO film instead of SiO_2 layer is used in the multilayer structure [32]. All of these promising results have proved the first implemented all-silicon-based photonic device [33]. Nevertheless, despite all these promising results in luminescent silicon-based materials, the improvement of the efficiency of the light emitting devices is still necessary.

This chapter shows a review about our experience on the PL and EL properties of SRN and SRO films deposited by low-pressure chemical vapor deposition (LPCVD). The effect of the combination of the SRN and SRO luminescent properties is also analyzed as an SRN/SRO structure. A study about the composition, structural, optical, and electro-optical properties of these films will be discussed. The study also includes the analysis of the charge transport mechanism through the SRO, SRN, and SRN/SRO films to understand their electroluminescence behavior and its correlation with the different luminescent centers (LCs) within the active material.

2. Experimental procedure

In this chapter, SRN, SRO, and SRN/SRO films were deposited in a homemade LPCVD hot-wall reactor. In these silicon-rich dielectrics materials, the Si content was controlled by a ratio of partial pressure of reactant gases; R_N and Ro for SRN and SRO, respectively.

The SRN films were deposited on N-type ((100)-oriented) Si wafers with a resistivity of 1–5 Ω-cm at 750°C using ammonia (NH_3) and 5% nitrogen (N_2)-diluted silane (SiH_4) as the reactant gases by the ratio $R_N = 20{*}P(NH_3)/P(SiH_4)$. SRN films with R_N values of 5, 20, and 80 were deposited with the parameters shown in **Table 1**. SRO films were deposited on N-type Si wafers with a resistivity of 5–10 Ω-cm ((100)-oriented) at a temperature of 720°C using pure nitrous oxide (N_2O) and 3.3% nitrogen (N_2)-diluted silane (SiH_4) as the reactant gases by the ratio $Ro = 30{*}P(N_2O)/P(SiH_4)$. SRO films with Ro values of 30 and 20 were deposited with the parameters shown in **Table 1**. Finally, SRN/SRO bilayers were deposited on P-type silicon substrates ((100)-oriented) with resistivity of 5–10 Ω-cm. $Ro = 20$ and 30 and $R_N = 80$ values were used for the SRO and SRN films, respectively. SRO and SRN films were deposited at 730 and 760°C, respectively. The deposition conditions are also summarized in **Table 1**. About 3.3% of nitrogen diluted silane was used for the bilayer structures.

After deposition, SRN, SRO and SRN/SRO samples were thermally annealed at 1100°C under nitrogen atmosphere conditions for 180 min.

	Sample name	R_N	R_O	Pressure of gases (Torr)			Time (min)	Thickness (nm)
				N_2O	NH_3	SiH_4		
SRO	M20		20	0.53		0.80	12	55.72 ± 5.0
	M30		30	0.80		0.80	15	64.00 ± 3.4
SRN	N5	5			0.22	0.85	10	102.83 ± 3.62
	N20	20			0.85	0.85	15	112.67 ± 6.19
	N80	80			2.00	0.50	13	66.93 ± 2.12
SRN/SRO	B20	80			1.08	0.41	4	16.32 ± 1.54
			20	0.53		0.80	15	55.72 ± 5.0
	B30	80			1.08	0.41	4	16.32 ± 1.54
			30	0.80		0.80	15	64.00 ± 3.4

Table 1. Process parameters of the SRO, SRN and SRN/SRO films deposited by LPCVD and thickness of the sample after thermal annealing.

For electrical and electroluminescence studies, light emitting capacitive (LEC) structures were fabricated. For SRN-LECs, a transparent 300-nm thick fluorine-doped tin oxide SnO_2:F (FTO) film was deposited onto the surface of the SRN by ultrasonic spray pyrolysis. Square-shaped patterns with 1 mm^2 area were defined by a photolithography process step to act as gate contact. For SRO-LECs, ~400-nm thick semitransparent n+ polycrystalline silicon (poly) gate was deposited onto the SRO film surface by LPCVD. After a photolithography process step, square-shaped gates of 4 mm^2 area were defined. For SRN/SRO-LECs, ~300-nm thick indium tin oxide (ITO) film was deposited by RF sputtering onto the surface of the films as gate contact. Square-shaped patterns with area of 1 mm^2 were defined by a photolithography process step to act as anode gate contact. Approximately 700-nm thick aluminum (Al) film was evaporated onto the backside of the silicon substrates as cathode contact in all of the LECs. A thermal annealing process at 460°C in N_2 atmosphere for 20 min was used to form the ohmic contact.

The thickness of thermally annealed SRO and SRN films was measured with a Gaertner L117 ellipsometer with a 70° incident laser with wavelength of 632.8 nm and is also shown in **Table 1**. Chemical bonding characteristics was analyzed by means of Fourier transform infra-red spectroscopy (FTIR) with a Brucker V22 equipment in the 4000–350 cm^{-1} range with a resolution of 5 cm^{-1}. The PL spectra were measured with a Fluoromax 3 of Horiba Jobin Yvon. The samples were excited using a 300 nm radiation, and the PL emission signal was collected from 400 to 900 nm with a resolution of 1 nm. The depth analysis profile of thermally annealed SRN, SRO, and SRN/SRO films was analyzed by means of X-ray photoelectron spectroscopy (XPS) Escalab 250Xi of Thermo Scientific equipment, with an Al Kα monochromated source. Current-voltage (I-V) measurements of SRN, SRO, and SRN/SRO-LECs were performed using a Keithley 4200-SCS parameter analyzer at the same time that the EL was collected with an optical fiber, which was located right on the surface of the device and connected to an Ocean Optics QE-65000 spectrometer.

3. Composition

The composition of SRN and SRO films play an important role in order to understand their luminescence, electrical, and electro-optical properties. In this sense, some techniques such as FTIR and XPS spectroscopies have been used.

3.1. Silicon-rich nitride (SRN) film

The Si-N bonds of SRN films were determined by FTIR measurements. **Figure 1(a)** shows the IR spectra measured from SRN films with $R_N = 5$ (N5), 20 (N20), and 80 (N80) before (B-TA) and after (A-TA) thermal annealing.

IR peaks at 460 and 840 cm^{-1} ascribed to Si-N wagging and stretching modes, respectively, were observed for all samples [35–37]. An IR band appears at 1080 cm^{-1} after thermal annealing, being more evident in the N80 sample. The presence of this peak has been observed before and attributed to a reordering in the films toward a-Si_3N_4 bonding configuration [38, 39]. Nevertheless, it could be related to the Si–O stretching mode due the oxygen incorporation in the samples.

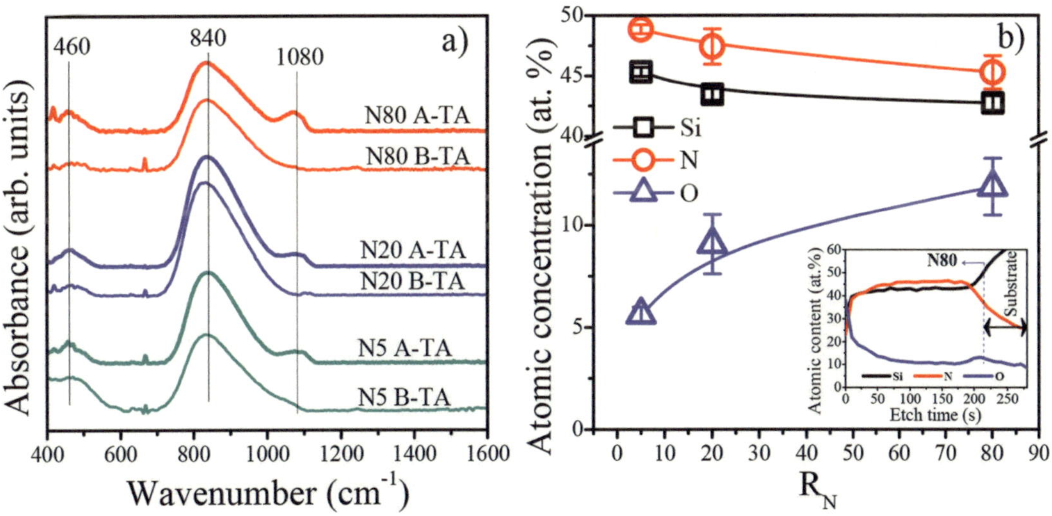

Figure 1. (a) FTIR spectra of SRN films with $R_N = 5$ (N5), 20 (N20) and 80 (N80) before and after thermal annealing, and (b) atomic concentration (at.%) of Si, O, and N as a function of R_N. Inset shows the XPS depth profile of SRN film with $R_N = 80$ (N80). From Cabañas-Tay et al. [34].

In order to comprehend the stoichiometry and the presence of some oxygen into the SRN films, analysis of their composition was performed by means of XPS. **Figure 1(b)** shows information about the chemical composition of the thermally annealed SRN samples. The inset of **Figure 1(b)** exhibits the depth profile composition of the thermally annealed $R_N = 80$ (N80) sample. As can be observed, some oxygen is present at the outmost part of the layer through the film depth, being present mainly at the film surface. This behavior is similar for different R_N, but with different concentrations. **Figure 1(b)** shows the atomic concentration of Si, O, and N as a function of R_N. Mean silicon content values along the layer of 45.55 ± 0.38, 43.58 ± 0.48, and 42.88 ± 1.39 at.% were obtained for the SRN films with $R_N = 5$, 20, and 80, respectively. Moreover, the presence of oxygen increases from 5.6 to 11.8 at.% as the R_N value increases from 5 to 80, respectively.

3.2. Silicon-rich oxide (SRO) film

The Si–O bonds of SRO films were also determined by FTIR measurements. The IR spectra measured from SRO films with *R*o = 20 (M20) and *R*o = 30 (M30) after thermal annealing are shown in **Figure 2(a)**. Typical vibration bands at 460, 810, and 1070 cm^{-1} related to Si–O rocking, Si–O bending, and Si–O stretching modes, respectively, were present in both SRO films [40, 41]. The shoulder from ~1100 to ~1300 cm^{-1} observed in both SRO films has been attributed to Si–O stretching out of phase [42]. It has been reported that this shoulder is less pronounced for the suboxides compared to the stoichiometric oxide [43, 44]. The absorption at 610 cm^{-1} due to unsaturated Si–Si bonds (phonon-phonon interactions) was observed only in M20 (SRO film with higher proportion of silicon precursor) showing the presence of structural imperfections at the SiO_2/silicon nanoparticles (Si-nps) interface [45].

Figure 2(b) exhibits the depth profile composition of the thermally annealed SRO films. Mean silicon content values of 41.85 ± 1.1 and 39.98 ± 0.8 at.% were obtained for the SRO films with

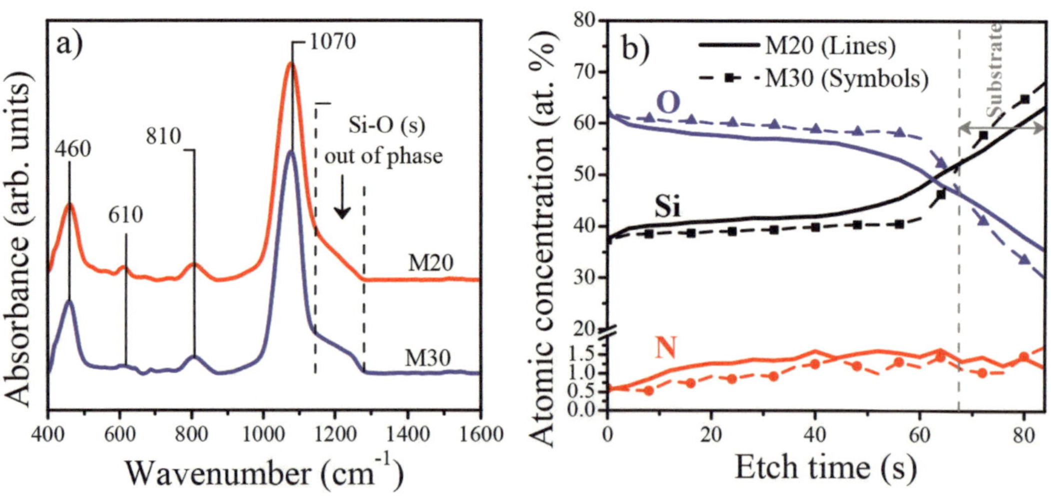

Figure 2. (a) FTIR spectra of SRO films with *R*o = 20 (M20) and 30 (M30) after thermal annealing, and (b) XPS depth profile of SRO films with *R*o = 20 (M20) and 30 (M30).

http://dx.doi.org/10.5772/64894

Ro = 20 (M20) and 30 (M30), respectively. The presence of nitrogen was also observed in both SRO films being slightly more evident for M20. An analysis of the Si2p XPS signal of SRO film with *Ro* = 30 thermally annealed at 1100°C (not showed here) [46] shows that the microstructure of this film is almost a stoichiometric SiO_2 (Si^{4+}) and that the silicon excess is present as Si–O compounds, which could include the neutral oxygen vacancy (NOV) and non-bridging oxygen hole center (NBOHC) defects. As expected by the Si2p XPS signal, crystalline silicon nanoparticles were not observed in this films; however, energy-filtered transmission electron microscopy (EFTEM) analysis has shown (although not clearly) the presence of silicon nanoparticles with a mean size of 1.5 nm and density of ~1.1×10^{12} cm^{-2} as reported in [47, 48]. In fact, it has been reported that a largely disordered system is expected when the size of Si-nps is between 1 and 2 nm. Thus, the small Si-nps are no longer crystalline but amorphous like, as also indicated by theoretical calculations [49, 50]. Then, the dispersed Si atoms in the M30 films are preferentially arranged as very small Si-nps (E'_{δ}) and some defects like oxygen-deficiency center (ODC) and NBOHC as reported for thin SiOx films (<100 nm) [51].

The presence of the Si^0 peak in the Si2p XPS signal correlates well with the presence of silicon nanocrystal observed by high-resolution transmission electron microscopy (HRTEM, not shown here) in the M20 film. HRTEM reveals Si nanocrystals with an average size (and density) of 2.91 ± 0.40 nm (8.66 x 10^{11} cm^{-2}) [46].

3.3. Silicon-rich nitride/silicon-rich oxide (SRN/SRO) bilayer

The Si–N and Si–O bonds of SRN/SRO bilayers with *Ro* = 20 (B20) and *Ro* = 30 (B30) were also determined. **Figure 3(a)** shows the FTIR spectra of SRO and SRN/SRO films with *Ro* = 20 (M20 and B20, respectively) after thermal annealing. The IR peak at 820 cm^{-1} observed only in the B20 bilayer is ascribed to Si–N stretching present in SRN films [45]. The intensity of a shoulder

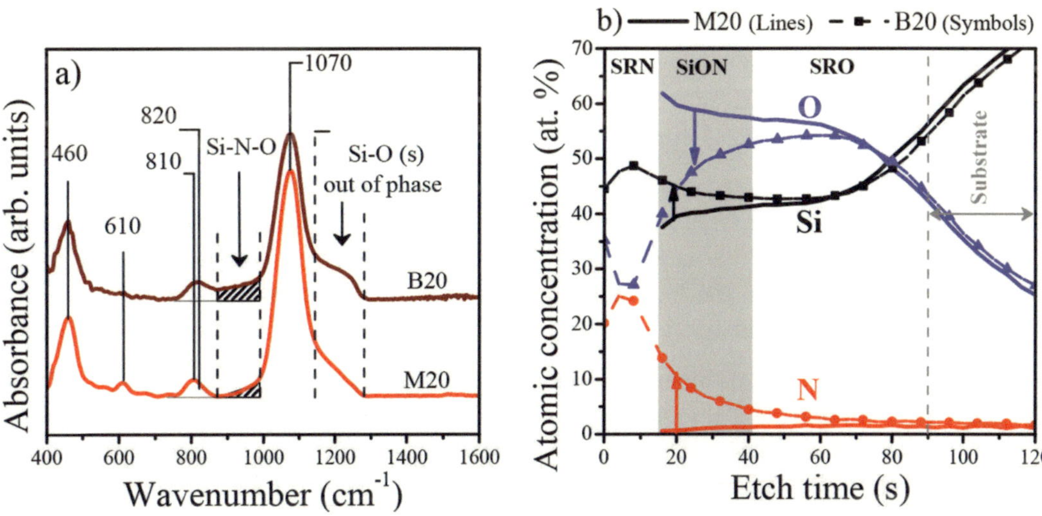

Figure 3. (a) FTIR spectra of SRO and SRN/SRO films with *Ro* = 20 (M20 and B20, respectively) after thermal annealing, and (b) XPS depth profile of SRO and SRN/SRO films with *Ro* = 20 (M20 and B30, respectively) after thermal annealing.

(shadow area) from 900 to 1000 cm^{-1} increased when the B20 bilayer was formed, and it was attributed to the formation of Si–N–O bonds in silicon oxynitride [45].

As shown in **Figure 3(a)**, the shoulder from ~1100 to ~1300 cm^{-1} was observed in both samples (M20 and B20) and it was attributed to Si–O stretching out of phase [42]. The IR peak at 610 cm^{-1} observed in the SRO monolayer (M20) disappeared for the bilayer (B20) and it could be related to the nitrogen incorporation within the SRO that creates Si–N–O–Si bridges. The presence of these bridges decreased the quantity of strained bonds and Si dangling bonds at the SiO_2/Si-np interface [45]. The SRN/SRO film with $Ro = 30$ (B30) showed the same IR peaks than the B20 bilayer, except for the 610 cm^{-1} IR peak.

Figure 3(b) showed the depth profile composition of the thermally annealed SRO monolayer and SRN/SRO bilayer, both with $Ro = 20$ (M20 and B20, respectively). The average Si content within the SRO monolayer was about 41.85 ± 1.1 at.%. When the SRN/SRO bilayer was formed, the oxide-nitride interface became an imprecise oxynitride (SiON) layer almost for 30 s etching time (shadow area in **Figure 3(b)**). In that region, a gradual increasing of both nitrogen and silicon was observed toward the interface. The average Si content slightly increased up to ~43.21 ± 0.7 at.% in the SRO layer, whereas the oxygen content in the SRO film was reduced from 56.78 ± 1.3 to 53.08 ± 1.0 at.%. The Si, O, and N diffusion could be enhanced by the high annealing temperature. This is supported by two facts: first, the presence of a nitrogen concentration within the oxide layer which goes from 2 to 10%, and second, the oxygen presence in the nitride layer (~25%). The nitrogen content inside of the SRO in the SRN/SRO bilayer structure could modify its optical and structural properties. It has been reported that the nitrogen hinder the diffusion of Si atoms and prevents the phase separation in the amorphous SiOx:N films [52]. Thus, the mobility of the Si atoms is smaller and the growth of the Si-nps during the thermal diffusion process is reduced, giving as a result smaller Si-nps. A similar behavior was observed for the B30 bilayer, but with different concentration.

4. Photoluminescence

4.1. Silicon-rich nitride (SRN) film

The PL spectra of SRN films before and after thermal annealing are shown in **Figure 4**. The PL intensity was normalized to the thickness of each SRN film. The as-deposited SRN film with $R_N = 80$ (N80) emit a broad PL band with the main peak at 490 nm, as shown in **Figure 4(a)**. As the R_N value decreases, the PL emission band becomes narrower and shifts to a higher wavelength reaching 590 nm for $R_N = 5$ (N5, higher silicon excess) but with a lower intensity.

The PL band of the SRN films blueshifts after thermal annealing, particularly for $R_N = 80$ where the main band shifts from 490 to 420 nm, as shown in **Figure 4(b)**. It is well known that the silicon excess in silicon-rich dielectrics agglomerates forming Si-nps after high temperature annealing, resulting in a redshift of the PL band due to the quantum confinement effects [53–57]. However, contrary to this assumption, in this work, the PL blueshifts after the annealing, as shown in **Figure 4(b)**. This effect has been observed before and it has been ascribed to compositional-dependent changes in the concentration of defect states within the samples [58].

http://dx.doi.org/10.5772/64894

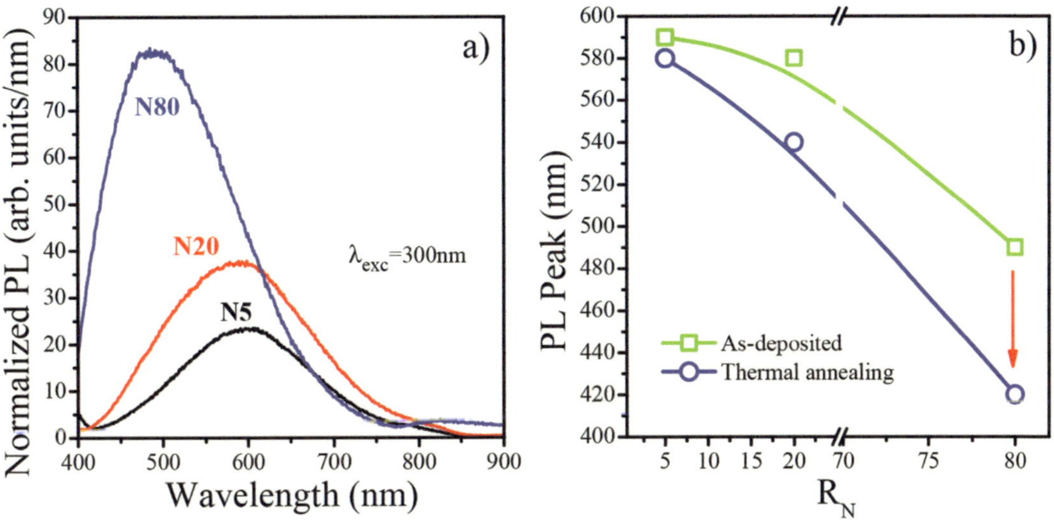

Figure 4. (a) Normalized PL spectra of as-deposited, and (b) maximum PL peak position before and after thermally annealed, SRN films with $R_N = 5$ (N5), 20 (N20), and 80 (N80). Lines are plotted as an eye-guide. From Cabañas-Tay et al. [34].

Therefore, defects are created rather than Si-nps, after thermal annealing. Moreover, due to the low diffusivity of silicon atoms in Si_3N_4, a high Si content (>52 at.%) is needed to form Si-nps [59, 60]. Because of the low Si content present in the SRN films from this work (<46 at.%), as shown above in the XPS results, these films can be explained as a sub-stoichiometric nitride with structural defects, as reported in the study of Cabañas-Tay et al. [34].

PL bands between 380 and 600 nm (2.0–3.2 eV) have been observed before in SRN films, and they have been ascribed to the radiative recombination of carriers in band tail states, which are related to defect energy levels within the gap of amorphous silicon nitride [61–64]. Therefore, the PL bands emitted by the SRN films from this work can be explained by the excitation of different defects as discussed in the study of Cabañas-Tay et al. [34]. As observed in **Figure 4(a)**, the as-deposited SRN films emit at 590, 580, and 490 nm, whereas the thermally annealed films emit at 580, 540, and 420 nm for R_N of 5, 20, and 80, respectively. PL emission bands observed at 580–590 nm (~2.1 eV) have been related to electronic transitions from the conduction band minimum (CBM) to K^0 centers located near of the mid-gap, the emission band at 540 nm (~2.3 eV) is related to electronic transition from the K^0 centers to $=N^-$ centers located near of the valence band maximum (VBM), and the emission band at 420 nm (~2.9 eV) has been related to electronic transitions from the K^0 centers to VBM. Hence, the PL bands observed in this work could have a similar origin, as discussed in the study of Cabañas-Tay et al. [34].

Some studies have shown that electronic transition related with K^0 centers to =Si–O–Si states are observed when oxygen is incorporated in the SRN film [65, 66]. The presence of oxygen in SRN films creates a gap state of Si–O above the VBM, giving rise to the 485 nm (~2.55 eV) emission. The XPS analysis demonstrated that SRN films contain oxygen, being the higher concentration for $R_N = 80$ (11.8 at.%). Therefore, the PL emission band at 490 nm (~2.53 eV) observed in the as-deposited sample with $R_N = 80$ could be related to the electronic transition from the K^0 centers to =Si–O–Si states.

In summary, the analysis of the PL emission observed in the SRN films before and after thermal annealing indicates that it could be mainly originated from the radiative recombination via luminescent Si dangling bonds, N dangling bonds, and Si–O bonds existing in the silicon nitride matrix.

4.2. Silicon-rich oxide (SRO) film

Figure 5 shows the normalized PL spectra of SRO films before and after thermal annealing. The as-deposited SRO film with $Ro = 30$ (M30) emits a broad PL band with the main peak at 460 nm. After thermal annealing, the PL emission band becomes narrower and redshifts reaching 690 nm, as shown in **Figure 5(a)**. As the *Ro* value decreases, the PL emission band becomes wider and shifts to a higher wavelength reaching 660 nm for $Ro = 20$ (M20, higher silicon excess), as shown in **Figure 5(b)**. After thermal annealing, the PL of M20 redshifts, reaching 740 nm.

The PL emitted by M30 exhibit a significant redshift of the main peak after thermal annealing as shown in **Figure 5(a)**. Nevertheless, when the silicon content is increased (M20), the PL band appears mainly at the red side of the spectrum. The redshift of the main PL peak (after thermal annealing) has been widely observed and ascribed to the agglomeration of silicon excess and a subsequent silicon nanoparticle formation as a result of thermal annealing process [67–71]. Nevertheless, some point defects are also present within the SRO films. It is widely accepted that violet-blue (400–460 nm), green (520), and even red (630) emission bands, obtained from the deconvolution of the PL spectrum [72], can be related with oxygen

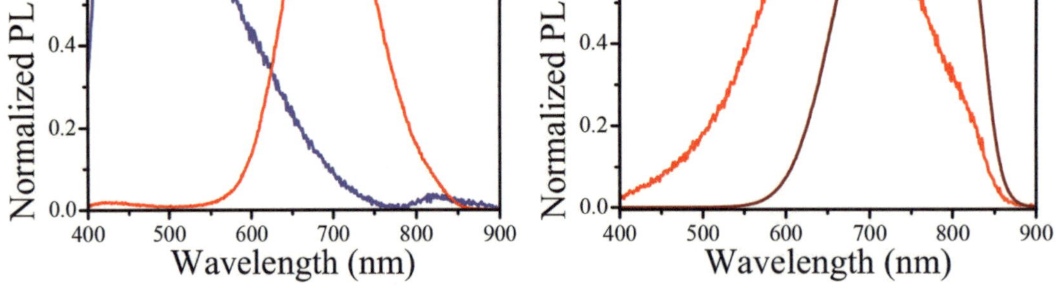

Figure 5. Normalized PL spectra of SRO monolayer (a) with $Ro = 30$ (M30) and (b) $Ro = 20$ (M20) before and after thermal annealing. From Palacios-Huerta et al. [46].

defect centers (ODC), E'_δ (Si↑Si=Si) and NBOHC defects, respectively [67–71]. The E'_δ center is one of the at least four different E′ centers [43], which comprises an unpaired spin delocalized over five silicon atoms and suggest the presence of very small Si-nps in the films.

4.3. Silicon-rich nitride/silicon-rich oxide (SRN/SRO) bilayers

In previous studies, it has been reported that the combination of Si_3N_4/SRO structure improves luminescent emission properties [73, 74]. Previous studies have also shown that a Si_3N_4-SRO bilayer structure improves the operation of light-emitting devices, such as a reduced leakage current, a reduced electric field on the oxide layer, and results an improvement in efficiency and a longer device life [73–76]. In this chapter, the effect of a SRN film on a SRO film (SRN/SRO bilayers) on their optical properties is analyzed. SRN films, deposited by LPCVD with $R_N = 80$ (N80), show luminescence in the blue region, while the SRO films emit in the red region, as observed in **Figures 4** and **5**, respectively. Therefore, a combination of SRN with $R_N = 80$ and SRO films could allow reaching a broad emission spectrum.

Figure 6 shows the PL spectra of SRN/SRO bilayers with $Ro = 20$ (B20) and 30 (B30) after thermal annealing. The emission spectra of the SRN (N80) and SRO monolayer films with $Ro = 20$ (M20) and 30 (M30) are shown as references, respectively. The emission spectra of the samples were normalized to the SRO thickness. Each PL spectrum of SRN and SRO

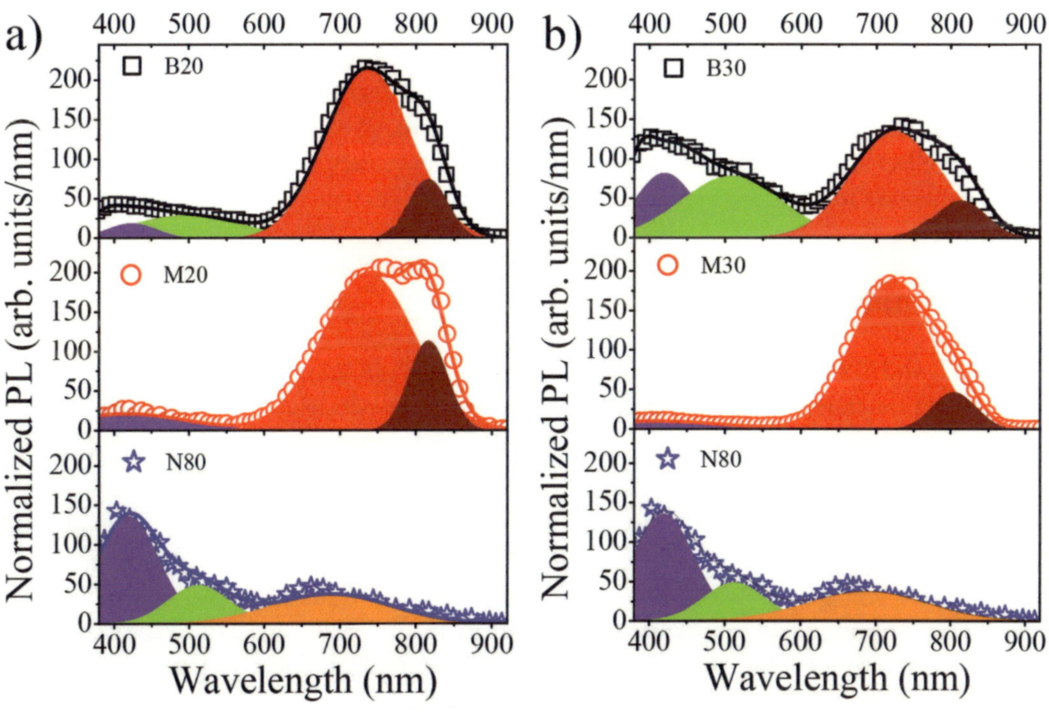

Figure 6. Normalized PL spectra after thermal annealing of (a) SRN/SRO film with $Ro = 20$ (B20), SRN film with $R_N = 80$ (N80), and SRO with $Ro = 20$ (M20), and (b) SRN/SRO film with $Ro = 30$ (B30), SRN film with $R_N = 80$ (N80), and SRO with $Ro = 30$ (M30).

monolayers was fitted three Gaussian, while four Gaussians were used to fit the PL spectrum of the SRO/SRN bilayer.

For the SRN film (N80), PL bands at about ~420, 505, and 680 nm were identified, which are related to electronic transitions from the K^0 centers to the VBM, K^0 centers to =Si–O–Si, and states of defects nitrogen (N^0_2), respectively [64]. For the SRO film with *R*o = 20 (M20), PL bands at ~420, 735, and 820 nm were identified and have been widely related to weak oxygen bond (WOB) defects, interactions that take place at the Si-np/SiOx interface, and quantum confinement in silicon nanocrystals, respectively [72]. The PL spectrum of the SRN/SRO bilayer with *R*o = 20 (B20) exhibits the same PL bands of the SRO films at ~420, ~735, and 812 nm, but with an additional band at ~505 nm (green band), which is related to transitions from K^0 centers to =Si–O–Si featured of the SRN film, as shown in **Figure 6(a)**.

As observed in **Figure 6(b)**, the SRO monolayer with *R*o = 30 (M30) exhibits PL bands at ~420, 720, and 805 nm, which have been also linked to WOB defects, interactions that take place at the Si-nps/SiOx interface, and silicon nanocrystals (Si-ncs) through quantum confinement, respectively [72]. The PL spectrum of the SRN/SRO with *R*o = 30 (B30) is composed by the same luminescent bands than that observed in the bilayer with *R*o = 20 (B20), but with an increased intensity of the bands at ~420 and ~505 nm related with transitions from K^0 centers to VBM (violet band), transitions from K^0 centers to =Si–O–Si (green band), respectively.

In summary, it is observed that the emission intensity in the blue and green bands (~420 and 505 nm) is improved when a SRN/SRO bilayer is formed compared to SRO monolayers, being higher when a *R*o = 30 (B30) is used. The XPS results show a diffusion of oxygen from the SRO layer to the SRN layer and nitrogen from the SRN layer to the SRO layer during SRN deposition (onto SRO layer). Therefore, some defects could be passivated into the SRO layer and others generated into the SRN layer like WOB or =Si–O–Si defects.

5. Electroluminescence

SRN, SRO, and SRN/SRO films exhibit intense and visible photoluminescence. In this section, luminescent characteristics of the samples are present but under electrical excitation.

5.1. Silicon-rich nitride (SRN)-LECs

Figure 7 shows the EL spectra of the light emitting capacitors using the as-deposited SRN film with R_N = 20 as an active dielectric layer at different injected currents. At forward bias (FB) (positive voltage to FTO contact respect to Si substrate), the SRN-LEC shows a broad spectrum with the maximum emission centered at around 580 nm, and it remains at the same wavelengths for different voltages as shown in **Figure 7(a)**. This luminescence has been reported as a characteristic behavior of defect-related EL [67]. In fact, the luminescent properties of SRN films in this chapter are related to the presence of defects. Therefore, the EL in SRN-LECs may originate from the same electronic transitions of the CBM to K^0 centers,

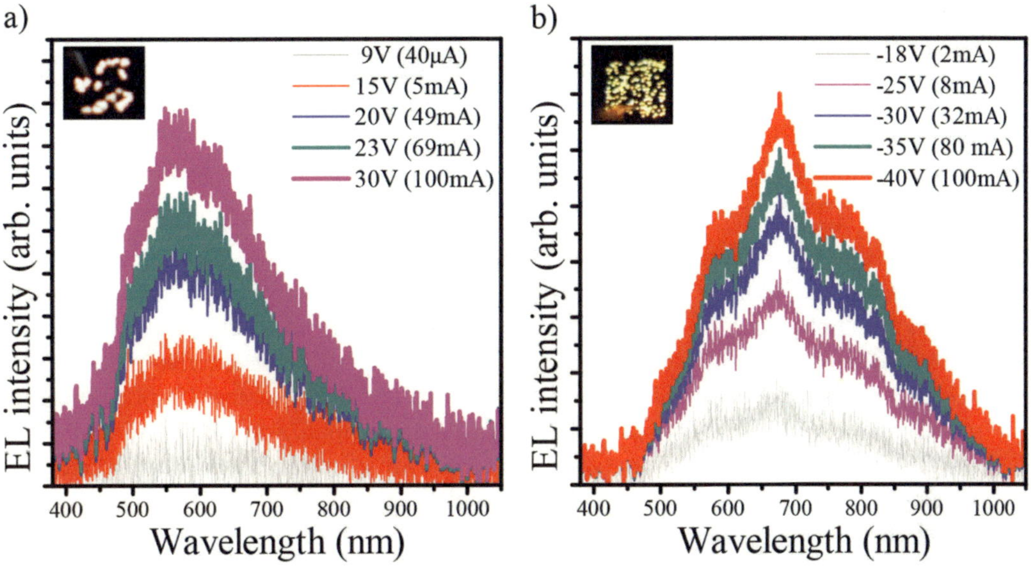

Figure 7. EL spectra of the as-deposited SRN LEC with $R_N = 20$ for different injected currents at (a) forward bias and (b) reverse bias. In the inset, pictures for the device with 100 mA of injected current for each bias. FTO was used as gate electrode. From Cabañas-Tay et al. [34].

as observed in PL [77, 78]. This behavior is similar for all SRN films in both as-deposited and after thermal annealing, as reported in the study of Cabañas-Tay et al. [34].

On the other hand, at reverse bias (RB), the EL spectrum changes and now three main emission bands are observed at around 600, 680, and 780 nm, as observed in **Figure 7(b)**. Namely, the maximum emission band blueshifts when the polarity is changed from RB to FB. The EL bands at 600, 680, and 780 nm also remains at the same wavelength when the voltage is increased indicating the EL emission is also produced by defects. These EL bands have been ascribed to electronic transitions from the K^0 centers to valence band tail states [58, 77]. The EL emission for all the SRN-LECs in both polarities is through shine dots as shown in the inset of **Figure 7(a)** and **(b)**. The device's area is covered with more shine dots when the current increases. The EL emission composed by shining spots is attributed to the formation of a finite number of preferential conductive paths within the SRN films, which connect the top and bottom electrodes, as discussed in the study of Cabañas-Tay et al. [34].

Negligible spectral shift is observed between the EL at forward bias and PL spectra of SRN films with $R_N = 5$ and 20 before and after thermal annealing, indicating that both PL and EL emissions are originated from the same radiative centers [34]. Nevertheless, a long spectral shift is observed for the SRN films with $R_N = 80$ (before and after thermal annealing). This behavior was explained as a significant smaller electrical pumping of electrons allowing the holes to relax to the lowest defect states before recombining. For the EL at RB, a similar behavior is presented but the electrical pumping of electrons is slightly improved allowing only few of the holes to relax to the lowest defect states before recombining [34].

5.2. Silicon-rich oxide (SRO)-LECs

The presence of defects including the Si-nps, either crystalline or amorphous, and their density and size in silicon-rich dielectric materials affect clearly the current transport, and therefore the EL, as in the SRO case. **Figure 8** shows J-E curves of SRO-LECs with *R*o = 30 and 20; samples were thermally annealed at 1100°C. SRO-LECs are forwardly biased (accumulation mode) considering the substrate as reference.

SRO-LECs with *R*o = 30 (M30) films show a high current state (HCS) at low electric fields, then after the applied voltage increased, the current was switched to an LCS, as shown in **Figure 8**. This behavior was observed by our group before in SRO films with *R*o =30, and it was related to the creation and annihilation of preferential conductive paths generated possibly by adjacent stable Si-nps and unstable silicon nanoclusters (Si-ncls) through structural changes and by the possible creation of defects (breaking off Si–Si bonds) [70, 79, 80]. Indeed, a clear correlation between current jumps/drops and EL dots appearing/disappearing on the LEC surface was observed [70, 79, 80]. This RS behavior is independent on the thermal annealing temperature [46]. Recent studies regarding the same electrical RS behavior in SRO films was observed and related with a conductive filament formed by Si-nps [81–84], which undergoes structural changes through a crystallization and amorphization process of the Si-nps, as discussed in [82]. Such observations are in agreement with the presented asseverations about the behavior of the SRO with *R*o = 30 (M30)-based LECs.

Once the current fluctuations disappear, through the electrical annealing, the current behavior stabilized (see I–V curve marked as M30-after in dark blue line, **Figure 8**) and EL on the whole area (WA EL) was observed at higher electric fields. On the other hand, the electrical behavior of SRO-LEC with *R*o = 20 (M20) did not show current fluctuations. The latter could

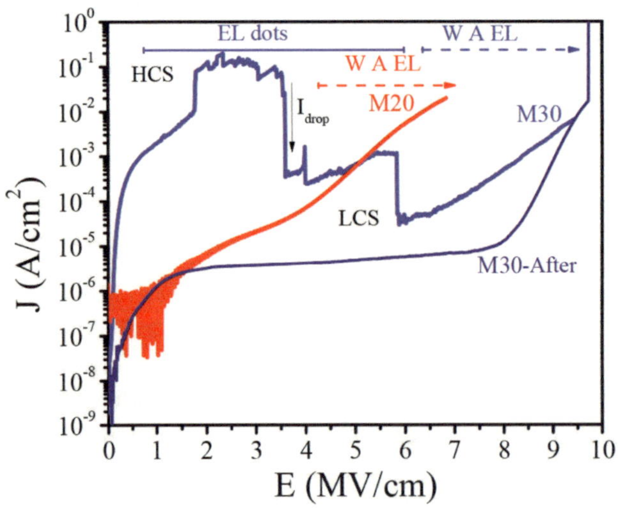

Figure 8. J-E curve of SRO-LECs with *R*o = 20 (M20) and 30 (M30) thermally annealed at 1100°C. From Palacios-Huerta et al. [46].

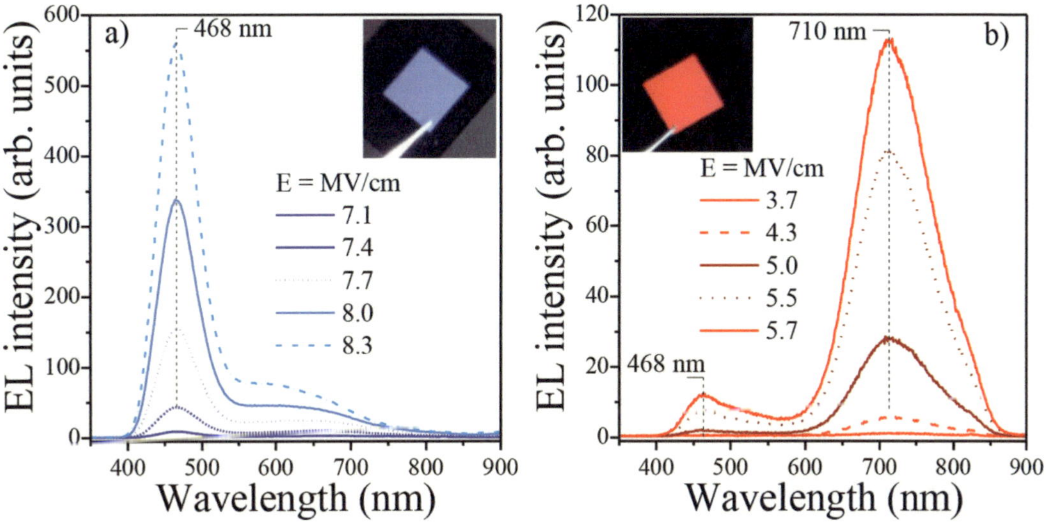

Figure 9. EL spectra and images of SRO-LECs with (a) *R*o = 20 (M20) and (b) 30 (M30) annealed at 1100°C. Semitransparent n+ polycrystalline silicon was used as gate electrode. From Palacios-Huerta et al. [46].

be related to the presence of well-separated and crystalline silicon nanoparticles, mainly on the density of Si-nps, as shown by transmission electron microscopy (TEM) analysis [46]. Therefore, a uniform network of conductive paths becomes possible as the Si-nps density increases, allowing a uniform charge flow through the whole capacitor area.

Figure 9 shows the EL spectra from the SRO-LECs. Blue EL in the whole area was observed with *R*o = 30 (M30) only after the current drop. The main EL peak remains at 468 nm when the electric field increases, as observed in **Figure 9(a)**. When the silicon content is increased (*R*o = 20), the LECs emit a broad EL spectrum in the red region, as observed in **Figure 9(b)**. The EL peak is placed at 710 nm. This value is blueshifted with respect to the value found in PL at 740 nm. Differences in the PL and EL peak wavelength have been related to the transmittance of the top electrode, which influences the real light emission from the active layer [85]. Indeed, the process of carrier injection and the several transport mechanism taking place through the active luminescent materials produces a complicated understanding of the EL process, as compared to the PL one [86]. Nevertheless, both EL and PL spectra appear in the same red region, which could indicate that the emission is originated by the same radiative centers. Images of LECs are shown as insets in **Figure 9**. As can be observed, the LEC devices emit an intense whole area EL.

5.3. Silicon-rich nitride/silicon-rich oxide (SRN/SRO)-LECs

Figure 10 shows the J(E) characteristic and the EL spectra of SRN/SRO-LECs at forward bias considering the substrate like reference. SRN/SRO-LECs show a high current state (HCS) at low electric fields, and then after the applied voltage increases, the current is switched to a LCS, as shown in **Figure 10(a)**. The resistive switching in the B20 bilayer occurs at a lower electric field (~2 MV/cm) compared to the B30 bilayer. When SRN/SRO-LECs, both with *R*o = 20

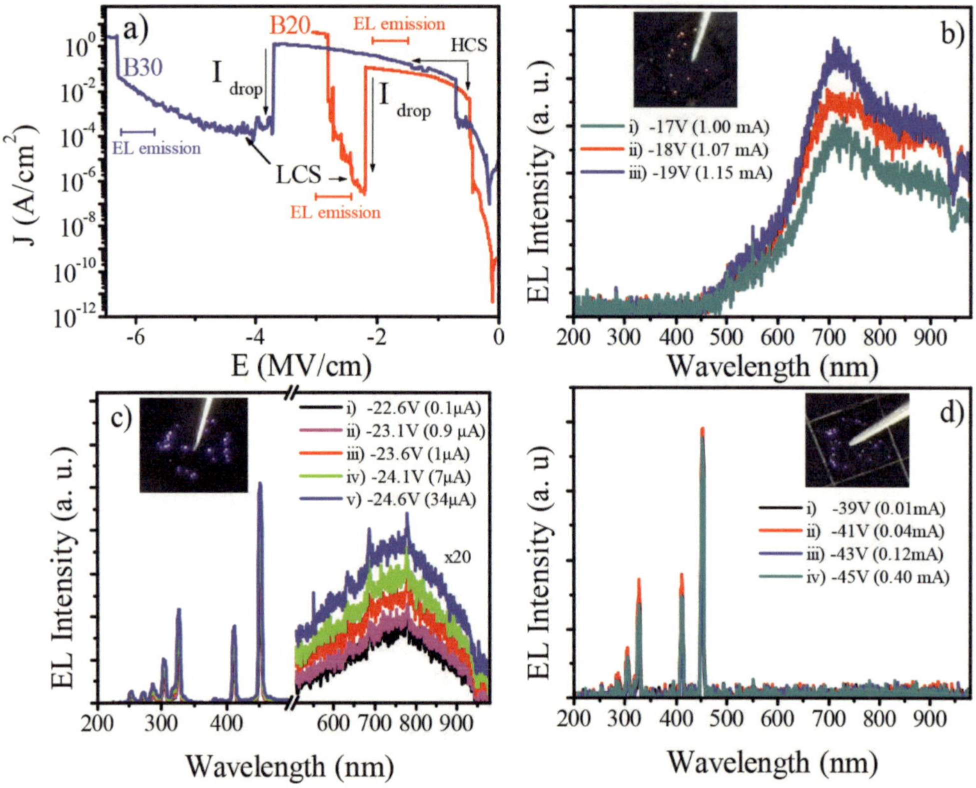

Figure 10. (a) J-E curve of SRN/SRO-LECs with *Ro* = 20 (B20) and 30 (B30) and EL spectra of SRN/SRO LECs at forward bias for different currents applied with (b) *Ro* = 20 (B20) in the HCS, (c) *Ro* = 20 (B20) in the LCS, and (d) *Ro* = 30 (B30) in the LCS. Inset of each graph are shown photographs of the LEC. ITO was used as gate contact.

(B20) and *Ro* = 30 (B30), are in the HCS (<20V), a broad spectrum with the maximum emission centered at about 730 nm and a shoulder at ~900 nm is observed. The 730 nm emission increases as the voltage also increases, as observed in **Figure 10(b)**. Nevertheless, first the shoulder at 900 nm increases its intensity when the voltage is between −17 and −18 V and then remains at the same value for larger voltages. As we can see in **Figure 10(b)**, the EL is observed as randomly scattered luminescent spots (mainly red) over the entire LEC surface. The spectrum exhibits that the EL emission follows the PL, showing that the emission is caused by the same luminescent centers.

After current switching, in the LCS (>20V), the EL spectra changes, as observed in **Figure 10(c)**. Also, a narrow (width of 7 ± 0.6 nm) and highly intense UV EL peaks appear at ~250, 270, 285, 305, 325, and 415 nm. A narrow blue EL peak with the highest intensity is observed at ~450 nm. All EL peaks remain at the same wavelength as the current increases. The red EL band observed in the HCS is still present at the LCS, but with a slight redshift and with a low intensity. The emission intensity of the blue EL is about 20 times higher than the red emission. The EL emission is observed as luminescent blue-violet dots randomly distributed on the surface of the contact, as shown in the insets of **Figure 10(c)**.

Figure 10(d) shows the EL emission of the SRN/SRO-LEC with *R*o = 30 (B30) in the range of LCS (>35V). The same narrow and highly intense UV EL peaks at ~250, 270, 285, 305, 325, 415, and 450 nm with an average width of 7 ± 0.6 nm are also observed in these devices. Nevertheless, these LECs do not show EL emission in the red-near infrared region.

Four narrow UV EL peaks have been reported at 293.78, 316.10, 403.07, and 444.82 nm, with an average width at half peak of 4 nm, in ITO/Y_2O_3/Ag EL devices. This emission was attributed to the characteristic radiation of indium ions [87]. However, as the best of our knowledge, narrow UV EL peaks in silicon rich dielectric materials have not been reported before. The most intense EL peaks emitted by the present SRN/SRO-LECS (~305, 325, 415, and 450 nm) are very similar to those obtained in reference [87], but displaced ~10 nm toward longer wavelengths. Thus, they could have a similar origin; however, a deeper analysis of these narrow emission bands needs to be done.

6. Conduction mechanisms

The origin of the EL emission from the SRO, SRN, and SRN/SRO-LECs can be determined by identifying the charge transport mechanism that takes place within the different silicon-rich dielectric materials as discussed below.

6.1. Silicon-rich nitride (SRN)-LECs

Figure 11 shows the J(E) dependency for the SRN-LECs with R_N = 5, 20, and 80. The conduction mechanism of the carriers in the SRN films was studied by analyzing the different reported mechanism including P-F conduction [88, 89], TAT [90], and space charge limited current (SCLC) [91]. As observed in **Figure 11**, the TAT mechanism is dominating the carrier transport at both forward and reverse bias, and it can be ascribed to the defect states generated during the SRN deposition. A model based on the trap-assisted tunneling carrier transport is shown with the EL radiative recombination process in the reference [34].

6.2. Silicon-rich oxide (SRO)-LECs

Figure 12 shows the experimental J-E data from SRO-LECs with *R*o = 20 and 30 fitted to the carrier transport mechanisms including Poole-Frenkel (P-F) conduction [88, 89] and trap-assisted tunneling (TAT) [90]. As can be observed, the TAT conduction mechanism predominates in the SRO-LEC with *R*o = 30 (M30).

The trap energy was estimated to be $\phi_t \cong 1.75$ eV, if a uniformly distributed trap concentration is assumed. It is worth to mention that these trap energy levels correlates well with the one obtained for Er-doped SRO/SiO_2 superlattice (2.1 eV) and Er-doped SiO_2 (1.9 eV), which are assumed to be deep traps inherent to the SiO_2 [92] possibly neutral oxygen vacancy [93]. Mehonic et al. have reported that the barrier height of the trap is modified by the concentration of oxygen vacancies in the connecting tissue; lower barrier heights correspond to a higher concentration of oxygen vacancies, which defines the device resistance [84]. Moreover, it has been reported that a poor quality of nanocrystalline composite material (formed by Si

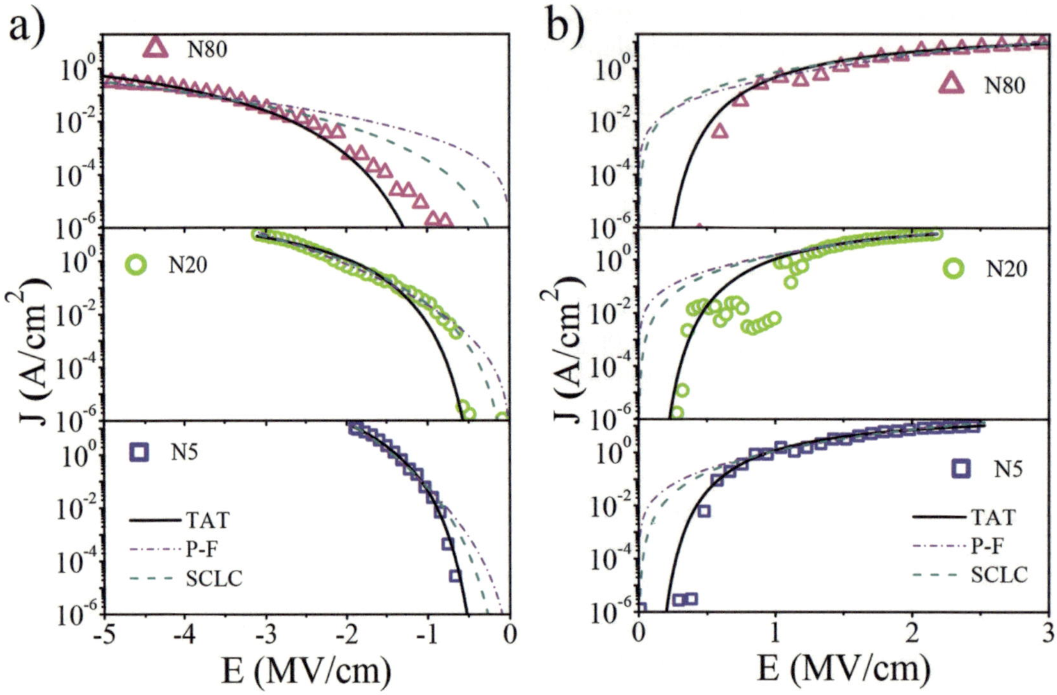

Figure 11. J-E characteristic and conduction mechanism fitting of SRN-LECs with $R_N = 5$ (N5), 20 (N20), and 80 (N80) after thermal annealing at: (a) reverse bias (RB) and (b) forward bias (FB). Symbols are the experimental current data, black line represents the TAT conduction, green dash line SCLC, and violet dash dot line the Poole-Frenkel conduction. From Cabañas-Tay et al. [34].

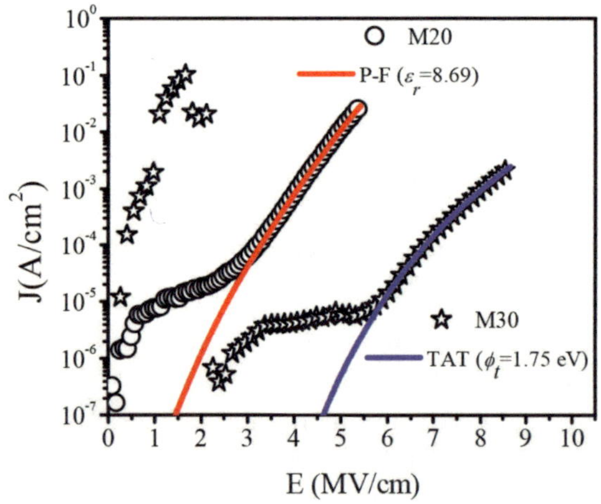

Figure 12. J-E characteristic and conduction mechanism fitting of SRO-LECs with $Ro = 30$ (M30) and $Ro = 20$ (M20). Red line represents the TAT conduction and blue line the Poole-Frenkel conduction. Symbols are the experimental data. From Palacios-Huerta et al. [46].

nanocrystals and SiOx) yields parasitic current paths, while low density of nanocrystals turns difficult the direct charge injection into the nanocrystals [94]. A similar behavior is obtained for SRO-LECs with *R*o = 30 (M30), where the annihilation of conductive paths, created by adjacent stable Si-nps and unstable silicon nanoclusters (Si-ncls), produce structural changes including the possible creation of defects. Then, it can be assumed that M30 films exhibit a poor quality of the silicon oxide resulting from the phase separation and the low silicon content. This behavior is well correlated to the blue EL emission observed in the M30-based LECs.

On other hand, the P-F conduction fits well the charge transport in SRO-LECs with *R*o = 20 (M20), as observed in **Figure 12**. A relative permittivity (ε_r) value of 9.16 was obtained from the P-F fit for M20, which is closer to the relative permittivity of silicon (ε_{Si} = 11.9), similar to other reports [92]. Relatively high permittivity values are a good indication of the large amount of silicon present as Si-nps within the M20 films. Moreover, the Si-nps size calculated by using the P-F estimate relative permittivity are very close to that obtained through the TEM analysis, as reported in [46].

6.3. Silicon-rich nitride/silicon-rich oxide (SRN/SRO)-LECs

When the SRN/SRO bilayer structure is used as active layer in LECs, several charge transport mechanisms are present, as observed in **Figure 13**. For the SRN/SRO-LEC with *R*o = 20 (B20), and in the HCS where red EL emission is observed, the hopping tunneling is the predominant mechanism (pink line), as shown in **Figure 13**(a).

The average distance between traps (*a*) was estimated to be about 0.25 nm. When the current switches to the low conduction state, the tunneling mechanism changes, then the P-F conduction

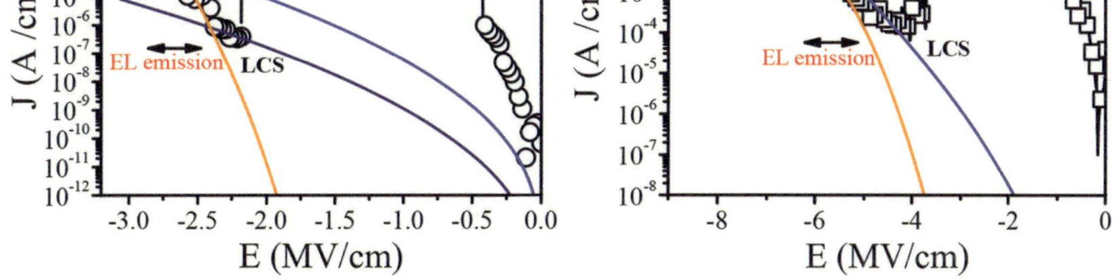

Figure 13. J-E characteristic and conduction mechanism fitting of SRN/SRO-LECs with (a) *R*o = 20 (B20) and (b) *R*o = 30 (B30) thermally annealed at 1100°C. Pink line represents the hopping conduction, orange line the TAT conduction, and blue lines the Poole-Frenkel conduction. Symbols are the experimental data.

dominates (dark blue line). The dielectric constant (ε_r) obtained by the P-F fit was about 7.96. However, when the deep UV EL emission starts, the conduction transport is dominated by two different mechanisms: TAT (orange line) and P-F (blue line). In this region, the traps depth obtained by the TAT fit was about 2.06 eV, and the dielectric constant (ε_r) obtained by the P-F fit was about 7.81.

The SRN/SRO-LECs with $Ro = 30$ (B30) behaves similar to the B20 bilayer, as observed in **Figure 13(b)**. In the high conduction state, the dominant mechanism is hopping tunneling (pink line), where the average distance between traps is about 0.15 nm, indicating a higher density of traps compared to the B20 sample (0.25 nm). When the current drops to the low conduction state, the dominating mechanism changes to the P-F conduction (blue line). The dielectric constant (ε_r) was estimated at 5.98, indicating a lower content of silicon related to B20 sample (8.61), as expected. However, when the EL emission starts, the conduction begins to be dominated by TAT (orange line). The traps depth was estimated at about 2.15 eV, very similar to that obtained in B20 sample (2.06) and very close to the depth at which are the centers K^0 (~2.1 eV) [64].

In summary, the conduction mechanism in the region of low electric fields at HCS is hopping for SRN/SRO bilayers. It was observed that when the nitrogen content in the SRO layer increases, the average distance between defects also increases either due to the defects passivation or due to the reduction of the Si-nps size. In the region of high electric fields at low conduction, both TAT and P-F mechanisms take place simultaneously. The trap depth, obtained by the TAT fit, was ~2.1 eV from the minimum of the conduction band for both bilayers (B20 and B30), which relates to the location of the centers K^0 in SRN film.

7. Conclusions

The compositional, structural, optical, and electro-optical properties of SRO, SRN films as well as combination of SRN/SRO bilayers deposited by LPCVD were studied. The EL of the SRN-LECs showed a broad emission spectrum where the maximum peak blueshifts when the polarity changed from reverse to forward bias. The EL spectrum was nearly similar to that of PL when LECs were forwardly biased and the silicon excess was increased. Analyzing the current-voltage characteristics, it was found that TAT was the main carrier transport mechanism in SRN films in both biases, where typical EL was observed. In SRO films, it was demonstrated that the silicon content affects the luminescence centers density obtaining the EL emission at lower electric field as the silicon excess increased. SRO with $Ro = 30$ films exhibited the presence of structural defects like NOV, NBOHC, and E'_δ (Si↑Si=Si) centers, which produced blue to red PL band after thermal annealing. SRO with $Ro = 20$ films showed mainly red PL produced by the presence of well-defined Si-ncs. It was found that SRO-LECs with $Ro = 30$ shows an intense blue EL, after a resistance switching behavior was reached. The RS was originated by the annihilation of preferential conductive paths created by Si-nps and silicon nanoclusters through structural changes that result in a long spectral blueshift (~222 nm) between the PL and EL. On the other hand, the RS was not observed in

SRO-LECs with *R*o = 20 due to the well-defined Si-nps with 2.7 nm size. These well-defined Si-nps produce a better charge injection through multiple conductive paths enhancing an intense red EL, which is related mainly to defects at the Si-nps/SiOx interface. The transport mechanism was affected by the composition and structure of the films, being TAT and PF the dominant transport mechanism in SRO with *R*o = 30 and 20, respectively. When the SRN was deposited onto the SRO to form the bilayer, compositional and electro-optical changes were observed. The SRN/SRO layers showed a broad PL emission respect to a simple SRO or SRN layer. In addition, multiple defects or centers were responsible of PL emission. The same luminescent centers in SRN/SRO were present in SRO and SRN monolayers. SRN/SRO-LECs showed an intense UV EL at high electric fields related to indium ions radiation. The conduction mechanism in SRN/SRO bilayers was found to be dominated by Pool-Frenkel and TAT mechanisms at LCS, and hopping at HCS.

Acknowledgements

This work has been partially supported by the project CONACyT-180992. The authors acknowledge technicians Pablo Alarcon, Armando Hernández, and Victor Aca from INAOE and Luis Gerardo Silva from CIMAV.

Author details

Santiago A. Cabañas-Tay[1], Liliana Palacios-Huerta[2], Mariano Aceves-Mijares[2], Antonio Coyopol[1], Sergio A. Pérez-García[1], Liliana Licea-Jiménez[1], Carlos Domínguez[3] and Alfredo Morales-Sánchez[1]*

*Address all correspondence to: alfredo.morales@cimav.edu.mx

1 Advanced Materials Research Center (CIMAV), Monterrey-PIIT Apodaca, Nuevo León, México

2 Electronics Department, National Institute of Astrophysics Optics and Electronics Tonantzintla, Puebla, México

3 Microelectronics Institute of Barcelona (IMB-CNM, CSIC) Bellaterra, Barcelona, Spain

References

[1] Tseng C.-K., Lee M.-C. M., Hung H.-W., Huang J.-R., Lee K.-Y., Shieh J.-M., and Lin G.-R. Silicon-nanocrystal resonant-cavity light emitting devices for color tailoring. Journal of Applied Physics. 2012;111(7): 074512. DOI: 10.1063/1.3702793

[2] Zhiping Z., Bing Y., and Jurgen M. On-chip light sources for silicon photonics. Light: Science & Applications. 2015;4: e358. DOI: 10.1038/lsa.2015.131

[3] Berencén Y., Illera S., Rebohle L., Ramírez J. M., Wutzler R., Cirera A., Hiller D., Rodríguez J. A., Skorupa W., and Garrido B. Luminescence mechanism for Er^{3+} ions in a silicon-rich nitride host under electrical pumping. Journal of Physics D: Applied Physics. 2016;49(8): 085106. DOI: 10.1088/0022-3727/49/8/085106

[4] Canham L. T. Silicon quantum wire array fabrication by electrochemical and chemical dissolution of wafers. Applied Physics Letters. 1990;57:1046–1048. DOI: 10.1063/1.103561

[5] Delerue C., Allan G., and Lannoo M. Theoretical aspects of the luminescence of porous silicon. Physical Review B. 1993;48(15):11024–11036. DOI: 10.1103/PhysRevB.48.11024

[6] Barreto J., Morales A., Perálvarez M., Garrido B., and Domínguez C. Stoichiometry of silicon-rich dielectrics for silicon nanocluster formation. Physica Status Solidi (C). 2011;8(3):1610–1642. DOI: 10.1002/pssc.201000363

[7] Shih C.-F., Hsiao C.-Y., and Su K.-W. Enhanced white photoluminescence in silicon-rich oxide/SiO_2 superlattices by low-energy ion-beam treatment. Optics Express. 2013;21(13): 15888–15895. DOI: 10.1364/OE.21.015888

[8] Mohammed S., Nimmo M. T., Malko A. V., and Hinkle C. L. Chemical bonding and defect states of LPCVD grown silicon-rich Si_3N_4 for quantum dot applications. Journal of Vacuum Science & Technology A. 2014;32(2): 021507. DOI: 10.1116/1.4861338

[9] Liao W., Zeng X., Wen X. and Wen Y. Tunable photoluminescence of Si-rich nitride films with silicon quantum dots by changing the total pressure. International Photonics and Optoelectronics. 2015; JW3A.33. DOI: 10.1364/OEDI.2015.JW3A.33

[10] Vlasukova L. A., Komarov F. F., Parkhomenko I. N., Milchanin O. V., Makhavikou M. A., Mudryi A. V., Żuk, J., Kopychiński P., and Togambayeva A. K. Visible photoluminescence of non-stoichiometric silicon nitride films: The effect of annealing temperature and atmosphere. Journal of Applied Spectroscopy. 2015;82(3): 386–389. DOI: 10.1007/s10812-015-0117-9

[11] Lin G.-R., Lin C.-J., and Lin C.-K. Enhanced Fowler-Nordheim tunneling effect in nanocrystallite Si based LED with interfacial Si nano-pyramids. Optics Express. 2007;15(5):2555–2563. DOI: 10.1364/OE.15.002555

[12] Lin Y.-H., Wu C.-L., Pai Y.-H., and Lin G.-R. A 533-nm self-luminescent Si-rich SiNx/SiOx distributed Bragg reflector. Optics Express. 2011;19(7):6563–6570. DOI: 10.1364/OE.19.006563

[13] Nishi H., Tsuchizawa T., Shinojima H., Watanabe T., Itabashi S-I., Kou R., Fukuda H., and Yamada K. Low-polarization-dependent silica waveguide monolithically integrated on SOI photonic platform. Journal of Lightwave Technology. 2013;31(11):1821–1827. DOI: 10.1109/JLT.2013.2256880

[14] Cheng C.-H., Lien Y.-C., and Wu C.-L., and Lin G.-R. Mutlicolor electroluminescent Si quantum dots embedded in SiOx thin film MOSLED with 2.4% external quantum efficiency. Optics Express. 2013;21(1): 391–403. DOI: 10.1364/OE.21.000391

[15] Zimmermann H., editors. Integrated Silicon Optoelectronics. 2nd ed. Germany: Springer-Verlag Berlin Heidelberg; 2010. 406 p. DOI: 10.1007/978-3-642-01521-2

[16] Pavesi L., Gaponenko S., and Dal Negro L., editors. Towards the First Silicon Laser. 1st ed. Dordrecht: Kluwer Academic Publishers; 2003. 490 p. DOI: 10.1007/978-94-010-0149-6

[17] Klak M. M., Zatryb G., Wojcik J., Misiewicz J., Mascher P., and Podhorodecki A. Mechanism of enhanced photoluminescence of Tb ions in hydrogenated silicon-rich silicon oxide films. Thin Solid Films. 2016;611: 62–67. DOI: 10.1016/j.tsf.2016.04.050

[18] Barbagiovanni E. G., Lockwood D. J., Simpson P. J., and Goncharova, L. V. Quantum confinement in Si and Ge nanostructures. Journal of Applied Physics. 2012;111(3): 034307. DOI: 10.1063/1.3680884

[19] Iacona F., Franzò G., and Spinella C. Correlation between luminescence and structural properties of Si nanocrystals. Journal of Applied Physics. 2000;87(3): 1295–1303. DOI: 10.1063/1.372013

[20] Kim T.-W., Cho C.-H., Kim B.-H., and Park S.-J. Quantum confinement effect in crystalline silicon quantum dots in silicon nitride grown using SiH_4 and NH_3. Applied Physics Letters. 2006;88(12): 123102. DOI: 10.1063/1.2187434

[21] Qin G. G., Li A. P., Zhang B. R., and Li B.-C. Visible electroluminescence from semitransparent Au film/extra thin Si-rich silicon oxide film/p-Si structure. Journal of Applied Physics. 1995;78(3): 2006–2009. DOI: 10.1063/1.360175

[22] Lin G.-R., and Lin C.-J. Improved blue-green electroluminescence of metal-oxide-semiconductor diode fabricated on multirecipe Si-implanted and annealed SiO_2/Si substrate. Journal of Applied Physics. 2004;95(12): 8484–8486. DOI: 10.1063/1.1739283

[23] Liu Y., Chen T. P., Ding L., Yang M., Wong J. I., Ng C. Y., Yu S. F., Li Z. X., Yuen C., Zhu F. R., Tan M. C., and Fung S. Influence of charge trapping on electroluminescence from Si-nanocrystal light emitting structure. Journal of Applied Physics. 2007;101(10): 104306. DOI: 10.1063/1.2713946

[24] Liao L.-S., Bao X.-M., Li N.-S., Zheng X.-Q., and Min N.-B. Visible electroluminescence from Si^+-implanted SiO_2 films thermally grown on crystalline Si. Solid State Communications. 1996;97(12): 1039–1042. DOI: 10.1016/0038-1098(95)00846-2

[25] Wang T., Wei D. Y., Sun H. C., Liu Y., Chen D. Y., Chen G. R., Xu J., Li W., Ma Z. Y., Xu L., and Chen K. J. Electrically driven luminescence of nanocrystalline Si/SiO_2 multilayers on various substrates. Physica E: Low-dimensional Systems and Nanostructures. 2009;41(6): 923–926. DOI: 10.1016/j.physe.2008.08.001

[26] López-Vidrier J., Berencén Y., Hernández S., Blázquez O., Gutsch S., Laube J., Hiller D., Löper P., Schnabel M., Janz S., Zacharias M., and Garrido B. Charge transport and electroluminescence of silicon nanocrystals/SiO_2 superlattices. Journal of Applied Physics. 2013;114(16): 163701. DOI: 10.1063/1.4826898

[27] Cho K. S., Park N.-M., Kim T.-Y., Kim K.-H., Shin J.-H., Sung G. Y., and Shin J. H. High efficiency visible electroluminescence from silicon nanocrystals embedded in silicon nitride using a transparent doping layer. Applied Physics Letters. 2006;88(20): 209904. DOI: 10.1063/1.2205754

[28] Huang R., Chen K., Qian B., Chen S., Li W., Xu J., Ma Z., and Huang X. Oxygen induced strong green light emission from low-temperature grown amorphous silicon nitride films. Applied Physics Letters. 2006;89(22): 221120. DOI: 10.1063/1.2399393

[29] Huang R., Chen K., Dong H., Wang D., Ding H., Li W., Xu J., Ma Z., and Xu L. Enhanced electroluminescence efficiency of oxidized amorphous silicon nitride light-emitting devices by modulating Si/N ratio. Applied Physics Letters. 2007;91(11): 111104. DOI: 10.1063/1.2783271

[30] Cen Z. H., Chen T. P., Liu Z., Liu Y., Ding L., Yang M., Wong J. I., Yu S. F., and Goh W. P. Electrically tunable white-color electroluminescence from Si-implanted silicon nitride thin film. Optics Express. 2010;18(19): 20439–20444. DOI: 10.1364/OE.18.020439

[31] Warga J., Li R., Basu S. N., and Dal Negro L. Electroluminescence from silicon-rich nitride/silicon superlattice structures. Applied Physics Letters. 2008;93(15): 151116. DOI: 10.1063/1.3003867

[32] Tai H. Y., Lin Y. H., and Lin G. R. Wavelength-shifted yellow electroluminescence of Si quantum-dot embedded 20-pair SiNx/SiOx superlattice by Ostwald ripening effect. IEEE Photonics Journal. 2013;5(1): 6600110. DOI: 10.1109/JPHOT.2012.2232285

[33] González-Fernández A. A., Juvert J., Aceves-Mijares M., and Domínguez C. Monolithic integration of a silicon-based photonic transceiver in a CMOS Process. IEEE Photonics Journal. 2016;8(1):1–13. DOI: 10.1109/JPHOT.2015.2505144

[34] Cabañas-Tay S. A., Palacios-Huerta L., Luna-López J. A., Aceves-Mijares M., Alcántara-Iniesta S., Pérez-García S. A., and Morales-Sánchez A. Analysis of the luminescent centers in silicon rich silicon nitride light-emitting capacitors. Semiconductor Science and Technology. 2015;30(6):065009. DOI: 10.1088/0268-1242/30/6/065009

[35] Tsu D. V., Lucovsky G., Mantini M. J., and Chao S. S. Deposition of silicon oxynitride thin films by remote plasma enhanced chemical vapor deposition. Journal of Vacuum Science & Technology A. 1987;5(4):1998–2002. DOI: 10.1116/1.574902

[36] Vuillod J. Preparation and characterization of plasma enhanced chemical vapor deposited silicon nitride and oxynitride films. Journal of Vacuum Science & Technology A. 1987;5(4):1675–1679. DOI: 10.1116/1.574545

[37] Debieu O., Nalini R. P., Cardin J., Portier X., Perrière J., and Gourbilleau F. Structural and optical characterization of pure Si-rich nitride thin films. Nanoscale Research Letters. 2013;8(1): 1–13. DOI: 10.1186/1556-276X-8-31

[38] Lin K-C, and Lee S-C. The structural and optical properties of a-SiNx:H prepared by plasma-enhanced chemical-vapor deposition. Journal of Applied Physics. 1992;72(11):5474–5482. DOI: 10.1063/1.351992

[39] Scardera G., Puzzer T., Conibeer G., and Green M. A. Fourier transform infrared spectroscopy of annealed silicon-rich silicon nitride thin films. Journal of Applied Physics. 2008;104(10):104310. DOI: 10.1063/1.3021158

[40] Tsu D. V., Lucovsky G., and Mantini M. J. Local atomic structure in thin films of silicon nitride and silicon diimide produced by remote plasma-enhanced chemical-vapor deposition. Physical Review B. 1986;33(10):7069–7076. DOI: 10.1103/PhysRevB.33.7069

[41] Aceves-Mijares M., Espinosa-Torres N. D., Flores-Gracia F., González-Fernández A. A., López-Estopier R., Román-López S., Pedraza G., Domínguez C., Morales A., and Falcony C. Composition and emission characterization and computational simulation of silicon rich oxide films obtained by LPCVD. Surface and Interface Analysis. 2014;46: 216–223. DOI:10.1002/sia.5212

[42] Fazio E., Barletta E., Barreca F., Neri F., and Trusso S. Investigation of a nanocrystalline silicon phase embedded in SiOx thin films grown by pulsed laser deposition. Journal of Vacuum Science & Technology B. 2005;23(2):519–524. DOI: 10.1116/1.1880252

[43] Schliwinski H. –J., Schnakenberg U., Windbracke W., Neff H., and Lange P. Thermal annealing effects on the mechanical properties of plasma-enhanced chemical vapor deposited silicon oxide films. Journal of the Electrochemical Society. 1992;139(6):1730–1735. DOI: 10.1149/1.2069484

[44] Pai P. G., Chao S. S., Takagi Y., and Lucovsky G. Infrared spectroscopic study of SiOx films produced by plasma enhanced chemical vapor deposition. Journal of Vacuum Science & Technology A. 1986;4(3):689–694. DOI: 10.1116/1.573833

[45] Diniz J. A., Tatsch P. J., and Pudenzi M. A. A. Oxynitride films formed by low energy NO^+ implantation into silicon. Applied Physics Letters. 1996;69(15):2214–2215. DOI: 10.1063/1.117169

[46] Palacios-Huerta L., Cabañas-Tay S. A., Luna-López J. A., Aceves-Mijares M., Coyopol A., and Morales-Sánchez A. Effect of the structure on luminescent characteristics of SRO-based light emitting capacitors. Nanotechnology. 2015;26(39):395202. DOI: 10.1088/0957-4484/26/39/395202

[47] Morales-Sánchez A., Monfil-Leyva K., González A. A., Aceves-Mijares M., Carrillo J., Luna-López J. A., Domínguez C., Barreto J., and Flores-Gracia F. J. Strong blue and red luminescence in silicon nanoparticles based light emitting capacitors. Applied Physics Letters. 2011;99(17):171102. DOI: 10.1063/1.3655997

[48] Lin G.-R., Lin C.-J., and Kuo H.-C. Improving carrier transport and light emission in a silicon-nanocrystal based MOS light-emitting diode on silicon nanopillar array. Applied Physics Letters. 2007;91(9):093122. DOI: 10.1063/1.2778352

[49] Dohnalová K., Ondič L., Kusová K., Pelant I., Rehspringer J. L., and Mafouana R.-R. White-emitting oxidized silicon nanocrystals: Discontinuity in spectral development with reducing size. Journal of Applied Physics. 2010;107(5):053102. DOI: 10.1063/1.3289719

[50] Wang X., Zhang R. Q., Lee S. T., Niehaus T. A., and Frauenheim T. Unusual size dependence of the optical emission gap in small hydrogenated silicon nanoparticles. Applied Physics Letters. 2007;90(12):123116. DOI: 10.1063/1.2715101

[51] Cueff S., Labbé C., Jambois O., Berencén Y., Kenyon A. J., Garrido B., and Rizk R. Structural factors impacting carrier transport and electroluminescence from Si nanocluster-sensitized Er ions. Optics Express. 2012;20(20): 22490–22502. DOI: 10.1364/OE.20.022490

[52] Kiebach R., Yu Z., Aceves-Mijares M., Bian D., and Du J. The deposition and control of self-assembled silicon nano islands on crystalline silicon. International Journal of High Speed Electronics and Systems. 2008;18(4):901–910. DOI: 10.1142/S0129156408005862

[53] Park N.-M., Choi C.-J., Seong T.-Y., and Park S.-J. Quantum confinement in amorphous silicon quantum dots embedded in silicon nitride. Physical Review Letters. 2001;86(7):1355–1357. DOI: 10.1103/PhysRevLett.86.1355

[54] Wang Y. Q., Wang Y. G., Cao L., and Cao Z. X. High-efficiency visible photoluminescence from amorphous silicon nanoparticles embedded in silicon nitride. Applied Physics Letters. 2003;83(17):3474–3476. DOI: 10.1063/1.1621462

[55] Park N.-M., Kim T.-S., and Park S.-J. Band gap engineering of amorphous silicon quantum dots for light-emitting diodes. Applied Physics Letters. 2001;78(17):2575–2577. DOI: 10.1063/1.1367277

[56] Mercaldo L. V., Veneri P. D., Esposito E., Massera E., Usatii I., and Privato C. PECVD in-situ growth of silicon quantum dots in silicon nitride from silane and nitrogen. Materials Science and Engineering B. 2009;159–160:77–79. DOI: 10.1016/j.mseb.2008.09.029

[57] Liao W., and Zeng X., Wen X., Chen X., and Wang W. Annealing and excitation dependent photoluminescence of silicon rich silicon nitride films with silicon quantum dots. Vacuum. 2015;121: 147–151. DOI: 10.1016/j.vacuum.2015.08.002

[58] Xie M., Li D., Wang F., and Yang D. Luminescence properties of silicon-rich silicon nitride films and light emitting devices. ECS Transactions. 2011;35(18):3–19. DOI: 10.1149/1.3647900

[59] Dal Negro L., Li R., Warga J., and Basu S. N. Sensitized erbium emission from silicon-rich nitride/silicon superlattice structures. Applied Physics Letters. 2008;92(18):181105. DOI: 10.1063/1.2920435

[60] Sahu B. S., Delachat F., Slaoui A., Carrada M., Ferblantier G., Muller D. Effect of annealing treatments on photoluminescence and charge storage mechanism in silicon-rich SiNx:H films. Nanoscale Research Letters. 2011;6(1):1–10. DOI: 10.1186/1556-276X-6-178

[61] Giorgis F., Vinegoni C., and Pavesi L. Optical absorption and photoluminescence properties of a-Si_{1-x}:H films deposited by plasma-enhanced cvd. Physical Review B. 2000;61(7):4693–4698. DOI: 10.1103/PhysRevB.61.4693

[62] Warren W. L., Robertson J., and Kanicki J. Si and N dangling bond creation in silicon nitride thin films. Applied Physics Letters. 1993;63(19):2685–2687. DOI: 10.1063/1.110420

[63] Warren W. L., Lenahan P. M., and Curry S. E. First observation of paramagnetic nitrogen dangling-bond centers in silicon nitride. Physical Review Letters. 1990;65(2):207–210. DOI: 10.1103/PhysRevLett.65.207

[64] Robertson J., and Powell M. J. Gap states in silicon nitride. Applied Physics Letters. 1984;44(4):415–417. DOI: 10.1063/1.94794

[65] Liu Y., Zhou Y., Shi W., Zhao L., Sun B., and Ye T. Study of photoluminescence spectra of Si-rich SiNx films. Materials Letters. 2004;58(19):2397–2400. DOI: 10.1016/j.matlet.2004.02.015

[66] Nguyen P. D., Kepaptsoglou D. M., Ramasse Q. M., Sunding M. F., Vestland L. O., Finstad T. G., and Olsen A. Impact of oxygen bonding on the atomic structure and photoluminescence properties of Si-rich silicon nitride thin films. Journal of Applied Physics. 2012;112(7):073514. DOI: 10.1063/1.4756998

[67] Hao X. J., Podhorodecki A. P., Shen Y. S., Zatryb G., Misiewicz J., and Green M. A. Effects of Si-rich oxide layer stoichiometry on the structural and optical properties of Si QD/SiO_2 multilayer films. Nanotechnology. 2009;20(48):485703. DOI: 10.1088/0957-4484/20/48/485703

[68] Lai B. H., Cheng C. H., and Lin G. R.. Multicolor ITO/SiOx/p-Si/Al Light Emitting Diodes With Improved Emission Efficiency by Small Si Quantum Dots. IEEE Journal of Quantum Electronics. 2011;47(5):698–704. DOI: 10.1109/JQE.2011.2109699

[69] González-Fernández A. A., Aceves-Mijares M., Morales-Sánchez A., and Monfil-Leyva K. Intense whole area electroluminescence from low pressure chemical vapor deposition-silicon-rich oxide based light emitting capacitors. Journal of Applied Physics. 2010;108(4):043105. DOI: 10.1063/1.3465335

[70] Salh R. Defect Related Luminescence in Silicon Dioxide Network: A Review. In: Prof. Sukumar Basu, editor. Crystalline Silicon—Properties and Uses. Umeå, Sweden: InTech; 2011. DOI: 10.5772/22607

[71] Rodríguez J. A., Vásquez-Agustín M. A., Morales-Sánchez A., and Aceves-Mijares M. Emission mechanisms of Si nanocrystals and defects in SiO_2 materials. Journal of Nanomaterials. 2014;2014:409482. DOI: 10.1155/2014/409482

[72] González-Fernández A. A., Juvert J., Aceves-Mijares M., Llobera A., and Domínguez C. Influence of silicon binding energy on photoluminescence of Si-implanted silicon dioxide. ECS Transactions. 2012;49(1):307–314. DOI: 10.1149/04901.0307ecst

[73] Perálvarez M., Carreras J., Barreto J., Morales A., Domínguez C., and Garrido B. Efficiency and reliability enhancement of silicon nanocrystal field-effect luminescence from nitride-oxide gate stacks. Applied Physics Letters. 2008;92(24):241104. DOI: 10.1063/1.2939562

[74] Parker C. G., Lucovsky G., and Hauser J. R. Ultrathin oxide-nitride gate dielectric MOSFET's. IEEE Electron Device Letters. 1998;19(4):106–108. DOI: 10.1109/55.663529

[75] Wang X., Huang R., Song C., Guo Y., and Song J. Effect of barrier layers on electroluminescence from Si/SiOxNy multilayer structures. Applied Physics Letters. 2013;102(8):081114. DOI: 10.1063/1.4794079

[76] Berencén Y., Ramírez J. M., Jambois O., Domínguez C., Rodríguez J. A., and Garrido B. Correlation between charge transport and electroluminescence properties of Si-rich oxide/nitride/oxide-based light emitting capacitors. Journal of Applied Physics. 2012;112(10):109901. DOI: 10.1063/1.4768249

[77] Pei Z., Chang Y. R., and Hwang H. L. White electroluminescence from hydrogenated amorphous-SiNx thin films. Applied Physics Letters. 2002;80(16):2839–2841. DOI: 10.1063/1.1473230

[78] Li D., Huang J., and Yang D. Enhanced electroluminescence of silicon-rich silicon nitride light-emitting devices by NH_3 plasma and annealing treatment. Physica E: Low-dimensional Systems and Nanostructures. 2009;41(6):920–922. DOI: 10.1016/j.physe.2008.08.024

[79] Morales-Sánchez A., Barreto J., Domínguez C., Aceves-Mijares M., and Luna-López J. A. The mechanism of electrical annihilation of conductive paths and charge trapping in silicon-rich oxides. Nanotechnology. 2009;20(4):045201. DOI: 10.1088/0957-4484/20/4/045201

[80] Morales-Sánchez A., Barreto J., Domínguez C., Aceves-Mijares M., Perálvarez M., Garrido B., and Luna-López J. A. DC and AC electroluminescence in silicon nanoparticles embedded in silicon-rich oxide films. Nanotechnology. 2010;21(8):085710. DOI: 10.1088/0957-4484/21/8/085710

[81] Yao J., Sun Z., Zhong L., Natelson D., and Tour J. M. Resistive switches and memories from silicon oxide. Nano Letters. 2010;10(10):4105–4110. DOI: 10.1021/nl102255r

[82] Yao J., Zhong L., Natelson D., and Tour J. M. In situ imaging of the conducting filament in a silicon oxide resistive switch. Scientific Reports. 2012;2:242. DOI: 10.1038/srep00242

[83] Mehonic A., Vrajitoarea A., Cueff S., Hudziak S., Howe H., Labbé C., Rizk R., Pepper M., and Kenyon A. J. Quantum conductance in silicon oxide resistive memory devices. Scientific Reports. 2013;3:2708. DOI: 10.1038/srep02708

[84] Mehonic A., Cueff S., Wojdak M., Hudziak S., Labbé C., Rizk R., and Kenyon A. J. Electrically tailored resistance switching in silicon oxide. Nanotechnology. 2012;23(45):455201. DOI: 10.1088/0957-4484/23/45/455201

[85] González-Fernández A. A., Juvert J., Aceves-Mijares M., Llobera A., and Domínguez C. Influence by layer structure on the output EL of CMOS compatible silicon-based light emitters. IEEE Transactions on Electron Devices. 20123;60(6): 1971–1974. DOI: 10.1109/TED.2013.2258158

[86] Wang D.-C., Chen J.-R., Zhu J., Lu C.-T., and Lu M. On the spectral difference between electroluminescence and photoluminescence of Si nanocrystals: a mechanism study of

electroluminescence. Journal of Nanoparticle Research. 2013;15(11):1–7. DOI: 10.1007/s11051-013-2063-x

[87] Yin X., Wang S., Li L., Mu G., Tang Y., Duan W., and Yi L. Intense deep-blue electroluminescence from ITO/Y_2O_3/Ag structure. Optics Express. 2015;23(14):18092–18097. DOI: 10.1364/OE.23.018092

[88] Huang R., Song J., Wang X., Guo Y. Q., Song C., Zheng Z. H., Wu X. L., and Chu P. Origin of strong white electroluminescence from dense Si nanodots embedded in silicon nitride. Optics Letters. 2012;37(4):692–694. DOI: 10.1364/OL.37.000692

[89] Tao Y., Zheng J., Zuo Y., Xue C., Cheng B., and Wang Q. Enhanced current transportation in silicon-riched nitride (SRN)/silicon-riched oxide (SRO) multilayer nanostructure. Nano-Micro Letters. 2012;4(4): 202–207. DOI: 10.1007/BF03353715

[90] Huang R., Wang D. Q., Ding H. L., Wang X., Chen K. J., Xu J., Guo Y. Q., Song J., and Ma Z. Y. Enhanced electroluminescence from SiN-based multilayer structure by laser crystallization of ultrathin amorphous Si-rich SiN layers. Optics Express. 2010;18(2):1144–1150. DOI: 10.1364/OE.18.001144

[91] Vila M., Prieto C., and RamÍrez R. Electrical behavior of silicon nitride sputtered thin films. Thin Solid Films. 2004;459(1–2):195–199. DOI: 10.1016/j.tsf.2003.12.082

[92] Ramírez J. M., Berencén Y., López-Conesa L., Rebled J. M., Peiró F., and Garrido B. Carrier transport and electroluminescence efficiency of erbium-doped silicon nanocrystal superlattices. Applied Physics Letters. 2013;103(8):081102. DOI: 10.1063/1.4818758

[93] Perera R., Ikeda A., Hattori R., and Kuroki Y. Trap assisted leakage current conduction in thin silicon oxynitride films grown by rapid thermal oxidation combined microwave excited plasma nitridation. Microelectronic Engineering. 2003;65(4):357–370. DOI: 10.1016/S0167-9317(02)01025-0

[94] Anopchenko A., Marconi A., Moser E., Prezioso S., Wang M., Pavesi L., Pucker G., and Bellutti P. Low-voltage onset of electroluminescence in nanocrystalline-Si/SiO2 multilayers. Journal of Applied Physics. 2009;106(3):033104. DOI: 10.1063/1.3194315

Section 3

Thermoluminescence Dating: From Theory to Applications

Chapter 9

Recombination Radiation in the Diamond

Evgeniy Igorevich Lipatov,
Dmitriy Evgen'evich Genin,
Denis Valer'evich Grigor'ev and
Victor Fedotovich Tarasenko

Additional information is available at the end of the chapter

http://dx.doi.org/10.5772/65064

Abstract

Original experimental data on the recombination radiation of free excitons, the band-A of luminescence, and the recombination radiation of electron-hole liquid in a diamond have been presented. A review of the literature on the recombination radiation in diamond and its application was performed. There was no displacement of free-exciton band at 5.275 eV in the temperature range of 80–300 K. At low excitation levels, the temperature dependence of free-exciton band intensity had the maximum at ~150 K. The band-A of luminescence, due to defects containing sp^2-hybridizedcarbon bonds, is located in the spectral range 350–650 nm with a maximum at ~440 nm and is characterized by the decay time of 8–19 m sec in the temperature range of 80–300 K. The electron-hole liquid recombination radiation in the diamond was observed at temperatures of <200 K and at peak densities of charge carriers of $\geq(0.3–1.0)\times10^{18}$ cm^{-3}. Condensation of electron-hole liquid implies the displacement of the free-exciton intensity maximum on the temperature dependence to higher temperatures. The critical temperature of electron-hole liquid condensation takes values in the range of 160–220 K. The literature data on the diamond light-emitting devices are discussed.

Keywords: diamond, recombination, electron-hole pair, exciton, electron-hole liquid, electron-hole plasma, photoluminescence, cathodoluminescence, KrCl-laser, light-emitting device

1. Introduction

Recombination radiation in semiconductors appears due to radiative recombination of nonequilibrium charge carriers, which are generated in the processes of photon and high-energy particles absorption, or are injected through the contacts [1]. Usually, recombination luminescence is observed in the spectra of photoluminescence (PL), cathodoluminescence (CL) and electroluminescence (EL).

For simple indirect-gap semiconductors (germanium, silicon, diamond), plasma band recombination of electron-hole pairs (at least at low excitation levels) is not observed in the luminescence spectra. Depending on the defect-impurity composition, temperature and density of nonequilibrium carriers, phonon components of radiative recombination bands of free (FE) and bound (BE) excitons, exciton complexes (EC), the electron-hole liquid (EHL) and electron-hole plasma (EHP) are observed in the luminescence spectra of the sample near the fundamental absorption edge. In the latter case, there is a bound state of nonequilibrium charge carriers arising when the Mott transition is fulfilled (phase transition of the second kind)—transition gas FE–EHP. In addition to the processes mentioned above, the radiative recombination of nonequilibrium charge carriers can occur through the impurity-defect allowed states located in the band gap. In this case, recombination emission bands are observed in the spectral region close to the half-width of the energy gap.

Due to the short covalent C–C bond, the low dielectric constant and high Debye temperature diamond excel other semiconductors. **Table 1** shows comparative data on diamond and other semiconductor-related recombination radiation.

Semiconductor	Electric constant, ε	Debye temperature, T [K]	Binding energy of free exciton, E_x [meV]	Radius of free exciton, a_x [nm]	Critical EHL carrier density, n [cm^{-3}]	Critical EHL temperature, T [K]
Germanium, Ge	15.6	374	3.2	15.37	8.9×10^{16}	6.7
Silicon, Si	11.4	640	14.3	4.15	1.2×10^{18}	28
Silicon carbide, 3C-SiC	9.7	1200	13.5	3	2.3×10^{18}	41
Gallium phosphide, GaP	9.1	445	10	5	2.6×10^{18}	45
Diamond, C	5.7	1860	80.5	1.26	3×10^{19}	>165

Table 1. Characteristics of some indirect-gap semiconductors associated with the recombination radiation: electric constant, Debye temperature, the binding energy of exciton, the free-exciton radius, critical temperature, and concentration for formation of EHL.

Diamond is characterized by high exciton binding energy (80.5 meV) [2], which corresponds to a temperature of 660°C. At room temperature, the thermal energy of the charge carriers is less than twice the binding energy of FE. At the temperature of liquid nitrogen, binding energy of FE in the diamond exceeds the thermal energy of lattice on an order of magnitude. Therefore, nonequilibrium charge carriers in diamond form FEs, because it is energetically favorable for them. These temperatures are not extremely low (helium). At the same time, the high density of trapping/recombination centers of charge carriers hinders the formation of FE. However, in the diamond samples containing the impurity-defect centers of $<10^{18}$ cm^{-3}, nonequilibrium charge carriers being generated or injected (at small values of the electric field) form excitons, which are then involved in the radiative and nonradiative recombination, the formation of the EC, condensation and evaporation of EHL.

This chapter of the book is devoted to the review process of radiative recombination FE, EHL, and EHP bands, as well as the so-called band-A of luminescence. We discuss the observed effects, the outstanding issues and possible research directions.

2. Experimental

We investigated the properties of several diamond samples, whose main characteristics are given in **Table 2**.

Sign	Synthesis method	Monocrystal/ polycrystalline	Size [mm]	Transparency	FE-band	Band-A	EHL-band
C4	Natural	Mono	∅4 × 0.25	Yes	No	Strong	No
C5	Chemical vapor deposition (CVD)	Poly	10 × 10 × 0.5	Yes	Strong	Very weak	Strong
C6 (fragments)	CVD	Poly	10 × 10 × 0.1	Yes	Weak	Very weak	No
C10	CVD	Mono	5 × 5 × 0.25	Yes	Strong	No	Strong
C11	CVD	Poly	5 × 5 × 0.25	No	No	Very weak	No
C12	High-pressure high-temperature (HPHT)	Mono	5 × 5 × 0.25	Yes	Strong	No	Weak

Table 2. General characteristics of diamond samples: synthesis method, monocrystal/polycrystalline, size, transparency, observation of spectral FE-, band-A, or EHL-components.

The densities of impurities and defects in samples were less than 10^{18} cm^{-3}. IR absorption spectra of samples did not reveal any peculiarities in one-phonon absorption in the range of 7.5–15 microns (see Ref. [3]), due to the impurity-defect centers.

IR absorption spectra were measured at room temperature (RT) using infrared spectrometer Nicolet 5700 with Fourier transformation in the spectral region from 2.5 to 15 microns. Raman spectra of samples in the wave number range of 100–3700 cm^{-1} were obtained at RT using the same infrared spectrometer equipped with the Raman attachment NXR FT-Raman. Spectral resolution was 2 cm^{-1}.

The PL spectra of diamond samples were measured at temperatures from 80 to 300 K using spectrometer Ocean Optics HR 4000 in the spectral range of 200–300 nm with spectral resolution of 0.4 nm and Stellar-Net EPP-2000C in the spectral range of 200–850 nm with resolution of 1.5 nm. Before measurements, diamond sample was mounted on a cooled copper holder, placed in a vacuum chamber. The sample temperature was measured with a platinum thermoresistor Heraeus Pt1000. Excitation of PL was provided by pulsed KrCl-laser radiation ($\lambda = 222$ nm) with pulse duration 18 ns (FWHM) or ArF-laser radiation ($\lambda = 193$ nm) with pulse duration of 8 ns (FWHM). Laser pulse energy was 30 mJ. The radiation was focused on a sample by a quartz cylindrical lens with a focal length of 65 mm. The peak intensity of the laser radiation on the surface of the sample reached 13 and 10 MW/cm^2 at 222 and 193 nm, respectively. PL radiation from the opposite side of the diamond sample (excitation relative) was transported to the spectrometer through the optical fiber. The incident energy on the sample was measured by laser radiation pyroelectric sensor Ophir PE50-BB. The experimental scheme is described in detail in Ref. [3].

Figure 1 shows the evacuated chamber design, which was used to research the pulse PL of diamond samples at cooling from room temperature (RT) to the temperature of liquid nitrogen. The design of this camera also enables to conduct the measurement of diamond photoconductivity.

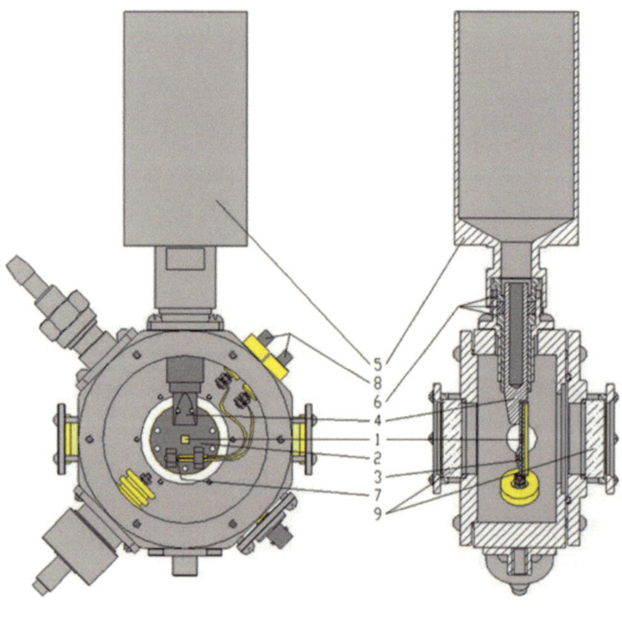

Figure 1. Vacuum chamber. **1**-diamond sample; **2, 3**-copper round plates; **4**-hollow copper heat sink; **5**-metal cup with liquid nitrogen; **6**-seals; **7**-platinum thermistor; **8**-vacuum-tight electrical bushings; **9**-fused silica windows.

Diamond sample **1** was placed between two copper round plates **2** and **3**. In the center of each plate, there was a square hole 3 × 3 mm^2. Plates were fastened to the hollow copper heat sink **4**. The heat sink **4** was out of the evacuated chamber in a metal cup **5** with liquid nitrogen. The vacuum seal and partial heat insulation were provided with seals **6** (polyamide, vacuum rubber). To measure sample temperature, the platinum thermistor **7** (Heraeus Pt1000) has been placed on the copper plate **2** in contact in the copper holder. The resistance of the thermistor **7** was measured with a multimeter (MY-64 S-line) through vacuum-tight electrical bushings **8**. The temperature of the diamond sample **1** was considered equal to the temperature of the thermistor **7** and was determined from the temperature dependence of thermistor resistance provided from the device data sheet. In the experiment, the temperature range of the thermistor resistance decreases from 1100 to 225 ohms. Thus, the minimum temperature recorded was 80 K at a fore-vacuum pump (~10^{-2} torr).

Thermal contact of the sample **1**, copper plates **2** and **3**, heat sink **4** and the thermistor **7** was provided by thermopaste KPT-8 with a specific electrical resistance >10^{12} Om cm. The laser beam fell into the sample through a quartz window **9**. The chamber pumping was made by the backing pump 2NVR-5DM.

The absorption of photons in the diamond sample leads to a nonuniform distribution of nonequilibrium charge carriers. Reduction in the peak intensity of the laser radiation with the depth of the sample (taking into account the reflections from the rear face) is calculated by the formula [4]:

$$I(x) = I_0 \cdot \frac{(1-r)\cdot\left(e^{-\alpha\cdot x} - r\cdot e^{-\alpha\cdot(2\cdot d - x)}\right)}{1 - r^2\cdot e^{-2\cdot\alpha\cdot d}} \tag{1}$$

where x [cm]—depth, I_0 [W/cm^2]—the peak intensity of the laser radiation on the surface of the sample, r—reflection coefficient of diamond at the laser wavelength ($r_{222} = 0.21031$, $r_{193} = 0.24432$ [5]), α [cm^{-1}]—the absorption coefficient of the sample at the laser wavelength ($\alpha_{222} = 400$ and $\alpha_{193} = 3250$ cm^{-1}), d [cm]—thickness of the sample.

Profiles of charge carrier peak density over the sample thickness were calculated using the formula [6]:

$$n(x) = \frac{I(x)\cdot\tau}{h\nu\cdot x} \tag{2}$$

where $I(x)$ [W/cm^2]—profile of the laser radiation peak intensity with the depth of the sample, calculated according to the formula (1), τ [s]—lifetime of nonequilibrium charge carriers, $h\nu$ [J]—the energy of photons.

Figure 2 shows the distribution of the laser peak intensity and the peak carrier density in the sample, calculated by the formulas (1) and (2), at peak laser intensity $I_{222} = 13$ MW/cm^2 and $I_{193} = 10$ MW/cm^2.

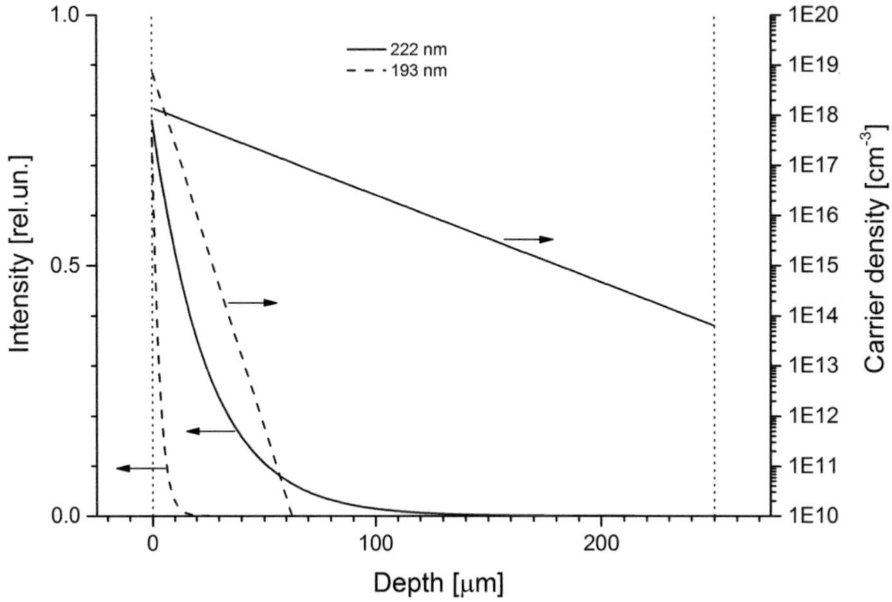

Figure 2. Laser intensity and charge carrier density in the diamond sample C5 versus the depth of sample for laser excitation on 222 nm (solid curves) and 193 nm (dashed curves).

Thus, the photons at 222 and 193 nm created in the samples the maximum nonequilibrium charge carrier density of $n_{222} \sim 1.1\times10^{17}$ and $n_{193} \sim 5.6\times10^{17}$ cm^{-3}, respectively. The carrier density reduced exponentially by two orders of magnitude in the diamond layer thickness $d_{222} \sim 110$ microns and $d_{193} \sim 13$ microns, respectively.

CL spectra were measured using another vacuum chamber, a similar device. The camera was attached directly to the electron accelerator. The luminescence radiation of a sample was transported to a spectrometer through the optical fiber. The technique of CL measurements is described more detailed in [7].

3. Free exciton recombination band

In the diamonds with the densities of impurity-defect centers of less than 10^{18} cm^{-3}, the recombination FE band is characterized by considerable intensity even at RT. In Ref. [8], observation of the FE luminescence band of the diamond sample doped with boron at temperatures up to 277°C was reported. Presumably, in luminescence spectra of undoped ultrapure samples, the FE recombination band can be observed up to temperatures of ~660°C.

http://dx.doi.org/10.5772/65064

Due to the indirect zone structure in the diamond, the FE radiative recombination arises with the generation of momentum-conserving phonons. Recombination radiation of FEs in diamond is accompanied by the generation of the three main phonon components—the dominant TO (transverse optical phonon 141 meV) at 5.275 eV, TA (transverse acoustic phonon 87 meV) at 5.325 eV and TO+O^{Γ} (two-phonon process involving TO + Raman phonon 141 + 159 meV) at 5.117 eV, as shown in **Figure 3**. The LO-component (longitudinal optical phonon 156 meV) at 5.254 eV is characterized by low intensity and is indistinguishable on the background of intense temperature-broadened TO-components at temperatures higher than 100 K [9].

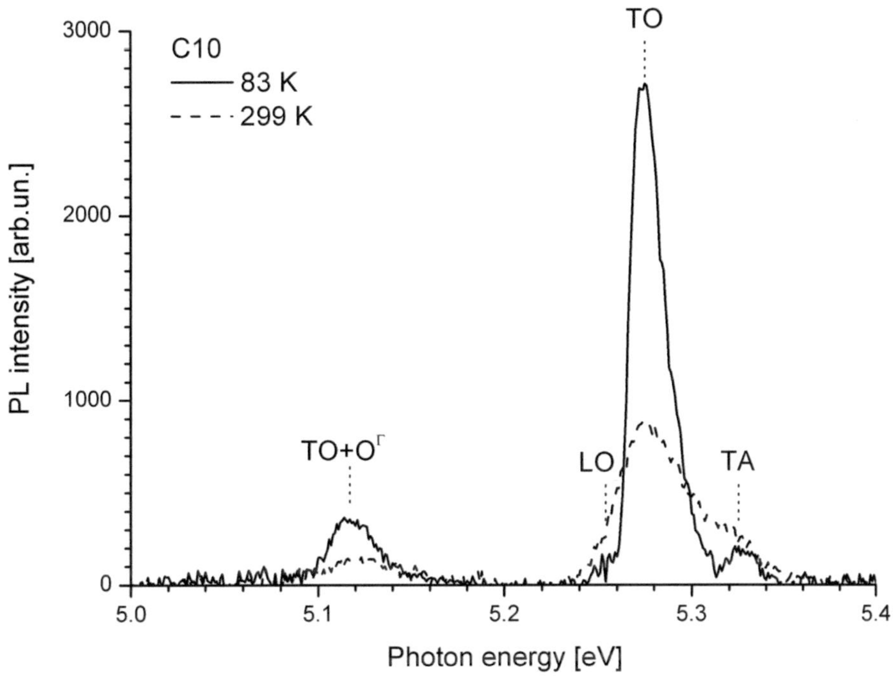

Figure 3. PL integrated spectra of diamond monocrystal C10 under excitation at 222 nm at temperatures of 83 K (solid curve) and 299 K (dashed curve). The spectrum contains the phonon components: TA (87 meV) at 5.325 eV, TO (141 meV) at 5.275 eV, LO (156 meV) at 5.254 eV and TO + O^{Γ} (141 + 159 meV) at 5.117 eV.

The spectral position of the phonon component of the recombination FE band $h\nu$ was determined by the band gap E_g, the binding energy of FE E_{FE}, and the corresponding phonon energy $h\omega$:

$$h\nu = E_g - E_{FE} - h\omega \tag{3}$$

Depending on experimental conditions, all three terms in the right side of the expression (3) vary slightly, causing the displacement of the spectral components.

It is known that the band gap decreases with increasing temperature due to increase in thermal energy of the atomic lattice [10]. In Ref. [11], the difference between the band gap E_g and exciton binding energy E_x in the 80–620 K temperature range was measured by finding the position of the phonon component of exciton absorption threshold at 5.258 and 5.314 eV for several natural diamonds. In the temperature range 80–300 K, the reduction of value $E_g - E_x$ was ~15 meV (see. **Figure 4**). According to Eq. (3), this reduction, $E_g - E_x$, should cause a corresponding shift of the maxima of phonon FE component. In Ref. [12], the position of TO-component maximum of FE recombination was measured in the temperature range 80–250 K for CVD diamond samples (see **Figure 4**). The FE band maximum shifted to shorter wavelengths by ~16 meV as the temperature increases from 80 to 200 K. This observation contradicts to the effect of band gap reduction with increasing temperature.

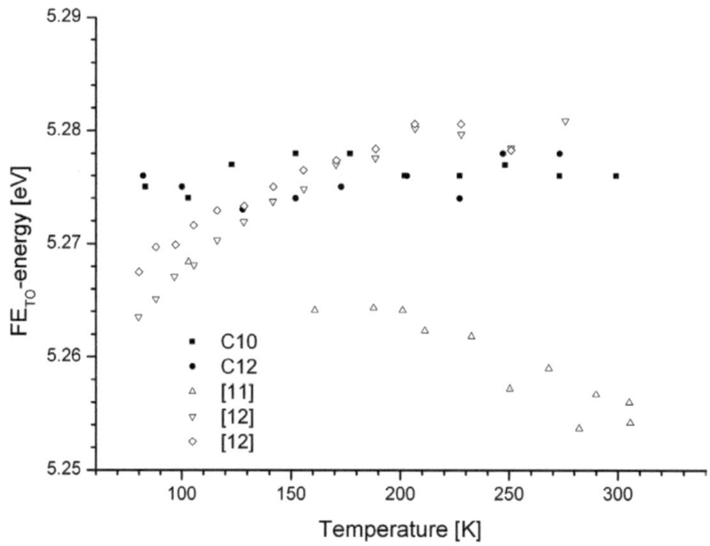

Figure 4. Positions of indirect optical transitions depending on diamond sample temperature. Excitation was provided at 222 nm with peak intensity 1 MW/cm^2. Black squares-sample C10, black circles-sample C12, open triangles-data from Ref. [11], open turned triangles and open rhombs-data from Ref. [12].

We have measured the position of dominant TO-component maximum of FE recombination depending on the temperature for the samples C10 and C12 at low level of excitation at 222 nm. The measured values are shown in **Figure 4**. The observed change in the maximum position of radiative FE recombination is about 3–5 meV, which lies within the measurement error, and is significantly less than one measured in [11]. In Ref. [12], there was another trend of TO-component position shifting to shorter wavelengths with increasing temperature. Thus, the data are inconsistent, and the observed phenomena deserve separate studies.

At low excitation levels, the binding energy of FE in diamond is 80.5 meV [2]. As the FE density approaching the Mott transition level or EHL condensation critical density (phase transition of the first kind), which is characterized by a decrease in the average distance between FEs to a value $r_s = 1–2$ (in FE radius size), the FE binding energy should decrease to zero [13]. In the literature, there is no information for diamonds about the FE binding energy measurements under such conditions.

For various diamond samples with various defect-impurity compositions, different phonon energy is mentioned for indirect optical transitions of radiative recombination. **Table 3** shows the values of TA-, TO-, LO- and O^{Γ}-phonon energies to various literature references.

References	Phonon energy, meV			
	TA	TO	LO	O^{Γ}
[11]	83	143	132	167
[2]	83.6	135.8	156	159.2
[9]	87	141	163	165
[14]	87	140	–	151
[15]	88	140	162.4	164.2

Table 3. Energies of TA-, TO-, LO- and O^{Γ}-phonons from a variety of published data.

Differences in phonon energies arise from different calculation methods, measurement errors and experimental conditions. Experimental conditions are the most important thing, because it is necessary to use the phonon energy as the varying values for the best fit of the experimental and calculated data. This indicates an incomplete understanding of observed phenomena.

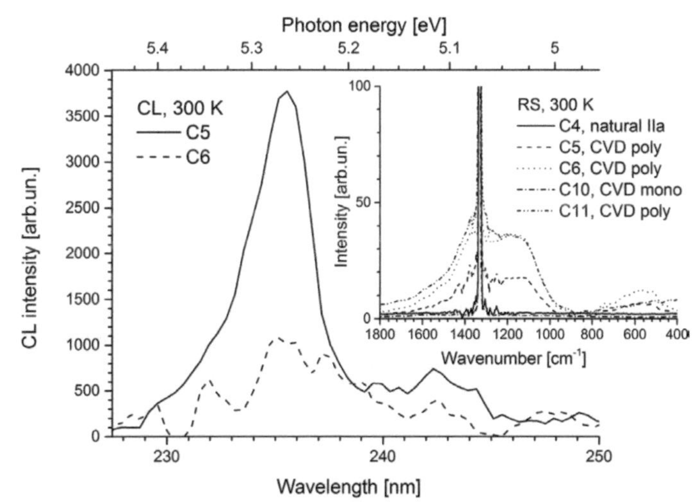

Figure 5. CL integrated spectra of polycrystalline CVD diamond samples C5 and C6 at room temperature. The CL excitation was provided by high-current electron beam (220 keV) with pulse duration of 2 ns. On the inset Raman spectra of several diamond samples at room temperature are shown.

The presence of defects and impurities in the sample increases the probability of nonradiative recombination processes of FEs. CL spectra at room temperature of two different polycrystalline CVD diamond samples C5 and C6 are shown in **Figure 5**. FE recombination band dominated in both spectra. The inset in **Figure 5** shows the Raman spectra of several diamond samples, including C5 and C6. For comparison, the spectra of natural (C4) and CVD single crystals (C10), as well as a fully opaque CVD polycrystalline sample (C11), are presented. In contrast to sample C5, the Raman spectrum of sample C6 is characterized by high intensity "non-diamond" bands: amorphous carbon (570–580 and 1540 cm^{-1}), disoriented graphite (1350 cm^{-1}) and microcrystalline diamond (1140 cm^{-1}). These features are reflected in the intensity of FE band. Thus, the presence of large density of states of "non-diamond" carbon increases the probability of nonradiative recombination and reduces the probability of optical transitions in FE band. Samples C5 and C6—diamonds of "optical quality"— had no measurable absorption bands in the IR range, had low intensity of luminescence band-A (see Section 4), showed no other luminescence bands (except the band-A and the FE band) and revealed the clear edge of fundamental absorption [3]. In addition, sample C5 demonstrated the radiative recombination of EHL at higher excitation levels [3, 16], which will be discussed in Section 5.

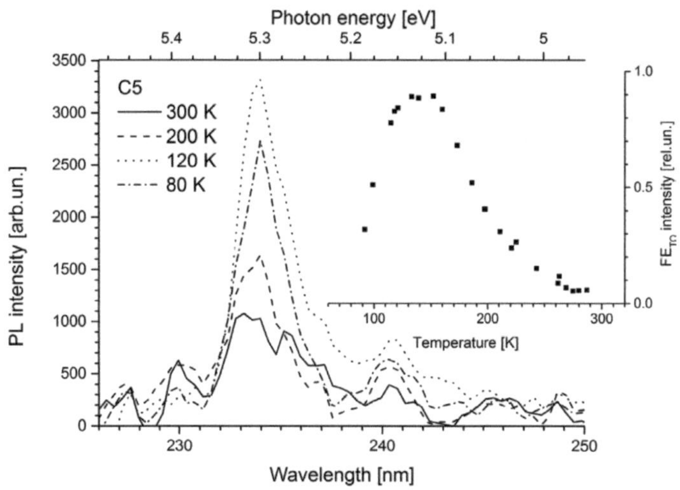

Figure 6. PL integrated spectra of polycrystalline CVD diamond sample C5 at different temperatures. The PL excitation was provided by KrCl-laser (222 nm) with pulse duration of 18 ns. On the inset there is the temperature dependence of FE band intensity of diamond sample C5.

The intensity of FE recombination band varies with the temperature of the sample as shown in **Figure 6**. At low excitation levels, this dependence has the form as shown in the inset of **Figure 6** and is determined by the FE lifetime of charge carriers in diamond [17]:

$$\frac{1}{\tau_{FE}} = P_{FE} + F_{FE} \cdot \exp\left(-\frac{E_{FE}}{kT}\right) + R_T \cdot \frac{P_{BE}}{P_{BE} + F_{FE} \cdot \exp\left(-\frac{E_{FE}}{kT}\right)} \tag{4}$$

where P_{FE}—annihilation rate FE, the second term—the rate of FE thermal ionization with frequency F_{FE} and binding energy E_{FE}, the third term—the localization rate of bound excitons (BE) on shallow traps, R_T—capture speed of FE, R_{BE}—the BE annihilation rate, F_{BE}—frequency of BE thermal dissociation on a FE and a trap, E_{BE}—binding energy of BE.

The intensity maximum of TO-component of recombination FE band usually locates at temperature about 150 K for all diamond samples and excitation methods [12, 17, 18]. The exception is the PL at high excitation levels (see Section 5).

Thus, the FE recombination radiation in the diamond at low excitation levels has been studied quite well. However, there are following unresolved aspects: the absence of spectral shifting of dominant TO-component, when the temperature changes; the different phonon energies of FE recombination component in different experiments; and the unknown change in FE binding energy by increasing the charge density to values $\geq 10^{19}$ cm^{-3}.

4. The band-A of luminescence

Diamond samples of any synthesis method show the blue luminescence band (350–650 nm) in the visible region of luminescence spectra. This so-called band-A is observed at any excitation method [19] (See **Figures 7–9**). There is a competition between the FE recombination band and band-A [8]. More defective samples show intense band-A, and purer samples—intensive FE band. Thus, the centers of band-A are effective FE radiative recombination centers.

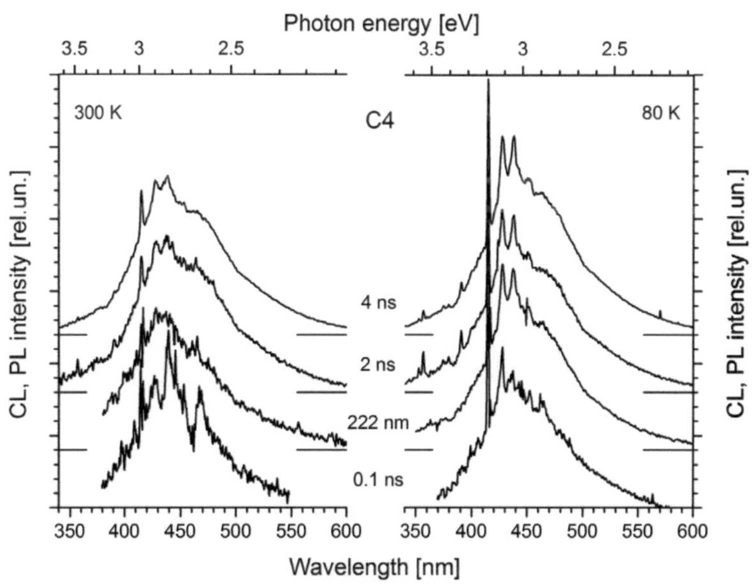

Figure 7. CL and PL integrated spectra of natural diamond IIa-type sample C4 at temperatures of 300 and 80 K. The CL excitation was provided by electron beams with durations of 0.1, 2, and 4 ns. The PL excitation was provided by KrCl-laser with pulse duration of 18 ns.

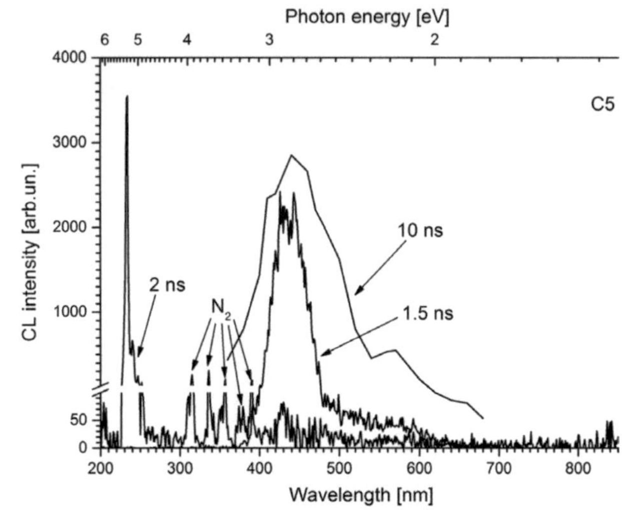

Figure 8. CL integrated spectra of polycrystalline CVD diamond sample C5 at room temperature. The CL excitation was provided by several electron beams with pulse duration of 1.5, 2, and 10 ns. The integrated spectrum obtained at 1.5 ns excitation was magnified in 30 times compared with the spectrum obtained at 2 ns excitation. The spectrum obtained at 10 ns excitation was reconstructed from the maximum of CL emission signals at different wavelength (the method see in Ref. [29]) and was normalized to intensity of CL spectrum obtained at 1.5 ns excitation.

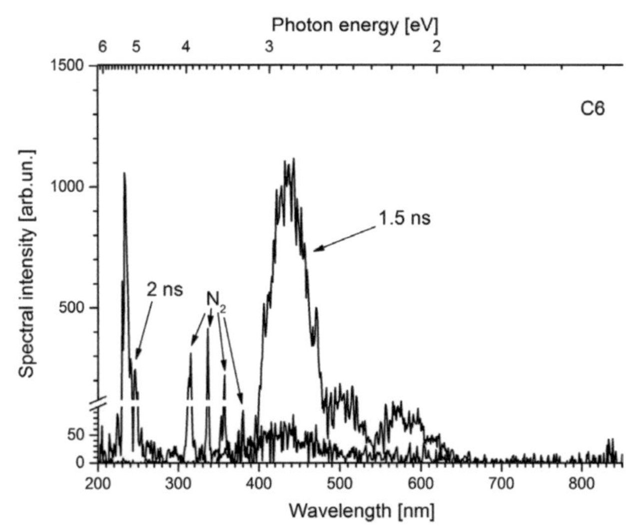

Figure 9. CL integrated spectra of polycrystalline CVD diamond sample C6 at room temperature. The CL excitation was provided by electron beams with pulse duration of 1.5 and 2 ns. The integrated spectrum obtained at 1.5 ns excitation was magnified in 30 times compared with the spectrum obtained at 2 ns excitation.

There is no common opinion about the origin of luminescence band-A. Initially, luminescence analysis objects of the diamond structure were natural crystals and specimens grown in conditions of high pressure and high temperature (HPHT). Such samples always contain nitrogen in significant densities (10^{17}–10^{20} cm^{-3}) in the dispersed form or in polyatomic complexes. In Ref. [20], it was suggested that the band-A is the result of recombination of electrons and holes on closely spaced donor-acceptor pairs. However, intensive band-A was observed in samples containing impurities below 10^{17} cm^{-3} [21]. In [19], band-A has been attributed to optical transitions in N_4V centers. However, a clear correlation between the intensities of the band-A and N9 system (associated with N_4V centers) has not been identified [22]. In Refs. [8, 12, 23, 24], it has been proposed that band-A is associated with dislocations. However, not all of the dislocations were luminesced [19, 23], a clear correlation with the type of dislocation was not observed [25], and band-A radiation may come from sample areas, which are free from dislocations [23]. Sometimes, the observed emission of band-A appeared from a single dislocation, which has led to assume about the optical transitions in defect centers, decorating dislocation and dangling bonds in dislocations.

The development of CVD technology was enabled to receive diamond samples of large sizes (up to Ø120 × 3 mm^3) [26]. Untreated CVD diamonds contain no nitrogen with aggregation degree of more than 2, because the synthesis gas receives nitrogen from air only. The observation of band-A in luminescence spectra of undoped CVD diamonds has led to speculation that the band-A is a result of recombination of electron-hole pairs on lattice defects, containing the sp^2-hybridized carbon bonds [27]. In natural diamonds subjected to natural HPHT treatment, this type of defects exists throughout the whole volume of sample [19].

The band-A is practically immeasurable in CVD single crystals [3]. Intrinsic defects, containing sp^2-hybridized carbon bonds, are concentrated in grain boundaries of polycrystalline CVD diamonds [28]. In Ref. [21], it was found that the band-A of CVD polycrystalline diamond luminescence is observed from defective disoriented areas—grain boundaries. Calculations [28] showed that the grain boundaries contain ~40% of amorphous carbon with sp^3-hybridized carbon bonds, preserving the short-range order of diamond lattice only, ~10% of dangling bonds and ~30% of sp^2-hybridized carbon dimers, which create allowed states in the band gap. The radiative recombination of FEs in the spectral region of band-A occurs through allowed states in the band gap.

In Ref. [8], there was competition between the FE recombination band and band-A in the CVD diamond films. The increase in boron impurity density led to the emergence of "non-diamond" bands in Raman spectra, reduction of FE band intensity and increase in the band-A intensity and vice versa. The FE binding energy in the diamond is about two times higher than heat energy at room temperature. There is a competition between the processes of spontaneous FE radiative recombination and FEs capture on band-A defects.

Natural samples (e.g., C4), containing a small amount of N_2 defects, exhibit intensive band-A of luminescence in the spectral range of 350–650 nm (see **Figure 7**). The luminescence spectra of polycrystalline CVD samples of "optical quality" contain the dominant FE recombination band at 235.2 nm. At the same time, the band-A is characterized by low intensity (see **Figures 8** and **9**).

The natural diamonds and synthetic samples, which were subjected to HPHT treatment, demonstrate the band-A of luminescence that overlapped the bands of other defects [22]. These defects (N_3V, N_2V and NV centers) show vibronic bands of luminescence with the zero-phonon lines (ZPL) at 415, 503 and 575 nm, respectively. For this reason, the integrated luminescence spectra of the diamonds in the visible region can have multiple peaks and exhibit shifts from sample to sample. **Figure 7** shows the CL and PL spectra of the natural type IIa specimen (C4). CL was excited by electron beam with pulse duration of 0.1, 2 and 4 ns; for PL excitation, the pulsed laser radiation at 222 nm with a pulse duration of 18 ns (FWHM) was used. Besides band-A, the luminescence spectra exhibit the so-called N3 system with ZPL at 415.2 nm due to N_3V centers.

Measurements of luminescence spectra with time resolution allow to resolve the band-A and superimposed bands in time. **Figure 10** shows the change in the CL spectrum of natural sample C4 in time. The CL spectrum in the maximum of optical signal consisted of system N3 with ZFL 415.2 nm and its phonon replicas (427, 438, 450 and 463 nm), and the wide anti-Stokes (with respect to the ZPL) component at 350–410 nm. At 30 ns after starting of excitation pulse, the CL intensity decreased by seven times, the position of the ZPL and phonon replicas did not change, and anti-Stokes luminescence component drastically decreased. At 2 ms after starting of excitation pulse, the ZPL 415.2 nm and its phonon replicas completely disappeared, and the intensity was reduced by three orders of magnitude, maximum shifted to 450–480 nm.

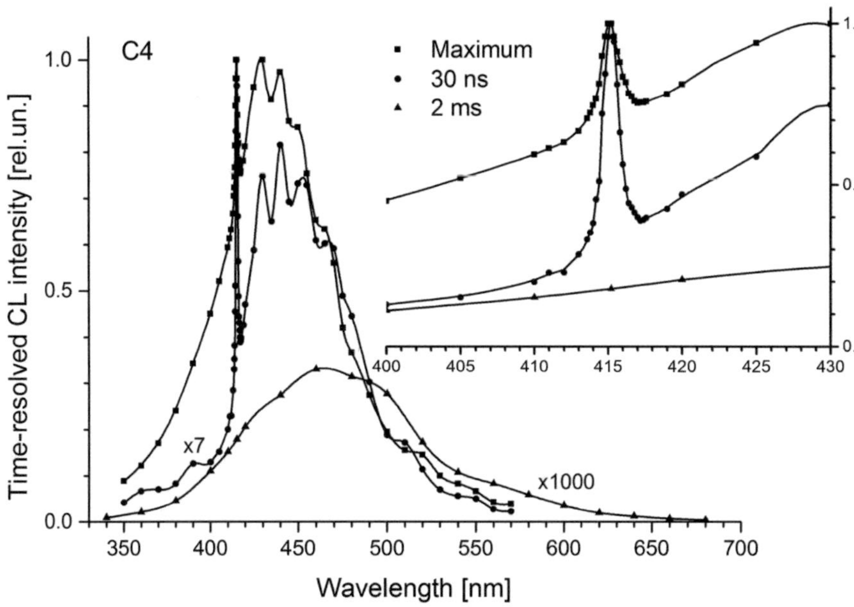

Figure 10. CL spectra of natural diamond sample C4 at room temperature obtained at 10 ns excitation. CL spectra were reconstructed at different wavelength (the method see in Ref. [29]) from the maximum of CL emission signals (squares), after 30 ns from the excitation starting (circles) and after 2 ms from the excitation starting (triangles). On the inset, there are CL spectra in the magnified spectral scale.

Thus, the band-A is characterized by prolonged weak emission in the 350–650 nm range with decay time 8–9 ms at room temperature, as shown in **Figure 11**. At temperature ~80 K of

http://dx.doi.org/10.5772/65064

diamond sample, the decay time increased up to 18–19 ms [7, 29]. The temperature dependence of the band-A was investigated in detail earlier [23]. The sample heating from 30 to 110°C resulted in complete quenching of the band-A. In Ref. [22], the observation of band-A at temperatures up to 125°C was reported. So the activation energy of nonradiative recombination of band-A centers is about 35 meV.

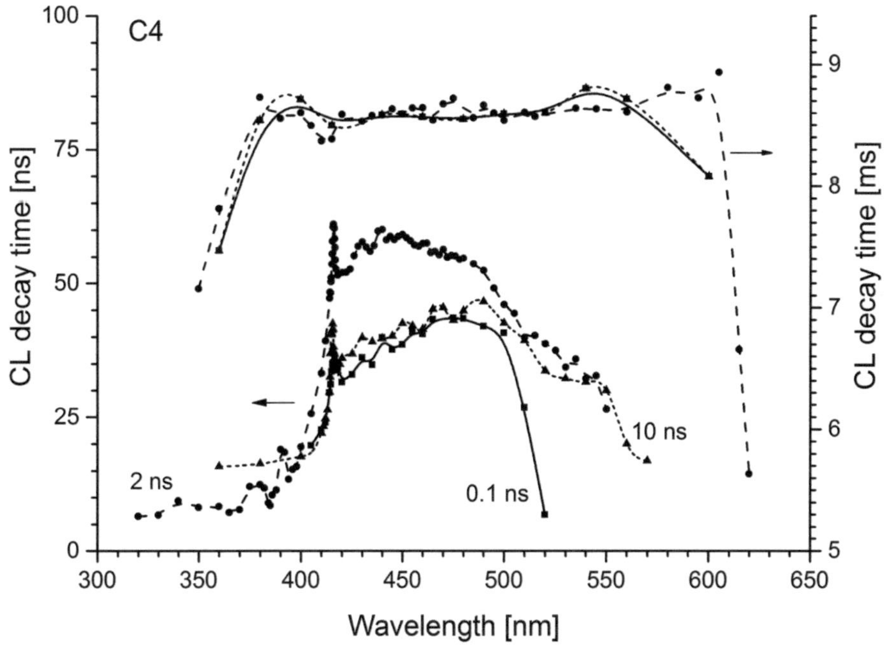

Figure 11. Spectra of CL decay time of natural diamond sample C4 at room temperature in nanosecond and millisecond time scale. The vibronic N3 system (N_3V centers) is characterized by the nanosecond decay time of CL. The band-A (sp^2-hybridized bonds) is characterized by the millisecond decay time of CL. Note that the CL decay time spectrum of band-A did not changed at different durations of electron beams and was 8–9 ms in the whole emission range. Squares-e-beam pulse duration of 0.1 ns, circles-2 ns, triangles-10 ns.

The observed temperature quenching of band-A in nonradiative recombination processes showed the complex structure of energy levels of corresponding defects. Temperature quenching was observed at temperatures above 80 K. There are no references in the available literature about the dependence of band-A intensity at temperature less than 80 K.

Thus, the experimental data indicate that the recombination band-A of luminescence in the diamond is due to the intrinsic defects of the diamond lattice, containing the sp^2-hybridized carbon bonds. The lifetime of the metastable excited level centers of band-A is ~8–19 ms in the temperature range 300–80 K. The activation energy of temperature quenching of band-A is ≥35 meV. At the same time, the atomic structure and the position of energy levels of band-A centers are not known exactly. Research in this area will improve the understanding of defect formation in diamond and advance in the synthesis of ultra-pure diamond samples for optoelectronic applications.

5. Electron-hole liquid recombination band

The electron-hole liquid (EHL) is a condensed state of FE gas in semiconductors [13]. For the condensation of EHL, it is necessary to provide the temperature below the critical point and high charge carrier density. The EHL exists in the form of droplets with size 10^{-8}–10^{-4} m depending on the material. The lifetime of the EHL is determined by the charge carrier lifetime and the processes of droplet condensation/evaporation. The EHL is observed in PL and CL spectra in the form of recombination band shifted toward longer wavelengths relative to FE recombination band.

EHL properties in germanium and silicon are well understood [30]. For them, the critical constitute 6.7 K and 24.5 K, respectively. Diamond is the third one-component indirect-gap semiconductor. Due to the high FE binding energy of 80.5 meV and the high Debye temperature of 1860 K, the critical temperature of EHL existence in diamond is more than 160 K [31–33].

The first reports on the EHL observation in diamond have been made quite recently [33, 34] as improving CVD synthesis technology. Observation of EHL in luminescence spectra of natural diamonds (as well as FEs) is difficult because of high level of doping with nitrogen [31]. These defect centers act as nonradiative (N_S, N_2, N_4V) and radiative (N_3V, N_2V, NV, sp^2-bonds) centers of FE recombination. Currently, the EHL properties in diamond attract some interest [3, 35].

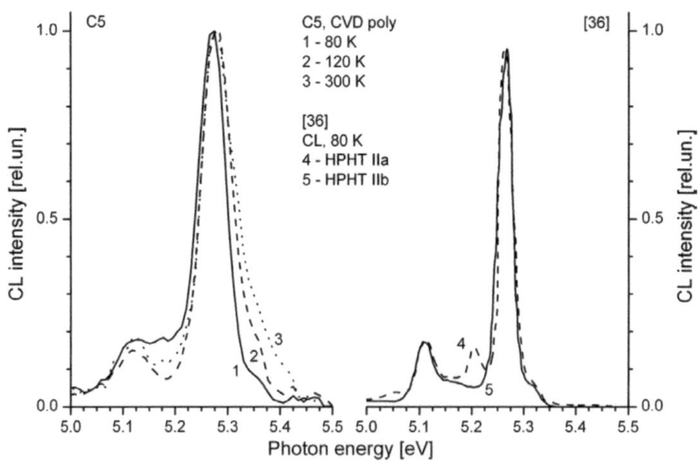

Figure 12. CL spectra of polycrystalline CVD diamond sample C5 at several temperatures from 80 up to 300 K at excitation by 2-ns electron beam (left part). CL spectra of two HPHT samples at 80 K from Ref. [36].

The observation of EHL recombination band in CL spectra of the diamond is complicated by heating the area of electron beam exposure. An accelerated electron generates tens-thousands of "hot" electron-hole pairs, which then thermalize in cascades of phonon emission, causing the heating of lattice and the "phonon wind," making difficulties to EHL condensation. In CL spectra, the EHL recombination band is observed in the form of low-intensity band at 5.16–

5.23 eV range and between TO- and TO+O$^{\Gamma}$-components of FE recombination band (see **Figure 12**, CVD sample C5). In addition to this work in the available literature, it was reported only once about the EHL recombination band in CL spectra of the HPHT IIa type (pure) and IIb type (doped with boron) diamonds [36].

The excitation of nonequilibrium charge carriers by laser radiation provides less heating of the lattice, as an absorbed photon produces an electron-hole pair. Moreover, if the excitation of nonequilibrium carriers by laser radiation takes place at the fundamental absorption edge, the generated carriers will have a minimum excess energy, and consequently, less laser energy is converted to the lattice heating and the "phonon wind."

The threshold laser intensity for the observation of EHL recombination band in the PL spectra depends on the fundamental properties of the material, as well as the impurity-defective composition of a particular sample. The higher the recombination center density in the sample is, the higher the threshold intensity to compensate the exciton recombination should be. On the other hand, some impurity-defect centers can act as condensation nuclei of EHL drops, which should reduce the threshold intensity, as condensation on the irregularities is thermodynamically favorable than the condensation in the homogeneous vapor [13, 37].

Figure 13 shows the PL spectra of polycrystalline diamond sample C5 in the temperature range 90–235 K with excitation at 222 nm with peak intensity of 3 MW/cm^2 [38]. The EHL recombination band is observed in the form of long-wavelength shoulder (5.2–5.16 eV) of TO-component of FE band. The EHL recombination radiation arises from the central regions of crystallites, but not from the crystallite boundaries. **Figure 14** shows a micrograph of sample C5 with magnification ×4. Crystallite sizes range from 50 to 400 microns, which is significantly higher than the FE diffusion length in diamond (1–10 microns) at RT [21, 39, 40].

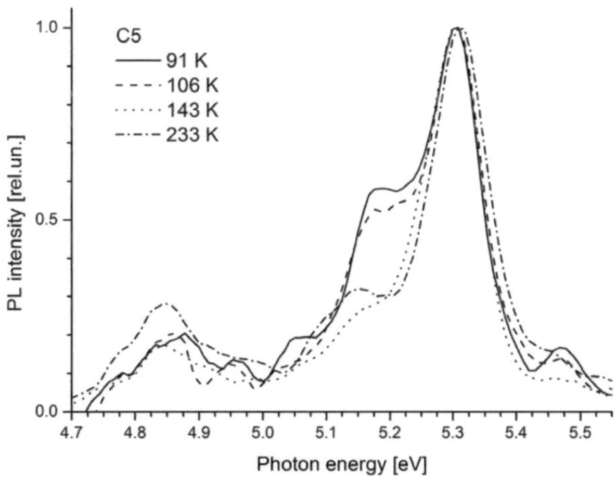

Figure 13. PL spectra of polycrystalline CVD diamond sample C5 at several temperatures from 80 K up to 300 K at excitation by 18-ns laser pulses at 222 nm with peak intensity 3 MW/cm^2. The TO-component of FE band at 5.27 eV dominated in PL spectra. The TO-component of EHL band in the spectral range 5.1–5.2 eV was appeared on the long-wavelength edge of FE band.

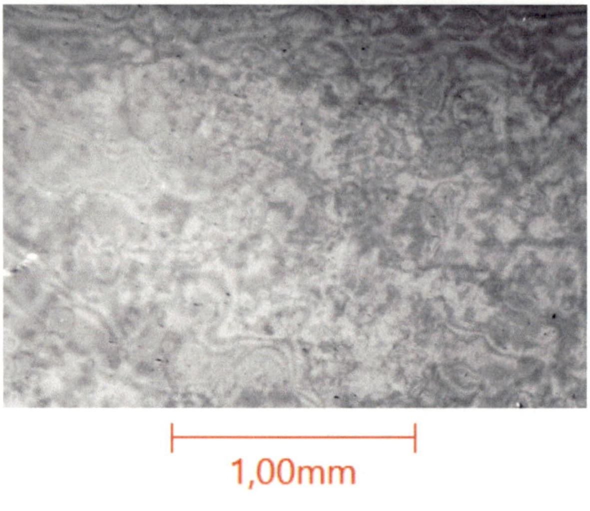

Figure 14. The microphotograph of surface of polycrystalline CVD diamond sample C5 with magnification ×4. The crystallite sizes take values from 50 up to 400 μm.

For a single-crystal sample C10 with excitation at 222 nm, the threshold intensity of EHL recombination band was about 7 MW/cm², which is more than two times higher than the threshold intensity for sample C5. The reason for this effect has not yet been established.

In addition to changing the threshold intensity for observation EHL recombination radiation from sample to sample due to its different impurity-defective structure, the threshold intensity also depends on the excitation wavelength. As mentioned above, the higher photon energy of excitation exceeds the band gap, and more laser energy is converted into thermal energy of lattice. The spectra of edge luminescence of the sample C10 are shown in **Figures 15** and **16** with excitation at 222 and 193 nm, respectively. Spectra are marked with numbers, which correspond to the peak intensity of laser radiation in units of MW/cm².

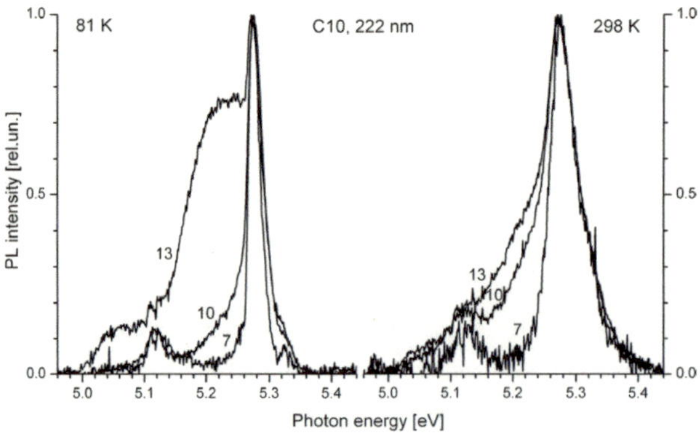

Figure 15. PL spectra of monocrystal CVD diamond sample C10 under the excitation at 222 nm at temperatures of 81 and 298 K. The PL spectra marked by peak intensities [7, 10, and 13 MW/cm²] of laser radiation on the sample surface.

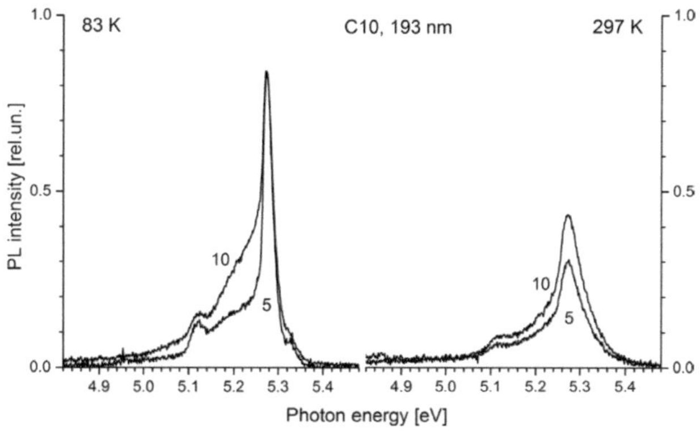

Figure 16. PL spectra of monocrystal CVD diamond sample C10 under the excitation at 193 nm at temperatures of 83 and 297 K. The PL spectra marked by peak intensities [5 and 10 MW/cm^2] of laser radiation on the sample surface.

According to the expression (2) at 222 nm excitation with peak intensities of 7, 10 and 13 MW/cm^2, the peak densities of non-equilibrium carriers were (0.72, 1.04 and 1.35)×10^{18} cm^{-3}, respectively. At 193 nm excitation with peak intensities of 5 and 10 MW/cm^2, the peak densities were (3.47 and 6.94)×10^{18} cm^{-3}, respectively. Thus, for comparable peak densities, the PL spectra, shown in **Figures 15** and **16**, were significantly different.

At 222 nm excitation with the peak intensity 7 MW/cm^2, the FE recombination band was observed only at temperatures of 81 and 298 K in the PL spectrum (see **Figure 15**). At 298 K, there was a considerable thermal broadening of the FE band components. At 10 MW/cm^2 on the long-wavelength edge of the TO-component of FE band, the shoulder appeared due to the EHL radiative recombination (and EC, possibly). When the peak intensity achieved of 13 MW/cm^2 and the sample temperature was 81 K, the EHL band was expressed explicitly as the TO-component at 5.16–5.22 eV and the TO+O$^\Gamma$-component at 5.0–5.09 eV. There was possible contribution of TA-component in the spectral range of 5.18–5.26 eV and of EC band in the spectral range of 5.20–5.26 eV. The broadening of short-wavelength edge of TO-component of FE band at 81 K in the spectral range of 5.18–5.34 eV was also due to the contribution of TA-component of EHL band. At the temperature of 298 K, the short-wavelength edge of TO-component of FE band practically did not change with increasing of laser intensity. Note that at the temperatures of liquid nitrogen and of RT, the long-wavelength shoulder of TO-component of FE band was presented in the PL spectrum. At RT, this broadening cannot be described in terms of thermal broadening.

Figure 16 shows the PL spectra excited at 193 nm with a peak intensity of 5 and 10 MW/cm^2 for temperatures 83 and 297 K. At 193 nm PL excitation, the peak density of electron-hole pairs was higher by 3–10 times than at 222 nm excitation. The radiation at 193 nm is absorbed in layer ~13 μm. It was an order of magnitude less than that at 222 nm excitation. However, at 193 nm excitation the PL spectra at 83 K showed a smaller contribution EHL recombination band than at 222 nm excitation. At excitation on the same wavelength, the charge carrier density increase was resulted in the increase in the contribution of the EHL band. Presumably, there is a high density of structural defects in a surface layer, which are nonradiative recombination centers. It results in an increase in threshold FE density for EHL condensation. At the same time, the envelope of PL spectra essentially unchanged at RT with different peak intensities. Unlike excitation at 222 nm in PL spectra at 193 nm excitation, "tails" in the spectral region of <5.08 eV were observed that for the case of silicon was attributed to the EHP recombination [13].

Figure 17 shows the edge luminescence spectra of the HPHT single crystal C12 at 222 nm excitation. The EHL recombination band was not expressed clearly, and FE band was three times less intensive with the same peak intensities as for sample C10. The spectrum envelopes did not change during the peak intensity growth for both temperatures. This indicates a much greater density of nonradiative recombination centers in the sample C12, as compared to C10, which makes it difficult for EHL condensation in sample C12.

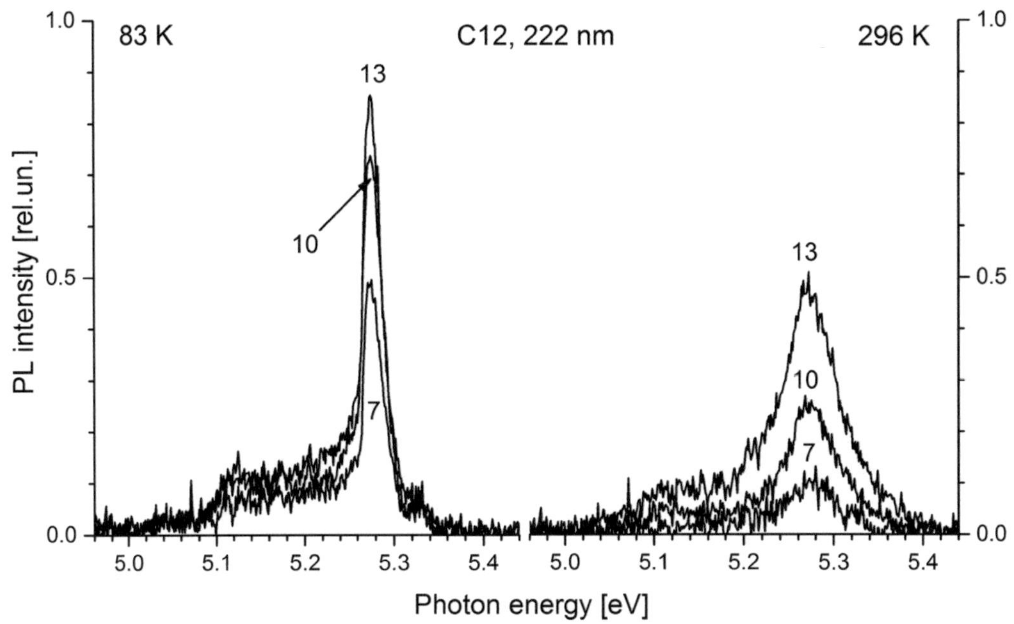

Figure 17. PL spectra of monocrystal HPHT diamond sample C12 under the excitation at 222 nm at temperatures of 83 K and 296 K. The PL spectra peaks are marked by peak intensities [7, 10 and 13 MW/cm^2] of laser radiation on the sample surface.

The EHL condensation was evident also in the temperature dependence of intensity of the dominant TO-component of FE recombination band, as shown in **Figure 18**. In conditions where condensation EHL was insignificant, the maximum of temperature dependence of FE band intensity was close to 150 K (see **Figure 18** for C10, 7 MW/cm^2). The peak position arises from the FE lifetime temperature dependence (Eq. 4). As the EHL band intensity increases with the increase in laser intensity (10 MW/cm^2), the additional maximum appeared at 190–220 K on the temperature dependence. As the intensity increases further (13 MW/cm^2), the maximum at 190–220 K becomes dominant.

http://dx.doi.org/10.5772/65064

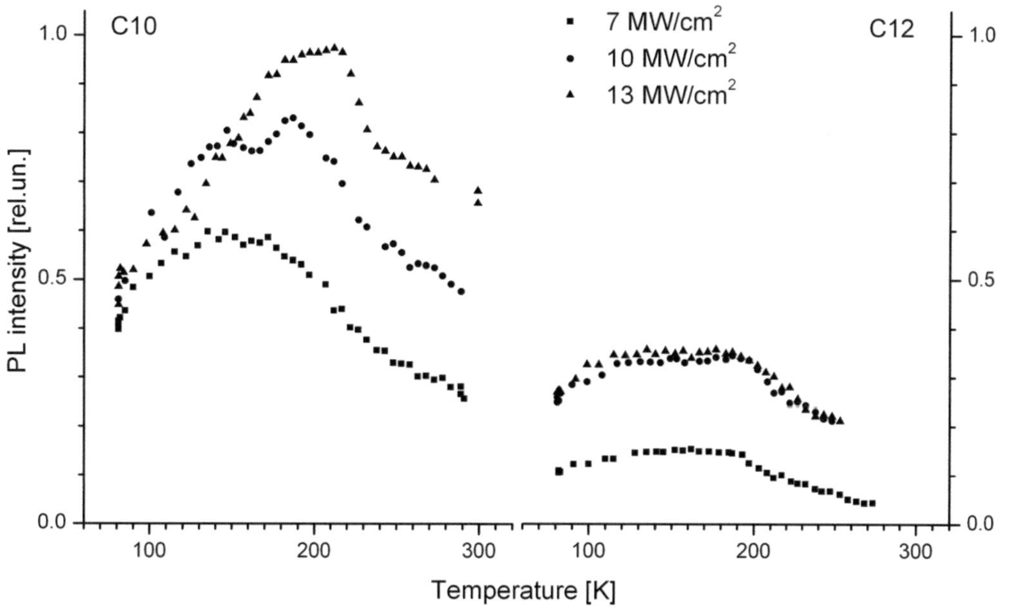

Figure 18. Temperature dependencies of FE recombination band intensities of two monocrystal diamond samples C10 and C12 at different excitation intensities at 222 nm. Squares-laser intensity 7 MW/cm^2, circles-10 MW/cm^2, triangles-13 MW/cm^2.

Note that for the C12 sample (**Figure 18**), the temperature dependences were differed. The FE band intensity remained practically unchanged in the temperature range of 130–190 K, i.e., the temperature maximum has been extended essentially, in contrast to the dependencies for the sample C10. With an increase in the peak laser intensity from 10 to 13 MW/cm^2, the FE recombination intensity remained approximately unchanged. The additional maximum at 190–220 K was not clearly manifested.

The analysis of dependencies in **Figures 17** and **18** leads to the conclusion that:

1. When EHL droplet size increases, the FE density decreases due to their loss in the process of EHL condensation (FE intensity maximum shifts on the temperature dependence);
2. The FE condensation can take place up to temperatures of ~220 K (EHL or EHP according to Mott);
3. The interaction cross-section of nonradiative recombination centers with EHL drops is higher than with FEs (no intensity increase for sample C12 with the peak laser intensity increase from 10 to 13 MW/cm^2).

The plot of phase diagram of electron-hole system in the diamond as the temperature dependence on the charge carrier density is an important problem, which is not completely solved yet. The results of the plot of phase diagram are summarized in **Figure 19**. Experimental data and theoretical models were used in [14, 41, 42].

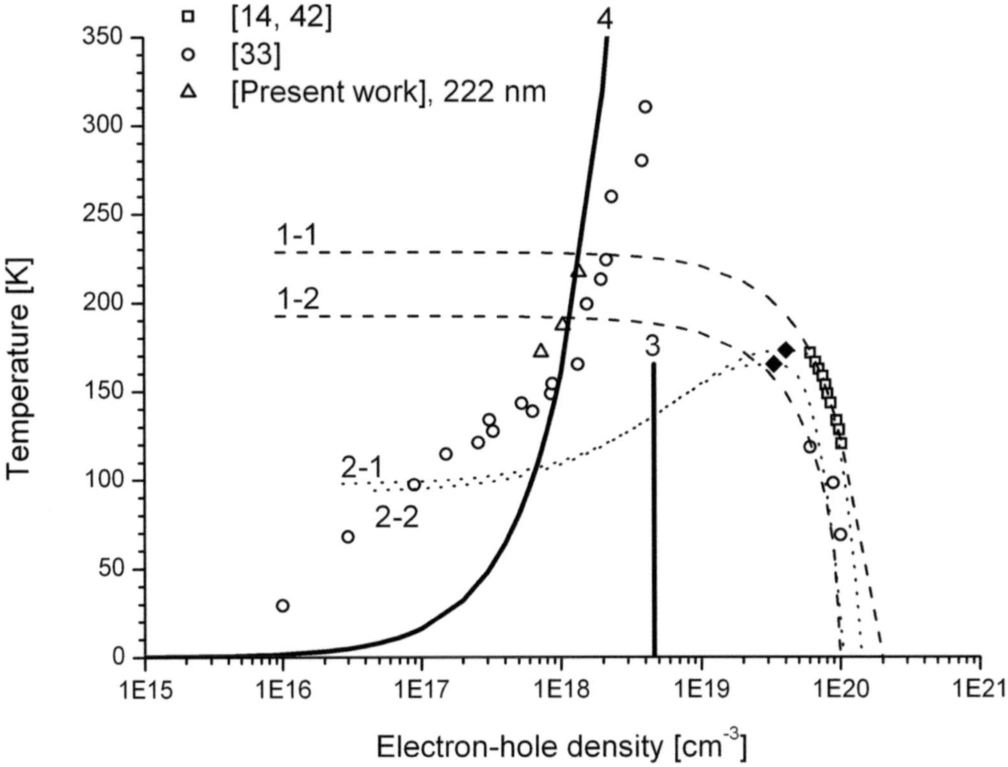

Figure 19. The phase diagram of the electron-hole system in diamond. White squares are experimental data from [14, 42]. White circles are experimental data from Ref. [33]. 1-1-degenerate plasma model $T_{crit} = 173$ K, $n_{crit} = 4.05 \times 10^{19}$ cm^{-3} from [42], 1-2-degenerate plasma model $T_{crit} = 165$ K, $n_{crit} = 3.33 \times 10^{19}$ cm^{-3} from [33], 2-1-liquid-gas transition boundary in Guggenheim model $T_{crit} = 173$ K, $n_0 = 1.41 \times 10^{20}$ cm^{-3} from [42], 2-2—liquid-gas transition boundary in Guggenheim model $T_{crit} = 165$ K, $n_0 = 1.0 \times 10^{20}$ cm^{-3} from [33], 3-Mott criterion in Thomas-Fermi approximation [33, 42], 4-Mott criterion in Debye-Huckel approximation [33, 42]. Black rhombs are two sets of critical temperature and EHP density: $T_{crit} = 160$ и 173 K, $n_{crit} = (3.33$ и $4.05) \times 10^{19}$ cm^{-3}. White triangles are experimental data of present work.

The experimental data [14, 42] were approximated (white squares in **Figure 19**) in the model of a degenerate plasma with the critical temperature $T_{crit} = 173$ K and the critical density $n_{crit} = 4.05\ 10^{19}$ cm^{-3} (curve 1-1 in **Figure 19**) and in the model of simple liquid near the critical point with the critical temperature $T_{crit} = 173$ K and the charge carrier density at the absolute zero of temperature $n_0 = 1.0 \times 10^{20}$ cm^{-3} (curve 2-1 in **Figure 19**). Note that in [14, 42] the PL excitation of diamond samples was provided by laser radiation with duration of 5–8 ns at FWHM.

In Refs. [33] and [41], the model of a degenerate plasma was used also with the critical temperature $T_{crit} = 160$ K and the critical density $n_{crit} = 3.33 \times 10^{19}$ cm^{-3} (curve 1-2 in **Figure 19**) and the model of simple liquid near the critical point [43] with a critical value temperature $T_{crit} = 160$ K and the charge carrier density at the absolute zero of temperature $n_0 = 1.0 \times 10^{20}$

cm^{-3} (curve 2-2 in **Figure 19**). In Ref. [33], the laser pulses with subpicosecond duration (200 fs) were used for PL excitation. These experimental data (white circles in **Figure 8**) were obtained not only for right side of the phase diagram at densities higher than the critical, but at densities less than densities of the Mott transition. In Thomas-Fermi approximation, the Mott transition has a threshold (curve 3 in **Figure 19**), but in Debye-Huckel approximation the Mott transition criterion increases monotonically with the FE density (curve 4 in **Figure 19**).

Note that the values of the experimental densities in **Figure 19** were obtained [14, 33, 42] through the form factors analysis of spectral components. But the density range of 4.1×10^{18}–6.0×10^{19} cm^{-3} was not investigated.

Black rhombs in **Figure 19** mark two sets of critical temperature and density for the EHL condensation. Different values of these quantities define the different calculated values in a degenerate plasma model and a phenomenological model of the gas-liquid transition (curves 1-1, 1-2, 2-1, 2-2 in **Figure 19**). As shown in **Figure 20**, the several different values of critical temperature and density for the EHL condensation were defined or calculated for the diamond [16, 31–33] and other semiconductors [30, 32]. Furthermore, in [42] it was postulated that the critical temperature for the EHL condensation is in the range of 180–207 K at PL excitation by laser pulses of nanosecond duration. At the same time, in Ref. [43], the EHL was observed at temperatures below 120 K at PL excitation by subpicosecond laser pulses. Note that the critical EHL condensation temperatures were determined from the shapes of PL spectra, but the EHL critical densities were the calculated parameters. Significant differences between the nanosecond PL excitation [14, 42] and the femtosecond PL excitation [32, 33, 43] were caused by:

1. Different modes of excitation—stationary and pulsed, respectively. At the stationary mode of excitation, the pulse duration is significantly longer than the lifetime of EHL drops (less than 1 ns), which leads to the nucleation of drops during the excitation pulse. At the pulsed mode of excitation, the duration of laser radiation is shorter in three orders of magnitude than the lifetime of EHL drops, so the PL spectra of EHL recombination display drops formed around the same time.

2. Different peak intensities of laser excitation—a few to tens of MW/cm^2 at the stationary mode and a few to tens of GW/cm^2 at the pulsed mode. The peak intensity of laser radiation of gigawatt level leads to the generation of electron-hole plasma with an initial density of more than 10^{20} cm^{-3} and causes the lattice heating due to the thermalization process of hot carriers and to the Auger recombination. This heating leads to a decrease in the critical temperature for the EHL condensation. For example, the authors of [44] were forced to use the understated initial FE density of 10^{18} cm^{-3} to perform calculations.

3. Different excitation wavelengths—218–220 and 215 nm, respectively. Due to the sharp increase in the absorption coefficient at the fundamental absorption edge of diamond [3], the shorter wavelengths are characterized by the thinner diamond layer in which absorption occurs. Therefore, at the same radiation intensity the shortwave radiation creates the higher initial FE density (see **Figure 2**). Moreover, the FE generation by shortwave photons

creates "hot" carriers, which transmit more energy to the diamond lattice in the thermalization process, and it reduces the critical temperature.

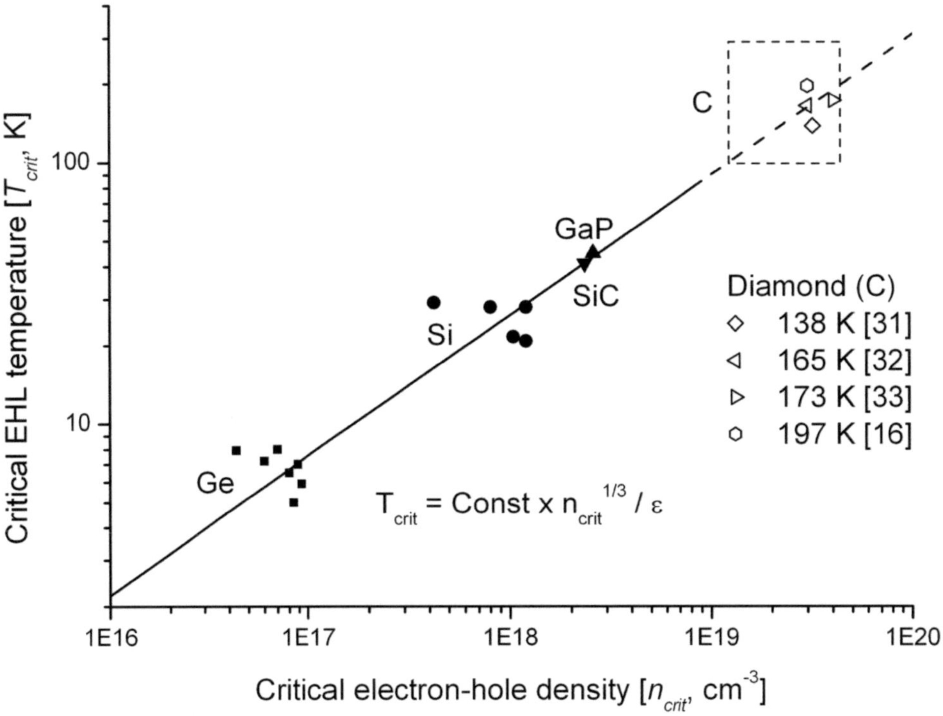

Figure 20. Critical EHL temperatures versus critical EHL densities in the several semiconductors. Black squares-germanium (Ge), black circles-silicon (Si), black down triangle-silicon carbide (SiC), black up triangle-gallium phosphite (GaP), white shapes-diamond (C). Data for Ge, Si, SiC and GaP referred to [30, 32], data for C (diamond)—[16, 31–33].

As it can be seen from **Figure 19**, the modeling of phase diagram of gas FE-EHL in the diamond describes satisfactorily the right branch, whereas, for the left branch the used models do not give a close description of experimental curves. The Mott criterion in Debye-Huckel approximation describes most closely the boundary of FE gas existence (left branch of the phase diagram) at the FE densities of $(1–4)\times10^{18}$ cm^{-3}. The values of critical temperatures and densities of experimental data of this work are marked as white triangles in **Figure 19**. Note that we have obtained the charge carrier densities according to expressions (1) and (2), but the values of critical temperature were taken from the positions of maxima on the temperature dependences of FE intensity (**Figure 18**).

Thus, the study of the EHL properties in the diamond has not yet responded to the key issues related to the conditions of condensation, evaporation and recombination. It remains still unclear what values of critical temperature and charge carrier density take place, and what their dependences on the impurity-defective composition of samples, the excitation energy of particles and the duration of excitation pulses are like. The phase diagram of EHL existence requires the additional data in the density range of 4.0×10^{18}–6.0×10^{19} cm^{-3}. An important aspect is to determine the temperature dependence of intensities of TA- and TO+O^{Γ}-components of EHL recombination band. Does EC band give a contribution to PL spectra (their position, intensity)? In addition, there are two different concentrated states of nonequilibrium charge carriers in the diamond—EHL and EHP—which can have the same densities under certain conditions. So, the question arises: what are the differences between these two states?

6. Applications of recombination radiation in the diamond

The most obvious application of recombination radiation in the diamond is producing of light emitting devices (LED) based on diamond. The recombination radiation in the luminescence spectrum of diamond is located in the spectral range of 230–245 nm near the fundamental absorption edge (225 nm) (see Section 3). The band-A of luminescence is located in the blue region of the visible spectrum with the maximum at 440 nm and has the recombination origin (see Section 4).

The LED's operating principle is based on the phenomenon of electroluminescence (EL)-radiative recombination of electron-hole pairs injected through contacts. The problem of making ohmic contacts to the diamond is solved in general [44]. Typically, the ohmic contacts to the diamond consist of three-layer metal deposition: carbide-former metal (Ti, Ta, Mo, W, Cr), interdiffusion-blocking layer (Pt) and protective/mounting layer (Au).

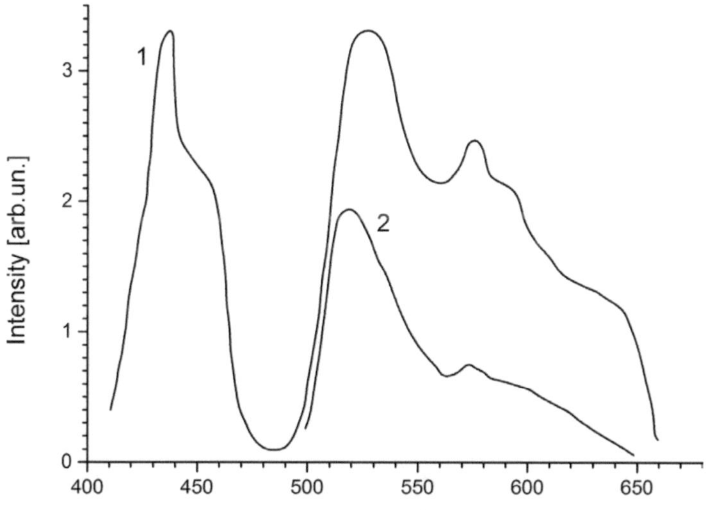

Figure 21. EL spectra of two natural diamond samples of IIb type referred to [46]. 1-sample without PL, 2-sample with yellow-green PL.

The first reports on EL in diamond were about semiconductor IIb type samples which were naturally doped with boron. The impurity of boron is an acceptor in diamond with an activation energy of 0.38 eV. In Ref. [45], the observation of the blue EL band of natural diamond sample with the maximum at 440 nm (the spectrum is not given) was reported. The EL was observed only at one polarity of bias of 50–100 B. In Ref. [46], a broadband EL (see **Figure 21**) was observed for several natural diamonds, presumably of IIb type (only a small fraction of the samples gave EL). In addition to the blue band with the maximum at 440 nm, the EL spectra of samples showed a broadband with several local maxima in the range from 500 to 650 nm. The green band of EL is appeared in the EL spectrum with a significant delay (tens of minutes). In general, the EL has demonstrated strong intensity fluctuations and has quenched when sample was cooled to a temperature below 100 K. This fact involves hopping conduction. In Ref. [47], the production of LED based on natural and HPHT diamonds was reported. The p- and n-types of conductivity were created by ion implantation of boron and lithium, respectively. The emission spectrum of resulting p-n transition is shown in **Figure 22**. The intensity of this transition decreases to zero with increasing temperature from -200 to 300°C, i.e., the band-A of luminescence took place.

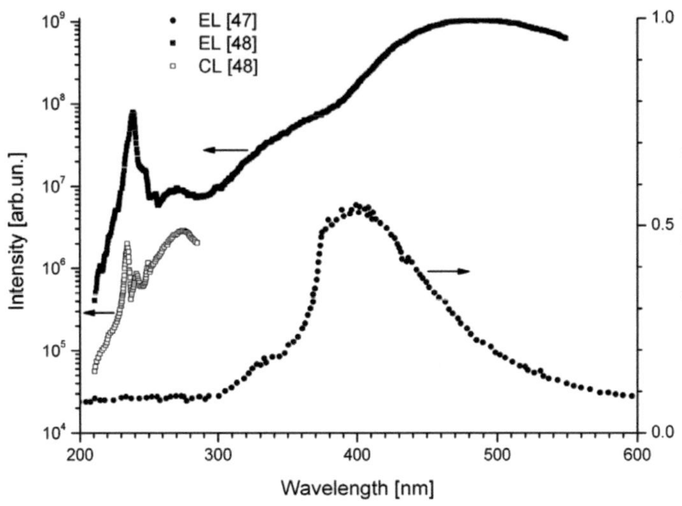

Figure 22. EL and CL luminescence spectra of diamond-based junctions referred to [47, 48]. The EL spectrum of p-n junction from [47] is showed by black circles. The p- and n-types of conductivity were created in the natural diamond by ion implantation of boron and lithium, respectively. The EL spectrum of diamond LED from [48] showed by black squares. The p- and n-types of conductivity were created in the CVD diamond during the synthesis process using admixtures of boron and phosphorus, respectively. The CL spectrum of the LED is shown as white squares.

The development of CVD-diamond synthesis and the ability to create p- (boron) and n- (0.47 eV phosphorus, lithium 0.29 eV) types of sample conductivity during the synthesis or during the subsequent ion implantation have led to reports on the EL research of CVD diamonds and the creation of LED based on various types of light-emitting junctions in diamond [48–62]. The voltage of only tens of volts was used.

In 2002, the first reports on diamond LED development, emitting in the UVC range at 235 nm, appeared. In Ref. [48], a diamond LED p-n junction was reported. The spectrum of LED emitted radiation is shown in **Figure 22**. The FE emission at 235 nm was characterized by intensity of an order of magnitude less than the intensity of band-A. So, the low intensity of FE band indicates the unsatisfactory quality of doped layers. In Ref. [49], a similar device showed the same characteristics: the presence of FE peak at 235 nm and the intense band-A.

In Refs. [50] and [51], the three-layer structure diamond:N-diamond:Ce-SiO_2 was studied. The EL spectrum showed the presence of wide and narrow bands in the spectral region 300–630 nm, arising from the band-A, transitions in cesium atoms and nitrogen complexes. The monotonically decreasing tail was observed from 630 nm to the IR region. This structure demonstrated the brightness of 3.5 cd/m^2 at RT and voltage of 150 V. In Ref. [52], the p-n junction was formed in a CVD single crystal by ion implantation of boron and lithium with subsequent ion implantation of xenon. The resulting device demonstrated the EL broadband in the spectral region from 450 to 850 nm at voltage 100 V and RT. There were the vibronic system of Xe centers with ZPL at 812.5 nm and the weak emission of N-V centers with ZPL at 575 nm.

In Ref. [53], a diamond-based p-i-p structure was described, which showed the EL emission at 235 nm. Later the same group produced a p-i-p-i-p device emitting FE recombination band [54]. In Ref. [55], a p-i-n transition doped with boron and phosphorus diamond layers was reported. At voltages of ±15 V, the p-i-n transition showed the dominant FE band at 235 nm, but the band-A was suppressed. In Ref. [56], the C-Si heterojunction demonstrated the red EL at RT. In Ref. [57], by ion implantation of B, As and P the p+-p-n structure was fabricated, which was based on p-type HPHT diamond. At RT the p+-p-n structure demonstrated the effect of superluminescence in the spectral region of band-A. In Ref. [58], the p-C/n-AlN-heterojunction showed the dominant FE recombination band at 235 nm in the diamond layer. In [59] the p-i-n structure based on boron and phosphorus (ion implanted) CVD diamond layers showed the EL of N-V defects at 575 nm. The authors of [60] suggested the method of creating graphite electrodes buried in a CVD diamond volume by the action of a focused MeV-ion beam (He^+). In the EL spectrum, there were the band-A, the ZPL at 575 nm of N-V centers, ZPL at 536.3 and 560.5 nm centers associated with helium and the broadband at 740 nm of Si-V or V centers.

We note that due to the superior thermal conductivity (higher than of copper in 5 times), the diamond samples are used also as passive heat sinks for LEDs based on other semiconductor materials [61, 62].

At the time being diamond LEDs are not widely used. The main reason is the need to improve the quality of the diamond structures. Another reason-the fundamental: diamond is a semiconductor with indirect-band structure, which reduces the efficiency of FE radiative recombination. At the same time, theoretical searches of ways to overcome the fundamental limitations of the diamond as a material for LEDs are being conducted. In [63] it was proposed to change the band structure of diamond to the direct-gap by using the linear compression of ~4% or lattice deformation by high doses of radiation damage. In [64] it was suggested to generate picosecond high-power pulses at 235 nm by using the superluminescence in diamond-cooperative spontaneous FE radiative recombination. The superluminescence is possible to be created in a ring resonator. At this condition, the Mott transition is not achieved.

Thus, the recombination radiation of diamond in the FE band (235 nm) or the band-A (400–650 nm) may be applied in LEDs. Despite the fact that the EHL is an interesting phenomenon in terms of quantum effects, its practical use as a radiation source or any other thing has not been found. The situation is the same for all semiconductors. At the same time, diamond has a highest critical temperature of EHL condensation (>160 K), which still gives hope for the practical application.

In Ref. [38], the strengthening of photoconductivity in the CVD diamond under conditions of EHL existing was reported. Unlike a single FE, drops of EHL are not electrically neutral particle clusters. Due to the different work function of electrons and holes (due to the different effective masses), the EHL drops have a noncompensated surface charge [13]. Consequently, the EHL drops may drift in the electric field and participate in the photocurrent. One can assume that the EHL drops in diamond could be used as controlled sources of photons, dynamic quantum dots, etc.

7. Conclusion

This chapter provides the original experimental data on the recombination radiation of natural and synthetic diamonds and the overview of the literature about.

The FE recombination radiation in the diamond is observed in CL, PL and EL spectra in the shape of the four-component band containing TA-, TO-, LO- and TO+O^{Γ}-components at 5.325, 5.275, 5.254 and 5.117 eV, respectively. The "non-diamond" carbon phase and impurity-defect centers have the negative effect on the intensity of FE recombination band. The spectral position of FE band maximum at 5.275 eV did not change within the measurement error in the temperature range of 80–300 K, which contradicts the phenomenon of band gap decreasing with the increase in temperature. At low excitation levels, the temperature dependence of the FE recombination band intensity has the maximum at ~150 K, which is in agreement with published data. The band-A, caused by intrinsic defects of the structure containing the sp^2-hybridized carbon bonds, is located in the spectral range of 350–650 nm with the maximum at about 440 nm and is characterized by the decay time of 8–19 ms in the temperature range of 80–300 K. The band-A reflects the perfection degree of diamond sample, i.e., the higher the band-A intensity is, the greater the structure defect density is. There is a competition between the band-A and FE recombination band in diamond. The emission spectrum of the band-A may be imposed by vibronic bands of N_3V, N_2V, NV centers, which are characterized by decay times in the nanosecond time range. Therefore, the measurement of luminescence emission spectra with time resolution allows to separate in time the spectra of band-A and other structural defects.

The recombination radiation of EHL drops in diamond is observed in CL and PL spectra at temperatures ≤200 K and peak carrier densities $\geq(0.3\text{–}1.0)\times10^{18}$ cm^{-3}. The EHL band consists of three phonon components: TA, TO and TO+O^{Γ} at 5.249, 5.211 and 5.075 eV, respectively. The dominant TO-component of EHL band begins to appear in the form of long-wavelength shoulder of TO-component of FE band when the charge carrier density reaches the critical value. A further increase in carrier density leads to an increase in EHL band intensity. The surface layer of diamond sample is more defective than the inner volume. Therefore, the absorption of exciting radiation in a thin layer hampers EHL condensation compared with volume absorption. The EHL condensation results in displacement of FE band intensity maximum on the temperature dependence to higher temperatures up to 220 K, which can be interpreted as a decline of excitons in the EHL condensation process at lower temperatures. In this context, the critical temperature of EHL condensation is not established and, presumably, takes values in the range of 160–220 K. In this regard, the phase diagram of EHL existence requires clarification.

Currently, fabricated LEDs based on p-n, p-i-n, p-i-p and other junctions in diamond structures have showed the possibility of spontaneous emission in the UV range at 235 nm and in the visible range of 400–800 nm. However, the diamond LEDs do not satisfy the requirements of efficiency and emission brightness. The main reason is the indirect-band zone structure of diamond, which leads to a long FE lifetime and low efficiency of the three-particle radiative recombination. To overcome this fundamental limitation, it was suggested to use the local deformation of the crystal, resulting in direct-gap zone structure in diamond and the super-luminescence in the ring resonator. The prospect of finding the solution to increase the efficiency and the emission brightness of the diamond-based LEDs is associated with the possibility of obtaining the compact devices that are emitting in the UV-C spectrum.

Acknowledgements

This report was prepared with the support of the Russian Foundation for Basic Research, Project No. 14-02-31132.

Author details

Evgeniy Igorevich Lipatov*, Dmitriy Evgen'evich Genin, Denis Valer'evich Grigor'ev and Victor Fedotovich Tarasenko

*Address all correspondence to: lipatov@loi.hcei.tsc.ru

Institute of High Current Electronics SB RAS, Tomsk, Russian Federation

References

[1] Pelant I, Valenta J. Luminescence spectroscopy of semiconductors. New York: Oxford University Press; 2012. 542 p. doi:10.1093/acprof:oso/9780199588336.001.0001

[2] Ruf T, Cardona M, Sternschulte H, et al. Cathodoluminescence investigation of isotope effects in diamond. Solid State Communications. 1998;105:311–316. doi:10.1016/S0038-1098(97)10196-X

[3] Lipatov EI, Genin DE, Tarasenko VF. Optical spectra and electron–hole liquid in diamond. Optics and Spectroscopy. 2015;119:918–923, doi:10.1134/S0030400X15120164

[4] Lipatov EI, Avdeev SM, Tarasenko VF. Photoluminescence and optical transmission of diamond and its imitators. Journal of Luminescence. 2010;130:2106. doi:10.1016/j.jlumin.2010.06.001

[5] Novikov N.V, editor. Handbook of Physical Properties of Diamond (in Russian). Kiev: Naukova dumka; 1987. 1132 p.

[6] Salvatori S., Pace E., Rossi MC, Galluzzi F. Photoelectrical characteristics of diamond UV detectors: dependence on device design and film quality. Diamond and Related Materials. 1997;6:361. doi:10.1016/S0925-9635(96)00757-1

[7] Lipatov EI, Lisitsyn VM, Oleshko VI, et al. Pulsed cathodoluminescence of natural and synthetic diamonds excited by nanosecond and subnanosecond electron beams. In: Yamamoto N, editor. Cathodoluminescence. Rijeka, Croatia: InTech; 2012. p. 51–70. doi: 10.5772/1989

[8] Yokota Y, Tachibana T, Miyata K, et al. Cathodoluminescence of boron-doped heteroepitaxial diamond films on platinum. Diamond and Related Materials. 1999;8:1587–1591. doi:10.1016/S0925-9635(99)00085-0

[9] Sharp SJ, Collins AT, Davies G, Joyce GS. Higher resolution studies of shallow bound exciton luminescence in diamond. Journal of Physics: Condensed Matter. 1997;9:L451–L455. doi:10.1088/0953-8984/9/33/003

[10] Gaman VI. Physics of semiconductor devices: Tutorial (in Russian). Tomsk: NTL Publishing House; 2000. 426 p.

[11] Clark CD, Dean PJ, Harris PV. Intrinsic edge absorption in diamond. Proceedings of the Royal Society of London A. 1964;277:312–329. doi:10.1098/rspa.1964.0025

[12] Robins LH, Farabaugh EN, Feldman A. Cathodoluminescence spectroscopy of free and bound excitons in chemical-vapor-deposited diamond. Physical Review B. 1993;48:14167–14181. doi:10.1103/PhysRevB.48.14167

[13] Keldysh LV. The electron-hole liquid in semiconductors. Contemporary Physics. 1986;27:395–428. doi:10.1080/00107518608211022

[14] Sauer R, Teofilov N, Thonke K. Exciton condensation in diamond. Diamond and Related Materials. 2004;13:691–699. doi:10.1016/j.diamond.2003.10.005

[15] Teofilov N, Sauer R, Thonke K, Koizumi S. Bound exciton luminescence related to phosphorus donors in CVD diamond. Physica B. 2003;340–342:99–105. doi:10.1016/j.physb.2003.09.011

[16] Lipatov EI, Genin DE, Tarasenko VF. Recombination radiation in synthetic and natural diamonds exposed to pulsed UV laser radiation. Russian Physics Journal. 2015;58:911–922. doi:10.1007/s11182-015-0590-x

[17] Takiyama K, Abd-Elrahman MI, Fujita T, Oda T. Photoluminescence and decay kinetics of indirect free excitons in diamonds under the near-resonant laser excitation. Solid State Communications. 1996;99:793–797. doi:10.1016/0038-1098(96)00309-2

[18] Fujii A, Takiyama K, Maki R, Fujita T. Lifetime and quantum efficiency of luminescence due to indirect excitons in a diamond. Journal of Luminescence. 2001;94–95:355–357. doi:10.1016/S0022-2313(01)00299-X

[19] Sobolev EV, Dubov YuI. On origin of diamond roentgenoluminescence (in Russian). Soviet Physics-Solid State. 1975;17:1142.

[20] Dean PJ. Bound excitons and donor-acceptor pairs in natural and synthetic diamond. Physical Review. 1965;139:588–602. doi:10.1103/PhysRev.139.A588

[21] Manfredotti C, Fizzotti F, Lo Giudice A, et al. Ion beam induced luminescence maps in CVD diamond as obtained by coincidence measurements. Diamond and Related Materials. 1999;8:1592–1596. doi:10.1016/S0925-9635(99)00051-5

[22] Walker J. Optical absorption and luminescence in diamond. Reports on Progress in Physics. 1979;42:1607–1659. doi:10.1088/0034-4885/42/10/001

[23] Prins JF. Recombination luminescence from defects in boron-ion implantation-doped diamond using low fluencies. Materials Research Innovations. 1998;1:243–253.

[24] Lawson SC, Kanda H, Kiyota H, et al. Cathodoluminescence from high-pressure synthetic and chemical-vapor-deposited diamond. Journal of Applied Physics. 1995;77:1729–1734. doi:10.1063/1.358865

[25] Nadolinny VA, Yureva OP, Eliseev AP, et al. The destruction of the nitrogen B1-centers in plastic deformation of natural diamonds of IaB type and the behavior of formed defects in the P,T-processing (in Russian). Doklady Akademii Nauk. 2004;399:532–536.

[26] Anoikin E, Muhr A, Bennett A, et al. Diamond optical components for high-power and high-energy laser applications. In: Proceedings of the SPIE; 2015. p. 93460T-1

[27] Takeuchi D, Watanabe H, Yamanaka S, et al. Origin of band-A emission in diamond thin films. Physical Review B. 2001;63:245328. doi:10.1103/PhysRevB.63.245328

[28] Cleri F, Keblinski P, Colombo L, et al. On the electrical activity of sp^2-bonded grain boundaries in nanocrystalline diamond. Europhysics Letters. 1999;46:671–677. doi:10.1209/epl/i1999-00318-5

[29] Lipatov EI, Tarasenko VF, Lisitsyn VM, Oleshko VI. Spectral and kinetic characteristics of the pulsed cathodoluminescence of a natural IIA-type diamond. Russian Physics Journal. 2007;50:52–57. doi:10.1007/s11182-007-0005-8

[30] Tikhodeev SG. The electron-hole liquid in a semiconductor, Soviet Physics Uspekhi. 1985;28:1–30.

[31] Vouk MA. Conditions necessary for the formation of the electron-hole liquid in diamond and calculation of its parameters. Journal of Physics C: Solid State Physics. 1979;12:2305–2312. doi:10.1088/0022-3719/12/12/016

[32] Shimano R, Nagai M, Horiuch K, Kuwata-Gonokami M. Formation of a high Tc electron-hole liquid in diamond. Physical Review Letters. 2002;88:057404. doi:10.1103/PhysRevLett.88.057404

[33] Nagai M, Shimano R, Horiuchi K, Kuwata-Gonokami M. Phase diagram of the quantum degenerate electron–hole system in diamond. Physica Status Solidi (b). 2003;238:509–512. doi:10.1002/pssb.200303171

[34] Thonke K, Schliesing R, Teofilov N, et al. Electron-hole drops in synthetic diamond. Diamond and Related Materials. 2000;9:428–431.doi:10.1016/S0925-9635(99)00315-5

[35] Kozak M, Trojanek F, Popelar T, Maly P. Dynamics of electron-hole liquid condensation in CVD diamond studied by femtosecond pump and probe spectroscopy. Diamond and Related Materials. 2013;34:13–18. doi:10.1016/j.diamond.2013.01.008

[36] Nakazawa K, Umezawa H, Tachiki M, Kawarada H. Non-linear increases in excitonic emission in synthetic type-IIa diamond. Diamond and Related Materials. 2003;12:1995–1998. doi:10.1016/S0925-9635(03)00194-8

[37] Saarela M, Taipaleenmaki T, Kusmartsev FV. Clustering of the electron-hole liquid. Physica E: Low-Dimensional Systems and Nanostructures. 2003;18:339–340. doi: 10.1016/S1386-9477(02)01078-0

[38] Lipatov EI, Genin DE, Tarasenko VF. Pulse photoconductivity of diamond at the quasi-stationary laser excitation at 222 nm under the conditions of existence of the electron-hole liquid. JETP Letters. 2016;103:663–668.

[39] Kozak M, Trojanek F, Maly P. Large prolongation of free-exciton photoluminescence decay in diamond by two-photon excitation. Optics Letters. 2012;37:2049–2051. doi: 10.1364/OL.37.002049

[40] Ivakin EV, Kisialiou IG, Ralchenko VG, et al. Investigation of free charge carrier dynamics in single-crystalline CVD diamond by two-photon absorption. Quantum Electronics. 2014;44:1055–1060. doi:10.1070/QE2014v044n11ABEH015540

[41] Nagai M, Shimano R, Horiuchi K, Kuwata-Gonokami M. Creation of supercooled exciton gas and transformation to electron-hole droplets in diamond. Physical Review B: Condensed Matter and Materials Physics. 2003;68:081202. doi:10.1103/PhysRevB.68.081202

[42] Teofilov N, Schliesing R, Thonke K, et al. Optical high excitation of diamond: phase diagram of excitons, electron-hole liquid and electron-hole plasma.

Diamond and Related Materials. 2003;12:636–641. doi:10.1016/S0925-9635(02)00386-2

[43] Naka N, Omachi J, Sumiya H, et al. Density-dependent exciton kinetics in synthetic diamond crystals. Physical Review B: Condensed Matter and Materials Physics. 2009;80:035201. doi:10.1103/PhysRevB.80.035201

[44] Werner M. Diamond metallization for device applications. Semiconductor Science and Technology. 2003;18:S41–S46. doi:10.1088/0268-1242/18/3/306

[45] Wolfe R, Woods J. Electroluminescence of semiconducting diamonds. Physical Review. 1957;105:921–922.

[46] Korsun' VM, Mal'tsev EK, Perekrestova LG, et al. Certain characteristics of electroluminescence in natural diamonds. Journal of Applied Spectroscopy. 1969;11:1211–1213.

[47] Melnikov AA, Denisenko AV, Zaitsev AM. Electrical and optical properties of light-emitting p-i-n diodes on diamond. Journal of Applied Physics. 1998;84:6127–6134. doi: 10.1063/1.368880

[48] Koizumi S., Watanabe K., Hasegawa M., Kanda H. Formation of diamond p-n junction and its optical emission characteristics. Diamond and Related Materials. 2002;11:307–311. doi:10.1016/S0925-9635(01)00537-4

[49] Wang WL, Liao KJ, Cai CZ, et al. Ultraviolet electroluminescence at room temperature from a pn junction of heteroepitaxial diamond film by CVD // Diamond and related materials. 2003;12:1385–1388. doi:10.1016/S0925-9635(03)00111-0

[50] Wang X, Wang L, Zhang Q, et al. Electroluminescence spectrum shift with switching behaviour of diamond thin films. Chinese Physics Letters. 2003;20:1868–1870.

[51] Wang X, Wang L, Zhang B, et al. Electroluminescence of diamond:Ce thin films. Semiconductor Science and Technology. 2003;18:144–146.

[52] Zaitsev AM, Bergman AA, Gorokhovsky AA, Huang M. Diamond light emitting diode activated with Xe optical centers. Physica Status Solidi. A: Applications and Materials Science. 2006;203:638–642. doi:10.1002/pssa.200521125

[53] Yamamoto M, Watanabe T, Hamada M, et al. Electrical properties of high-quality CVD diamond p-i-p structures at high electric fields. Applied Surface Science. 2005;244:310–313. doi:10.1016/j.apsusc.2004.10.136

[54] Watanabe T, Teraji T, Ito T. Fabrication of diamond p-i-p-i-p structures and their electrical and electroluminescence properties under high electric fields. Diamond and Related Materials. 2007;16:112–117. doi:10.1016/j.diamond.2006.04.003

[55] Makino T, Tokuda N, Kato H, et al. Electrical and light-emitting properties of (001)-oriented homoepitaxial diamond p-i-n junction. Diamond and Related Materials. 2007;16:1025–1028. doi:10.1016/j.diamond.2007.01.024

[56] Yang D, Chen J, Sekiguchi T, et al. Enhanced red electroluminescence from a polycrystalline diamond film/Si heterojunction structure. Applied Physics Letters. 2007;90:161123. doi:10.1063/1.2730584

[57] Buga SG, Blank VD, Bormashov VS, et al. P-n junction on high-pressure-high-temperature grown single crystal diamond: UV-emission spectra and electrical properties. Journal of Physics: Conference Series. 2008;121/3:032005. doi:10.1088/1742-6596/121/3/032005

[58] Kazuyuki H, Yoshitaka T, Makoto K. Electroluminescence and capacitance-voltage characteristics of single-crystal n-type AlN (0001)/p-type diamond (111) heterojunction diodes. Applied Physics Letters. 2011;98:011908–3. doi:10.1063/1.3533380

[59] Lohrmann A, Pezzagna S, Dobrinets I, et al. Diamond based light-emitting diode for visible single-photon emission at room temperature. Applied Physics Letters. 2011;99:251106. doi:10.1063/1.3670332

[60] Forneris J, Battiato A, Gatto Monticone D, et al. Electroluminescence from a diamond device with ion-beam-micromachined buried graphitic electrodes. Nuclear Instruments and Methods in Physics Research Section B. 2015;348:187–190. doi:10.1016/j.nimb.2014.12.036

[61] Man W, Sun L, Xie P, et al. Microwave CVD diamond thin film for led heat spreader. Diamond and Abrasives Engineering. 2008;2:1–4.

[62] Vilisov AA, Remnev GE, Linnik SA. Light-emitting diode with CVD diamond heat sink (in Russian). Russian Physics Journal. 2013;56:169–171.

[63] Hora H, Prelas MA. Theoretical aspects of diamond films and laser action. Diamond and Related Materials. 1995;4:1376–1382. doi:10.1016/0925-9635(95)00325-8

[64] Kukushkin VA. Simulation of ultraviolet- and Soft X-ray-pulse generation as a result of cooperative recombination of excitons in diamond nanocrystals embedded in a polymer film. Semiconductors. 2013;47:1442–1446. doi:10.1134/S1063782613110122

Chapter 10

Trap Level Measurements in Wide Band Gap Materials by Thermoluminescence

Pooneh Saadatkia, Chris Varney and Farida Selim

Additional information is available at the end of the chapter

http://dx.doi.org/10.5772/65306

Abstract

The electrical and optical properties of wide band gap materials are greatly affected by the presence of defects in the band gap. Identification and characterization of these defects that act as electron and hole traps are essential to understand charge carrier and exciton dynamics and ultimately control the electrical and optical properties of dielectrics and semiconductors. In this chapter, we will demonstrate how thermoluminescence (TL) spectroscopy can be used to characterize traps and measure their energy levels in the band gap. An advanced wavelength-based TL spectrometer will be presented, and its applications for the evaluation of trap levels and the characterization of donors and acceptors in semiconductors and dielectrics will be discussed.

Keywords: dielectrics and semiconductors, wide band gap materials, trap levels, exciton dynamics, luminescence

1. Introduction

The electrical and optical properties of materials are greatly affected by the presence of deep and shallow traps in the band gap. Deep traps capture charge carriers and prevent their recombination while shallow traps affect the transport of charge carriers and excitons in the lattice. Thus, identification and characterization of the electron and hole traps are essential to understand exciton and charge carrier dynamics and ultimately control the electrical and optical properties of dielectrics and semiconductors. Thermoluminescence (TL) spectroscopy can offer a powerful tool for the study of traps and the measurement of their depth in the band gap. It is commonly used for dosimetry and has also been applied for defect investigation and radiation damage studies. Recently, we have developed an advanced wavelength-based TL spectrometer for the evaluation of trap levels and the characterization of donors and acceptors and employed it to study a number of semiconductors and dielectrics.

This chapter explains in detail the basics of TL spectroscopy and introduces the reader to its use an effective method for trap and donor and acceptor characterization. Then it describes the experimental details of an advanced TL spectrometer, explains how to construct TL glow curves, and discusses different approaches for the calculations of trap levels in the band gap. The chapter also discusses in brief some of the complimentary methods that can be used in parallel with TL for the full characterization of traps including their structure, type, density, and energy level. Examples of these methods include positron annihilation and optical absorption spectroscopies. A few TL studies of traps in semiconductors and dielectrics are presented.

The chapter is organized as follows. It begins in Section 2 with a historical background about TL and its applications to radiation dosimetry and defect studies then it discusses the TL process and physics in Section 3. Section 4 gives a complete analysis for activation energy and trap level calculations including initial rise method, different heating rate method for first-order kinetics, and finally, it describes an analysis of general order kinetics. In Section 5, we describe the standard TL setup "TL reader", then introduce our advanced TL spectrometer that is especially useful for the study of traps in dielectrics and the characterization of donors and acceptors in semiconductors. Sections 6 and 7 discuss in detail the application of TL in the study of wide band gap materials by presenting our recent studies on yttrium aluminum garnet (YAG) as an example of dielectrics and ZnO as an example of wide band gap semiconductors.

2. Historical background

In Latin "lumen" means light, the term "luminescent" materials refer to materials that can emit light, especially in the visible range such as crystals, minerals, and chemical substances. Luminescence phenomena have been known for long times [1], it does not occur spontaneously, it needs some source of energy such as visible light, ultraviolet light, or X-rays in order to excite the sample [2]. Cascariolo [3] made the first attempt to create luminescence artificially from Bolognian stone in 1603. Afterwards, the stone and other luminescent materials were named "phosphor". In 1852, Stokes [4] formulated luminescence law, the energy of emission is typically less than the energy of absorption; therefore, luminescence occurs at lower energies or longer wavelengths. Later in 1888, Weidemann [5] used the term "luminescenz" for all phenomena of light not produced through increased temperature, i.e., incandescence. Luminescence is classified to many types such as photoluminescence, cathodoluminescence, thermoluminescence, and X-ray luminescence [6, 7].

Measuring material properties as a function of temperature is called thermally stimulated processes (TSPs) that are usually as follows: first, the sample is perturbed, in most cases by exposure to some source of ionizing radiation, then by elevating the temperature, it is thermally stimulated back to equilibrium that releases accumulated energy as light emission in TL or as current in thermally stimulated conductivity [8]. We will limit our discussion and literature review in this chapter to TL spectroscopy. Existence of thermoluminescence in crystals is dependent on the presence of crystal defects that are capable of absorbing some part of incident energy during irradiation, and only emit light after heating the material [9–13].

Different types of irradiation sources can be used to excite the samples in TL measurement such as γ-rays, X-rays, and UV radiation. The dependence of energy absorbed during irradiation, which is termed "radiation dose" and defined as dose rate × time of exposure makes TL very effective in radiation dosimetric applications [14, 15] in medicine, industry, biology and agriculture, and also in dating of ancient pottery sherds and certain rocks [16, 17]. Besides dosimetric applications, thermoluminescence spectroscopy is a versatile tool for defect studies in crystals and trapping phenomena at the grain boundaries in nanostructures and soft materials.

The TL glow curve represents the intensity of emitted light versus temperature; each glow peak is associated with recombination center and related to specific trap. Activation energy and escape frequency factor can be calculated from the glow curve [18]. As mentioned in the introduction, TL is an important technique for investigating of nature and depth of traps in the different type of solids [19, 20], and various methods can be used to calculate trap parameters based on the kinetics order of glow peaks such as initial rise method and variable heating rates that will be discussed later. Heating rate has critical effect on the intensity and peak area of thermoluminescence glow peaks [21]. Variable heating rate techniques including very slow heating rate down to 0.0008°C/s were used to investigate the TL of $CaSO_4$: Tm phosphor. Using slow heating rates helped to resolve eight narrow components in the TL glow curve [22]. Because of inner shell electronic transitions, rare-earth ions and transition metal ions are considered excellent luminescence centers [23–28]. For example, TL measurement was applied to investigate the luminescence efficiency of rare-earth ions doped $SrTiO_3:Pr^{3+}$ [28]. The authors used a 100 W deuterium lamp for excitation and measured emission from as-fired $SrTiO_3:Pr^{3+}$ and annealed samples in reduced atmosphere (5% H_2 and 95% Ar). They observed a considerable decrease in the TL intensity of the annealed samples, which were attributed to the suppression of traps. They also tried to increase the luminescence efficiency of $SrtiO_3:Pr^{3+}$ by adding impurities. After adding Al, new peaks were appeared, it seems that above 150 k, Al^{3+} ions are responsible for providing free carriers, which help to improve the luminescence efficiency of Pr^{3+} [28].

Recently, nanosized phosphor has fascinated researchers, and its applications have been emerged in different fields of materials especially luminescence area [29]. They have great potential for high dose detection in dosimetry applications [30, 31]. ZrO_2 nanophosphor is a good example where TL spectroscopy was successfully applied to study its properties [30]. Different concentrations of titanium were used as dopants to study TL behavior of ZrO_2 nanophosphors [32]. Samples were exposed to 254 nm UV for variable times in order to study the effect of time exposure and dose on TL intensity. Trapping parameters, i.e., activation energy and frequency factor were also calculated with different heating rate and peak shape methods. By increasing the heating rate, peaks position were shifted to higher temperature side. The sample was found to be useful for TL dosimetry applications due to linear dose response, high stability, and low fading phenomenon. Two peaks were observed at 167 and 376°C with fixed positions, and enhancing or quenching of intensity was found to be strongly dependent on dopant concentrations. Similarly, Tamrakar et al. [33] used TL spectroscopy to study the effect of different concentration of Er^{3+} as dopant for gadolinium oxide nanophosphor ($Gd_2O_3:Er^{3+}$ phosphor). They found shallow and deep traps when using

254 nm UV irradiation and Co^{60} gamma irradiation, respectively. In addition, gamma-irradiated samples indicated low fading and higher stability in comparison with samples exposed to UV. However, UV-irradiated Mn-doped $CaYAl_3O_7$ phosphor indicates higher intensity compared to the samples irradiated by Gamma and megavoltage beam, indicating that by exposing to lower energy radiations, large numbers of shallower traps were formed [34]. Gamma rays have also been applied to irradiate pure LiF crystals to investigate trap levels and their connections to color centers generated during ionization process. The authors reported the relation of four color centers to four resolved and six unresolved glow peaks, which shows that TL spectroscopy is quite useful [35].

It is important to mention that the physical and optical properties of individual nanosize phosphor materials are often not the same as those of their bulk counterparts [36–39]. For example, activation energy and trap parameters were calculated by Tamrakar et al. [37, 40] from TL glow curve of bulk CdS and CdS: Cu phosphor nanoparticles using Chen's peak shape methods [8, 41–43]. The former was found to exhibit first-order kinetics while the second one showed general order kinetics in TL emission implying the presence of charge retrapping in the nanoparticles. They also found that the level of excitation, presence of other traps, and preionized luminescence centers can affect the shape, intensity, and position of a glow curve.

3. Basics of TL process

In a typical TL measurement, a sample is irradiated with high-energy light for a duration, which excites electrons from the valence band or from the dopant levels within the band gap to the conduction band. The excited state is not stable, so after some time the charge carriers recombine. A fraction of charge carriers may be trapped at a defect or impurity and cannot recombine with their counterpart. After removal of the excitation, heating injects thermal energy into the system, providing the trapped charge carriers with the energy necessary to escape their traps. These released charge carriers can either recombine with their charge carrier counterpart at luminescence centers, giving rise to TL emission, or they can become trapped at a deeper, higher-energy trap. They possess too much energy at this point to retrap at the same type of trap they just escaped or any shallower trap. The probability of detrapping increases with raising temperature. **Figure 1** presents a schematic diagram of the process of thermal stimulation and recombination of trapped charge carriers. TL emission intrinsically occurs at one or few luminescence centers. Thus, TL measurements provide a reliable method by which to record small luminescence intensities. This makes TL an ideal method for the characterization of trap levels in insulating crystals such as yttrium aluminum garnet ($Y_3Al_5O_{12}$) [44–58].

4. Activation energy and trap level calculations

A plot of thermoluminescence intensity versus temperature is known as a glow curve. Analysis of glow curves can provide quantitative information about trap levels. Most notably, one may use glow curve analysis to calculate the depth of the charge carrier traps. The calculated

http://dx.doi.org/10.5772/65306

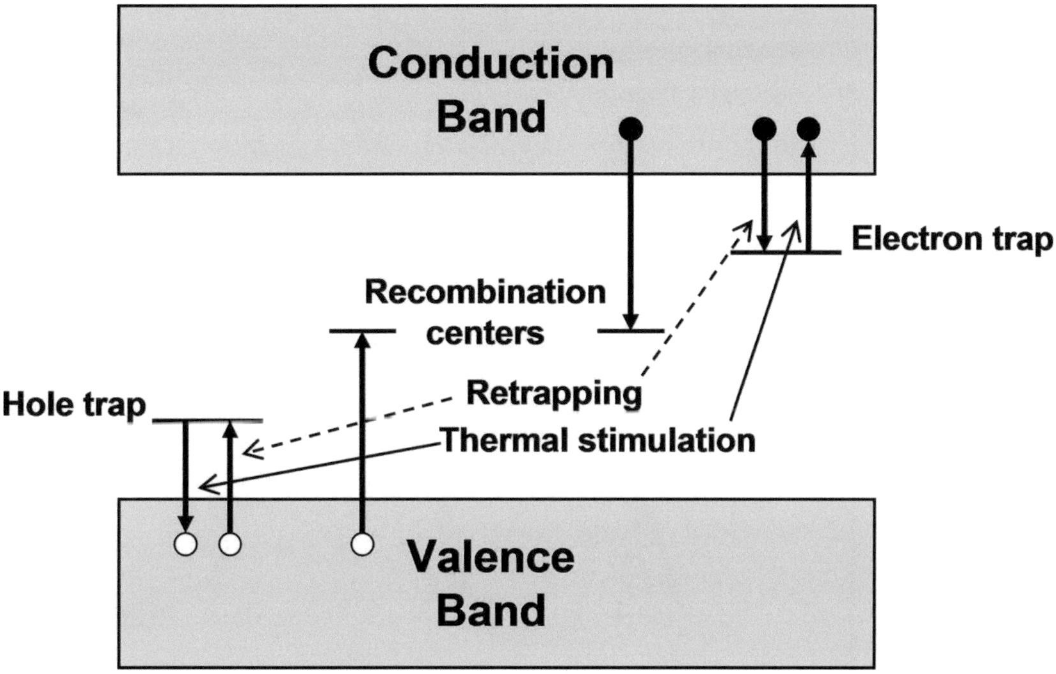

Figure 1. Schematic diagram of the TL process after the thermal activation of trapped charge carriers [55].

depth of a trap is referred to as its activation energy. To explain TL phenomena, Randall and Wilkins presented a model based on a single defect level within the band gap and assumed negligible retrapping, in which case the charge carrier can go straight to a luminescence center [59–61]. This is based on first-order kinetics of TL. Higher order kinetics assumes higher degrees of retrapping.

There are several methods to calculate activation energy from glow curves based on TL kinetics. The initial rise method calculates the activation energy directly by fitting the beginning of TL peak and is applicable to any order kinetics but can only be applied for nonoverlapping glow curves, where the peak being studied initially rises from the baseline [2, 41, 55, 62–64]. Other methods of activation energy calculation rely on conducting the experiment multiple times with different heating rates. At a faster heating rate, a glow curve shifts to higher temperature [2, 41, 65]. This effect is shown in **Figure 2** for glow curves of YAG doped with Ce to 0.2% concentration (Ce:YAG 0.2%) at heating rates of 5, 10, 20, 30, 40, 50 and 60°C/min. The method of two heating rates uses two separate constant heating rates to cancel unknown terms in kinetics calculations, thus rendering activation energy calculation trivial [2, 41, 55, 65–66]. This method is limited by peak fitting capabilities, and since peak temperature is usually not very sensitive to the heating rate, can give large uncertainty.

This method is typically applied to first-order kinetics, but it is possible to apply this method to any order kinetics. Following the same principles as the method of two heating rates is the variable heating rate method, where several glow curves using different heating rates are constructed [2, 55]. As with the method of two heating rates, unknown terms cancel in kinetics

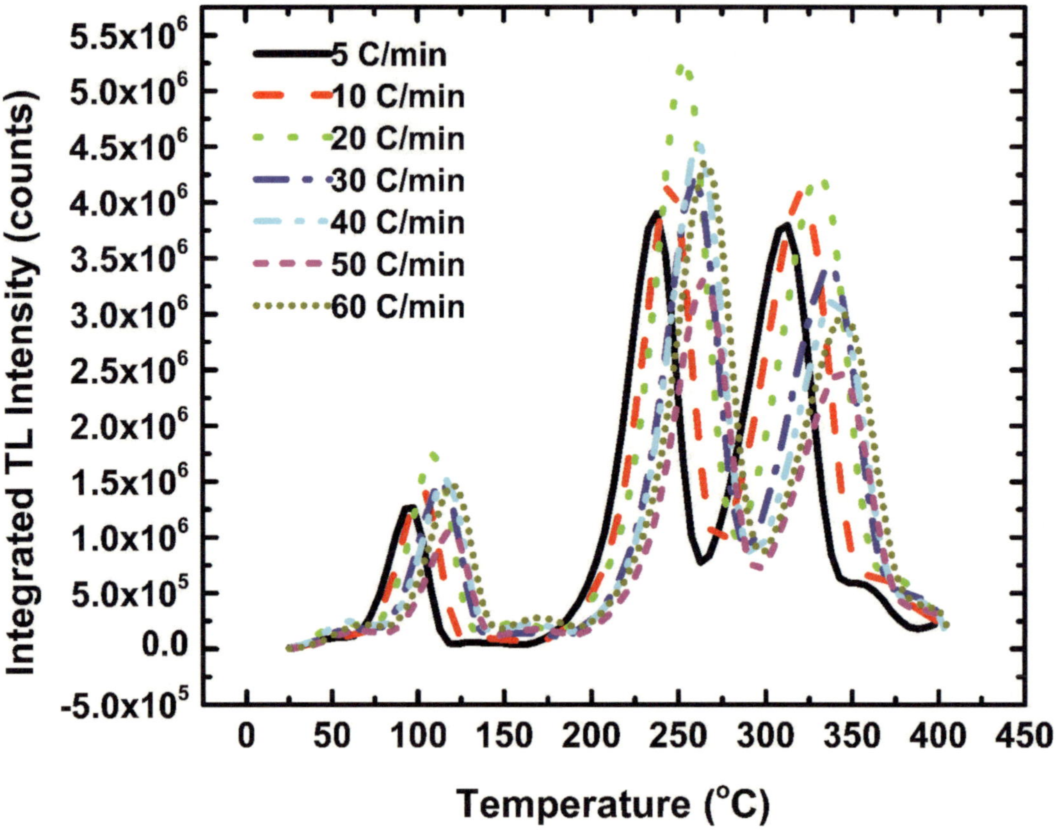

Figure 2. Glow curves of Ce:YAG 0.2% at multiple heating rates, showing the effects of heating rate on glow curve peaks.

equations, and in this case the activation energy can be determined graphically. The benefit of this method over the method of two heating rates is that the greater number of data points decreases uncertainty. This method is typically applied to first-order kinetics, but it can be generalized to any order kinetics. Other methods, such as the half-width method or direct computational glow curve fitting with the activation energy as a fitting parameter, are not explored in this chapter.

4.1. Initial rise method

Let us now derive each of the discussed methods for determining activation energies. The probability p of an electron escaping from a trap of depth at temperature T can be described by the Arrhenius Equations [60, 65, 67].

$$p = s\exp\left(-\frac{E}{k_B T}\right) \tag{1}$$

where k_B is the Boltzmann constant and s, known as the frequency factor, is given by [67]

http://dx.doi.org/10.5772/65306

$$s = N_B v \sigma_t \tag{2}$$

where N_B is the density of states in the band in which the carriers escape, v is the carrier's thermal velocity, and σ_t is the trap's capture cross section. The frequency factor s has units of time^{-1} and may vary weakly with temperature. The rate of detrapping $-dn/dt$ at temperature T can be expressed as

$$-\frac{dn}{dt} = s\,n\,\exp\left(-\frac{E}{k_B T}\right) \tag{3}$$

where n is the concentration of trapped charge carriers. Determining $n(T)$ follows from Eq. (3) by

$$\frac{dn}{n} = -s\,\exp\left(-\frac{E}{k_B T}\right)dt = -\frac{s}{q}\,\exp\left(-\frac{E}{k_B T}\right)dT \tag{4}$$

where $q = dT/dt$ is the heating rate, which for simplicity we shall assume is linear. Integration gives

$$\ln(n) = -\int \frac{s}{q}\,\exp\left(-\frac{E}{k_B T'}\right)dT' + \text{constant} \tag{5}$$

The prime in T' is introduced to distinguish between the variable and the upper limit of integration. When $t = 0$, let $n = n_0$ and $T = T_i$ = some initial temperature, then the constant is equal to n_0 and we have

$$\begin{gathered} \ln\left(\frac{n}{n_0}\right) = -\int_{T_i}^{T} \frac{s}{q}\,\exp\left(-\frac{E}{k_B T'}\right)dT' \\ \text{or} \\ n = n_0 \exp\left[-\int_{T_i}^{T} \frac{s}{q}\,\exp\left(-\frac{E}{k_B T'}\right)dT'\right] \end{gathered} \tag{6}$$

The thermoluminescence intensity I is proportional to the rate that charge carriers are supplied to luminescence centers [2, 41, 60, 64–65, 68–69]:

$$\begin{aligned} I(T) &= -c\frac{dn}{dt} \\ &= c\,s\,n\exp\left(-\frac{E}{k_B T}\right). \end{aligned} \tag{7}$$

Combining this with Eq. (6),

$$I(T) = c\,s\,n_0\exp\left(-\frac{E}{k_B T}\right)\exp\left[-\int_{T_i}^{T} \frac{s}{q}\exp\left(-\frac{E}{k_B T}\right)dT'\right] \tag{8}$$

which is known as the Randall-Wilkins first-order expression of a single glow peak [60]. For low temperatures ($T \approx T_i$), we may estimate $n \approx n_0$ and thus write

$$I \approx c\, s\, n_0 \exp\left(-\frac{E}{k_B T}\right) \tag{9}$$

The initial rise method, introduced by Garlick and Gibson [64], is based on Eq. (9). To calculate the activation energy by the initial rise method, we may rewrite Eq. (9) in a more useful form

$$\ln(I) = \ln(c\, s\, n_0) - \frac{E}{k_B T} \tag{10}$$

Plotting ln(I) versus $1/k_B T$ gives the slope = E.

The initial rise method is applicable to first, second, and even general order cases, but it is limited to the region where $I_i \leq I << I_m$.

Bräunlich [62] showed that the values found for E using the initial rise method will be smaller than the actual values when the retrapping factor $R[h(T_i)/f(T_i)] >> 1$, where R is given by

$$R = \frac{\text{probability of retrapping}}{\text{probability of recombination}} \tag{11}$$

and $h(T_i)$ and $f(T_i)$ are traps with electrons and the initial concentration of unoccupied recombination centers, respectively (at the temperature T_i). This suggests that the initial rise method actually gives a poor estimate for E for high order kinetics.

Christodoulides [63] estimated a correction to the initial rise method that may be used if the intensity of the range used for calculation is larger than a small fraction of the peak height. The activation energy calculated by the initial rise method EIR can be adjusted to the corrected value of the activation energy E_C by

$$E_C = (1 + 0.74a_1 + 0.082a_2)E_{IR} - (2a_1 + 0.22a_2)\frac{T_m}{11{,}605} \tag{12}$$

The factors a_1 and a_2 are the fraction of the peak value of the TL intensity observed at temperatures T_1 and T_2, respectively, marked as the lower and upper bounds of the fit, as demonstrated in **Figure 3**.

4.2. Methods of variable heating rates

The initial rise method is applicable to general order kinetics, but with decreased accuracy for higher order kinetics. However, if peaks are overlapped it is not possible to apply this method since the beginning of a peak is not available for analysis. The methods of two and variable heating rates only require that the peak temperature can be determined for calculation of E. These methods can be used with general order kinetics. The methods of two and variable heating rates can be derived from first-order kinetics or simplified from general order kinetics. We will first derive the method for general order kinetics and then apply it to the case of

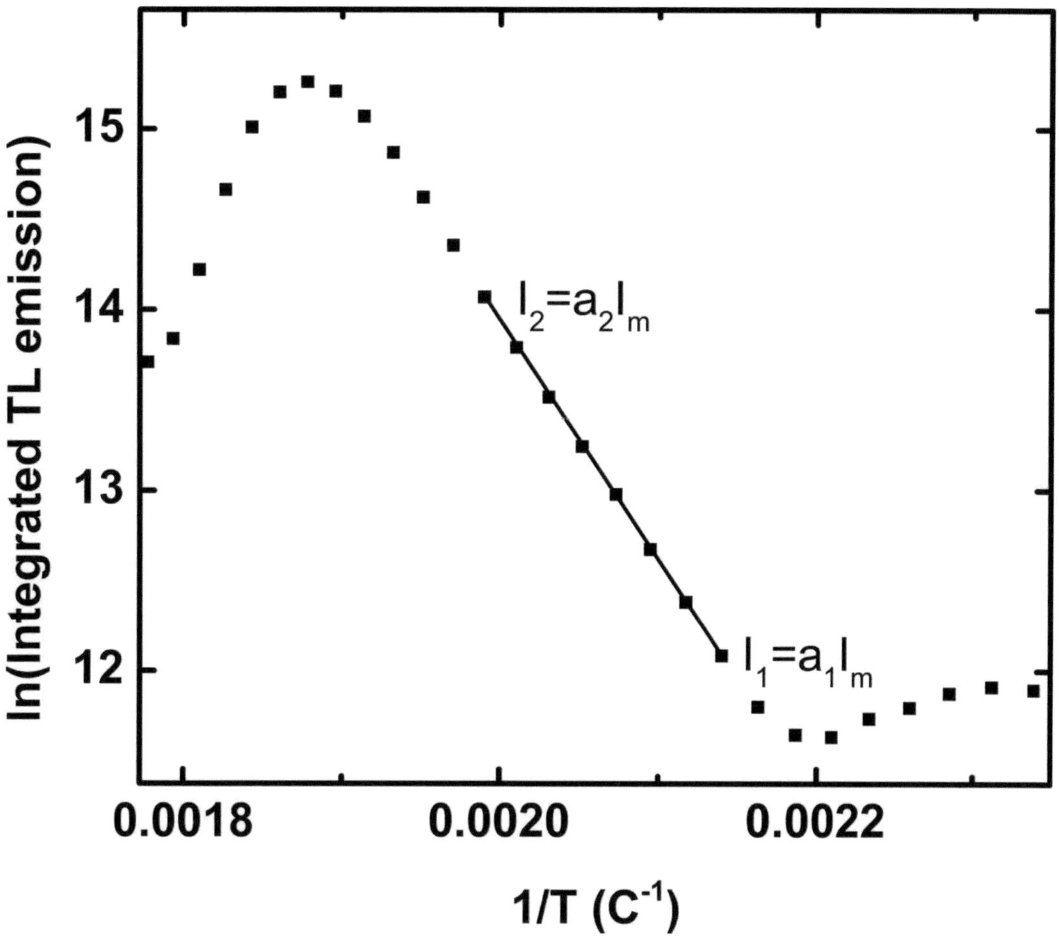

Figure 3. TL glow curve of Ce:YAG 0.2% plotted as ln(I) versus 1/T for calculation of the activation energy by the initial rise method, demonstrating the definitions of a_1 and a_2.

first-order kinetics. It will be necessary to assume that s is independent of temperature to proceed with this derivation. To general order of kinetics, Eq. (7) becomes [2, 41, 65, 69–71]

$$I(T) = -c\frac{dn}{dt} = c\,s^{'}n^{b}\exp\left(-\frac{E}{k_B T}\right) \quad (13)$$

where s' is the general order preexponential factor, be represents the order of kinetics, and $s' = s$ for $b = 1$. Eq. (13) can be integrated

$$\int_{n_0}^{n} (n^{'})^{b} = -\int s^{'}\exp\left(-\frac{E}{k_B T}\right)dt \quad (14)$$

This has the solution

$$n^{1-b} = n_0^{1-b}\left[1 + \frac{s'(b-1)n_0^{b-1}}{q}\int_{Ti}^{T}\exp\left(-\frac{E}{k_B T'}\right)dT'\right]$$
or
$$n = n_0\left[1 + \frac{s'(b-1)n_0^{b-1}}{q}\int_{T_i}^{T}\exp\left(-\frac{E}{k_B T'}\right)dT'\right]^{\frac{1}{1-b}} \tag{15}$$

The intensity of Eq. (13) then becomes

$$I(T) = c\,s'\,\exp\left(-\frac{E}{k_B T}\right)n_0^b\left[1 + \frac{s'(b-1)n_0^{b-1}}{q}\int_{T_i}^{T}\exp\left(-\frac{E}{k_B T'}\right)dT'\right]^{\frac{b}{b-1}} \tag{16}$$

The maximum intensity of a peak occurs at temperature T_m. At this point, the derivative of the intensity is zero, as is the derivative of the logarithm of the intensity. Thus, we obtain at $T = T_m$ the relationship

$$\frac{d\{\ln[I(T)]\}}{dT}\Big|_{T=T_m} = 0 = \frac{E}{k_B T_m^2} - \frac{b}{b-1}\left[1 + \frac{s'(b-1)n_0^{b-1}}{q}\int_{T_i}^{T}\exp\left(-\frac{E}{k_B T'}\right)dT'\right]_{T=T_m}^{-1} \times\left(\frac{s'(b-1)n_0^{b-1}}{q}\exp\left(-\frac{E}{k_B T_m}\right)\right) \tag{17}$$

This yields

$$1 + \frac{s'(b-1)n_0^{b-1}}{q}\int_{T_i}^{T_m}\exp\left(-\frac{E}{k_B T}\right)dT = \frac{b\,s'\,n_0^{b-1}k_B T_m^2}{q\,E}\exp\left(-\frac{E}{k_B T_m}\right) \tag{18}$$

The following approximation can be made

$$\int_{T_i}^{T}\exp\left(-\frac{E}{k_B T'}dT'\right) \cong T\exp\left(-\frac{E}{k_B T}\right)\sum_{n=1}\left(\frac{k_B T}{E}\right)^n(-1)^{n-1}n! \tag{19}$$

which Kitis and Pagonis determined is usually “a very good numerical approximation” [69]. The series converges quickly, so only the first two terms need to be taken. Inserting Eq. (19) into (18) yields

$$1 + \frac{s'(b-1)n_0^{b-1}}{q}\left[T\exp\left(-\frac{E}{k_B T_m}\right)\left(\frac{kT}{E} - \frac{2k^2T_m^2}{E^2}\right)\right] = \frac{b\,s'\,n_0^{b-1}k_B T_m^2}{qE}\exp\left(-\frac{E}{k_B T_m}\right),$$
which becomes
$$\frac{q\,E}{k_B T_m^2} = s'\,n_0^{b-1}\exp\left(-\frac{E}{k_B T_m}\right)\left[1 + (b-1)\frac{2k_B T_m}{E}\right] \tag{20}$$

At this point, we can calculate the first-order solutions for the method of two heating rates and the variable heating rate method by setting $b = 1$ and $s' = s$

$$\frac{q\,E}{k_B T_m^2} = s\exp\left(-\frac{E}{k_B T_m}\right) \tag{21}$$

Conducting a pair of TL measurements and recording the maximum temperature of a peak in both glow curves gives two such equations with different q and T_m, but with the same E and s. Division of one equation by the other gives

$$\frac{q_1 T_{m2}^2}{q_2 T_{m1}^2} = \exp\left(\frac{E}{k_B T_{m2}} - \frac{E}{k_B T_{m1}}\right) \tag{22}$$

which can be solved for E

$$E = \frac{k_B T_{m1} T_{m2}}{T_{m1} - T_{m2}} \ln\left(\frac{q_1 T_{m2}^2}{q_2 T_{m1}^2}\right) \tag{23}$$

which represents the activation energy to first-order kinetics using the method of two heating rates [2, 41, 65–66, 70]. The frequency factor can also be found by Eq. (21) once we have obtained E.

The variable heating rate method adapts the method of two heating rates to the case of many heating rates, to be solved graphically. From Eq. (12), for first-order kinetics

$$\ln\left(\frac{T_m^2}{q}\right) = E\left(\frac{1}{k_B T_m}\right) - \ln\left(\frac{s\,k_B}{E}\right) \tag{24}$$

Data can easily be plotted in this way to give the activation energy as its slope. **Figure 4** plots the data obtained from **Figure 2** for calculation of E and s using Eq. (24). Once E is calculated, s may be calculated from the y-intercept.

Activation energies calculated using the initial rise method, corrected initial rise method [63], and the method of variable heating rates is compared and presented in **Table 1**.

Note that the activation energies found for Peaks 2 and 3 match closely between the method of multiple heating rates and the corrected initial rise method and that the initial rise method in those peaks gives a significantly smaller result than the method of two heating rates. The latter point suggests that we may see the effects of retrapping proposed by Braunlich [62].

4.3. General order method of variable heating rates

The general order variable heating rate method and the method of multiple heating rates are as of yet poorly documented in the literature [2]. The solution of the general order variable heating rate method begins by taking the logarithm of Eq. (20)

$$\frac{E}{k_B T_m} = \ln\left\{\frac{s'\,n_0^{b-1} k_B T_m^2}{q\,E}\left[1 + (b-1)\frac{2k_B T_m}{E}\right]\right\}, \tag{25}$$

which can be separated into the parts

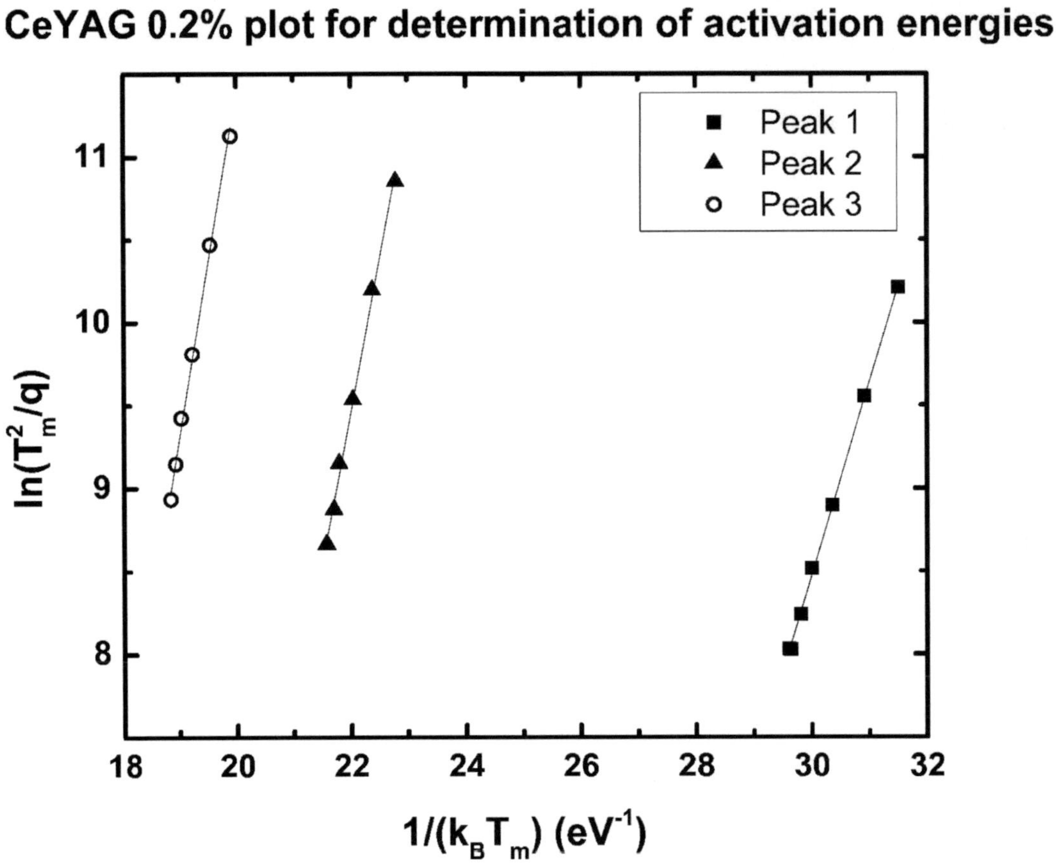

Figure 4. Data and fits for all three peaks using multiple heating rates to get the results.

Peak 1 (470–720 nm, ~400 K)	**Activation energy (eV)**
Method of multiple heating rates	1.16 ± 0.01
Initial rise method	0.88 ± 0.06
Corrected initial rise	1.46 ± 0.05
Peak 2 (470–720 nm, ~550 K)	
Method of multiple heating rates	1.80 ± 0.05
Initial rise method	1.15 ± 0.06
Corrected initial rise	1.83 ± 0.10
Peak 3 (470–720 nm, ~600 K)	
Method of multiple heating rates	2.08 ± 0.05
Initial rise method	1.49 ± 0.07
Corrected initial rise	1.97 ± 0.05

Table 1. Calculated activation energies for the three major peaks found in Ce:YAG 0.2% thermoluminescence measurements from room temperature to 400°C using all methods previously discussed.

$$\frac{E}{k_B T_m} = \ln\left(\frac{T_m^2}{q}\right) + \ln\left[1 + (b-1)\frac{2k_B T_m}{E}\right] + \ln\left(\frac{s' n_0^{b-1} k_B}{E}\right) \tag{26}$$

The activation energy is typically at least an order of magnitude larger than $k_B T_m$ [2], so we can apply the series expansion for $\ln(1 + x)$ for $x \leq 1$ to get

$$\ln\left(\frac{T_m^2}{q}\right) \approx \ln\left(\frac{E}{s' n_0^{b-1} k_B}\right) + \frac{E}{k_B T_m} - 2(b-1)\frac{k_B T_m}{E} + 2(b-1)^2 \frac{k_B^2 T_m^2}{E^2} - \frac{8}{3}(b-1)^3 \frac{k_B^3 T_m^3}{E^3} + \ldots \tag{27}$$

If we let $y = \ln(T_m^2/q)$ and $x = k_B T_m$ and designate y_0 as the logarithmic term independent of both q and T_m, Eq. (27) becomes

$$y = y_0 + \frac{E}{x} - 2(b-1)\frac{x}{E} + (b-1)^2 \frac{x^2}{E^2} - \frac{8}{3}(b-1)^3 \frac{x^3}{E^3} + \ldots \tag{28}$$

which can be plotted and fit using a simple fitting program. Mahesh et al. [2] solve this in a very different way:

$$\ln\left[I_m^{b-1}\left(\frac{T_m^2}{q}\right)^b\right] = b\ln E + \ln\left[\frac{\left(c\, s'\, n_0^{b-1}\right)^{b-1}}{\left(s'\, b\, k_B\right)^b}\right] + \frac{E}{k_B T_m} \tag{29}$$

Eq. (29) shows dependence on intensity, which adds a further source of uncertainty. Detector zero drift, overlapping glow curve peaks, nonlinear intensity response by the detector, or other misleading intensity measurements will add error to calculations. Thus, Eq. (27) is preferred over Eq. (29) since it contains less factors that may create further sources of error.

The general order method of two heating rates can be solved by subtracting Eq. (27) for two different measurements

$$\begin{aligned} 0 = &\frac{E}{k_B}\left(\frac{1}{T_{m1}} - \frac{1}{T_{m2}}\right) + \ln\left(\frac{q_1 T_{m2}^2}{q_2 T_{m1}^2}\right) + \frac{2(b-1)k_B}{E}\left(T_{m2} - T_{m1})\right) - \frac{2(b-1)^2 k_B^2}{E^2}\left(T_{m2}^2 - T_{m1}^2\right) \\ &+ \frac{8(b-1)^3 k_B^3}{3E}\left(T_{m2}^3 - T_{m1}^3\right) - \ldots \end{aligned} \tag{30}$$

This is easiest solved graphically. After inputting all the known values for T_{mi} and q_i and the constants and letting $x = E$, we can write Eq. (30) as

$$y = Ax + B + \frac{C}{x} - \frac{D}{x^2} + \frac{E}{x^3} - \ldots \tag{31}$$

which can be plotted to find the x-intercept = E. The substitutions made are as follows:

$$
\begin{aligned}
A &= \frac{1}{k_B}\left(\frac{1}{T_{m1}} - \frac{1}{T_{m2}}\right),\\
B &= \ln\left(\frac{q_1 T_{m2}^2}{q_2 T_{m1}^2}\right),\\
C &= 2(b-1)k_B(T_{m2}-T_{m1}),\\
D &= 2(b-1)^2 k_B^2(T_{m2}^2-T_{m1}^2),\\
E &= \frac{8}{3}(b-1)^3 k_B^3(T_{m2}^3-T_{m1}^3).
\end{aligned}
\tag{32}
$$

We may check our work here by setting $b = 1$ in Eq. (30), and we see that we return the first-order kinetics solution of Eq. (22).

The obvious challenge with this method, however, is the required guesswork for the order b.

The solution presented by Mahesh et al. [2] for the general order method of two heating rates is

$$
\frac{T_{m1}^2}{q_1}\left(\frac{1}{b-1} + \frac{2k_B T_{m1}}{E}\right)\exp\left(-\frac{E}{k_B T_{m1}}\right) = \frac{T_{m2}^2}{q_2}\left(\frac{1}{b-1} + \frac{2k_B T_{m2}}{E}\right)\exp\left(-\frac{E}{k_B T_{m2}}\right) \tag{33}
$$

which is not so easily solved for E and still requires inputting an assumed b. If $b = 1$, this can also be shown to return the first-order kinetics solution of Eq. (22).

Thus, we have formulated the theory for the initial rise method (Eq. 10), which can be applied to general order kinetics, the first-order method of two heating rates (Eq. 22) and variable heating rate method (Eq. 24), and general order method of two heating rates (Eq. 30) and variable heating rate method (Eq. 27).

5. Experimental setup

A schematic of a standard TL dosimetry reader (TLD) is shown in **Figure 5**. The irradiated dosimeter is mounted on a tray located within a readout chamber. Two different temperatures are used for heating the dosimeter, first one is preheating to clear noisy peaks and second one is readout temperature used to thermally stimulate the sample and collect data. Nitrogen gas is continuously pumped to decrease spurious phenomena [72] and reduce the background.

The absorbed dose is proportional to the output current received from photomultiplier tube [73].

We have developed an advanced TL setup that uses a special heating stage with an optical window that transmits light and blocks heat to achieve perfect temperature control during the measurement. Solarized fiber optics connected to collimation lenses are used to collect the emitted light (**Figure 6**). The TL spectra are recorded by a CCD detector with a 200 μm slit. The spectral range of the detector can be adjusted from 200 to 800 nm or from 350 to 1100 nm with 1 nm resolution. The optical components are water cooled to allow for high temperature

http://dx.doi.org/10.5772/65306

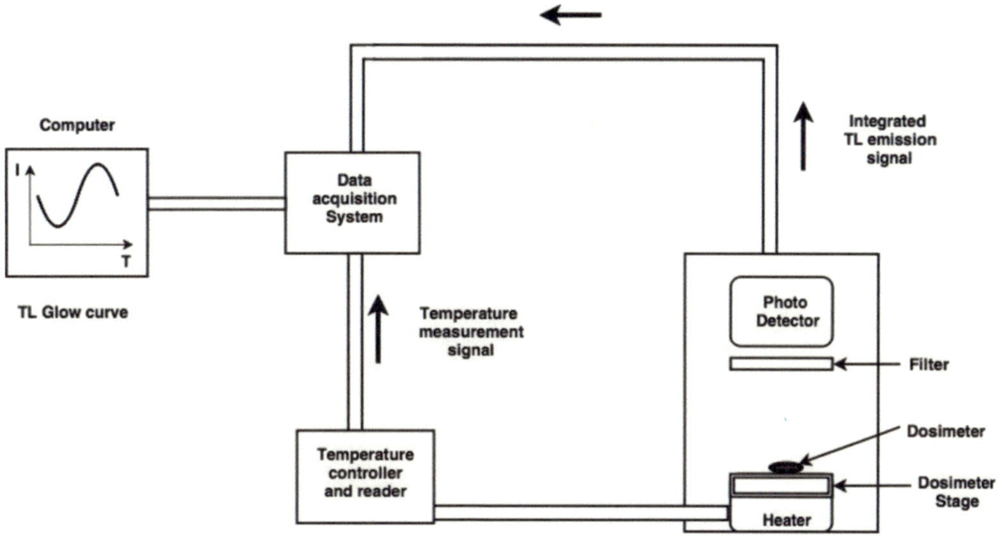

Figure 5. Schematic of TLD reader setup [73].

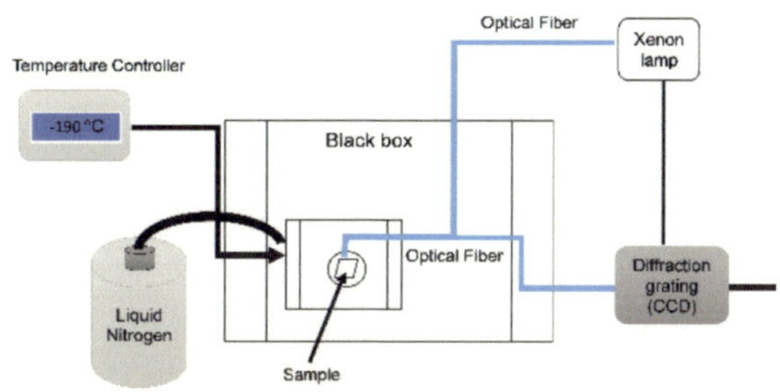

Figure 6. Schematic diagram of a typical TL setup. The sample is heated on a heating stage at a constant rate and the frame temperature is regulated by a water pump. A collimated lens collect emitted light from the sample and carries it to a spectrometer.

(up to 400°C) TL measurements. Before TL experiment, we often irradiate the samples in dark for certain time using xenon lamp, light emitting diodes (LEDs), or 254 nm irradiation panel. The spectral range of the xenon lamp is 220–750 nm. This setup has many advantages including:

1. Precise control and measurements of temperature, 1 mK precision and high stability.
2. Adjustable heating/cooling rate, can go up to 1000°C/min. In most measurements, it is operated at 60°C/min and 120°C/min. Using different heating rates is required to calculate trap levels by variable heating rate methods.
3. TL can be measured as a function of temperature and wavelength simultaneously as shown in (**Figure 7**).

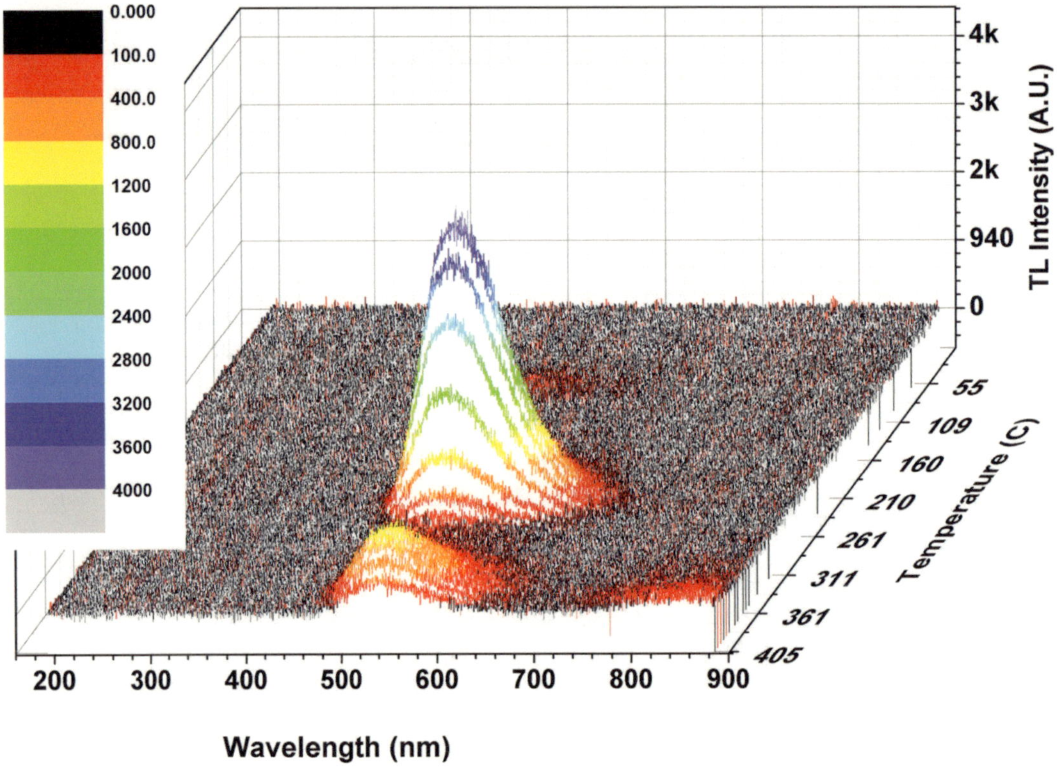

Figure 7. TL intensity versus temperature and wavelength.

A TL measurement is performed in a dark box in a temperature-insulated heating stage as discussed above and as shown in the figure. This is because the TL emission is relatively weak, so it is necessary to isolate the measurement from ambient lighting that would greatly interfere with it. The heating stage must be also temperature isolated for two reasons. First, it makes it possible to fully control the temperature of the sample instead of potentially creating a gradient of temperature across the thickness of the sample. Second, it separates the optical components used for excitation and measurement so that they may be operated independently.

6. TL applied to dielectrics

Dielectric materials possess a large band gap separating the valence and conduction bands. This large gap determines the characteristic insulating properties. Dopants, defects, and impurities alter the band structure of a dielectric material by adding energy levels and electron and electron hole traps within the band gap. Charge carrier traps can be vacancy, interstitial, substitutional, or antisite defects. Typically, charge carriers can be trapped at defects of the opposite charge. For instance, an anion vacancy is positively charged and would thus trap an electron. However, the behavior of some impurity traps depends on the electronic structure of the impurity [74, 75]. For instance, in YAG, Eu^{3+}, and Yb^{3+} impurities have been found to act as stable electron traps at room temperature due to their large electron affinity. Meanwhile, Pr^{3+}, Tb^{3+}, and Ce^{3+} were found to act as hole traps, stable at room temperature [45, 57]. At low temperatures, other traps stabilize, such as Sm^{3+} electron traps.

TL can be used to characterize defects and provide quantitative information about charge carrier traps. It shares many similarities to color center measurements [76]; however, it has more advantages in evaluating the trap depths in the band gap. In color center measurements, a sample is irradiated for a period of time with an excitation source, similar to the first step for a TL measurement. Absorption spectra of this sample are recorded before and after excitation. As discussed previously, this excitation leads to the trapping of some charge carriers at defects. These trapped charge carriers carry their own optical properties, which are observable in absorption measurements. On top of that, an irradiated sample may be visibly discolored. Heating to sufficiently high temperature bleaches this discoloring and returns the trapped charge carriers to recombination centers. TL is the qualitative study of this effect. **Figure 8** shows the absorption spectra of undoped YAG before and after irradiation and after heating to 400°C. This sample will be also used later for TL glow curves presented in **Figure 12**. Some qualitative information can be gained from this figure in conjunction with prior knowledge about the sample, such as ascribing the peak at 256 nm to Fe impurities [77, 78]. These plots give no information about the depth of these traps other than the fact that heating to 400°C seemed to clear them up which implies that they are deep centers. TL provides the more qualitative analysis.

We will discuss our TL study on Ce:YAG as an example of dielectric materials because of its great importance in a wide range of applications from phosphors to scintillation. In general, YAG is one of the most important optical materials, with Nd:YAG as the most widely used laser crystal. As it stands now, Ce:YAG is an efficient luminescent material, but its performance as a scintillator is often hindered by the presence of traps.

6.1. YAG TL

In our TL measurements on YAG crystals [55], samples were cooled or heated within a temperature range of −190 to 400°C and TL spectra are recorded as described in the experimental section. Glow curves were constructed by integrating each spectrum over a specific range of wavelengths. In Ce-doped YAG, Ce ions provide the recombination centers and thus all TL emission are shown to have the luminescence profile characteristic of Ce ions in YAG, as shown in **Figure 9**. For the sake of demonstration, a contour plot of temperature versus wavelength for the same sample is provided in **Figure 10**, demonstrating only one luminescence center.

Sometimes care had to be taken when integrating over wavelengths to distinguish between different luminescence centers activating at different temperatures. **Figure 11** shows the contour plot of temperature versus wavelength for an undoped YAG sample, originally grown in a pure Ar atmosphere, where the luminescence centers are comprised of different trace impurities. Something else to note in **Figure 10** is the onset of IR emission near 400°C due to blackbody radiation. The wavelength of this IR emission overlapped slightly with the integration range of one of the peaks and was visible in the constructed glow curve. Also note the

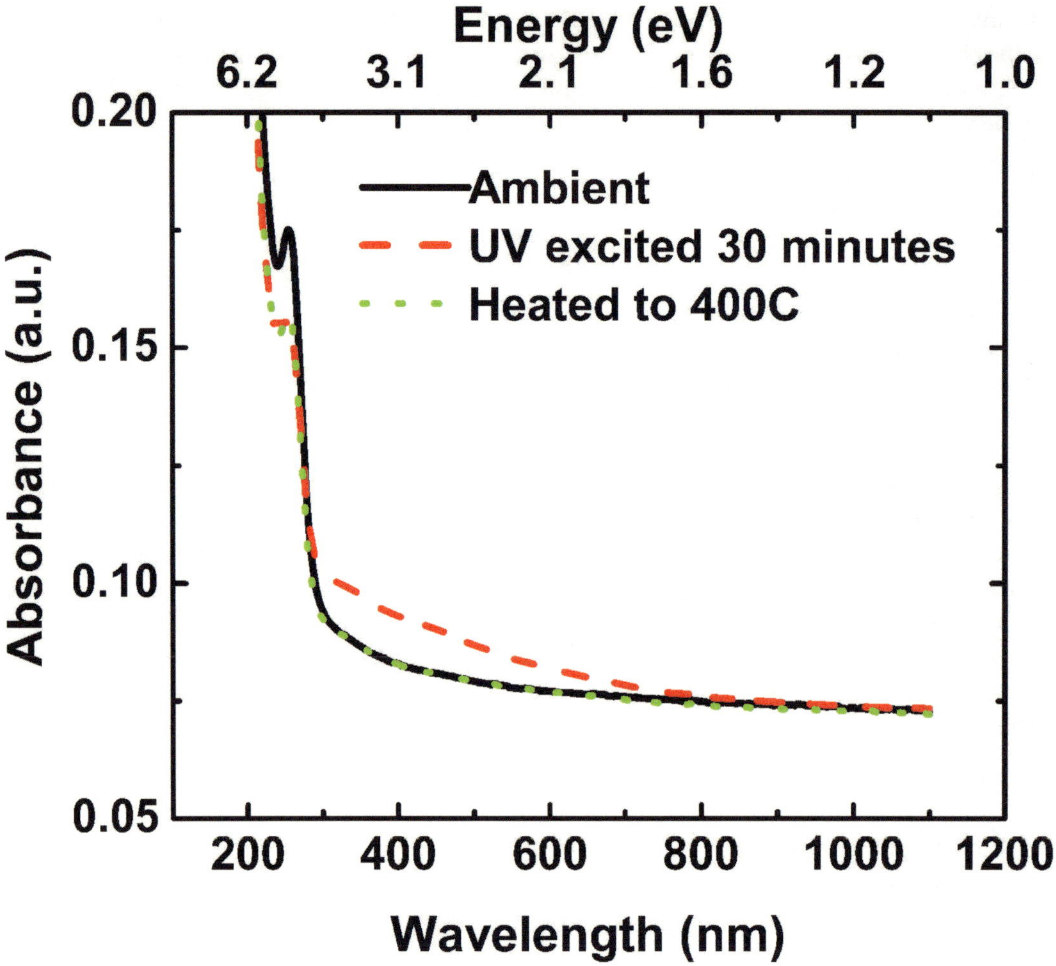

Figure 8. Air annealed Ar-grown YAG absorption spectra after UV excitation and heating.

much weaker emission, as each contour now corresponds to a step of 200 counts, whereas in **Figure 10**, it corresponded to 1500 counts. The IR radiation was too weak to show on the contour plot of **Figure 10**. The Ar-grown undoped YAG sample was annealed at 1200° for 96 hours in air. **Figure 12** presents the two unique glow curves for this sample before and after annealing, separately integrating over the wavelength ranges 340–570 and 570–800 nm. The intensity of most peaks has been decreased suggesting decreased concentration of their corresponding defects. Also, the high-temperature peak in the high-wavelength plot shifts significantly, suggesting a change in activation energy. Activation energies were calculated for each of these peaks and frequency factors were calculated whenever possible for the as-grown sample. The small peak in the high-wavelength glow curve at 180°C had too high of uncertainty to get a reliable estimate for s. Results are tabulated in **Tables 2** and **3**.

Obviously, the activation energy increases with temperature, as charge carriers at deeper traps require more thermal energy for release. The most noticeable change is the large decrease in

http://dx.doi.org/10.5772/65306

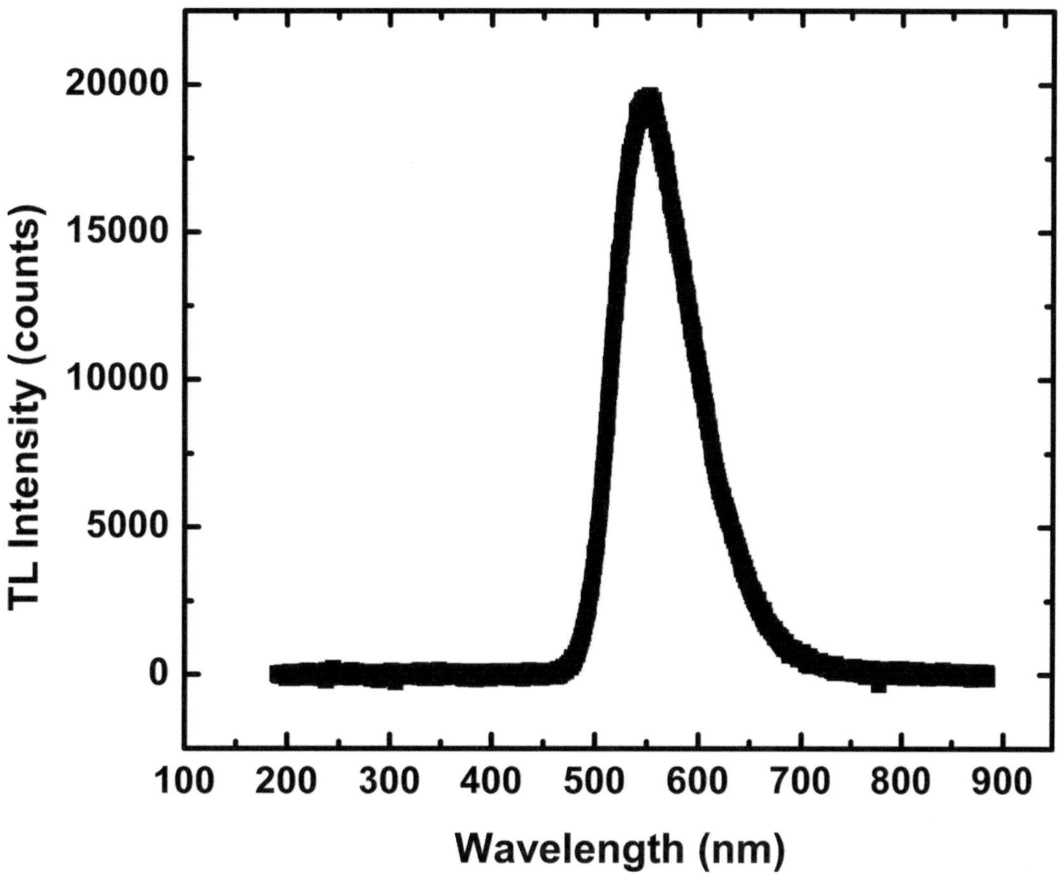

Figure 9. TL emission spectrum at 262°C for Ce:YAG 0.14%.

activation energy of the highest energy peak after annealing by about 0.9 eV. Due to the nature of anneal, it seems reasonable to attribute this defect partially to oxygen vacancies. However, the fact that the peak persists, just at a decreased activation energy, suggests that there still exists a defect, suggesting that the original defect was a complex defect consisting of both at least one Al vacancy and at least one O vacancy, and the anneal filled one O vacancy but left the Al vacancy. The depth of the low-wavelength peak at 250°C appears to decrease slightly after annealing, possibly due to the change in charge state of the impurity responsible for this luminescence and/or change of lattice defect structures surrounding them. No other activation energy appeared to change after annealing, so it would appear that annealing only decreased their concentration and not their depth.

Positron annihilation lifetime spectroscopy (PALS) measurements were also performed on these samples in combination with TL [79]. PALS works by injecting a positron into the crystal and timing how long it takes to annihilate with an electron within the sample. If the positron annihilates within the bulk of the sample, the process is very fast (of order 0.1 ns). If the positron becomes trapped at a vacancy or void, it takes longer to annihilate with an electron

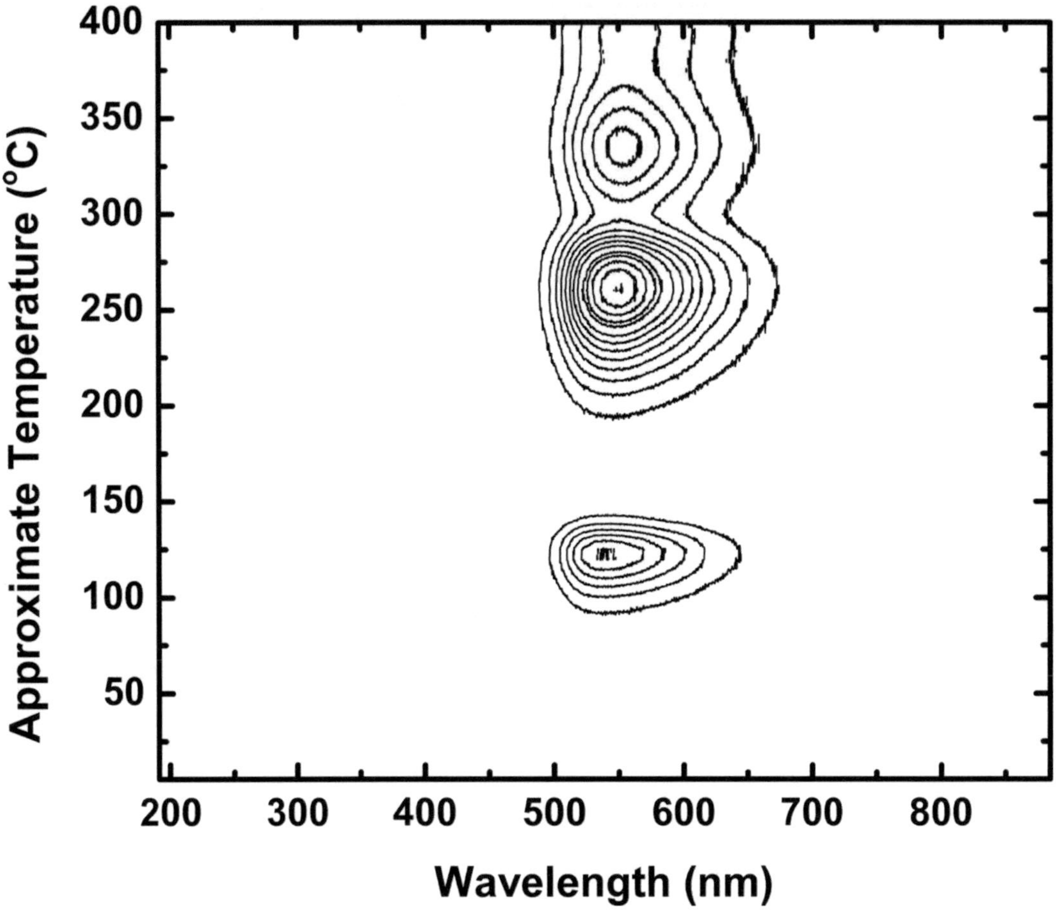

Figure 10. Contour plot of TL of Ce:YAG 0.14%, plotting luminescence intensity contours to a map of temperature versus wavelength. Each Contour represents 1500 counts [79].

due to the reduced number density of electrons at vacancies. Due to their positive charge, positrons typically trap at negatively charged vacancies or voids, such as cation vacancies, or neutral voids, such as cation + anion vacancy complexes. Typically, a larger vacancy means a lower number density of electrons and thus longer positron lifetime [80, 81].

PALS measurements on these YAG samples demonstrated a defect lifetime (i.e., the lifetime of a positron trapped at a defect) of 0.2931 ± 0.0032 ns for the as-grown sample and 0.2691 ± 0.0037 ns for the annealed sample, suggesting that annealing in air decreased the size of the defect but did not entirely eliminate it. This defect bears striking similarity to the peak in **Figure 12** that moved from 395 to 345°C after annealing in air and in fact the same conclusion may be drawn from this PALS result that this defect is due to an Al + O vacancy complex.

PALS measurements only identified one defect that trapped the injected positrons, suggesting that the remaining defects arose from either positively charged vacancies or nonvacancy

http://dx.doi.org/10.5772/65306

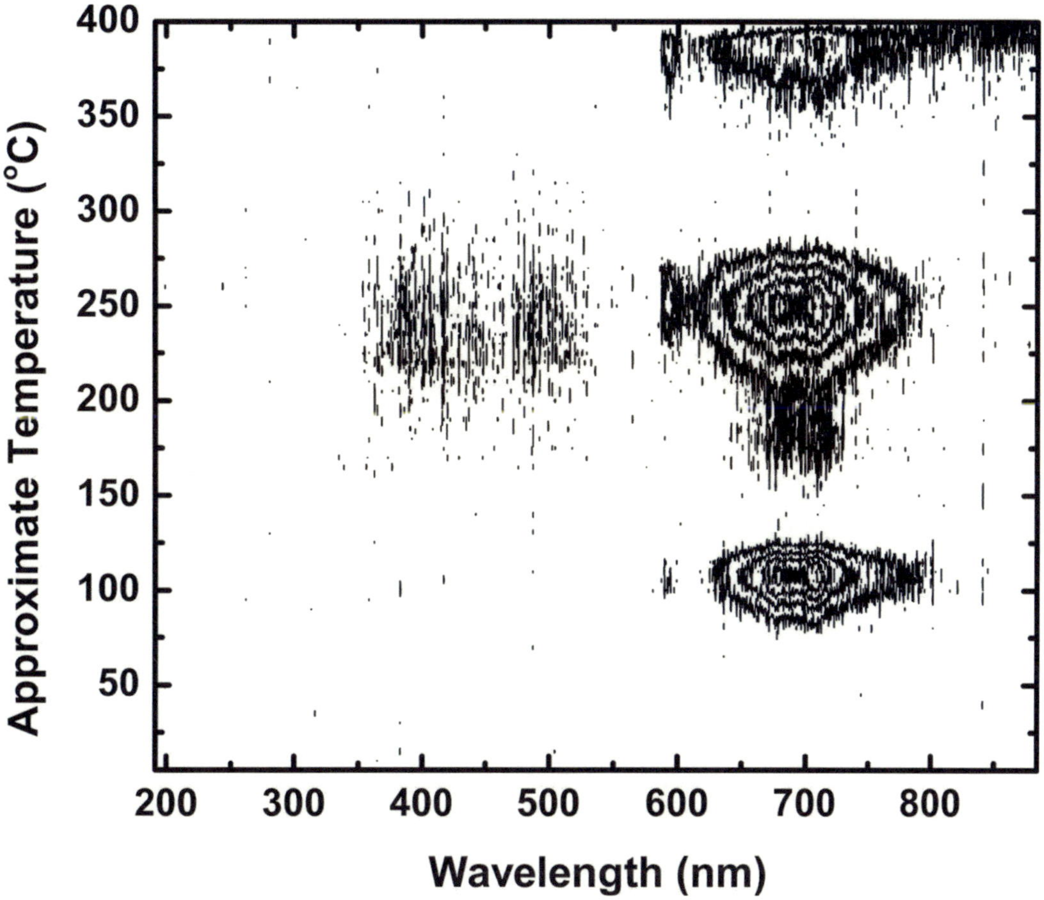

Figure 11. Contour plot of TL of undoped YAG grown in a pure Ar atmosphere, plotting luminescence intensity contours to a map of temperature versus wavelength. Each contour represents a step of 200 counts [79].

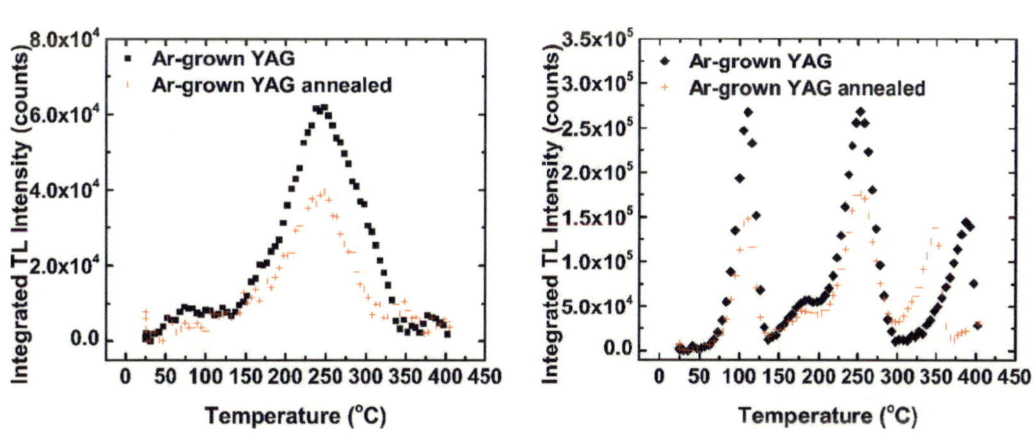

Figure 12. Glow curves for Ar-grown undoped YAG before and after annealing. The low wavelength curve (left) was integrated over 340–570 nm, the high wavelength curve (right) was integrated over 570–800 nm.

	Activation energy		Frequency factor
Approximate peak location (q = 60°C/min)	As grown	Air annealed	As grown
250°C	1.47 ± 0.23 eV	0.95 ± 0.09 eV	2.68 × 1010 s^{-1}

Table 2. Calculated activation energies of Ar-grown undoped YAG for the glow curves constructed using 340–570 nm integration.

	Activation energy		Frequency factor
Approximate peak location (q = 60°C/min)	As grown	Air annealed	As grown
110°C	1.65 ± 0.01 eV	1.63 ± 0.09 eV	6.47 × 1017 s^{-1}
180°C	3.04 ± 2.27 eV	1.44 ± 1.87 eV	(not calculated)
250°C	1.92 ± 0.09 eV	2.07 ± 0.15 eV	4.21 × 1019 s^{-1}
345°C	N/A	2.23 ± 0.55 eV	1.14 × 1022 s^{-1}
395°C	3.15 ± 0.25 eV	N/A	N/A

Table 3. Calculated activation energies of Ar-grown YAG for the glow curves constructed using 570–800 nm integration.

defects, such as impurities. Since a positively charged vacancy in this crystal means an O vacancy, annealing in air would be expected to eliminate or decrease this defect, and consequently any TL peaks associated with it. Other TL peaks that are not sensitive to air anneal can be ascribed to impurities.

6.2. Low-temperature TL of Ce:YAG

Low-temperature glow curves for Ce:YAG 0.14%, presented in **Figure 13**, provide an abundance of information. The sample was measured, then annealed in air for 48 hours at 1200°C and remeasured, then annealed in a vacuum for 24 hours at 800°C and measured once again. TL emission spectra were integrated over the wavelength range of 470–720 nm, which encompasses the strong and well-documented Ce luminescence peak in YAG peaked around 550 nm [51, 56, 57, 78, 82–106]. From figure, it can be seen that the highest intensity peak at −75°C is completely eradicated by annealing in air, but it returns after annealing in vacuum. With this information, it is easy to identify this trap as an oxygen vacancy. Upon eliminating this peak, TL emission can be distinguished near −100°C in the air-annealed sample. Curve fitting resolves that it is as two peaks centered at about −100 and −90°C. A weak peak at about −10°C appears to be eliminated by annealing in air.

Activation energies were calculated to compare the as-grown and air-annealed samples and are provided in **Table 4**. Interestingly, although all traps appear to decrease drastically in concentration, their depths uniformly increase.

These examples demonstrate some of the analytical capabilities of TL spectroscopy. The high-concentration oxygen vacancies observed in **Figure 13** at −75°C are not stable at room

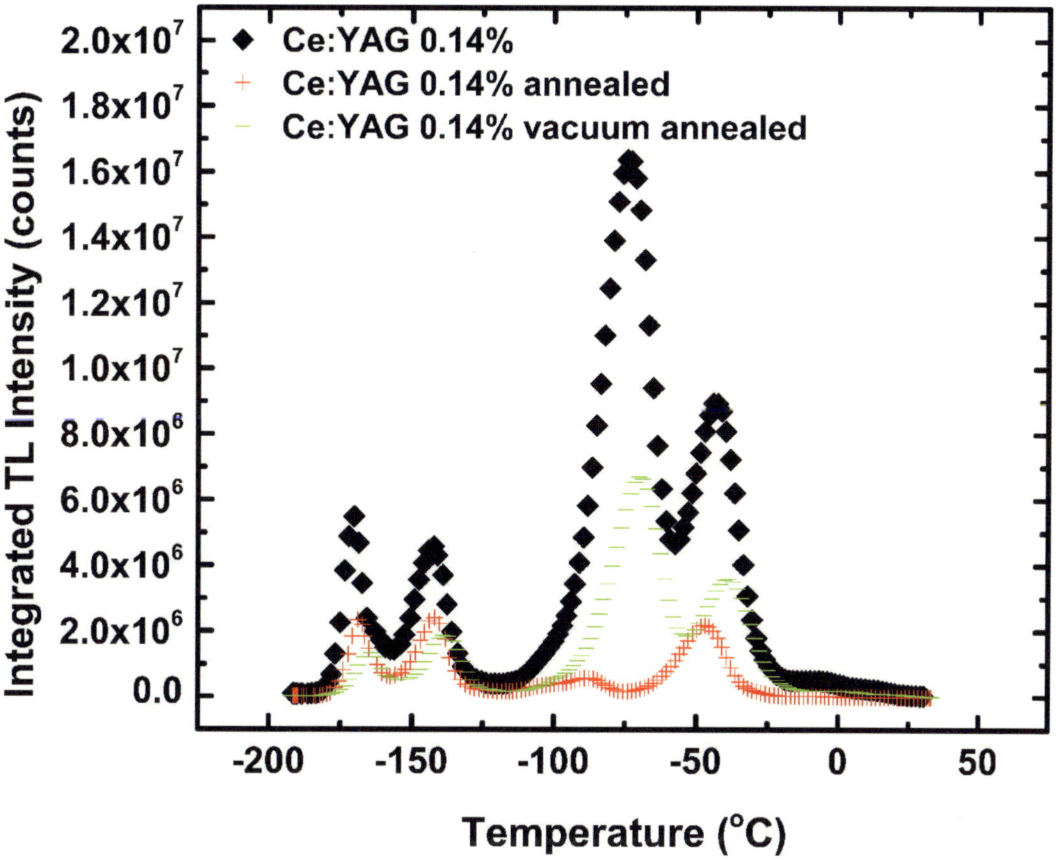

Figure 13. Low-temperature glow curves of Ce:YAG 0.14% as grown, annealed in air, and annealed in a vacuum using 470–720 nm integration.

	Activation energy	
Approximate peak location (q = 60°C/min)	As grown	Annealed in air
−170°C	0.20 ± 0.01 eV	0.37 ± 0.03 eV
−140°C	0.27 ± 0.01 eV	0.39 ± 0.02 eV
−100°C	N/A	0.38 ± 0.15 eV
−90°C	N/A	0.67 ± 0.06 eV
−75°C	0.55 ± 0.01 eV	N/A
−45°C	0.66 ± 0.01 eV	0.81 ± 0.01 eV
−10°C	0.90 ± 1.04 eV	N/A

Table 4. Calculated activation energies of Ce:YAG 0.14% for the glow curves constructed using 470–720 nm integration range.

temperature but would play a major role in any low-temperature processes involving Ce:YAG crystal. It also affects exciton dynamics in YAG crystal. It is clear how different processing affects the defect structure and concentration within these samples.

7. TL applied to semiconductors

7.1. Calculations of donor ionization energies in ZnO

In this section, we discuss a new application of TL that implies the use of low-temperature TL for the measurements of donor ionization energy in luminescent semiconductors. TL spectroscopy has been already applied in combination with other methods to identify defects in a few semiconductors as mentioned earlier in the chapter. However, the focus here is on its development as a method for donor ionization energy calculations. We have recently shown that it is possible to apply low-temperature TL to measure donor ionization energies in luminescent semiconductor. The study has been carried out on ZnO, one of the most important wide band gap semiconductors with many existing and future applications [107]. It is also considered one of the most

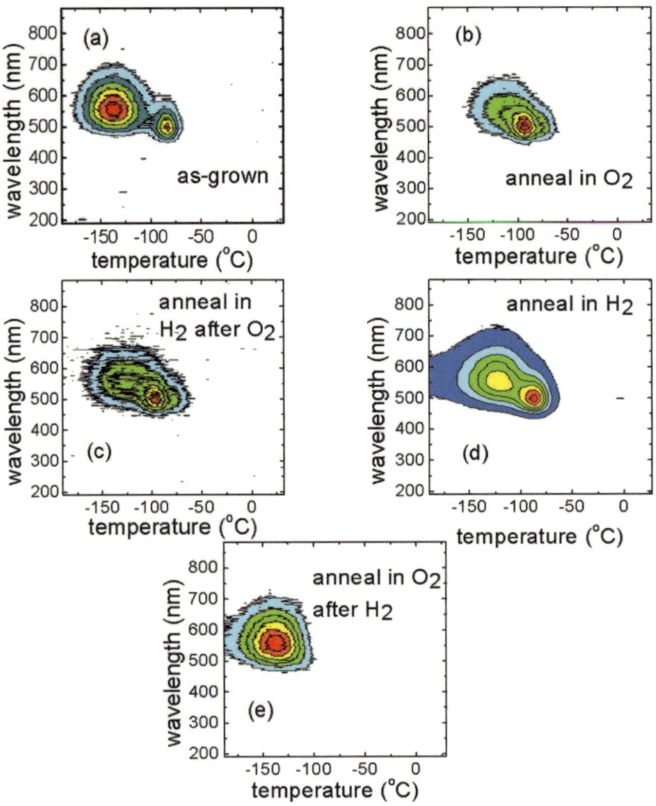

Figure 14. Contour plot of TL emission as a function of temperature and wavelength for as grown and annealed ZnO single crystals. 60°C/min was used for heating rate [116].

complex binary system in terms of defects and optoelectronic properties. There are several established methods for calculating donor and acceptor ionization energies in semiconductors. Temperature-dependent Hall-effect is the most common method, it is crucial for semiconductor characterization [108–110], however it suffers limitations in specific cases, and TL may provide an alternative method in these cases as explained below. Electron paramagnetic resonance (EPR) [110], and photoluminescence [111–115] have also been applied to measurements of donor ionization energies, however, EPR is very limited and PL alone may be not capable of measuring the donor ionization energy due to the collapse of exciton.

Low-temperature TL has been recently applied on as-grown and annealed ZnO single crystals [116]. Annealing was carried out in different atmospheres: (1) hydrogen atmosphere at 300°C for 1 hour. (2) Oxygen atmosphere at 1100°C for 1 hour. (3) Both atmospheres in different order. **Figure 14** represents the contour plots for as-grown and annealed samples, two peaks can be seen at 520 and 580 nm, which are slightly changed with the annealing atmospheres. **Figure 15** shows the glow curves for as-grown and annealed samples. Three donors were identified in the samples from the glow curves and their ionization energies were found to be 47 ± 3, 55 ± 5, and 36 ± 2 meV. The first two energies are in agreement with Lavrov [111, 112]

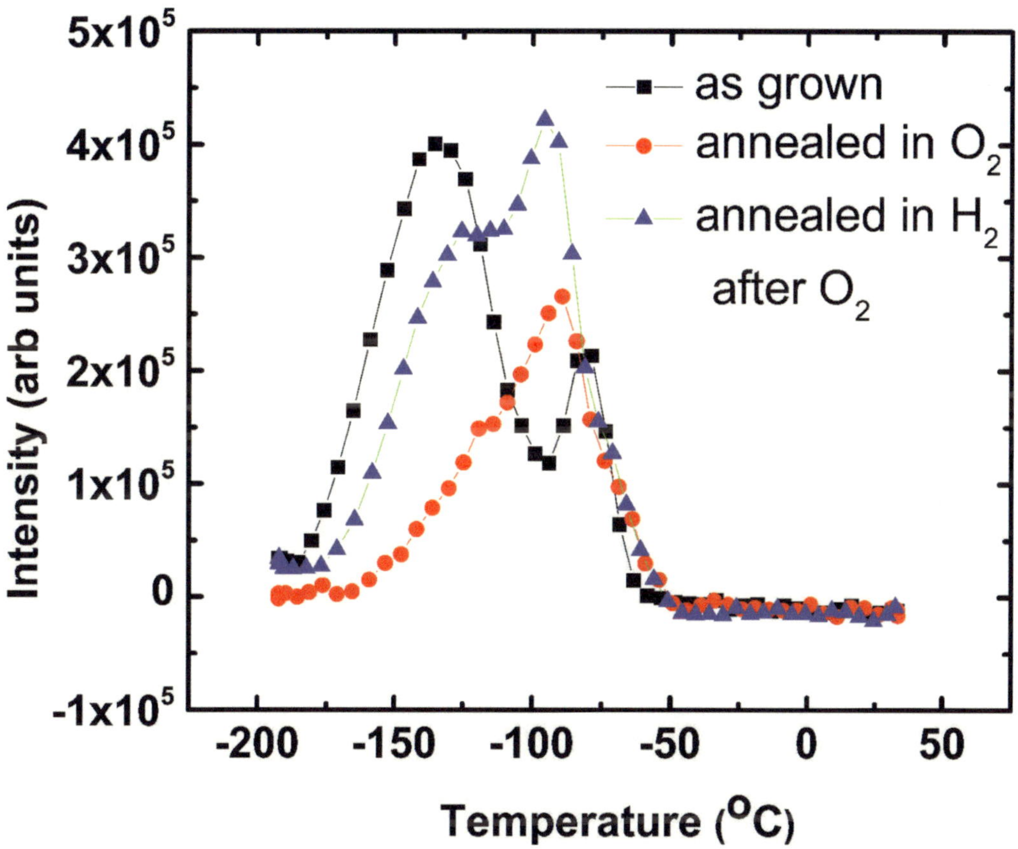

Figure 15. TL glow curve of ZnO single crystals produced from contour plots shown in **Figure 14 (a-c)** [116].

reports, the 47 meV was attributed to hydrogen bound in oxygen vacancy (H_o), and the 55 meV was related to hydrogen at a bound centered lattice site (H_{BC}). The 36 meV has also been reported for ZnO, and its origin is still in debate. **Figure 16** reveals the spectral range of TL emission that shows two peaks at 580 and 520 nm green luminescence centers, the ratio between them is dependent on the annealing conditions. The 580 nm emission peak is attributed to the Zn-O vacancy pairs which that have been modified after annealing in O_2 (**Figure 16b**). The 520 nm peak was reported as O-vacancy related defects, this peak was declined for the sample annealed in H_2 first and then annealed in O_2, where the 36 meV ionization energy has been measured. The reason behind it is that oxygen fills oxygen vacancies rather than Zn-O vacancies due to the formation of stable H-Zn vacancy complex defects, which supports the association of the 36 meV with "three or more hydrogen in Zn vacancy" donor. These TL measurements revealed the three known donors for ZnO in a single experiment, while each other method was only capable of measuring one type of the three donors. It should be noted that TL could be especially useful for small nonuniform samples because Hall-effect measurements require uniform sample and its resolution is limited to 5 × 5 mm.

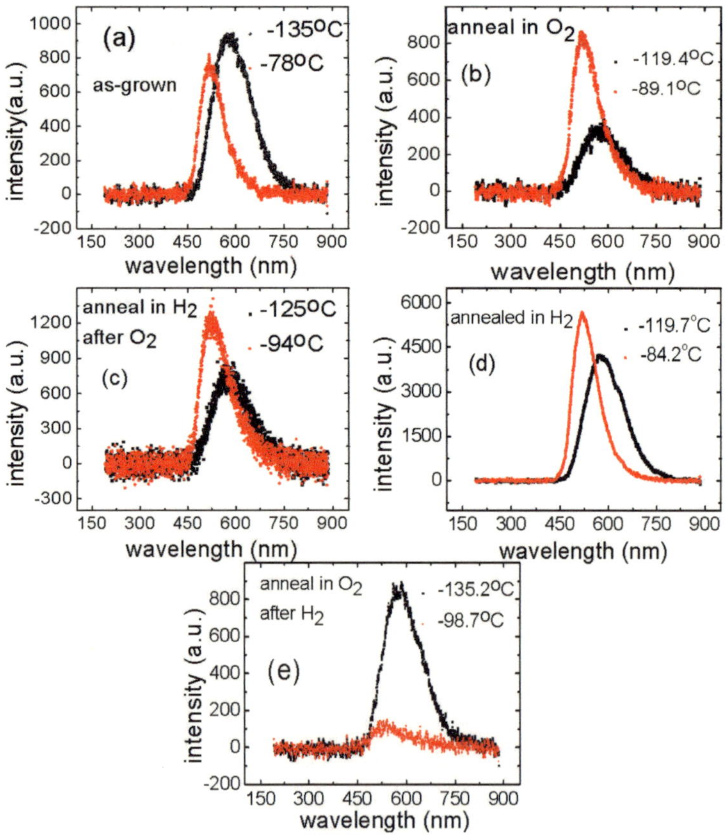

Figure 16. Emission peaks for as grown and annealed ZnO single crystals. Annealing atmosphere shows strong effects on the ratio of 580 and 520 nm [116].

Moreover, characterizing donors and acceptors in thin films on conductive layer using Hall-effect is a real challenge as the current diffuses to the conductive substrate and the measurements give a false indication for the electrical transport properties of the film. We expect that TL could be an alternative method for donor characterization in thin films on conductive layers

8. Conclusion

The chapter reviewed the basics of TL spectroscopy and its applications and described an advanced TL setup to extend its applications to the study and measurements of trap levels in semiconductors and dielectrics. Different methods for the analysis of TL glow curve and the calculations of activation energies were presented. By applying TL spectroscopy to the study of YAG and ZnO, we demonstrated the effectiveness of the technique in characterizing trap levels and measuring donor ionization energies in dielectrics and semiconductors, which is crucial to understand their electronic properties. By combining TL with other spectroscopies such as positron annihilation, Fourier transform infrared, and optical absorption spectroscopies, full characterization of trap levels in the band gap can be obtained. This is especially important for wide band gap materials where defects and dopants in the band gap determine most of the optical, electrical, and magnetic properties.

Author details

Pooneh Saadatkia[1], Chris Varney[2] and Farida Selim[1*]

*Address all correspondence to: faselim@bgsu.edu

1 Center for Photochemical Sciences, Bowling Green State University, Bowling Green, OH, USA

2 Seattle University, Seattle, WA, USA

References

[1] Harvey E. A history of luminescence. From early times until 1900. Baltimore, USA: The American Philosophical Society, JH Furst Co.; 1957.

[2] Mahesh K, Weng P, Furetta C. Thermoluminescence in solids and its applications. Ashford, Kent, England, UK: Nuclear Technology Pub.; 1989.

[3] Klingshirn CF. Semiconductor optics. Berlin: Springer Science & Business Media; 2012.

[4] Stokes GG. On the change of refrangibility of light. Philosophical Trans R Soc Lond 1852;142:463–562.

[5] Valeur B, Berberan-Santos MN. A brief history of fluorescence and phosphorescence before the emergence of quantum theory. J Chem Educ 2011;88(6):731–738.

[6] Selim FA. Luminescence based spectrometers. Patent 9261469; 2016.

[7] Varney C, Khamehchi M, Ji J, Selim F. X-ray luminescence based spectrometer for investigation of scintillation properties. Rev Sci Instrum 2012;83(10):103112.

[8] Chen R, McKeever SW. Theory of thermoluminescence and related phenomena. Singapore: World Scientific; 1997.

[9] Dincer O, Ege A. Synthesis and luminescence of Tb 3 doped lithium borate (LiBO 2). J Lumin 2013;138:174–178.

[10] Kucuk N, Kucuk I, Cakir M, Keles SK. Synthesis, thermoluminescence and dosimetric properties of La-doped zinc borates. J Lumin 2013;139:84–90.

[11] Sharma S, Pitale S, Malik M, Dubey R, Qureshi M, Vasisth S, et al. Synthesis and thermoluminescence studies of Sr4Al14O25: Eu2 phosphor. Indian J Eng Mater Sci 2009;16 (3):165.

[12] Meijerink A, Blasse G. Photoluminescence and thermoluminescence properties of Ca2PO4Cl: Eu2. J Phys Cond Matt 1990;2(15):3619.

[13] Tanno H, Zhang S, Shinoda T, Kajiyama H. Characteristics of photoluminescence, thermoluminescence and thermal degradation in Eu-doped BaMgAl 10 O 17 and SrMgAl 10 O 17. J Lumin 2010;130(1):82–86.

[14] McKeever S, Moscovitch M, Townsend P. Thermoluminescence and thermoluminescence dosimetry. Thermoluminescence dosimetry materials: properties and uses. Ashford: Nuclear Technology Publishing; 1995:15–44.

[15] McKinlay AF. Thermoluminescence dosimetry.United Kingdom: Adam Hilger; 1981.

[16] McDougall DJ. Thermoluminescence of geological materials. London, new York: Academic Press; 1968.

[17] Aitken M. Physics and archaeology. Oxford, UK: Oxford University; 1974.

[18] Vij D. Thermoluminescent materials. New Jersey: Prentice Hall; 1993.

[19] Falcony C, Garcia M, Ortiz A, Alonso J. Luminescent properties of ZnS: Mn films deposited by spray pyrolysis. J Appl Phys 1992;72(4):1525–1527.

[20] McKeever SW. Thermoluminescence of solids. Cambridge: Cambridge University Press; 1988.

[21] Rasheedy M, Zahran E. The effect of the heating rate on the characteristics of some experimental thermoluminescence glow curves. Phys Scripta 2005;73(1):98.

[22] Shinsho K, Suzuki Y, Yamamoto Y, Urushiyama A. The thermoluminescence activation energy and frequency factor of the main glow of $CaSO_4$: Tm phosphor determined by

heating rate method including very slow rates of heating. J Appl Phys 2005;97 (12):3523.

[23] Dubey V, Tiwari R, Pradhan MK, Rathore GS, Sharma C, Tamrakar RK. Photoluminescece and thermoluminescence behavior of Zn2SiO4: Mn2, Eu2 Phosphor. J Lumin 2014;1(1):30–39.

[24] Sohn K, Cho B, Park HD. Excitation energy-dependent photoluminescence behavior in Zn 2 SiO 4: Mn phosphors. Mater Lett 1999;41(6):303–308.

[25] Zhang H, Buddhudu S, Kam C, Zhou Y, Lam Y, Wong K, et al. Luminescence of Eu 3 and Tb 3 doped Zn 2 SiO 4 nanometer powder phosphors. Mater Chem Phys 2001;68(1):31–35.

[26] Reinhard C, Gerner P, Garcí S, Valiente R, Güdel HU. Near-infrared to green photon upconversion in Mn 2 and Yb 3 doped lattices. Chem Phys Lett 2004;386(1):132–136.

[27] Taghavinia N, Lerondel G, Makino H, Yamamoto A, Yao T, Kawazoe Y, et al. Growth of luminescent Zn 2 SiO 4: Mn 2 particles inside oxidized porous silicon: emergence of yellow luminescence. J Cryst Growth 2002;237:869–873.

[28] Yamamoto H, Okamoto S, Kobayashi H. Luminescence of rare-earth ions in perovskite-type oxides: from basic research to applications. J Lumin 2002;100(1):325–332.

[29] Nalwa HS. Handbook of nanostructured materials and nanotechnology, five-volume set. San Diego, San Francisco, New York, Boston, London, Sydney, Tokyo: Academic Press; 1999.

[30] Lochab S, Pandey A, Sahare P, Chauhan R, Salah N, Ranjan R. Nanocrystalline MgB4O7: dy for high dose measurement of gamma radiation. Phys Status Solidi (a) 2007;204 (7):2416–2425.

[31] Lochab S, Sahare P, Chauhan R, Salah N, Ranjan R, Pandey A. Thermoluminescence and photoluminescence study of nanocrystalline Ba 0.97 Ca 0.03 SO4. Eur J Phys D 2007;40 (5):1343.

[32] Tiwari N, Kuraria RK, Tamrakar RK. Thermoluminescence glow curve for UV induced ZrO2:Ti phosphor with variable concentration of dopant and various heating rate. J Radiat Res Appl Sci 2014;7(4):542–549.

[33] Tamrakar RK, Bisen DP, Sahu IP, Brahme N. UV and gamma ray induced thermoluminescence properties of cubic Gd 2 O 3: Er 3 phosphor. J Radiat Res App Scie 2014;7 (4):417–429.

[34] Pathak P, Selot A, Kurchania R. Thermoluminescence properties of Mn-doped CaYAl 3 O 7 phosphor irradiated with ultra-violet, mega-voltage and gamma radiation. Radiat Phys Chem 2014;99:26–29.

[35] Baldacchini G, Chiacchiaretta P, Gupta V, Kalinov V, Voitovich A. Thermoluminescence, glow curves, and carrier traps in colored and nominally pure LiF crystals. Phys Sol Stat 2008;50(9):1747–1755.

[36] Tamrakar RK. Studies on absorption spectra of Mn doped Cds nanoparticles: luminescence study and absorption spectra of CdS: Mn. Saarbrücken: Lambert Academic Publishing; 2012.

[37] Tamrakar RK, Bisen D. Optical and kinetic studies of CdS: Cu nanoparticles. Res Chem Intermed 2013;39(7):3043–3048.

[38] Tamrakar RK, Bisen D, Brahme N. Characterization and luminescence properties of Gd2O3 phosphor. Res Chem Intermediate 2014;40(5):1771–1779.

[39] Tamrakar RK, Bisen DP, Robinson CS, Sahu IP, Brahme N. Ytterbium doped gadolinium oxide (Gd 2 O 3: Yb 3). Indian J mat Sci 2014;2014:7 pages.

[40] Tamrakar RK. UV-irradiated thermoluminescence studies of bulk CdS with trap parameter. Res Chem Intermediate 2015;41(1):43–48.

[41] Chen R, Kirsh Y. The analysis of thermally stimulated processes. Oxford: Pergamon press; 2013.

[42] Chen R, Lawless J, Pagonis V. A model for explaining the concentration quenching of thermoluminescence. Radiat Measur 2011;46(12):1380–1384.

[43] Chen R, Pagonis V. Thermally and optically stimulated luminescence: a simulation approach. New Jersey: John Wiley 2011.

[44] Bagdasarov KS, Pasternak L, Sevast'yanov B. Radiation color centers in Y3Al5O12: Cr3 tals. Sov J Quant Electron 1977;7(8):965.

[45] Batygov SK, Radyukhi VS, Maier A, Denker B, Timoshec MI, Voronko Y, et al. Color centers in Y3Al5O12 crystals. Fizika Tverdogo Tela 1972;14(4):977.

[46] Guerassimova N, Dujardin C, Garnier N, Pedrini C, Petrosyan A, Kamenskikh I, et al. Charge-transfer luminescence and spectroscopic properties of Yb 3 in aluminium and gallium garnets. Nucl Instr Methods Phys Res A Accelerat Spectromet Detectors Assoc Equip 2002;486(1):278–282.

[47] Mackay D, Varney C, Buscher J, Selim F. Study of exciton dynamics in garnets by low temperature thermo-luminescence. J Appl Phys 2012;112(2):023522.

[48] Milliken E, Oliveira L, Denis G, Yukihara E. Testing a model-guided approach to the development of new thermoluminescent materials using YAG: Ln produced by solution combustion synthesis. J Lumin 2012;132(9):2495–2504.

[49] Mori K. Transient colour centres caused by UV light irradiation in yttrium aluminium garnet crystals. Phys Status Solidi (a) 1977;42(1):375–384.

[50] Murk V, Yaroshevich N. Exciton and recombination processes in YAG crystals. J Phys Condens Matt 1995;7(29):5857.

[51] Robbins D, Cockayne B, Glasper J, Lent B. The temperature dependence of rare-earth activated garnet phosphors I. Intensity and lifetime measurements on undoped and Ce-doped. J Electrochem Soc 1979;126(7):1213–1220.

[52] Selim F, Solodovnikov D, Weber M, Lynn K. Identification of defects in Y3Al5O12 crystals by positron annihilation spectroscopy. Appl Phys Lett 2007;91(10):4105.

[53] Smol'Skaya L, Martynovich E, Davydchenko A, Smirnova S. X-ray and thermally stimulated luminescence in YAG. J App Spectrosc 1987;46(1):44–46.

[54] Solodovnikov D, Weber M, Lynn K. Improvement in scintillation performance of Ce, Er codoped yttrium aluminum garnet crystals by means of a postgrowth treatment. Appl Phys Lett 2008;93(10):104102.

[55] Varney C, Mackay D, Pratt A, Reda S, Selim F. Energy levels of exciton traps in yttrium aluminum garnet single crystals. J Appl Phys 2012;111(6):063505.

[56] Vedda A, Di Martino D, Martini M, Mares J, Mihokova E, Nikl M, et al. Trap levels in Y-aluminum garnet scintillating crystals. Radiat Measur 2004;38(4):673–676.

[57] Zorenko Y, Zorenko T, Gorbenko V, Voznyak T, Savchyn V, Bilski P, et al. Peculiarities of luminescent and scintillation properties of YAG: Ce phosphor prepared in different crystalline forms. Optical Mat 2012;34(8):1314–1319.

[58] Zych E, Brecher C, Glodo J. Kinetics of cerium emission in a YAG: Ce single crystal: the role of traps. J Phys Condens Matter 2000;12(8):1947.

[59] Randall J, Wilkins M. The phosphorescence of various solids. Proc Royal Soc London A Math Phys Sci 1945;184(999):347–364.

[60] Randall J, Wilkins M. Phosphorescence and electron traps. I. The study of trap distributions. Proc Royal Soc London A Math Phys Eng Sci Royal Soc 1945;184:365-389.

[61] Randall J, Wilkins M. Phosphorescence and electron traps. II. The interpretation of long-period phosphorescence. Proc Royal Soc London A Math Phys Eng Sci Royal Soc 1945;184:390-407

[62] Bräunlich P. Comment on the initial-rise method for determining trap depths. J Appl Phys 1967;38(6):2516–2519.

[63] Christodoulides C. Errors involved in the determination of activation energies in TL and TSDC by the initial rise method. J Phys D 1985;18(8):1665.

[64] Garlick G, Gibson A. The electron trap mechanism of luminescence in sulphide and silicate phosphors. Proc Phys Soc 1948;60(6):574.

[65] Bos A. High sensitivity thermoluminescence dosimetry. Nucl Instrum Methods Phys Res B 2001;184(1):3–28.

[66] Kitis G, Tuyn J. A simple method to correct for the temperature lag in TL glow-curve measurements. J Phys D 1998;31(16):2065.

[67] Pankove JI. Optical processes in semiconductors. New York: Dover publications; 2012.

[68] Devault D, Govindjee, Arnold W. Energetics of photosynthetic glow peaks. Proc Natl Acad Sci U S A 1983;80(4):983–987.

[69] Kitis G, Pagonis V. Peak shape methods for general order thermoluminescence glow-peaks: a reappraisal. Nucl Instrum Methods Phys Res B 2007;262(2):313–322.

[70] Kitis G, Gomez-Ros J, Tuyn J. Thermoluminescence glow-curve deconvolution functions for first, second and general orders of kinetics. J Phys D 1998;31(19):2636.

[71] May C, Partridge J. Thermoluminescent kinetics of alpha-irradiated alkali halides. J Chem Phys 1964;40(5):1401–1409.

[72] Grandfond A, Gautier B, Militaru L, Albertini D, Descamps-Mandine A. Spurious phenomena occurring during current measurement on ultra-thin dielectric layers: from electro-thermal effects to surface damage. J Appl Phys 2014;115(13):134103.

[73] VanDam J, Marinello G. Methods for in vivo dosimetry in external radiotherapy. Belgium: Garant Publ.; 1994.

[74] Winarski D, Persson C, Selim F. Hydrogen in insulating oxide Y3Al5O12 strongly narrows the band gap. Appl Phys Lett 2014;105(22):221110.

[75] Reda S, Varney C, Selim F. Radio-luminescence and absence of trapping defects in Nd-doped YAG single crystals. Res Phys 2012;2:123–126.

[76] Varney CR, Selim FA. Color centers in YAG. AIMS Mater Sci 2015;2(4):560–572.

[77] Varney C, Reda S, Mackay D, Rowe M, Selim F. Strong visible and near infrared luminescence in undoped YAG single crystals. AIP Adv 2011;1(4):042170.

[78] Varney C, Mackay D, Reda S, Selim F. On the optical properties of undoped and rare-earth-doped yttrium aluminium garnet single crystals. J Phys D 2011;45(1):015103.

[79] Varney C, Selim F. Positron lifetime measurements of vacancy defects in complex oxides. Acta Phys Polonica A 2014;125(3):764–766.

[80] Selim F, Varney C, Tarun M, Rowe M, Collins G, McCluskey M. Positron lifetime measurements of hydrogen passivation of cation vacancies in yttrium aluminum oxide garnets. Phys Rev B 2013;88(17):174102.

[81] Selim F, Winarski D, Varney C, Tarun M, Ji J, McCluskey M. Generation and characterization of point defects in SrTiO 3 and Y 3 Al 5 O 12. Results Phys 2015;5:28–31.

[82] Babin V, Laguta V, Maaroos A, Makhov A, Nikl M, Zazubovich S. Luminescence of F-type centers in undoped Lu3Al5O12 single crystals. Physica Status Solidi (b) 2011;248 (1):239–242.

[83] Baciero A, Placentino L, McCarthy K, Barquero L, Ibarra A, Zurro B. Vacuum ultraviolet and x-ray luminescence efficiencies of Y3Al5O12: Ce phosphor screens. J Appl Phys 1999;85(9):6790–6796.

[84] Blasse G, Bril A. A new phosphor for flying-spot cathode-ray tubes for color television: yellow-emitting Y3Al5O12–Ce3. Appl Phys Lett 1967;11(2):53–55.

[85] Blasse G, Bril A. Investigation of some Ce3-activated phosphors. J Chem Phys 1967;47 (12):5139–5145.

[86] Blazek K, Krasnikov A, Nejezchleb K, Nikl M, Savikhina T, Zazubovich S. Luminescence and defects creation in Ce3-doped Lu3Al5O12 crystals. Phys Status Solidi (b) 2004;241 (5):1134–1140.

[87] Dong Y, Zhou G, Jun X, Zhao G, Su F, Su L, et al. Luminescence studies of Ce: YAG using vacuum ultraviolet synchrotron radiation. Mater Res Bull 2006;41(10):1959–1963.

[88] Fagundes-Peters D, Martynyuk N, Lünstedt K, Peters V, Petermann K, Huber G, et al. High quantum efficiency YbAG-crystals. J Lumin 2007;125(1):238–247.

[89] Gibbons E, Tien T, Delosh R, Zacmanidis P, Stadler H. Some factors influencing the luminous decay characteristics of Y 3Al5 O 12: Ce3. J Electrochem Soc 1973;120(6):835–837.

[90] Hamilton D, Gayen S, Pogatshnik G, Ghen R, Miniscalco W. Optical-absorption and photoionization measurements from the excited states of Ce 3: Y 3 Al 5 O 12. Phys Rev B 1989;39(13):8807.

[91] Kirm M, Lushchik A, Lushchik C, Zimmerer G. Investigation of luminescence properties of pure and Ce3+ doped Y3Al5O12 crystals using VUV radiation. Electrochem Soc Proc 1999; 99–40:113-22.

[92] Liu B, Gu M, Liu X, Huang S, Ni C. Formation energies of antisite defects in Y3Al5O12: a first-principles study. Appl Phys Lett 2009;94(12):121910.

[93] Meng JX, Li JQ, Shi ZP, Cheah KW. Efficient energy transfer for Ce to Nd in Nd/Ce codoped yttrium aluminum garnet. Appl Phys Lett 2008;93(22):1908.

[94] Miniscalco W, Pellegrino J, Yen W. Measurements of excited-state absorption in Ce3: YAG. J Appl Phys 1978;49(12):6109–6111.

[95] Moszyński M, Ludziejewski T, Wolski D, Klamra W, Norlin L. Properties of the YAG: Ce scintillator. Nucl Instrum Methods Phys Res A 1994;345(3):461–467.

[96] Pedrini C, Rogemond F, McClure D. Photoionization thresholds of rare-earth impurity ions. Eu2: CaF2, Ce3: YAG, and Sm2: CaF2. J Appl Phys 1986;59(4):1196–1201.

[97] Pédrini C. Electronic processes in rare earth activated wide gap materials. Phys Status Solidi (a) 2005;202(2):185–194.

[98] Robbins D, Cockayne B, Lent B, Duckworth C, Glasper J. Investigation of competitive recombination processes in rare-earth activated garnet phosphors. Phys Rev B 1979;19 (2):1254.

[99] Robbins D, Cockayne B, Glasper J, Lent B. The temperature dependence of rare-earth activated garnet phosphors II. A comparative study of Ce3, Eu3, Tb3, and Gd3. J Electrochem Soc 1979;126(7):1221–1228.

[100] Rotman S, Tuller H, Warde C. Defect-property correlations in garnet crystals. VI. The electrical conductivity, defect structure, and optical properties of luminescent calcium and cerium-doped yttrium aluminum garnet. J Appl Phys 1992;71(3):1209–1214.

[101] Suzuki Y, Sakuma T, Hirai M. UV emission from the second lowest 5d state in Ce3: YAG. Mater Sci Forum 1997;239–241:219–222.

[102] Tomiki T, Akamine H, Gushiken M, Kinjoh Y, Miyazato M, Miyazato T, et al. Ce3 centres in Y3Al5O12 (YAG) single crystals. J Physic Soc Japan 1991;60(7):2437–2445.

[103] Zorenko Y, Voloshinovskii A, Savchyn V, Voznyak T, Nikl M, Nejezchleb K, et al. Exciton and antisite defect-related luminescence in Lu3Al5O12 and Y3Al5O12 garnets. Phys Status Solidi (b) 2007;244(6):2180–2189.

[104] Zorenko Y, Mares J, Prusa P, Nikl M, Gorbenko V, Savchyn V, et al. Luminescence and scintillation characteristics of YAG: Ce single crystalline films and single crystals. Radiat Measur 2010;45(3):389–391.

[105] Zorenko Y, Voznyak T, Gorbenko V, Zych E, Nizankovski S, Dan'Ko A, et al. Luminescence properties of Y 3 Al 5 O 12: Ce nanoceramics. J Lumin 2011;131(1):17–21.

[106] Zych E, Brecher C, Glodo J. Kinetics of cerium emission in a YAG: Ce single crystal: the role of traps. J Phys 2000;12(8):1947.

[107] Selim F, Tarun M, Wall D, Boatner LA, McCluskey M. Cu-doping of ZnO by nuclear transmutation. Appl Phys Lett 2011;99(20):202109.

[108] Look DC, Hemsky JW, Sizelove J. Residual native shallow donor in ZnO. Phys Rev Lett 1999;82(12):2552.

[109] Look DC, Reynolds DC, Sizelove J, Jones R, Litton CW, Cantwell G, et al. Electrical properties of bulk ZnO. Solid State Commun 1998;105(6):399–401.

[110] Hofmann DM, Hofstaetter A, Leiter F, Zhou H, Henecker F, Meyer BK, et al. Hydrogen: a relevant shallow donor in zinc oxide. Phys Rev Lett 2002;88(4):045504.

[111] Lavrov E, Herklotz F, Weber J. Identification of two hydrogen donors in ZnO. Phys Rev B 2009;79(16):165210.

[112] Lavrov E. Hydrogen in ZnO. Phys B Condens Matt 2009;404(23):5075–5079.

[113] Meyer B, Sann J, Hofmann D, Neumann C, Zeuner A. Shallow donors and acceptors in ZnO. Semicond Sci Tech 2005;20(4):S62.

[114] Meyer B, Alves H, Hofmann D, Kriegseis W, Forster D, Bertram F, et al. Bound exciton and donor–acceptor pair recombinations in ZnO. Phys Status Solidi (b) 2004;241(2):231–260.

[115] Schildknecht A, Sauer R, Thonke K. Donor-related defect states in ZnO substrate material. Phys B Condens Matt 2003;340:205–209.

[116] Ji J, Boatner L, Selim F. Donor characterization in ZnO by thermally stimulated luminescence. Appl Phys Lett 2014;105(4):041102.

Chapter 11

Luminescence Dating: Applications in Earth Sciences and Archaeology

Ken Munyikwa

Additional information is available at the end of the chapter

http://dx.doi.org/10.5772/65119

Abstract

Over the last 60 years, luminescence dating has developed into a robust chronometer for applications in earth sciences and archaeology. The technique is particularly useful for dating materials ranging in age from a few decades to around 100,000–150,000 years. In this chapter, following a brief outline of the historical development of the dating method, basic principles behind the technique are discussed. This is followed by a look at measurement equipment that is employed in determining age and its operation. Luminescence properties of minerals used in dating are then examined after which procedures used in age calculation are looked at. Sample collection methods are also reviewed, as well as types of materials that can be dated. Continuing refinements in both methodology and equipment promise to yield luminescence chronologies with improved accuracy and extended dating range in the future and these are briefly discussed.

Keywords: luminescence dating, thermoluminescence (TL), optically stimulated luminescence (OSL), paleodose, dose rate, earth sciences, archaeology

1. Introduction

Luminescence dating refers to age-dating methods that employ the phenomenon of luminescence to determine the amount of time that has elapsed since the occurrence of a given event. In this chapter, the application of luminescence techniques in dating geological and archaeological events is examined. Generally, the term luminescence dating is a collective reference to numerical age-dating methods that include thermoluminescence (TL) and optically stimulated luminescence (OSL) dating techniques. Other terms used to describe OSL include optical dating [1] and photon-stimulated luminescence dating or photoluminescence dating [2]. Luminescence dating methods are based on the ability of some dielectric and semiconducting materials to absorb and store energy from environmental ionizing radiation. In earth sciences and archaeological applications, the dielectric materials are usually minerals such as feldspar and quartz. The materials are also sometimes referred to as dosimeters [3]. Environmental ionizing radiation for earth science and archaeological applications typically comes from radioactive elements within the immediate surroundings of the mineral grains as well as from cosmic radiation. When the minerals are stimulated, they release the stored energy in the form of light, from which the term luminescence is derived. Stimulation of energy release by heating is referred to as TL. When light is used, on the other hand, the technique is described as OSL. Measuring the amount of energy released in conjunction with a determination of the rate at which the energy was accumulated allows an age to be calculated, indicating time that has elapsed since the storage of energy began. Luminescence methods can generally be used to date materials that range in age from a few decades to about 100,000 years. However, ages of up to several hundred thousands of years have been reported in some studies [4]. Therefore, the method can be used for dating events of Late Pleistocene and Holocene age (ca. < 126 000 years). Radiocarbon dating is a technique that can also be used to date some materials of Late Pleistocene or Holocene age if they bear carbon. Hence, when dealing with materials that do not contain organic carbon, luminescence dating can serve as an optional chronometer to radiocarbon dating. Furthermore, the age range that can be dated using luminescence techniques is greater than that of radiocarbon dating, and this provides researchers with a viable alternative dating method.

This chapter presents a brief examination of how luminescence is used in earth sciences and archaeology to measure time. In both fields of study, the imperativeness of assigning a temporal scale to events and processes is an inherent aspect of the discipline. Hence, the role played by dating cannot be overstated. The chapter is not intended to be used as a practice manual. Rather, the aim is to provide a primer that acquaints scholars who may be familiar with the science of luminescence but are not accustomed to its application as a dating method. Thus, the chapter comprises eight sections. Following an introductory look at the development of luminescence dating in Section 1, principles of luminescence dating are examined in Section 2. In Section 3, basic luminescence measurement equipment and sample stimulation mechanisms during measurement are explored after which luminescence properties of some common minerals are examined. The determination of parameters used in the luminescence age equation is discussed in Section 5, and this is followed by a look at materials that can be dated with luminescence methods in Section 6. Section 7 looks at methods used in sample collection and preparation prior to the analysis. In Section 8, the chapter concludes with a look at future developments in luminescence dating.

1.1. Historical development of luminescence dating

The ability of some minerals to luminesce when stimulated is not a recent finding. The earliest recorded observation of the behavior in minerals has been attributed [5] to Robert Boyle, who in 1664 recounted his discovery [6] that a diamond that had been loaned to him could emit light when heated. Subsequently, Boyle would describe the phenomenon as 'self-shining' [7]. Throughout the eighteenth, nineteenth and early parts of the twentieth centuries a lot of activity went into examining the effects of TL such that by the middle of the twentieth century,

http://dx.doi.org/10.5772/65119

physicists were fairly familiar with the phenomenon. The focus of this section, however, is not on the history of the science of luminescence but rather on the history of the emergence of the application of luminescence as a dating technique for geological and archeological materials. That inception can be traced back to approximately seven decades ago, to a period during which experiments were being conducted into applications of TL [8, 9]. Zeller et al. [10] noted that research into applying TL to determine geological ages of rocks using minerals began at the University of Wisconsin around 1950. Progress in dating geological materials, however, was hampered by a limited understanding of the luminescence process in rocks and accurate ages were hard to come by [11]. It is in the dating of heated archaeological artifacts that progress was realized. In 1953, Daniels et al. [9] had proposed that TL from ancient pottery could be used to determine its antiquity [12]. The idea was premised on the recognition that heating the pottery to high temperatures (>500°C) during production had the effect of erasing any previously accumulated energy from constituent mineral grains. Subsequently, Kennedy and Knoff [13] provided some basic aspects of dating heated archaeological materials using TL, though the approach had not been tested on actual samples as yet. Around the same time, Grogler et al. [14] provided additional experimental details on how a dating experiment measuring TL of pottery could be conducted. It was in 1964 that Aitken et al. [12] designed a study that sought to determine the relationship between TL glow of an artifact and its archaeological age. Results demonstrated that the luminescence intensity emitted by the samples was linearly proportional to radiocarbon ages of sediments from which the samples were obtained. The positive results noted in these early studies gave impetus for further refinements in TL protocols, and throughout the rest of the 1960s and 1970s, the dating method gained a foothold in archaeological studies.

A major development in luminescence studies occurred when TL dating was extended to determining burial ages of unheated sediments. The evolution appears to have followed two parallel paths, one in the West and another in former Eastern Bloc countries. However, it seems there was minimal interaction between the two geographical regions, especially in the early stages. In the West, some of the earliest work includes a study [15] that looked at TL signals of deep-sea sediment that mostly comprised foraminiferal shells. The investigators [15] considered the signals to be from calcite and noticed that the TL intensity increased with depth. Later, another study [16] investigated a deep-sea sediment core that comprised predominantly siliceous plankton and reported results similar to those presented earlier [15]. Subsequent studies by Wintle and Huntley [17] provided additional TL data from deep-sea sediments that also showed increasing signal intensities with depth. However, it was later observed [17] that the TL signals from deep-sea cores actually came from detrital minerals that were mixed with the plankton. Significantly, it was also suggested [17] that what was being dated was the last time the ocean sediments had been exposed to sunlight. Though it was some time before researchers fully understood the zeroing mechanism for unheated sediment [18], that discovery was extremely important for sediment dating since it meant that the exposure of sediments to sunlight had the same zeroing effect of accumulated energy in sediment grains as did heating in pottery.

In former Eastern Bloc countries, the analysis of unheated terrestrial sediments using TL appears to have begun sometime during the 1960s. Early published reports from the former

Soviet Union include a study [19] that examined TL signals of Quaternary deposits. A few years later, Morozov [20] presented relative ages of Quaternary sediments from Ukraine that had been dated using TL methods. The study was mainly based on the recognition that luminescence signal intensities increased with depth, which was interpreted as commensurate with age. Morozov [20] also suggested that the signal was coming from quartz in the sediments. Shortly afterwards, Shelkoplyas [21] reported a range of Quaternary TL ages obtained from soils and loess deposits [18]. Throughout the 1970s, researchers in Eastern Bloc countries [22–29] as well in China [30] reported studies in which TL was used to date Quaternary deposits. The accuracy of some of these early ages, however, is questionable [18], not least because zeroing mechanisms were not well-understood at the time. In other studies, efforts were directed at understanding TL characteristics of dosimeters, especially quartz [31–33].

Around the time Wintle and Huntley [17] discovered that TL signals in their deep-sea cores were coming from detrital mineral grains mixed with the plankton, they became aware of the work by Eastern Bloc researchers who had worked extensively with terrestrial sediments. Ultimately, they realized that TL dating could be applied much more broadly to date Quaternary deposits [18]. These developments resulted in the landmark publications by Wintle and Huntley [17, 34, 35] in which TL dating of sediments was outlined. With increased research throughout the early 1980s, dating procedures improved as efforts were made to standardize procedures [18]. However, optimal conditions for solar resetting of sediments remained unclear to researchers and this hampered the accuracy of TL ages. Researchers who examined the problem include Huntley [36] who investigated solar resetting of sediments from various environments and proposed a method to address the issue of inadequate zeroing. By 1985, another monumental step in luminescence dating would be realized when

Authors	Topic	Source
Murray and Olley (2002)	Luminescence dating of sedimentary quartz	[40]
Feathers (2003)	Luminescence dating in archaeology	[41]
Lian (2007)	Optically stimulated luminescence dating	[42]
Duller (2008)	Luminescence dating guidelines for archaeology	[43]
Fuchs and Owen (2008)	Luminescence dating of glacial sediments	[44]
Preusser et al. (2008)	Luminescence dating: principles and applications	[45]
Rittenour (2008)	Luminescence dating of fluvial deposits	[46]
Singhvi and Porat (2008)	Luminescence dating in geomorphology	[47]
Wintle (2008)	History of luminescence dating over the last 50 years	[39]
Wintle (2008)	Luminescence dating: history and prospects	[3]
Rhodes (2011)	Luminescence dating of sediments	[48]
Liritzis et al. (2013)	Luminescence dating in archaeology	[49]

Table 1. Recent review studies examining luminescence dating and its applications.

Huntley et al. [11] reported that a light source could be used to stimulate energy release from a dosimeter during measurement instead of heating as was used in TL. This led to the development of OSL dating which offered a number of advantages over TL methods when dating unheated sediments. With further equipment and methodological refinements, there was a burgeoning of OSL dating studies of Quaternary sediments throughout the 1990s that saw luminescence dating emerge as a robust dating technique. Over the last two decades, the technique has developed further [37–39], and today, it is the method of choice for dating detrital sediments of Late Pleistocene and Holocene age as well as previously heated archaeological artifacts. **Table 1** presents some recent review studies that have examined various aspects of luminescence dating and its applications.

2. Principles of luminescence dating

2.1. Luminescence behavior of dielectric materials used in dating

Many minerals such as quartz, feldspar, calcite and zircon are dielectric materials and, when subjected to ionizing radiation, they are able to store energy in their crystal lattices. In natural geological and archeological settings, the ionizing radiation emanates naturally from the immediate surroundings of the minerals. Cosmic radiation may also contribute a small component. If the minerals used in dating are stimulated, they release the energy by luminescing and, within certain constraints, the energy released is proportional to the stored energy.

2.2. Luminescence dating equation

In luminescence dating, the energy given out by the minerals or dosimeters following stimulation is measured using appropriate instrumentation. This energy is referred to as the paleodose [50]. In order to determine an age, the rate at which the energy was accumulated by the dosimeter, or the dose date, is also ascertained. The quotient of the paleodose and the dose rate, as indicated in Eq. (1), is the luminescence age.

$$Luminescence\,age = \frac{Paleodose}{Dose\,rate} \tag{1}$$

If the mineral grains were emptied of all previously accumulated energy prior to the latest energy storage episode, the age obtained will denote time that has elapsed since the start of that episode. Hence, both in geology and archaeology, the luminescence age simply connotes time that has passed since the occurrence of a specific energy zeroing event. In geology, this might be a geomorphic event that exposed sediment to sunlight. Zeroing by sunlight is also sometimes referred to as optical bleaching [3]. In pottery, zeroing would normally occur during a firing event associated with the manufacture.

2.3. Electron trapping mechanisms

Mechanisms by which minerals store energy in their crystal lattices as a result of ionizing radiation are complex [50–52]. However, it is thought that ionizing radiation drives mineral crystals into a metastable state where electrons are displaced from their parent nuclei. The positions from which the electrons have been evicted act as holes. The electrons and holes then diffuse within the mineral crystals and become trapped separately at lattice defects. Examples of common defects include a negative ion missing from its lattice position, a negative ion positioned in an interstitial site or the presence of impurity atoms in the lattice through substitution [52]. Other more complex trap types exist [52]. **Figure 1** depicts an energy level diagram that is used to visualize the trapping mechanisms involved in luminescence in crystalline materials. The depth of the trap (T) below the conduction band, indicated by 'E' (**Figure 1**) is a reflection of the efficacy of a given trap. Stable traps are those that can withstand perturbations such as lattice vibrations that could dislodge the electrons from their traps. If the crystal lattice is stimulated using an appropriate mechanism, for example, by heating to an adequately high temperature or by exposure to an optical source with a suitable wavelength, trapped electrons will be evicted out of the traps. Once evicted, the electrons diffuse within the crystal lattice until they reach a site that is attractive to electrons. Such sites are referred to as recombination centers. Some recombination centers emit energy in the form of light when they capture electrons. Where stimulation is conducted by heating, the effect would be referred to as TL. When stimulation is by optical means, OSL will be obtained. The diffusion of evicted electrons to their recombination centers occurs fairly rapidly to the extent that the time between stimulation and recombination can be viewed as instantaneous. Effective recombination centers are usually those sites in the lattice where electrons are missing. These are the holes created when the materials are exposed to ionizing radiation (**Figure 1a**). The intensity of the luminescence that is obtained when a material is stimulated is proportional to the number of electrons that are trapped in the material which, in turn, is commensurate with the energy absorbed from the ionizing radiation [50, 51]. However, despite the energy

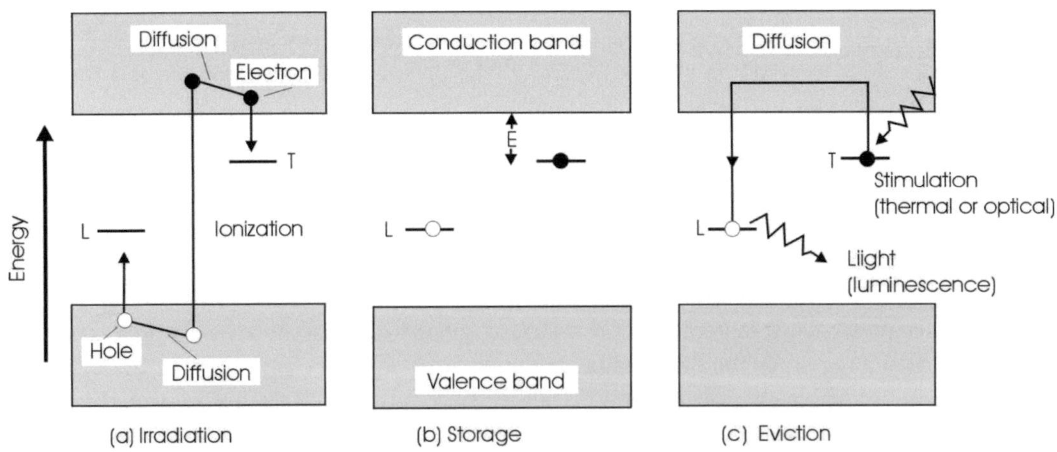

Figure 1. An energy level diagram that illustrates how ionizing radiation creates luminescence centers in crystal lattices (redrawn from Ref. [50]). (a) Following irradiation, electrons are expelled from their original sites leaving holes behind. Both electrons and holes diffuse within the lattice. (b) At appropriate sites, electrons are trapped while holes may become localized. (c) Thermal (TL) or optical (OSL) stimulation of the material results in electrons being evicted from the traps. Evicted electrons that reach luminescence centers result in light being emitted.

http://dx.doi.org/10.5772/65119

storage mechanism being the same for a given mineral, the sensitivity to radiation may vary greatly between samples, an aspect that has important implications for dating procedures as will be shown later.

Importantly, for dating purposes, the number of electron traps within any mineral lattice is finite. As a result, when minerals are exposed to ionizing radiation for an extended period, the traps become exhausted such that energy can no longer be stored efficiently. This effect is referred to as saturation. In dating, saturation determines the upper limit beyond which samples cannot be dated using luminescence techniques. Materials that are subjected to very high dose rates will have the number of traps exhausted more rapidly such that the specific age representing the upper age limit will depend on both the number of traps present as well as the dose rate.

2.4. Natural sources of ionizing radiation

In geological and archaeological dating applications, natural sources of ionizing radiation that contribute to the trapped energy in mineral grains include isotopes of uranium (^{238}U and ^{235}U) and thorium (^{232}Th) decay chains, potassium (^{40}K) and rubidium (^{87}Rb). The elements are found in natural materials in very low concentrations (about 3–10 parts per million for uranium and thorium and less than 5% for potassium, where ^{40}K is one part in 10,000). Despite the low concentrations, these radioactive isotopes collectively emit enough radiation to induce luminescence that is detectable for dating purposes. The radiation emitted includes alpha and beta particles as well as gamma radiation. Beta particles and gamma rays have penetration ranges of about 0.02 cm and 20 cm in earth materials, respectively, whereas alpha particles penetrate about 0.02 mm [51].

An additional though smaller radiation component received by earth materials comes from cosmic radiation. Cosmic rays from outer space consist of a soft and a hard component. On earth, surface substrate absorbs the soft component such that it cannot penetrate deeper than 50 cm. The hard component, however, largely comprising muons, penetrates deeper and is lightly ionizing. Hence, only the hard component is relevant to luminescence dating. On earth, the intensity of the hard component is also influenced by both latitude and altitude. Special formulae for evaluating cosmic ray contribution to dose rate have been developed for luminescence dating [53].

3. Basic luminescence measurement equipment and sample stimulation mechanisms

For dating studies, the primary aim of TL and OSL measurements is to determine the amount of energy that has been stored in the mineral grains of a given material since the start of the event that is being investigated. As outlined above, two main methods are used to stimulate energy release in luminescence studies. Heating allows TL to be measured, whereas stimulation using a light source is used for OSL measurements.

3.1. The luminescence reader

The basic layout of equipment used to measure luminescence in geological and archaeological dating is shown in **Figure 2**. Modern luminescence dating systems commonly possess both TL and OSL measurement capabilities [54, 55]. To conduct a measurement, samples are usually loaded on discs about 1 cm in diameter that sit on an appropriate sample holder in multiples. These are then introduced into the device, commonly referred to as a luminescence reader [44] and selectively moved into position for measurement. The luminescence signal from the sample is captured by a photon detector system [1] for example, photomultiplier tube (PMT) or charge-coupled device (CCD) camera after passing through optical filters. When conducting TL measurements, the filters exclude infrared signals from the heating but permit blue or violet emissions to pass through. In OSL measurements, wavelengths used for stimulation are rejected by the filters, whereas violet and near ultraviolet wavelengths are usually transmitted. Output from a TL measurement is distinctly different from that obtained using OSL stimulation (**Figure 2**).

3.2. Thermal stimulation

When using thermal stimulation, samples are heated at rates approaching 20°C per second. On reaching a temperature that corresponds to the trap depth 'E' (**Figure 1**), usually characteristic

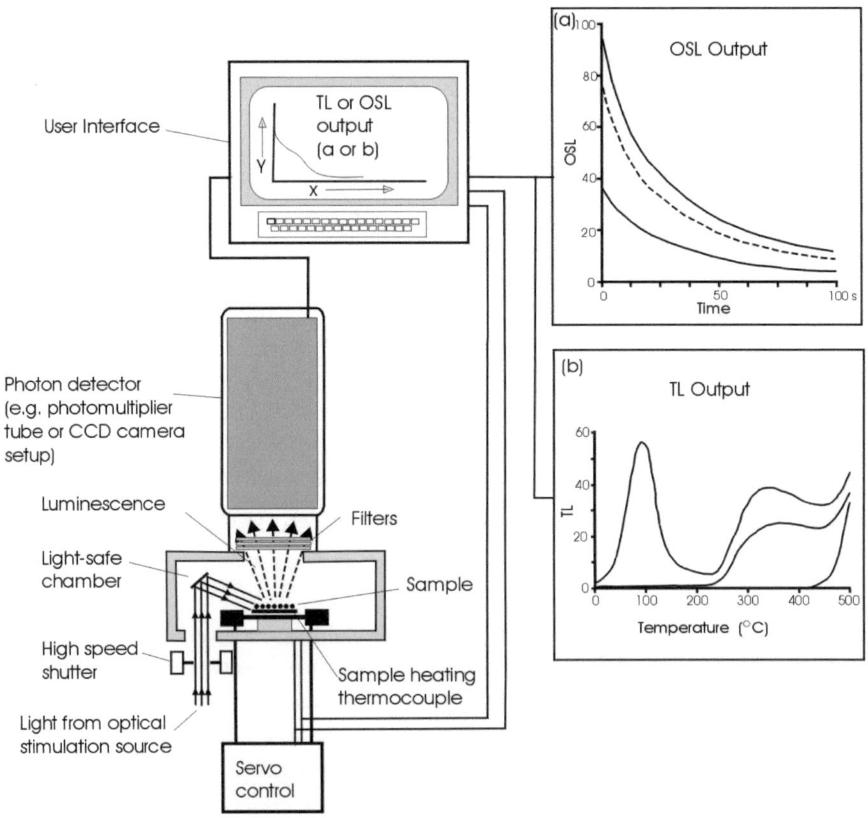

Figure 2. Main components of a luminescence reader that measures both TL and OSL signals (modified after Ref. [56]).

of a given trap type, trapped electrons are rapidly evicted [51] into the conduction band. The equation below gives the probability per second at temperature T that the energy provided is sufficient to evict a trapped electron at depth 'E' into the conduction band:

$$\rho(T) = s(T)\exp(-E/kT) \tag{2}$$

where $s(T)$ is a temperature dependent factor associated with the lattice vibrational frequency and change in entropy and k is Boltzmann's constant [52]. Once evicted, electrons are free to be re-trapped at the same site, be trapped at a different site or get to a recombination site where luminescence occurs. The eviction temperature is depicted by a peak in emission on a plot of the luminescence signal versus temperature which is referred to as a TL glow-curve. If the heating continues, all the electron traps will be emptied. A glow-curve that is obtained after the first heating of a sample is given in **Figure 3**. Heating the sample again soon after the first heating will produce a different curve. The second curve corresponds to incandescence that is usually observed when any material is adequately heated to an elevated temperature. Hence, from this second heating, there will be no luminescence from trapped electrons that had accumulated from ionizing radiation since the last zeroing event.

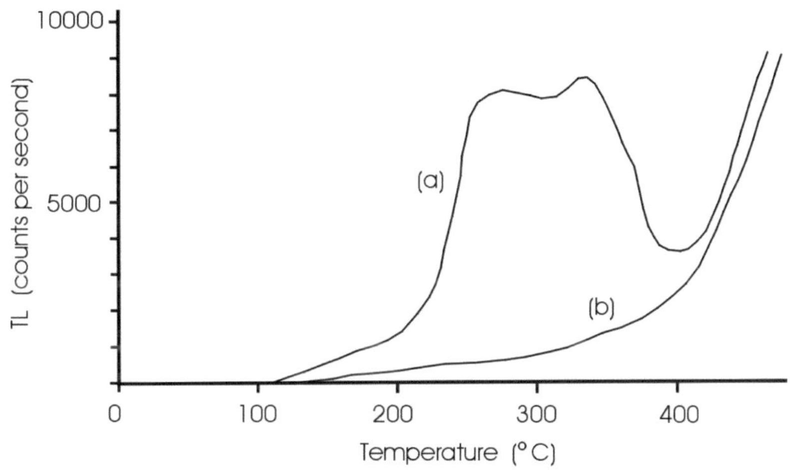

Figure 3. Illustration of glow-curves obtained following thermal stimulation. (a) A signal yielded after the first heating of a quartz sample with trapped electrons (paleodose). (b) A glow-curve from incandescence resulting from the second heating of the sample to a high temperature. It is important to note that incandescence is also realized above 400°C during the first heating (redrawn from Ref. [56]).

3.3. Optical stimulation

In optical stimulation, electrons are expelled from their traps using a source of a chosen wavelength. Commonly used sources in luminescence dating include blue, green or near-infrared wavelengths. The rate at which trapped electrons are evicted is influenced by the rate at which stimulating photons are emitted by the source as well as by the sensitivity of the trap types to optical stimulation. Generally, however, starting with a concentration of 'n' trapped electrons and if '*p*' is the probability for electrons to be evicted by an optical source to the conduction band per unit time, 'n' will change according to Eq. (3):

$$\frac{dn}{dt} = -np \tag{3}$$

assuming that no re-trapping of freed electron occurs. Resolving Eq. (3) shows that as traps are emptied, the concentration of trapped electrons decays exponentially as given in Eq. (4) [5].

$$n(t) = n_0 e^{-pt} \tag{4}$$

where n_0 is the concentration of trapped electrons prior to the stimulation. If all freed electrons reach recombination sites instantaneously, the luminescence intensity will be proportional to electrons being evicted from the traps. Thus from Eq. (3) and Eq. (4):

$$I_{OSl}(t) \alpha \left| \frac{dn}{dt} \right| = n_0 p e^{-pt} \tag{5}$$

The exponentially decaying emission curve obtained is referred to in OSL dating as a shine-down curve (**Figure 4**). With continued stimulation, a point is reached where all trapped electrons that are susceptible to optical stimulation are depleted. If all the photons released during stimulation are integrated, the total luminescence energy released by the mineral can be ascertained.

Factors that influence the sensitivity of a trap type to optical eviction include characteristics of the trap as well as the wavelength of the optical source. Generally, however, eviction rates are faster when shorter wavelengths are employed. Electron eviction from some traps could require more energy than that provided by an optical source. To circumvent that limitation,

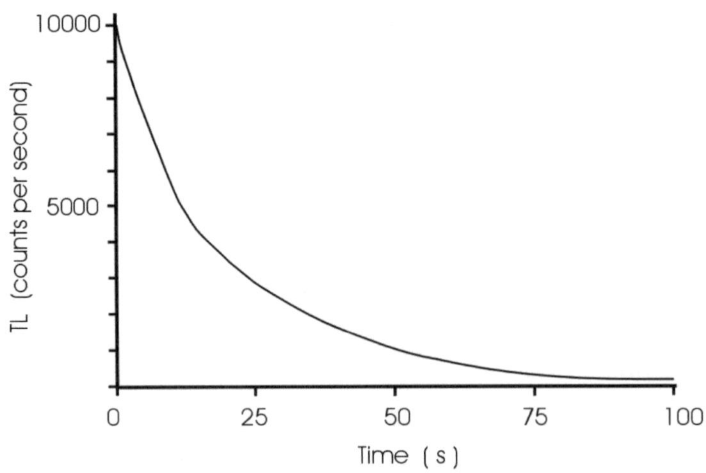

Figure 4. A shine-down curve obtained following the optical stimulation (OSL) of a hypothetical mineral sample for about 100 s (modified after Ref. [51]).

thermal assistance is used to attain the energy threshold required for eviction. This allows longer wavelengths that would not normally be employed for optical stimulation to be used in dating [50].

Generally, sources for optical stimulation are selected such that separation can be made between wavelengths of the source used for stimulation and those of signals emitted by the minerals being analyzed. That separation is usually aided by the use of optical filters. As an illustration, the main emissions for quartz and feldspar are in the near-ultraviolet (356 nm) and violet (410 nm) regions of the electromagnetic spectrum. Thus, filters that are employed when analyzing quartz and feldspars have windows in those respective regions but exclude wavelengths used for stimulation, for example, blue for quartz and near infrared for feldspar.

3.4. Advantages of OSL over TL

OSL dating has a number of inherent advantages compared with TL when analyzing sediments that have been zeroed by exposure to sunlight. Investigations have shown that solar bleaching of electron traps that are stimulated by TL proceeds more slowly than with traps that are sensitive to OSL [50]. In the study summarized in **Figure 5**, after 20 h of exposure to

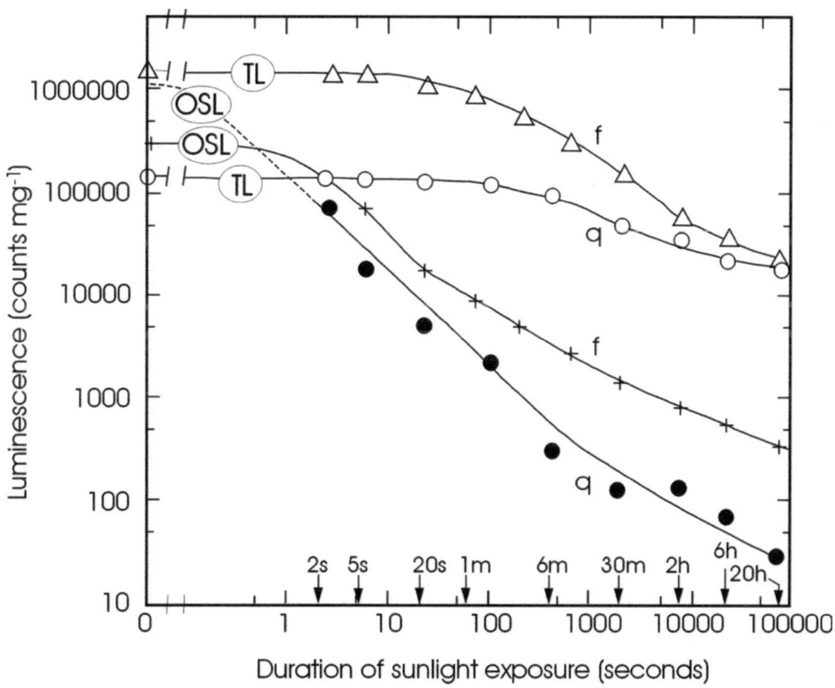

Figure 5. Bleaching rates of quartz (q) and feldspar (f) stimulated using TL and OSL (green light) conducted by Godfrey-Smith et al. [57] (redrawn from Ref. [50]). The slower bleaching curves are from TL signals. For quartz, TL measurement focused on a peak between 320 and 330°C whereas for feldspar, a peak at 310–320°C was analyzed. Detection of TL signals was through a window with a center at 400 nm (violet). For the OSL, both quartz and feldspar used a green laser and a detection window of 380 nm which is violet to near-UV [50].

sunlight, both quartz and feldspar were shown to have less than 0.1% of the signal originally present in the mineral remaining [57]. The TL signal that remained following the same period of bleaching, on the other hand, was a few orders of magnitude higher than the OSL signal. In practical terms, the slower bleaching of TL traps by solar energy means that higher residual signals will be found in unheated sediments analyzed using TL, to the extent that it is difficult to date very young samples using TL [11, 50]. Thus, OSL analysis is generally preferred for dating sediments that were not reset by heating [50]. For dating materials that have previously been zeroed by heating, however, such as archeological artifacts, TL remains an appropriate stimulation mechanism.

4. Luminescence properties of some common minerals

Many minerals will luminesce when stimulated using an appropriate source following a period of exposure to ionizing radiation. However, not all such minerals are suitable for use in luminescence dating. Today, luminescence dating primarily employs quartz and feldspar. Zircon and calcite have been tried in some studies but both minerals are associated with a number of complications. As a result, they are not commonly used in luminescence dating at present. This section examines the luminescence properties of the four minerals. In the discussions below, natural dose refers to energy acquired from natural radiation sources by a mineral grain in its field setting. This is differentiated from an artificial dose that a sample would obtain when irradiated using an artificial source in a laboratory setting.

4.1. Quartz

Quartz is the most commonly used mineral in luminescence dating because it offers a number of advantages when contrasted with alternatives. Due to its resistance to both chemical weathering and mechanical abrasion, it is very stable at the earth's surface. As a result, it is one of the most abundant minerals in clastic depositional environments. Quartz's luminescence properties are also very stable. Additionally, it does not have an internal source of radiation as a major element of its composition. Thus, the ionizing radiation that quartz grains receive in nature is usually from an external source, which simplifies dose rate calculation procedures. Situations exist, however, where quartz grains may contain very low levels of uranium but these are rare [51].

4.1.1. Quartz TL properties

Quartz that has a natural dose displays TL peaks at 325°C and 375°C [58]. TL dating usually employs the 375°C peak which is very stable and is thought to be the result of AlO_4 lattice impurities that serve as hole traps [59]. Under sunlight, the 325°C peak bleaches much more rapidly than the 375°C peak [60]. Quartz that has been irradiated artificially also shows a peak at 110°C [**Figure 6**]. With regards to emissions, heating quartz above 300°C shows a natural TL emission band around 460–480 nm (blue) and another in the region of 610–630 nm (orange). Laboratory irradiated quartz has a TL emission band below 300°C in the region of 360–420 nm.

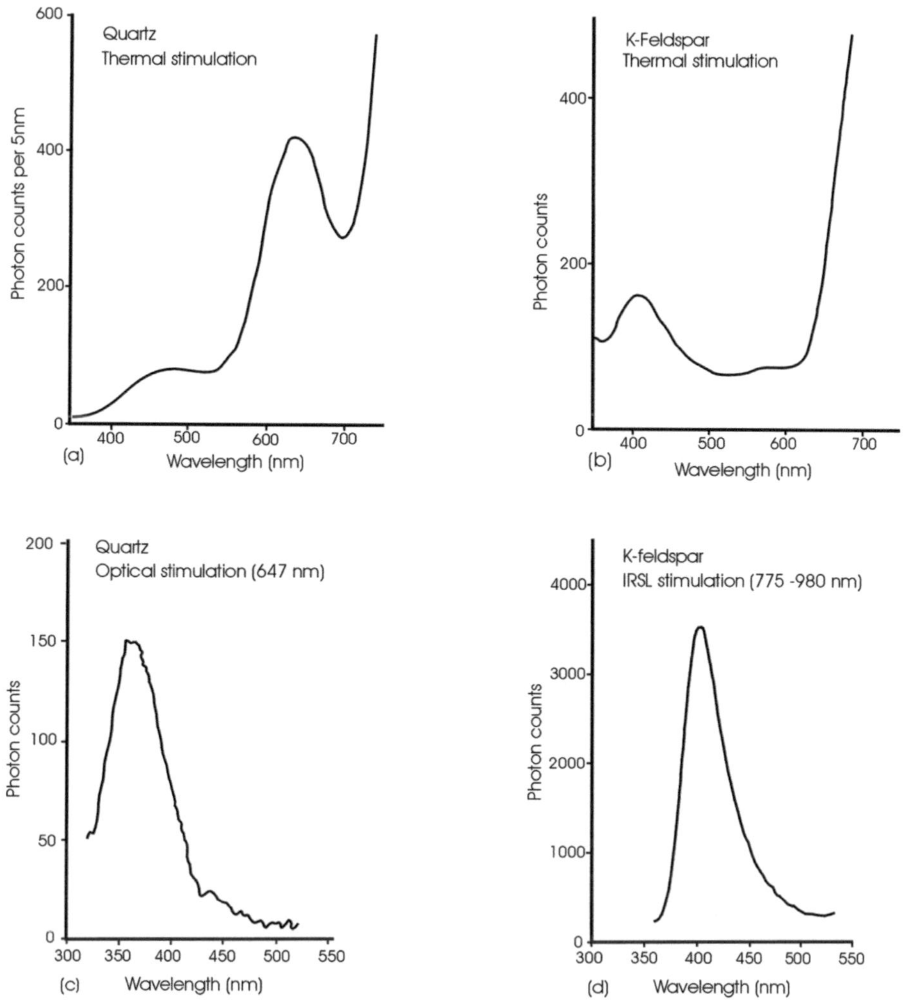

Figure 6. Examples of TL (a, b) and OSL (c, d) emission spectra (adapted with permission from Ref. [60]). For TL spectra, the sharp rise in emissions beyond 650 nm is largely from incandescence (rather than from electrons evicted from traps).

4.1.2. *Quartz OSL properties*

Quartz has been shown to luminesce when stimulated by wavelengths from any part of the visible spectrum [60]. Most current OSL studies, however, prefer using blue light for stimulation because of the higher OSL intensities it yields [55]. Investigations have demonstrated that the OSL signal of quartz can consist of at least three or more components that are referred to as fast, medium and slow, based on their decay rates [61, 62]. To separate the components during stimulation, constantly increasing power is used to give linearly modulated OSL (LM-OSL) [63]. Most regular dating procedures, however, employ a constant power (continuous wave–CW) and are unable to resolve the components. Through the use of heat treatments or stimulation for limited times (to exclude the slower components), desired signals can be targeted when using CW stimulation. Emission bands that are observed in quartz OSL are in the range of 360–420 nm (**Figure 6**), which corresponds to emissions that are seen in quartz TL at temperatures below 300°C [60] (**Figure 7**).

4.2. Feldspar

Feldspar is another widely used mineral in OSL dating. It constitutes about 60% of the earth's crust and even though it weathers more rapidly than quartz, it is also a very common mineral at the earth's surface. In terms of chemistry, feldspars are aluminosilicates that form solid solution series with potassium (K) calcium (Ca) and sodium (Na) as end members of a ternary system. Since potassium has an isotope that contributes ionizing radiation in luminescence dating, the potassium in K-feldspars has to be treated as a source of internal dose, in addition to dose contributions from sources external to the grains. As a result, when dating feldspars, it is necessary to separate K-feldspars from Ca and Na-feldspars and analyze them separately.

Compared with quartz, feldspar has a number of attractive luminescence features. First, feldspar emissions are generally brighter than those from quartz which produces stronger signals. This means that smaller doses can be measured during analysis. Second, the internal dose from potassium is not susceptible to external influences such as variations in pore water and this allows dose rates to be ascertained more accurately. Third, feldspar can be stimulated using infrared radiation which allows effective separation to be made between the stimulation source and emission wavelengths. The main drawback for feldspar, however, is its susceptibility to anomalous fading [64]. Anomalous fading occurs when trapped electrons reside in their traps for shorter periods than what would be predicted by physical models such that the luminescence intensity drops over time from the time of irradiation. Ultimately, the result of anomalous fading is that most feldspar grains yield equivalent doses that are slightly lower than they would in the absence of fading.

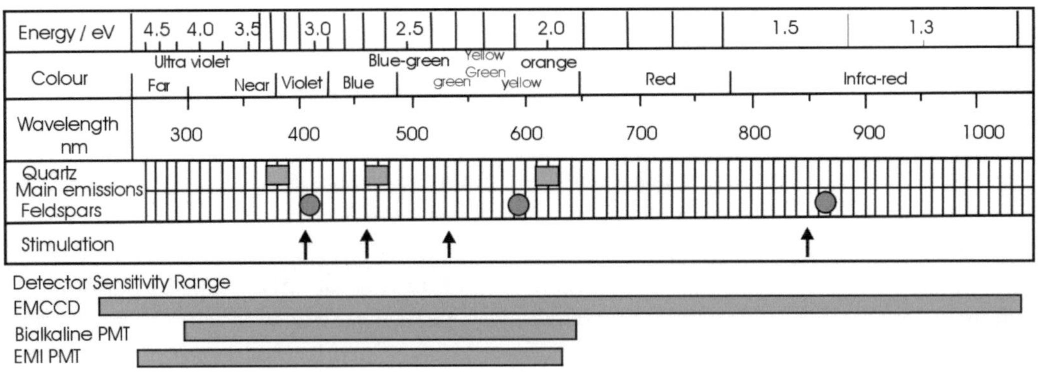

Figure 7. Main emission wavelengths for quartz and feldspars used in luminescence dating as well as wavelengths employed for stimulation. Sensitivity ranges for some detectors are also shown. Standard bialkaline photomultipliers (PMT) and EMI 9235QB PMTs do not detect emissions beyond 650 nm, whereas electron multiplying CCD (EMCCD) cameras can capture the whole range (modified after Ref. [54]).

Correction methods have been developed for dealing with anomalous fading when dating feldspars [65, 66].

4.2.1. Feldspar TL properties

Various studies have shown that K-feldspars extracted from sediments yield natural TL signals that display peaks at 280 and 330°C [60]. In terms of emission wavelengths, K-rich feldspars have been reported [67] to show maximums in the range of 390–440 nm (violet to blue). Conversely, emissions for some plagioclase feldspars have been reported to appear in the range of 550–560 nm (blue-green). Other studies, however, have intimated at a more complex emission pattern for feldspars [68].

4.2.2. Feldspar OSL properties

Optical stimulation of luminescence from feldspars has been investigated using visible light. Early studies employed lasers which included the 514.5 nm wavelength from argon and the 633 nm (red) wavelength from krypton. The emissions were then monitored at shorter wavelengths [1, 57] and shown to be centered around 400 nm [69]. The application of OSL stimulation in dating feldspars, however, has been relatively limited because near-infrared stimulation (discussed below) has been shown to be a more desirable approach. Nonetheless, a study [70] comparing green light stimulation (GLSL) of feldspars with infrared stimulated luminescence (IRSL) reported data suggesting that at 10°C, GLSL signals were more stable than IRSL signals. This would indicate that different trap types might be involved [50]. Apart from green and red stimulation, luminescence in feldspar has been demonstrated using a range of other wavelengths in the region spanning 380–1020 nm [71].

4.2.3. Feldspar IRSL properties

As mentioned above, wavelengths in the near infrared region (peaking around 880 nm) can also be used to induce luminescence in feldspars. Since this effect was first noticed [72], most research in optical dating of feldspars has focused on IRSL stimulation. The main advantage of using IRSL is that the rest of the visible spectrum can then be used for emission detection. Fine-grained sediments containing mixtures of both plagioclase and K-feldspars have also been demonstrated to display a major stimulation peak around 845 nm as well as a weaker one at 775 nm [73]. Today, most studies that use feldspar for optical dating employ light emitting diodes (LEDs) for stimulation that have emission peaks at around 880 ± 40 nm. LEDs are much cheaper than lasers and are widely available, making them a desirable alternative. Sedimentary K-feldspars stimulated using IRSL show a major natural emission peak at 410 nm (**Figure 6d**) and another minor peak in the range of 300–350 nm [74]. Some IRSL studies [75] have reported additional natural emission maxima for K-feldspars at 280, 560 and 700 nm. With plagioclase feldspar, an IRSL emission peak has been identified at 570 nm.

Feldspars stimulated using IRSL following the administration of a laboratory dose also exhibit an emission peak at 290 nm. That peak is not observed in feldspars that have a natural signal. When not required during dating, the peak can be removed by preheating the sample to an appropriate temperature.

4.3. Calcite

Thermally stimulated calcite has an emission maximum at 570 nm [60]. However, efforts to use the mineral in luminescence dating have been encumbered by the limited environmental occurrence of calcite. Calcite also tends to concentrate uranium in its lattice and this complicates dose rate calculations since isotopic disequilibrium of uranium has to be taken into account. Worth noting is that uranium disequilibrium dating can yield ages from calcite that are more reliable than those obtained using luminescence techniques. As a result, the incentive to employ luminescence methods in dating calcite has been small. It should be mentioned that some of the earliest, albeit unsuccessful, TL studies that tried to date rocks employed calcite [10]. Other attempts to use calcite in archaeological dating include a report by Ugumori and Ikeya [76].

4.4. Zircon

Zircon is an attractive dosimeter because it usually has a relatively high concentration of uranium. Consequently, the dose from the mineral grain's interior is far greater than that originating from outside. This yields a dose rate that is relatively constant since it is not susceptible to variations arising from external effects such as changes in water content or burial depth. An associated drawback, however, is that the uranium content of zircon varies between individual grains. Consequently, measurements for dose rate are made on single grains. Also notably, zircon crystal lattices often have natural inhomogeneities that make it difficult to make comparisons between artificial irradiation administered in the laboratory with natural doses originating from within the grain. As outlined in Section 5, such comparisons are the standard approach for determining the paleodose when dating quartz or feldspar. To address that problem, zircon dating uses a technique called autoregeneration. With autoregeneration, after the natural signal from the zircon grains is measured, the grains are stored for a few months to allow a new dose to accrue. Measuring the signal at the end of the storage period and comparing it to the natural signal obtained from the initial measurement allows a calibration to be made that yields an age of the natural signal.

Analysis of zircon using TL includes a study by Huntley et al. [67] and by Templer and Smith [77]. OSL studies using zircon include investigations by Smith [78].

5. Paleodose and dose rate determination

The age equation introduced in Section 1 (Eq. 1) shows that two parameters need to be determined before a luminescence age can be calculated: the paleodose and the dose rate. This section examines methods that are used to determine the two variables.

5.1. Paleodose determination

The start of the accumulation of the paleodose should typically coincide with a geomorphic or archeological event that emptied (or zeroed) any previously accumulated energy in the sample grains. The dose refers to the energy absorbed per kilogram of material, and it is measured using units of gray (Gy), where 1 Gy = 1 J/kg. For materials that were previously zeroed by heating or firing, the start of the accumulation of the paleodose would correspond with the last time the material was heated to a temperature high enough to expel electrons from their traps. In the case of sediments that were zeroed by exposure to sunlight, the start of the accumulation of the paleodose would correspond to the last time that a sample was subjected to the bleaching effects of the sun for a period long enough to evict all trapped electrons. As indicated earlier, the natural signal refers to the luminescence signal yielded by a sample collected from the field. In order to determine the paleodose of a sample of unknown age, the natural signal is measured first after which the sample is irradiated artificially using a well-calibrated laboratory-based source. The signals from the artificial dose are then measured and compared with the natural dose signals in order to determine an artificial dose that gives a signal similar to that of the natural dose. This is referred to as the equivalent dose (D_e). Investigators use two main methods to determine D_e: the additive dose and the regenerative dose (or regeneration) methods [2, 50].

5.1.1. *Additive dose method*

When determining luminescence ages using the additive dose method, a sample of unknown age is split into two sets of aliquots. The natural signal for one set is measured first after which the second set is irradiated with incremental doses using an artificial source and also measured. Plotting the artificial signals against the dose administered produces a dose–response curve depicting the luminescence signal against the laboratory dose (**Figure 8a**). The curve is also known as a growth curve. The natural signal is also plotted on the growth curve against zero dose (**Figure 8a**). Extrapolating the curve backwards until it intercepts the horizontal axis at zero signal intensity provides D_e (**Figure 8a**) [51]. Growth curves are unique to each sample because luminescence sensitivity of mineral grains can vary from sample to sample. As a result, a new growth curve has to be constructed for each sample whose age is being determined. When dating materials that were zeroed by the sun using TL, the residual TL signal that is noted following solar bleaching would have to be taken into account when extrapolating the curve backwards. With both feldspar and quartz, the relationship between the luminescence signal and the laboratory radiation portrays a linear trend for low and moderate doses. At elevated doses, however, the growth curve plateaus, indicating that luminescence traps are getting exhausted, also referred to as saturation.

5.1.2. *Regenerative dose method*

The procedure used in the regeneration method is similar to that employed in the additive dose method apart from that, before the laboratory dose is applied, the sample aliquots in the regeneration method are first zeroed to remove any previously acquired dose. Incremental doses are then applied to the zeroed aliquots and measured. The acquired signals are plotted against the administered dose to give a regenerative dose growth curve. To get the equivalent dose, the signal from the sample of unknown age is interpolated into the growth curve (**Figure 8b**) [51]. Hence, when constructing the growth curve, the laboratory irradiation doses are selected such that the signals they produce lie above and below the signal obtained from the natural dose.

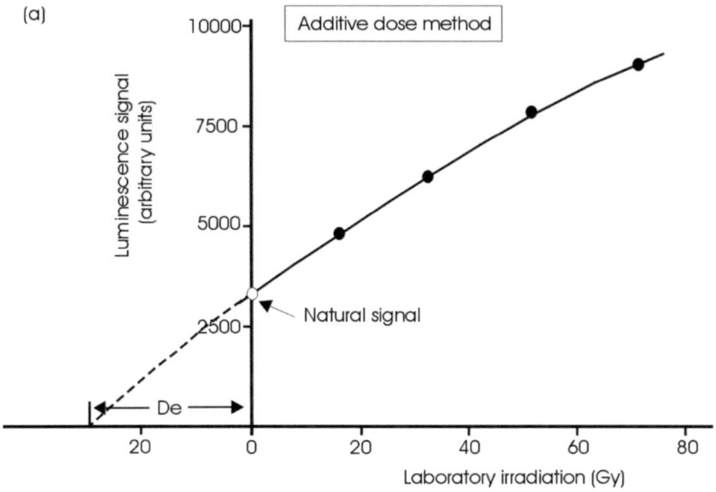

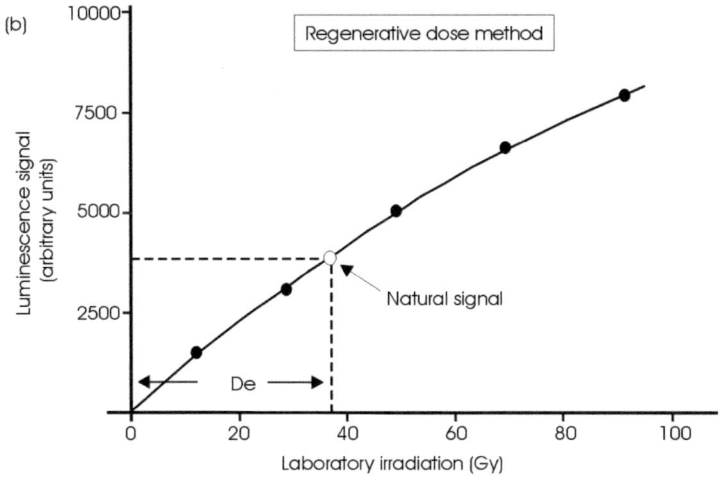

Figure 8. Main methods employed in determining the equivalent dose. (a) When using the additive dose method, incremental doses are administered on top of the natural dose and then measured. (b) With the regenerative dose approach, samples are first zeroed to remove any previously acquired dose and then given incremental doses, followed by measuring. Growth curves are unique to each sample being analyzed such that new measurements have to be made and a new curve constructed for every sample being dated.

5.1.3. Single aliquot methods versus multiple aliquot methods

Both the additive dose and regenerative dose methods employed multiple aliquots when they were originally developed for TL dating. Later, when OSL dating emerged, the possibility of using single aliquots only was brought up [1]. However, the concept did not take hold initially and multiple aliquots were also adopted for OSL dating. Generally, the use of multiple aliquots assumes that all aliquots of a given sample behave similarly to the dose administered.

However, in reality, inter-aliquot variations occur for a range of reasons that include changes in sensitivity [3, 79]. Normalization is used in some cases to try and reduce the effects of the variations. Nonetheless, the effects cannot be eliminated entirely such that uncertainties are contributed to the calculated ages. Ultimately, there was an incentive to devise an approach that only employed a single aliquot.

5.1.4. Single-aliquot regenerative-dose (SAR) protocol

When initially introduced for dating, single aliquot methods employed the additive dose method on feldspars [80, 81]. Quartz had not been used because sensitivity changes that it displays during repeat measurement cycles rendered single aliquot data unworkable. A few years later, however, following work carried out by Mejdhal and Bøtter-Jensen [82], a single aliquot regenerative (SAR) protocol for quartz that incorporated correction steps to address the changes in sensitivity was proposed. Subsequently, Murray and Wintle [37] presented an enhanced version of the SAR approach introduced by Murray and Roberts [83] in which a test dose was used to monitor sensitivity changes in the quartz. As initially proposed by Murray and Wintle [37], their SAR method entailed steps outlined in **Table 2**.

Minor modifications have been made to the SAR procedure presented in **Table 2** since its original inception [37, 38]. However, over the last two decades, the protocol has been widely adopted for routine dating of both sediments and heated materials using quartz and feldspar [84].

Step	Treatment[a]	Observed signal
1	Irradiate sample with artificial dose, D_i	–
2	Sample preheated (160–300°C) for 10s	–
3	Sample stimulated for 100 s at 125°C	L_i
4	Sample irradiated with test dose, D_t	–
5	Sample heated to 160°C	–
6	Sample stimulated for 100 s at 125°C[b]	T_i
7	Return to step 1 and repeat sequence[c]	–

D_i is the dose that yields the signal L_i while D_t is the test dose that produces the signal T_i. These values are used to plot a curve of Li/Ti vs. the regeneration dose D.
[a]For the natural sample, $i = 0$ and $D_0 = 0$ Gy.
[b]Stimulation time varies with the intensity of the light used for stimulation.
[c]Subsequent modifications to steps above include an additional step to help attenuate recuperation.

Table 2. Sequence of steps in the SAR procedure as initially proposed by Murray and Wintle [37].

5.1.5. Single grain analysis

SAR protocols have also been extended to determining paleodoses using individual mineral grains. The transition has been facilitated by the development of special equipment for loading and analyzing thousands of grains of sand size (<250 μm) relatively quickly [85]. The analysis of individual grains from the same sample is particularly useful for identifying differences in paleodose between grains [86]. Examples of cases where this may be expedient is when studying sediments deposited by rivers (fluvial) where well-bleached grains might be mixed with partially bleached fractions [87].

5.2. Determination of the dose rate

As outlined above, once the paleodose has been ascertained, the dose rate needs to be evaluated before an age can be determined. In Section 1, it was mentioned that ionizing radiation responsible for the energy accumulation in mineral grains in natural settings emanates from uranium and thorium decay chains as well as from potassium and rubidium isotopes. Cosmic radiation also contributes a minor component. A number of methods can be used to evaluate the total contributions of all these components. Using the concentration approach, levels of uranium, potassium and rubidium in a given sample are quantified with the help of an analytical procedure such as atomic absorption spectroscopy (AAS), neutron activation analysis (NAA), flame photometric detection (FPD), X-ray fluorescence (XRF) and inductively coupled plasma spectroscopy (ICPS). Once the concentrations have been measured, dose rate is determined using special conversion tables prepared for the purpose [50]. For uranium and thorium, however, isotopic disequilibrium could render the measurements using these analytical techniques unreliable [50].

As opposed to measuring elemental concentrations "using methods outlined" above, direct measurements of the activities of specific radionuclides can be achieved using methods such as alpha or gamma ray spectrometry. Despite being costly, these methods can provide accurate measurements, including from uranium and thorium decay chains and in cases where disequilibrium exists. Nonetheless, lengthy measurement times may be necessary [50].

Alternatively, errors from isotopic disequilibrium in the uranium and thorium decay chains can be reduced by measuring the contributions of uranium and thorium using thick source alpha counting (TSAC) after which potassium is measured using an analytical technique such as AAS, FPM, XRF or ICPMS. Yet another approach to determine the dose rate is to use TSAC to determine the alpha particle contribution after which a beta particle counter is employed to determine the beta contribution. The gamma dose rate is best determined in the field whenever possible. Highly sensitive portable gamma-ray spectrometers that make such onsite measurements possible are now available [44].

Measurement of the dose rate can also be conducted in the field using synthetic dosimeters such as α-Al_2O_3:C [44]. These are highly sensitive materials that are left in the field for a few weeks after which they are retrieved and analyzed.

The dose contribution from cosmic rays is usually minor. However, in settings where the radionuclide concentrations are low, the proportion from cosmic rays becomes significant. A methodology for calculating cosmic ray contribution to the luminescence dose rate was formulated by Prescott and Hutton [53].

Lastly, it is imperative to take the *in situ* moisture content of the material that is being dated into account when calculating the dose rate. This is because interstitial water absorbs part of the dose that should otherwise reach the dosimeter, with the attenuation of the dose rate intensifying as the moisture content increases.

http://dx.doi.org/10.5772/65119

5.3. Lower and upper age limits in luminescence dating

At present, luminescence dating methods can be used to date samples that are as young as a few decades [88]. Dating using the single grain approach can produce young ages that are relatively precise. When dating such young samples, it is desirable to use mineral grains characterized by a high luminescence sensitivity and for the grains to have been completely bleached prior to the burial [39]. Minimizing thermal charge transfer during measurement also improves the accuracy of the results.

Maximum ages that can be obtained using luminescence dating methods are ultimately controlled by the fact that the population of electron traps within any given dosimeter is fixed. As

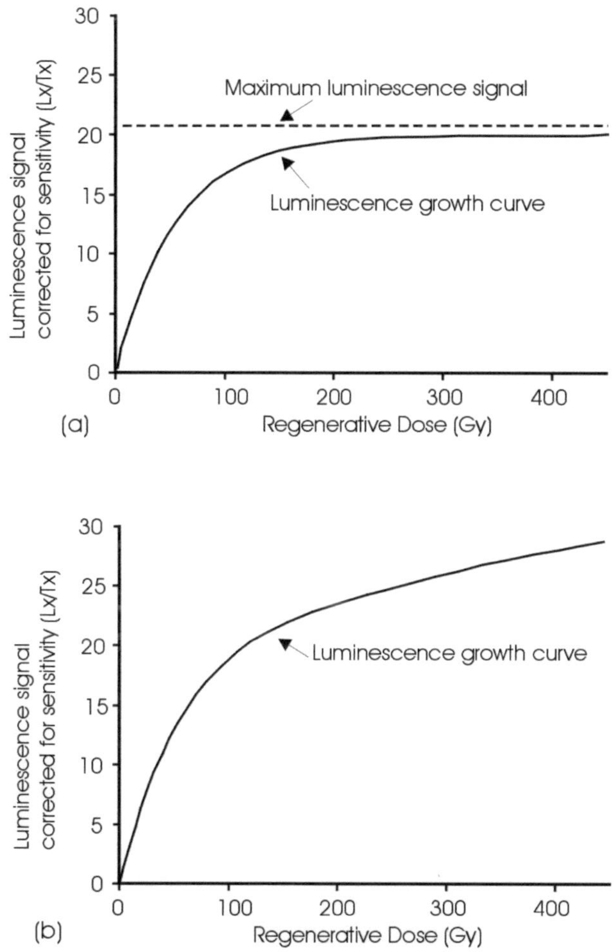

Figure 9. Examples of growth curves for quartz taken to saturation (redrawn after [3]). (a) Curve constructed using a saturating exponential function of the form $I(D) = I_0 (I - exp^{-D/D_0})$ where $I(D)$ is the luminescence signal produced by the dose D and I_0 is the maximum luminescence intensity produced. The parameter D_0 determines the shape of the curve. (b) Since curve (a) appeared to underestimate some older ages, the saturating function from (a) is combined with a linear function in (b) [3].

a result, the number electrons stored by trapping cannot increase indefinitely [50, 51, 56]. This is depicted in luminescence growth curves by a flattening of the signal obtained as the dose increases and is often expressed using a saturating exponential function. **Figure 9a** shows such a function expressing the fast component of a quartz signal. It shows that, once a certain dose is reached, the curve flattens. That dose is the upper limit above which the proportionality between the dose received and luminescence signal obtained breaks down. Quartz usually saturates with a dose of around 100–150 Gy. An approach used in some studies [4, 89] working with high doses is to model the growth curve by combining a linear function and a saturating function as shown in **Figure 9b** [3]. This approach has been used to report quartz ages in excess of 200 000–400 000 years. It should be noted that the limits of the ages that can be obtained are ultimately determined by the magnitude of the dose rate, with low dose rates giving higher age limits and vice versa. With feldspar, several studies have reported ages that exceed 100 000 years using IRSL [90]. Correction for anomalous fading of the feldspar at such high doses (>100–200 Gy) using standard methods [66], however, becomes problematic.

In essence, with both feldspar and quartz dating, there are maximum dose limits above which reliable ages cannot be produced as a result of electron trap exhaustion.

6. Sample collection and preparation for analysis

Preceding sections explored the basics of luminescence dating. A subject that now needs to be addressed is the nature of materials on which luminescence dating methods can be applied. Before that can be looked into, however, it is pertinent to examine the topic of sample grain size, since protocols employed in luminescence analysis are contingent upon the granulometry of the material being analyzed.

6.1. Sample grain size

There are generally two broad mineral grain size ranges that are employed in luminescence dating: coarse grains and fine grains. When using the coarse grain method, sand-sized particles in the diameter range 90–250 μm are extracted and analyzed. Grains in this size range normally receive ionizing radiation from alpha and beta particles as well as gamma and cosmic rays. Because of their size, alpha particles would only be able to penetrate the mineral grain's exterior 25 μm. When working with coarse grains, the outer rim that is affected by the alpha particles is removed by etching using hydrofluoric acid (HF). For this reason, coarse grain luminescence dating is sometimes described as inclusion dating [58, 91, 92]. When dating pottery using quartz inclusions, grains in the size range 90–125 μm are commonly selected. In sediment dating, the usual practice is to extract grain sizes that represent the modal size. Thus, for eolian dunes, grains in the size range 150–180 μm are often used [3]. Feldspar inclusion dating employs procedures similar to those used for quartz dating, including etching to remove the outer rim penetrated by alpha particles.

An important aspect of inclusion dating is that, since parts of the grain affected by alpha particles are removed prior to the performing luminescence measurements, only the contributions of beta, gamma and cosmic rays are taken into account when determining the dose rate. If Eq. (**1**) is modified to take those analytical considerations into account, the age equation becomes:

$$Luminescence\ Age = \frac{Paleodose}{0.90D_{\beta} + D_{\gamma} + D_{C}} \quad (6)$$

where D_{β}, D_{γ} and D_{c} are the doses from beta, gamma and cosmic ray contributions, respectively [50, 51]. If the dose rates are given as annual values, the equation yields the age in calendar years. The beta dose rate is factored by 0.9 as a correction for the grain size and the etching that disposes of the outer rim which also received a beta dose [50, 51].

Worth noting is that, on rare occasions, quartz may contain some uranium and thorium within its lattice. In such cases, Eq. (**6**) may need to be modified if coarse grains are used. Similarly, when dating feldspar using coarse grains, potassium-40 isotopes in K-feldspar constitute an additional source of beta particles that has to be factored into the dose rate calculation. As a result, feldspars are often separated into K-feldspar and low-potassium fractions that are then analyzed separately.

For fine grains, mineral grains in the size range 4–11 μm are usually extracted and mounted for analysis. When calculating the dose rate for such mineral grains, since their diameters are such that alpha particles can penetrate them entirely, contribution from alpha particles has to be included in the dose rate. Consequently, the age equation becomes:

$$Luminescence\ Age = \frac{Paleodose}{\mathrm{k}D_{\alpha} + D_{\beta} + D_{\gamma} + D_{C}} \quad (7)$$

where D_{α}, D_{β}, D_{γ} and D_{c} are the doses from alpha, beta, gamma and cosmic ray contributions, respectively.

Mineral grains in the diameter range 33–63 μm have been dated in other studies [93]. Because the size range is intermediate between coarse grain and fine grain categories, the analysis is sometimes referred to as middle-grain dating [50]. Etching with HF to remove the outer rim of the grains is not usually employed with middle-grain dating. Thus, the alpha particle contribution is also included when calculating the dose rate.

6.2. Sample collection and preparation

6.2.1. Sample collection

As outlined earlier, exposure of luminescence dating dosimeters to sunlight results in the zeroing of any previously acquired dose. Thus, a basic constraint when collecting samples is that they should be shielded from light until the time of analysis. Accordingly, when dealing with archaeological artifacts, sampling usually entails using a power drill in safe-light conditions to retrieve a representative sample from the interior of the artifact. When collecting sediments from their natural settings, on the other hand, sampling could be conducted at night [50]. However, working in the dark is inconvenient and chances of accidentally exposing the sample to light are also greater. Hence, night-time sampling is not the most popular approach. More often than not, sampling is conducted during the day. In cases where sediments have a firm consistency, it may be possible to cut out a block of sediment and transport it to the lab where the sample for dating is extracted from the block's interior. A more commonly used approach is to insert an opaque cylindrical pipe into a targeted geological unit. Both ends of the pipe are capped once it has been retrieved. At the laboratory a sample is analysis is removed from the center of the pipe. Samples for luminescence analysis can also be collected at depth by drilling vertically using sampling modules specially designed to avoid exposure of the sample to sunlight [94].

6.2.2. Sample preparation prior to the measurement

Collected samples have to be pretreated before luminescence analysis can be conducted. There are a number of standard procedures that have been developed for pretreating samples, depending on whether the grains targeted for analysis are fine or coarse. Analysis of coarse grains is usually performed on pure separates of either quartz or feldspar. Fine grains, however, are often analyzed as polymineralic fractions. The choice of what mineral or grain size to analyze will generally depend on the sample type. For instance, when dating loess, which is a silt-sized deposit, it is more appropriate to extract fine grains whereas using coarse grains would be more suitable for dating wind deposited dune sands. Where sediments have equal proportions of fine and coarse components, it is possible to validate results by dating both fractions.

In coarse grain dating, sample pretreatment entails first removing carbonates and organic material using dilute hydrochloric acid (HCl) and hydrogen peroxide (H_2O_2), respectively. Pure separates of quartz or feldspar are then extracted from the sample using a heavy liquid such as sodium polytungstate solution. Finally, the separates are mounted on discs (about 10 mm in diameter) before being analyzed. More details on sample extraction are given in [95].

When extracting fine grains for analysis, carbonates and organic materials are also first eliminated using dilute HCl and H_2O_2, respectively. To avoid flocculation of the grains, dilute sodium oxalate is added. After thorough mixing, a sedimentation column is then used to extract grains of the desired size range (usually 4–11 μm). If no further separation is required, the sample will be ready for mounting on discs. Occasionally, however, some studies do extract pure quartz from fine-grained samples. This is achieved by digesting feldspar from the polymineralic sample using fluorosilisic acid [96]. Additional details on separating and mounting fine grains can be found in [97].

7. On what materials can luminescence dating be applied?

Materials of geological or archaeological origin that are dated using luminescence methods have to satisfy a number of characteristics before they can be successfully analyzed. First, the material must be composed of a substance that behaves as a dosimeter. Second, the luminescence energy contained in the material must have been zeroed at some point that now serves as the starting point for counting time. Third, energy storage in the materials should not have reached saturation at the time of analysis, otherwise stored energy would cease to be

http://dx.doi.org/10.5772/65119

an accurate measure of time that elapsed. Within the context of these requirements, materials from which ages can be obtained using luminescence methods can be classified into two main categories: substances whose accrued energy has been reset by heating and sediment grains of sand and silt size whose previously acquired energy has been zeroed by exposure to sunlight while being transported.

7.1. Heated materials

The heating process employed in pottery production commonly fires earthenware to temperatures in excess of 800°C. Such elevated heat effectively evicts trapped electrons in constituent materials, providing a starting event for a luminescence age. As a result, archeological materials such as ancient pottery, bricks, terracotta figures and tiles are artifacts whose ages can be ascertained using luminescence methods, especially TL. This is because the artifacts commonly contain mineral grains such as quartz and feldspar. If the quartz or feldspar grains from the artifacts are extracted and dated using the inclusion dating approach (coarse grain technique), ages can be obtained. Alternatively, the fine-grained components can be extracted from the same materials and polymineralic fractions dated. Fine-grained quartz separates can also be isolated and dated similarly. It is thought that humans began making pottery about 30,000 years ago, during the Neolithic period [98]. By 10,000 years ago, the skill had spread globally [51]. Thus, most ancient pottery artifacts should be amenable to dating using luminescence methods.

In addition to pottery, other materials of archaeological context that could experience episodes of heating include flint and any other stones that may have been heated to high temperatures by fires associated with human settlements. These include residual rock chips from the production of tools by ancient humans that may have been heated either intentionally or otherwise. Dating any of these artifacts using luminescence techniques provides a timeline that approximates the age of the occupation of the archaeological site by humans [99].

Geological materials that have undergone heating to high temperatures at any point during the last about 100,000 years should also be datable using TL methods if they contain appropriate dosimeters. Materials that could fall into this category include sediments that border volcanic eruptions, as well as products of the volcanic eruptions themselves including lava and ash. In all instances, dating the materials using luminescence methods allows the age of the eruption to be approximated [100].

A final category of substances that can be dated using luminescence methods after being zeroed by elevated temperatures are materials associated with meteorite impacts. Such impacts create thermal shocks that erase any previously acquired dose in mineral grains contained in proximal geological structures, providing a convenient bleaching mechanism for dating the impact using luminescence methods [101].

7.2. Dating of sediments reset by sunlight

Granular sediments that contain dosimeters that have been bleached of all previously acquired energy by exposure to sunlight before being buried form an important class of materials that can be dated using luminescence methods. Stemming from studies [36] that showed that the exposure of some minerals to sunlight could evict electrons from traps in mineral lattices, luminescence methods have been successfully used to date sediment from a range of environments. While TL was used in the beginning, the subsequent development of OSL techniques was nurtured by the dating of this category of materials. The types of sediments that fall into this category include sediments transported by wind (eolian sediments), water-lain sediments, sediments associated with glaciers, as well as sediment related to earthquakes. The section below examines each of these sediment classes. In each case, luminescence ages denote time that has passed since the sediments were last zeroed by sunlight. Similarly, the ages correspond to time that has passed since the sediment was last shielded from sunlight (or buried) by overlying geological units.

7.2.1. Eolian deposits

Of all classes of geological materials that can be dated using luminescence methods, wind deposited (eolian) sediments are the most ideal. This is because the sediment grains are transported at the earth's surface or in mid-air, affording them adequate exposure to the sun to cause complete zeroing of any dose the sediment may have [50]. Consequently, the largest share of luminescence ages that have been published to date are from deposits of eolian origin. Eolian deposits generally fall into two main categories: sand-sized and silt-sized grains. Sands comprise sediment grains in the range 63–250 μm. The wind transports such particles via a process called saltation [102], which is a sequence of low altitude jumps, less than a foot from the surface of the depositional bed. Such grains usually form dunes. Silt-sized grains comprise particles in the size range 2–63 μm. These are transported in suspension and typically stay airborne for long periods of time [102]. In either case, eolian sands and silt grains are usually well-zeroed by the time they are buried by other sediments.

Eolian dune sediments have been dated in multiple studies [103] using quartz or feldspar extracts (mostly employing blue OSL and IRSL for stimulation, respectively) and produced ages that have been compared with radiocarbon ages from related sediments. In all cases, the results have validated luminescence dating as a competent dating method that provides reliable ages. Since eolian dune deposits are often treated as indicators of dry conditions, chronologies from eolian sands have been used to reconstruct environmental changes from the past [104–107]. Published ages span the last ca. 100,000 years. Hence, the method has emerged as an indispensable tool for the study of Late Pleistocene and Holocene paleoenvironments.

In addition to studying environmental change in inland deserts, luminescence chronologies from coarse-grained eolian sands have also been used to investigate depositional chronologies of coastal dune deposits [108]. Results from such studies have been used to develop temporal frameworks for processes in coastal environments, including sea level change.

Silt-sized eolian sediments can form extensive sequences of deposits referred to as loess [102]. Fine grain dating has been used to provide depositional chronologies of loess in numerous studies [107]. As in coarse grain dating, the chronologies have invariably been used to reconstruct paleoenvironments from the late Quaternary period. Luminescence analysis of fine grains can be conducted by targeting feldspar in polymineral fractions using IRSL. Alternatively, if quartz fines are extracted as outlined above, blue OSL is employed for stimulation [109].

7.2.2. *Water-lain deposits*

Coarse as well as fine-grained sediments that have been deposited by water have also been investigated using luminescence methods. However, studies have shown that sediments transported by rivers (fluvial deposits) are not always well-zeroed at deposition [46, 110]. As a result, single grain dating methods have been used to identify well-zeroed grains from such settings [46, 110]. Useful information on modern and ancient depositional rates of river systems have been acquired through luminescence studies of fluvial sediments [110]. Influences of tectonic and climatic effects on river systems have also been investigated using luminescence chronologies [110]. Other applications of luminescence ages from fluvial deposits include paleoseismic and archaeological studies. Additional information on dating fluvial deposits using luminescence methods can be found in [46, 110].

7.2.3. *Sediments of glacial origin*

As with water-lain deposits in fluvial settings, the greatest concern with deposits from glacial environments is the degree of bleaching that sediments experience prior to the burial. Sediments with the highest probability of being well-zeroed are deposits that meltwaters transport away from the glacier as outwash [44, 111]. Hence, chances for complete zeroing increases with distance from the glacier's front [111–113]. Nonetheless, even with outwash deposits, studies have shown that sediments are often partially zeroed [112]. As with fluvial deposits, single grain dating methods can be used to identify well-bleached grains in such instances. Some investigators have tried to identify signal components that bleach more rapidly for use when dating partially bleached glacial deposits [111]. More detailed discussions on luminescence dating of glaciofluvial deposits are found in [44, 111].

7.2.4. *Earthquake-related studies*

Tsunami episodes generated by earthquakes often displace sandy deposits from beach areas, depositing them further inland in tidal marshes, bogs and lakes that rim ocean margins. Afterwards, as part of the coastal sequence, other deposits that can include peat or mud may be emplaced atop the displaced sands, shielding them from sunlight. If the sands are dated using luminescence methods, the chronology they yield can be used to approximate recurrence rates of the tsunamis and, by extrapolation, of the earthquakes that cause the tsunamis. Such information would be critical when evaluating environmental hazards within the region [50]. Worth noting, however, is that tsunami events are rapid and often do not provide time for adequate bleaching of the sediments they mobilize. As a result, investigators target sediments that were exposed at the surface before the tsunami event and hence were already zeroed [50, 114]. Other deposits associated with earthquakes on which luminescence dating has been applied are sediments that occur on horizontal surfaces that have experienced differential vertical movement through earthquake-related faulting [49]. Sediments on the uplifted surfaces could be correlated by dating them using luminescence methods.

8. Current and future trends in luminescence dating

Luminescence dating has evolved into a full-fledged and robust chronometer over the last six decades. Concurrent with and driving this development has been the establishment of numerous luminescence dating laboratories throughout the world. To foster interaction between scholars, the Luminescence and Electron Spin Resonance Dating (LED) Conference is held triennially to examine a range of topics that encompass luminescence dating. In the years between the LED conferences, regional and national meetings are also held, including the UK Luminescence Dating Conference, the German Luminescence and ESR Conference or the North American New World Luminescence Dating Workshop. Topics explored at the meetings include fundamental research into luminescence dating as well as applications of luminescence chronologies in environmental, geological and archaeological sciences. Improvements in instrumentation for luminescence dating also constitute an important element at the meetings.

Current topics in fundamental research include investigations that aim to improve knowledge on luminescence characteristics of dosimeters such as feldspar and quartz. Ultimately, this should lead to the production of ages that are more accurate as well as the extension of maximum and minimum age limits that can be obtained using luminescence methods. Such efforts include studies that aim to understand the behavior of quartz at high doses [115, 116]. The information should lead to the formulation of protocols that extend the upper limit of ages obtainable using quartz [116].

Research on feldspar has included efforts to identify IRSL signals that are less inclined to experience anomalous fading [66]. Employing such signals would allow older ages to be determined using feldspar than are currently obtainable [117, 118]. Some studies have demonstrated that IRSL signals that are acquired by stimulating feldspar at a low temperature (for instance, 50°C) after which another measurement is immediately conducted at a higher temperature (for instance, 290°C), have a lower fading rate than those obtained using standard protocols [118]. This experimental procedure, termed post-IR IRSL [118] has yielded encouraging results that could potentially raise the upper limit of ages that can be obtained using current methods [118].

Improvements in instrumentation are also continuously being realized and advances include the development of luminescence readers equipped with broader spectral ranges for emission detection [54]. The recent development of portable OSL measurement systems that can be used to conduct rapid on-site measurements is also worth mentioning [119]. Compared to regular OSL readers, the portable systems have limited functions. However, if they were to be equipped with add-ons such as external X-ray sources for sample irradiation, practitioners would be faced with expanded options for signal collection.

Today, luminescence dating is a dynamic field of research that is set to continue witnessing ground-breaking developments in years to come. As both laboratory procedures and analytical equipment are improved, ages should be reported with higher precision and accuracy.

http://dx.doi.org/10.5772/65119

Author details

Ken Munyikwa

Address all correspondence to: kenm@athabascau.ca

Centre for Science, Athabasca University, Athabasca, Alberta, Canada

References

[1] Huntley DJ, Godfrey-Smith DI, Thewalt MLW. Optical dating of sediments. Nature. 1985;313:105–101.

[2] Aitken MJ. Luminescence dating: a guide for non-specialists. Archaeometry. 1989;31:147–159.

[3] Wintle AG. Luminescence dating: where it has been and where it is going. Boreas. 2008;37:471–482.

[4] Murray A, Buylaert JP, Henriksen M, Svendsen, JI, Mangerud, J. Testing the reliability of quartz OSL ages beyond the Eemian. Radiation Measurements. 2008;43:776–780.

[5] Yukihara EG, McKeever SWS. Optically Stimulated Luminescence: Fundamentals and Applications. Chichester: Wiley; 2011. 378 p.

[6] Boyle R Experiments and Considerations Upon Colors and Observations on a Diamond that Shines in the Dark. 1st ed. London: Henry Herringham; 1664.

[7] Hunter MCW. The Boyle Papers: Understanding the Manuscripts of Robert Boyle. 1st ed. Aldershot: Ashgate Publishing; 2007. 688 p.

[8] Daniels F, Boyd CA, Saunders DF. Thermoluminescence as a research tool. Abstracts of Papers presented at the 1949 Meeting. Science. 1949;109:440.

[9] Daniels F, Boyd CA, Saunders DF. Themoluminescence as a research tool. Science. 1953;117:343–349.

[10] Zeller EJ, Wray JL, Daniels F. Factors in age determination of carbonate sediments by thermoluminescence Bulletin of the American Association of Petroleum Geologists. 1957;41:121–129.

[11] McDougal DJ, editor. Themoluminescence of Geological Materials. New York: Academic Press; 1968.

[12] Aitken MJ, Tite MS, Reid J. Thermoluminescence dating of ancient ceramics. Nature. 1964;202:1032–1033.

[13] Kennedy GC, Knopff L. Dating by thermoluminescence. Archaeology. 1960;13:147–148.

[14] Grogler N, Houtermans FG, Stauffer H. About the dating of ceramics and bricks using thermoluminescence. Helvetica Physica Acta. 1960;(33):595–596.

[15] Bothner MH, Johnson NM. Natural thermoluminescent dosimetry of late Pleistocene perlagic sediments. Journal of Geophysical Research. 1969;74:5531–5338.

[16] Huntley DJ, Johnson HP. Thermoluminescence as a potential means for dating siliceous ocean sediments. Canadian Journal of Earth Sciences. 1976;13:593–596.

[17] Wintle AG, Huntley DJ. Thermoluminescence dating of a deep-sea sediment core. Nature. 1979;279:710–712.

[18] Wintle AG, Huntley DJ. Thermoluminescence dating of sediments. Quaternary Science Reviews. 1982;1:31–53.

[19] Shelkoplyas VN, Morozov GV. Some results of an investigation of Quaternary deposits by the thermoluminescence method. In: Materials on the Quaternary period of the Ukraine (for the VIIth International Quaternary Association Congress); 1965; Kiev: Nuakova Dumka; pp. 83–90.

[20] Morozov GV. The relative dating of Quaternary Ukranian sediments by the thermoluminescence method. In: Unpublished paper presented at the 8th International Quaternary Association Congress; Paris. Washington DC: US Geological Survey Library cat no. 208M8280.167; 1968.

[21] Shelkoplyas VN. Dating of the Quaternary deposits by means of thermoluminescence . In: Chronology of the Glacial Age. Leningrad: Geographic Society of the USSR, Pleistocene Commission; 1971. pp. 155–160.

[22] Shelkoplyas VN, Il'ichev VA, Svitoch AA. Thermoluminescence dating of recent sediments of the Ob Plateau and Gorray Altai. Doklady Academii Nauk SSSR. 1973;**212**:935–937.

[23] Dobredeyev OP, Il'ichev VA. Age of loess and fossil soils of the Russian Plain and their correlation with glacial events of the Pleistocene. Dolkady Akademii Nauk SSSR. 1974;**214**:401–403.

[24] Faustov SS, Il'ichev VA, Bol'shakov VA. Paleomagnetic and thermoluminescence analyses of the Likhvin section. Dolkady Akademii Nauk SSSR. 1974;**214**:1160–1162.

[25] Shelkoplyas VN. The age determination of loess deposits. In: Zubakov VA, editor. Geokronologiya SSSR 3. Leningrad;1974. p. 31–35.

[26] Hutt GI, Raukas AV. Perspectives for using the thermoluminescence method for determining the age of Quaternary deposits. Bulletin of the Commission for the Study of the Quaternary Period. 1977;**47**:77–83.

[27] Troitsky L, Punning JM, Hutt G, Rajamae R. Pleistocene glaciation chronology of Spitsbergen. Boreas. 1979;**8**:401–407.

[28] Borsy Z, Flerservalvi J, Szabo PP. Thermoluminescence dating of several layers of loess sequences at Paks Mende (Hungary). Acta Geologica of the Academy of Science, Hungary. 1979;**22**:451–459.

[29] Linder L, Proszyriski M. Geochronology of the Pleistocene deposits exposed at Washock, northern part of the Holy Cross Mtns. Acta Geologica Polonica. 1979;**29**:121–132.

[30] Li JL, Paie JX, Wang ZZ, Lu YC. A preliminary study of both the thermoluminescence of quartz powder in loess and the age determination of the loess layers. Kexue Tongbau. 1977;**22**:498–502.

[31] Hütt GI, Smirnov AV. Dosimetric properties of natural quartz and its possibilities in thermoluminescent dating of geological objects. Izvestiya Akademii Nauk SSSR, Seriya Fizichskaya. 1977;**41**:1367–1369.

[32] Hütt G, Vares K, Smirnov A. Thermoluminescent and dosimetric properties of quartz from Quaternary deposits. Academii Nauk ESSR, Khimiia-Geologia. 1977;**26**:275–283.

[33] Hütt G, Smirnov A, Tale I. On the application of thermoluminescence of natural quartz to the study of geochronology of sedimentary deposits. PACT. 1978;**3**:262–373.

[34] Wintle AG, Huntley DJ. Thermoluminescence dating of sediments. PACT. 1979;**3**:374–380.

[35] Wintle AG, Huntley DJ. Thermoluminescence dating of ocean sediments. Canadian Journal of Earth Science. 1980;**17**:348–690.

[36] Huntley DJ. On the zeroing of the thermoluminescence of sediments. Physics and Chemistry of Minerals. 1985;**12**:122–127.

[37] Murray AS, Wintle AG. Luminescence dating of quartz using an improved single-aliquot regenerative protocol. Radiation Measurements. 2000;**32**:57–73.

[38] Murray AS, Wintle AG. The single aliquot regenerative dose protocol: potential improvements in reliability. Radiation Measurements. 2003;**37**:377–381.

[39] Wintle AG. Fifty years of luminescence dating. Archaeometry. 2008;**50**:276–312.

[40] Murray AS, Olley JM. Precision and accuracy in the optically stimulated luminescence dating of sedimentary quartz: a status review. Geochronometria. 2002;**21**:1–16.

[41] Feathers JK. Use of luminescence dating in archaeology. Measurement Science and Technology. 2003;**14**:1493–1509.

[42] Lian OB. Optically-stimulated luminescence. In: Elias S, editor. Encyclopedia of Quaternary Science. 1st ed. Amsterdam: Elsevier Press; 2007. pp. 1480–1491.

[43] Duller GAT. Luminescence Dating: Guidelines on using Luminescence Dating in Archaeology. 1st ed. Swindon: English Heritage; 2008. 44 p.

[44] Fuchs M, Owen L. Luminescence dating of glacial and associated sediments: a review, recommendations and future directions. Boreas. 2008;**37**:636–659.

[45] Preusser F, Degering D, Fuchs M, Hilgers A, Kadereit A, Klasen N et al. Luminescence dating: basics, methods and applications. Eiszeitalter und Gegenwart Quaternary Science Journal. 2008;**57**(1–2):95–149.

[46] Rittenour TM. Luminescence dating of fluvial deposits: applications to geomorphic, palaeoseismic, and archaeological research. Boreas. 2008;**37**:613–635.

[47] Singhvi AK, Porat N. Impact of luminescence dating on geomorphological and palaeoclimate research in drylands. Boreas. 2008;**37**:536–558.

[48] Rhodes EJ. Luminescence dating of sediment over the past 200,000 years. Annual Reviews in Earth and Planetary Sciences. 2011;**39**:461–488.

[49] Liritzis I, Singhvi AK, Feathers JK, Wagner GA, Kadereit A, Zacharias N et al. Luminescence dating in archaeology, anthropology, and geoarchaeology. 1st ed. Berlin: Springer; 2013. 70 p.

[50] Aitken MJ. An Introduction to Optical Dating. 1st ed. Oxford: Oxford University Press; 1998. 267 p.

[51] Aitken MJ. Thermoluminescence Dating. 1st ed. London. Academic Press; 1985. 359 p.

[52] McKeever SWS, Moscovitch M, Townsend PD. Thermoluminescence Dosimetry: Properties and Uses. 1st ed. Ashford: Unclear Publishing Technology; 1995. 304 p.

[53] Prescott JR, Hutton JT. Cosmic ray contributions to dose rates for luminescence and ESR dating; large depths and long-term time variations. Radiation Measurements. 1994;**23**:497–500.

[54] Richter D, Richter A, Dornich K. Lexsyg–a new system for luminescence research. Geochronometria. 2011;**20**:220–228.

[55] Bøtter-Jensen L. Luminescence techniques: instrumentation and methods. Radiation Measurements. 1997;**17**:749–768.

[56] Stokes S. Luminescence dating applications in geomorphological research. Geomorphology. 1999;**59**:153–171.

[57] Godfrey-Smith DI, Huntley DJ, Chen WH. Optical dating studies of quartz and feldspar sediment extracts. Quaternary Science Reviews. 1988;**7**:373–380

[58] Fleming SJ. Thermoluminescence dating: refinement of the quartz inclusion method. Archaeometry. 1970;**12**:133–145.

[59] McKeever SWS. Thermoluminescence of Solids. 1st ed. Cambridge: Cambridge University Press; 1985. 390 p.

[60] Wintle AG. Luminescence dating: laboratory procedures and protocols. Radiation Measurements. 1997;**27**:769–817.

[61] Bailey RM, Smith BW, Rhodes EJ. Partial bleaching and the decay form characteristics of quartz OSL. Radiation Measurements. 1997;**27**:124–136

[62] Jain M, Murrary AS, Bøtter-Jensen L. Characterisation of blue-light stimulated luminescence components in different quartz samples: implications for dose measurement. Radiation Measurements. 2003;**37**:441–449.

[63] Bulur E. An alternative technique for optically stimulated luminescence (OSL) experiment. Radiation Measurements 1996; **26**: 701–709

[64] Wintle AG. Anomalous fading of thermoluminescence in mineral samples. Nature. 1973;**245**:143–144.

[65] Lamothe M, Auclair M. A solution to anomalous fading and age shortfalls in optical dating of feldspar minerals. Earth and Planetary Science Letters. 1999;**171**:319–323.

[66] Huntley DJ, Lamothe M. Ubiquity of anomalous fading in feldspars and the measurements and correction of it in optical dating. Canadian Journal of Earth Sciences. 2001;**38**:1093–1106.

[67] Huntley DJ, Godfrey-Smith DI, Thewalt MLW, Berger GW. Thermoluminescence spectra of some mineral samples relevant to thermoluminescence dating. Journal of Luminescence. 1988;**39**:123–136.

[68] Krbetschek MR, Gotze J, Dietrich A, Trautmann T. Spectral information from minerals. Radiation Measurements. 1997;**27**:495–748.

[69] Godfrey-Smith DI, Huntley DJ, Chen WH. Optical dating studies of quartz and feldspar sediment extracts. Quaternary Science Reviews. 1988;**7**:373–380.

[70] Li SH, Tso MYW. Lifetime determination of OSL signals from potassium feldspar. Radiation Measurements. 1997;**27**:119–121.

[71] Bøtter-Jensen L, Duller GAT, Poolton NRJ. Excitation and emission spectrometry of stimulated luminescence from quartz and feldspars. Radiation Measurement. 1994;**23**:613–616.

[72] Hutt G, Jack I, Tchonka J. Optical dating: K–feldspars optical response stimulation spectrum. Quaternary Science Reviews. 1988;**7**:381–386.

[73] Barnett SM, Baillif IK. Infrared stimulation spectra of sediment containing feldspar. Radiation Measurements. 1997;**27**:237–242.

[74] Huntley DJ, Godfrey-Smith DI, Haskell EH. Light-induced emission spectra from some quartz and feldspars. Nuclear Tracks and Radiation Measurements. 1991;**18**:127–131.

[75] Krbetscheck MR, Rieser U, Stloz W. Optical dating: some dosimetric properties of natural feldspars. Radiation Measurements. 1996;**27**:107–117.

[76] Ugumori T, Ikeya M. Luminescence of $CaCO_3$ under N_2 laser excitation and application to archaeological dating. Japanese Journal of Applied Physics. 1980;**19**:459–465.

[77] Templer RH, Smith BW. Auto-regenerative TL dating with zircon from fired materials. Nuclear Tracks and Radiation Measurements. 1988;**14**:329–332.

[78] Smith BW. Zircons from sediments: a combined OSL and TL auto-regenerative dating technique. Quaternary Science Reviews. Quaternary Science Reviews. 1988;**7**:401–406.

[79] Pietsch TJ, Olley JM, Nanson GC. Fluvial transport as a natural luminescence sensitizer of quartz. Quaternary Geochronology. 2008;**3**:365–376.

[80] Duller GAT. Equivalent dose determination using single aliquots. Nuclear Tracks and Radiation Measurements. 1991;**18**:371–378.

[81] Galloway RB. Equivalent dose determination using only one sample: alternative analysis of data obtained from infrared stimulated luminescence of feldspars. Radiation Measurements. 1996;**23**:103–106.

[82] Mejdhal V, Bøtter-Jensen L. Luminescence dating of archaeological materials using a new technique based on single aliquot measurements. Quaternary Science Reviews (Quaternary Geochronology). 1994;**13**:551–554.

[83] Murray AS, Roberts RG. Measurement of the equivalent dose in quartz using a regenerative-dose single aliquot protocol. Radiation Measurements. 1998;**29**:503–515.

[84] Wintle AG, Murray AS. A review of quartz optically stimulated luminescence characteristics and their relevance in single-aliquot regeneration dating protocols. Radiation Measurements. 2006;**41**:369–391.

[85] Duller GAT, Bøtter-Jensen L, Koshiek P, Murray AS. A high sensitivity optically stimulated luminescence scanning system for measurement of single sand-sized grains. Radiation Protection Dosimetry. 1999;84:325–330.

[86] Olley JM, Roberts RG, Yoshida, Bowler JM. Single-grain optical dating of grave infill associated with human burials at Lake Mungo, Australia. Quaternary Science Reviews. 2006;**25**:2469–2474.

[87] Olley JM, Catcheon GG, Roberts RG. The origin of dose distribution in fluvial sediments, and the prospect of dating single grains from fluvial deposits using optically stimulated luminescence. Radiation Measurements 1999; 30:207-217.

[88] Ballarini M, Wallinga J, Murray AS, van Heteren S, Oost AP, Bos AJJ et al Optical dating of young coastal dunes on a decadal time scale. Quaternary Science Reviews. 2003;**22**:1011–1017.

[89] Pawley SM, Bailey RM, Rose J, Moorlok BSP, Hamblin RJO, Booth SJ et al Age limits on Middle Pleistocene glacial sediments from OSL dating, north Norfolk, UK. Quaternary Science Reviews. 2008;**27**:1363–1377.

[90] Buylaert JP, Murray AS, Huot S. Optical dating of an Eemian site in northern Russia using K-feldspar. Radiation Measurements. 2008;**43**:505–512.

[91] Ichikawa Y. Dating of ancient ceramics by thermoluminescence. Bulletin of the Institute of Chemical Research, Kyoto University. 1965;**43**:1–6.

[92] Mejdahl V. TL dating based on feldspars. Nuclear Tracks and Radiation Measurements. 1985;10:133-136.

[93] Stevens T, Armitage SJ, Lu H, Thomas DSG. Examining the potential of high–resolution OSL dating of Chinese loess. Quaternary Geochronology. 2007;2:15-22.

[94] Munyikwa K, Telfer M, Baker I, Knight C. Core drilling of Quaternary sediments for luminescence dating using the Dormer DrillmiteTM. Ancient TL. 2011;29:15-22.

[95] Stokes S. Optical dating of young sediments using quartz. Quaternary Science Reviews (Quaternary Geochronology). 1992;11:153-159.

[96] Rees-Jones J. Dating young sediments using fine grained quartz. Ancient TL. 1995;13:9-14.

[97] Munyikwa K. Luminescence Chronology. In: Mörner NA, editor. Geochronology: Methods and Case Studies. 1st ed. Rijeka: Intech; 2014. pp. 31–71.

[98] Vandiver PB, Soffer O, Klima B, Svoboda J. The origins of ceramic technology at Dolni Vestonice, Czecholosvakia. Science. 1989;**246**:1002–1008.

[99] Mejdahl V. Feldspar inclusion dating of ceramics and burnt stones. PACT. 1983;**9**:351–364.

[100] Berger GW, Huntley DJ. Dating volcanic ash by thermoluminescence. PACT. 1983; **9**:581–592.

[101] Sutton SR. TL measurements on shock-metamorphosed sandstone and dolomite from Meteor Crater, Arizona: Pt. 1 shock dependence of TL properties. Journal of Geophysical Research. 1985;**9**:3683–3689.

[102] Pye K, Tsoar H. Aeolian Sand and Sand Dunes. 2nd ed. Heidelberg: Springer. 2009. 458 p.

[103] Stokes S, Gaylord DR. Optical dating of Holocene dune sands in the Ferris dune field, Wyoming. Quaternary Research. 1993;**39**:274–281.

[104] Munyikwa K. Synchrony of southern hemisphere late Pleistocene arid episodes: a review of luminescence chronologies from arid aeolian landscapes south of the Equator. Quaternary Science Reviews. 2005;**24**:2555–2583.

[105] Wolfe SA, Huntley DJ, Ollerhead J. Relict Late Wisconsinan dune fields of the Northern Great Plains. Canada. Geographie Physique et Quaternaire. 2004;**58**:323–336.

[106] Munyikwa K, Feathers J, Rittenour T, Shrimpton H. Constraining the Late Wisconsinan retreat of the Laurentide Ice Sheet from western Canada using luminescence ages from postglacial eolian dunes. Quaternary Geochronology. 2011;**6**:407–422.

[107] Frechen M. Luminescence dating of loessic sediments from the Loess Plateau, China. Geologische Rundschau. 1999;**87**:675–684.

[108] Munyikwa K, Choi JH, Choi KH, Byun JM, Kim JW, Park K. Coastal dune luminescence chronologies indicating a mid-Holocene highstand along the east coast of the Yellow Sea. Journal of Coastal Research. 2008;**24**:92–103.

[109] Lai Z, Fan A. Examining quartz OSL age underestimation for loess samples from Luochuan in the Chinese loess plateau. Geochronometria. 2013;**41**:57–64.

[110] Cunningham AC, Wallinga J. Realizing the potential of fluvial archives using robust OSL chronologies. Quaternary Geochronology. 2012;**12**:98–106.

[111] Thrasher IM, Mauz B, Chiverrell RC, Lang A. Luminescence dating of glaciofluvial deposits: a review. Earth Sciences Reviews. 2009;**97**:133–146.

[112] Rhodes EJ, Bailey RM. The effect of thermal transfer on the zeroing of the luminescence of quartz from recent glaciofluvial sediments. Quaternary Geochronology. 1997; **16**:291–298.

[113] Munyikwa K, Brown S, Kitabwala Z. Delineating stratigraphic breaks at the bases of postglacial dunes in central Alberta, Canada using a portable OSL reader. Earth Surface Processes and Landforms. 2012;**37**:1063–1614.

[114] Huntley DJ, Clague JJ. Optical dating of tsunami-laid sands. Quaternary Research. 1996;**46**:127–140.

[115] Lowick SE, Preusser F. Investigating age underestimation in the high dose region of optically stimulated luminescence using fine grain quartz. Quaternary Geochronology. 2011;**6**:33–41.

[116] Timar-Gabor A, Constanin D, Buylaert JP. Fundamental investigations of natural and laboratory generated SAR dose response curves for quartz OSL in the high dose range. Radiation Measurements. 2015;**81**:150–156.

[117] Thompsen KJ, Murray AS, Jain M, Bøtter-Jensen L. Laboratory fading rates of various luminescence signals from feldspar-rich sediment extracts. Radiation Measurements. 2008;**49**:1474–1486.

[118] Buylaert JP, Yeo EY, Thiel C, Yi S, Stevens T, Thompson W, Frechen M, et al A detailed post-IR IRSL chronology for the last interglacial soil at the Jingbian loess site (northern China) Quaternary Geochronology. 2015;**30**:194–199.

[119] Sanderson DCW, Murphy S. Using simple portable measurements and laboratory characterisation to help understand complex and heterogeneous sediment sequences for luminescence dating. Quaternary Geochronology. 2010;**5**:299–305.

Section 4

Bioluminescence Perspectives and Prospects

Chapter 12

Bioluminescent Fishes and their Eyes

José Paitio, Yuichi Oba and
Victor Benno Meyer-Rochow

Additional information is available at the end of the chapter

http://dx.doi.org/10.5772/65385

Abstract

What shaped the evolution of vision in fish more than anything else is the need to see, be it to avoid obstacles or find shelters, and recognize conspecifics, predators and prey. However, for vision to be effective, sufficient light has to be available. While there is no shortage of light in shallow water depths, at least during the day, the situation for species occurring at greater depths is a different one: they live in an environment where sunlight does not reach, but which nevertheless, is not totally devoid of light. Numerous marine organisms, including fishes of at least 46 families, possess the ability to 'bioluminesce', i.e. they can produce biological light. This chapter focuses on the interaction between bioluminescence and specific photoreceptor adaptations in fishes to detect the biological light.

Keywords: bioluminescence, light, sea, eye, vision, fish

1. Introduction

Bioluminescence is rare, but widespread. This seemingly contradictory statement refers to the fact that there is almost no animal phylum that does not have at least a few bioluminescent species, even if the vast majority are non-luminescent. Freshwater contains the smallest number of bioluminescent species; terrestrial environments possess a slightly greater number, but the oceans are inhabited by a wide variety of bioluminescent creatures and the vertebrate class of Pisces represents one of the best examples of this [1]. When one thinks of the evolutionary plasticity of fishes and the fact that most of the ocean is dark, it is not surprising that these animals show a high degree of remarkable luminous features. In fact, fishes present the most diverse and complex examples of bioluminescent adaptations in the world [2].

In an environment where organic light is used for different biological purposes [1, 3], light perception may be a determinant factor for the survival and success of the species involved [4]. The eyes of all animals on the planet are adapted to the light regimes of the habitat they are meant to function in [5] and bioluminescent fishes are no exception. The visual capacities of oceanic fishes are adapted not only to the spectral properties and intensity of downwelling sunlight at depths the fish inhabit, but also to the bioluminescence in their environment [6]. Vision and bioluminescence (at least in fishes) are likely to share a common evolutionary history, which is reflected in the bioluminescent fishes' ontogeny, behaviour and ecology.

2. Taxonomy of luminous fishes

While preparing this chapter, the authors found records of bioluminescence for around 1500 species of fish, but most likely this number is too low as some records of luminescent species may still be missing. The tally of luminescent fishes represents about 8% of the approximately 20,000 known species [7]. At least 43 families of 11 orders of bony fishes and 3 families of 1 order of sharks are luminous (**Table 1**). There are families with no luminescent representatives and there are families in which the luminescent species are more abundant than in some other families, which is often a reflection of the habitat that these species inhabit (see next section of this chapter).

Order (total number of families)	Family	Biolumine scent genus		Functions[a]
		Ratio[b]	Genus	
Cartilaginous fishes				
Squaliformes (7)	Dalatiidae	7/7	All [8]	L [8], CI [8]
	Etmopteridae	5/5	All [8]	M, CI, S[c] [92], A [67]
	Somniosidae	1/5	*Zameus squamulosus* [9]	?
Bony fishes				
Anguilliformes (16)	Congridae	1/30	*Lumiconger arafura* [16]	?
Aulopiformes (16)	Chlorophthalmidae	1/2	*Chlorophthalmus* [2]	S [60]
	Evermannellidae	2/3	*Cocorella atrata* [2], *Odontostomops normalops* [11]	?
	Paralepididae	2/12	*Lestidium, Lestroplepis* [23]	CI [4]
	Scopelarchidae	2/3	*Benthalbella, Scopelarchoides* [95]	R [95], CI [4]
Batrachoidiformes (1)	Batrachoididae	1/22	*Porichthys* [21]	M [4], L [21], CI, I [33], A [2]

Order (total number of families)	Family	Biolumine scent genus		Functions[a]
		Ratio[b]	Genus	
Beryciformes (7)	Anomalopidae	6/6	All [12]	M, L [33], S, I, Br [62]
	Monocentridae	2/2	All [26]	L, I [53]
	Trachichthyidae	1/18	*Aulotrachiththys* [2]	CI [21]
Clupeiformes (7)	Engraulidae	1/17	*Coilia dussumieri* [15]	CI [33]
Gadiformes (10)	Macrouridae	14/23	*Cetonurus, Coelorinchus, Haplomacrurus*[d], *Hymenocephalus* [23], *Kumba* [24], *Lepidorynchus* [23], *Lucigadus* [24], *Malacocephalus, Nezumia, Odontomacrurus, Sphagemacrurus* [23], *Spicomacrurus* [24], *Trachonurus, Ventrifossa* [23]	S [60], I [30]
	Merlucciidae		*Steindachneria argentea* [24]	?
	Moridae	5/11	*Gadella* [23], *Physiculus* [23], *Salilota* [24], *Tripterophycis* [23]	?
Lophiiformes (18) (Suborder Ceratioidei)	Centrophrynidae	1/1	All [14]	M, L, I [14], Sa, Ss [41]
	Ceratiidae	2/2	All[14]	
	Diceratiidae	2/2	All [14]	
	Gigantactinidae	1/2	*Gigantactis* [14]	
Lophiiformes (18) (Suborder Ceratioidei)	Himantolophidae	1/1	All [14]	
	Linophrynidae	5/5	All [14]	
	Melanocetidae	1/1	All [14]	M, L, I [14], Sa, Ss [41]
	Oneirodidae	16/16	All [14]	
	Thaumatichthyidae	2/2	All [14]	
(non-Ceratioidei)	Ogcocephalidae	1/10	*Dibranchus atlanticus* [17]	?
Myctophiformes (2)	Myctophidae	34/34	All [11]	R [32], M [4], S [59], CI [2], I [1], Sa [2]
	Neoscopelidae	1/1	*Neoscopelus* [15]	CI [11], L [20]
Osmeriformes (14)	Alepocephalidae	4/19	*Microphotolepis, Photostylus, Rouleina, Xenodermichthys* [23]	CI [57], I [30]

Order (total number of families)	Family	Biolumine scent genus		Functions[a]
		Ratio[b]	Genus	
	Microstomidae	1/3	*Nansenia* [100]	?
	Opisthoproctidae	3/8	*Opisthoproctus, Rhynchohyalus, Winteria* [23]	CI [21]
	Platytroctidae	13/13	All [13]	R, CI, Ss [13]
Perciformes (164)	Acropomatidae	2/8	*Acropoma* [23], *Synagrops* [17]	CI [21]
	Apogonidae	5/32	*Apogon, Archamia* [23], *Jaydia* [18], *Rhabdamia, Siphamia* [23]	M, L, CI [51], I [33]
	Chiasmodontidae	1/4	*Pseudoscopelus* [23]	CI [21]
	Epigonidae	3/7	*Epigonus, Florenciella, Rosenblattia* [23]	?, Ss [2]
	Howellidae	1/3	*Howella* [23]	?
	Leiognathidae	7/7	All [18]	R, M, L, S, CI, I, Sa [53]
	Pempheridae	2/2	*Pempheris, Parapriacanthus* [23]	CI [33]
	Scianidae	3/65	*Collichthys* [23], *Larimichthys, Sonorolux* [106]	CI [33]
Saccopharyngiformes (4)	Eurypharyngidae	1/1	All [2]	?
	Saccopharyngidae	1/1	All [22]	L [15]
Stomiiformes (4)	Gonostomatidae	8/8	All [22]	CI [11]
	Phosichthyidae	7/7	All [22]	?
	Sternoptychidae	10/10	All [22]	CI [2]
	Stomiidae	27/27	All [22]	R [2], M [46], L [4], S [46], CI [2], I [48], A [46], Sa, Br [4]

[a]R—recognition; M—mate; L—lure; S—school; CI—counterillumination; I—illumination; A—aposematism; Br—blink and run; Sa—startle predators; Ss—smokescreen; ?—unreported functions in the references.
[b]Ratio, the number of the genera containing luminous species/the total number of the genera in the family; All, all the species are luminous.
[c]Etmopterids also seem to use its photophore patterns to coordinate hunting in schools [8].
[d]Bioluminescence in *Haplomacrurus* is only stated by Herring [23] and no other references found to contradict or agree with this information. The taxonomy used in this table is in accordance with the WoRMS Editorial Board, 2016 [7]. The taxonomic data used in this table are included and available online in the website of the Living Light List Project (http://www3.chubu.ac.jp/faculty/oba_yuichi/Living%20Light%20List/).

Table 1. Bioluminescent families and genera of fishes and respective suggested ecological functions.

In cartilaginous fishes, bioluminescence is not as common as in bony fishes. Only three families of luminous sharks are known, Dalatiidae, Etmopteridae and Somniosidae [8, 9]. Luminescence in rays and skates as well as in Holocephali seems non-existent or has not been confirmed. In the luminescent sharks, numerous small light organs are predominantly present in the ventral region of the body, but lower densities can even reach the dorsal areas in dalatiid species [10]. Luminous sharks, like the majority of fish species, possess external light organs with intrinsic light production, which are denominated as photophores [8, 11].

In bony fishes, bioluminescence reaches its zenith in terms of complexity and diversity of light organs and their corresponding biological and ecological functions. It is common to find whole families of bioluminescent species, such as the Myctophidae [11], Leiognathidae, Anomalopidae [12], Platytroctidae [13] and several families of anglerfishes [14]. In such cases, the patterns or shapes of the light organs are generally species-specific and can even be used for taxonomic purposes. Nevertheless, the most common scenario is that only some genera of fishes are bioluminescent as, for example, in families of the order Perciformes with very few luminescent species or in Macrouridae with considerably more luminescent species [2]. An uncommon scenario is when there is only a single luminous species known in an entire order as is the case with *Coilia dussumieri* in the Clupeiformes [15] and *Lumiconger arafura* in the order Anguilliformes [16, 17]. Equally uncommon is it to find only one luminescent species within a family as for the merluciid *Steindachneria argentea* [2] and the ogcocephalid *Dibranchus atlanticus* [17]. On the other hand, we have the order Stomiiformes, in which all of its species produce their own light [1].

Generally speaking, the origin of the light emissions do not vary much within families, being either bacterial or intrinsic. Exceptions exist in the Apogonidae where only *Siphamia* species are known to have bacterial symbionts whereas the remainder of the luminescent species employ intrinsic bioluminescence with *Cypridina* luciferin [18]. A rare case is observed in anglerfishes of the families Lynophrinidae and Centrophrynidae which have an intrinsic bioluminescent barbel in addition to the symbiotic luminous lure of the first ray of the dorsal fin that projects from the dorsum of the anglerfishes and is known as the esca [14]. The location of the light organs is also generally similar within a family, but exceptions exist, namely in families where all the species are luminous. In Myctophidae, for example, only members of the genus *Diaphus* have head photophores and unlike fishes of some other genera do not possess caudal glands. Another example is that of luminous barbels, present in some species of Stomiidae [15]. Indirect bioluminescence is produced by internal luminous organs that depend on transparent tissues underneath to emit diffuse light from the ventral part of the body. Although this method is more common in the shallow-water species like those of the families Leiognathidae and Apogonidae [19], it can also be found in some fishes of greater depths as from the families of Opisthoproctidae [12] and Evermannellidae [2].

2.1. Dubious species

Misidentifications, confusion of non-luminous tissues and luminous bacteria on damaged specimens [20] are common sources of errors that led to reports of the luminous species, which do not exist or are not luminescent at all. Furthermore, unusual phylogenetic and habitat placements in records for certain species claimed to be luminous, apparently backed up with dubious references, are suspicious vis-à-vis the phylogeny and habitat preferences of the accepted luminous species and therefore cast doubt on the validity of some records. Another problematic issue are contradictory statements, when some authors report a species to be luminescent and others contradict such statements. As the authors of this chapter did not have the chance to observe and confirm the status of luminosity of all the species mentioned, they decided to be neutral and include in the chapter cases that could be dubious.

The batoid *Benthobatis moresbyi* [20] is one of the most doubtful cases of bioluminescence reported in fishes. Sharks of the genus *Somniosus* were erroneously categorized as bioluminescent [2]. The megamouth shark possesses reflective tissue along its upper jaw, which can be used to attract its prey through reflecting the bioluminescence of other animals [1].

In bony fishes, reports of luminescent members of opisthoproctids, *Dolichopteryx, Bathylychnops* and *Bathylagus* have been shown to be erroneous [2]. Some trachichthyid genera like *Hoplostethus* [21] and *Sorosichthys* [22] may have luminous members, but there is no convincing proof of their bioluminescence. In macrourids, the genera *Coryphaenoides, Mesobius, Pseudocetonurus, Macrourus* and *Idiolophorhynchus andriashevi* are stated as luminous by some, but not all investigators [21, 23, 24]. *Antimora* is another uncertain case among otherwise luminescent morids [2, 23].

Apart from the well-known anglerfishes, two unrelated species of shallower water lophiiformes, *Antennarius hispidus* and *Chaunax* [20, 25] are described as luminous based on a single reference. There are also some species of ogcocephalids, which have dubiously been reported as luminous [21, 23]. Bioluminescence in cetomimids has been stated by reference [21], but other authors disagree [26]. Reference [21] consider luminescence in the species of the families Halosauridae and Brotulidae and the gempylidae genus *Ruvettus* as dubious. The uncertainty of bioluminescence in *Kasidoron* and *Bathylagus* has been discussed [2] and possible luminescence in *Derichthys serpentines* has been muted [27] but definitive proof is lacking. The silaginid *Sillago bassensis* was found with bioluminescent bacteria [28] probably attached to its skin; this might also have been the explanation for the dorsal luminescence of a flying fish reported in Japan [29].

3. Habitat dispersal of luminous fishes

Light penetrates the sea and is then subjected to scattering by air bubbles and water molecules [3], phytoplankton and zooplankton as well as suspended particles [30]. This essentially imposes limits on the use of vision in the aquatic environment [31]. Major light attenuation occurs in the epipelagic zone where the planktonic biomass, dissolved nutrients and floating debris 'filter' most of the light, reducing its intensity and altering its spectral composition. However, the angle of this light is almost completely vertical apart from very near the surface [32]. Not many bioluminescent fishes inhabit the epipelagic zone (**Figure 1**) and in those that are present, photoreceptive structures and visually induced behaviours seem less complex than that encountered in species inhabiting deeper habitats [33]. This seems likely to be related

http://dx.doi.org/10.5772/65385

to two main factors. Primarily, at these shallow depths light intensities, especially during the day, are too high for luminescence [4, 31] to be useful for ecological purposes without investing a considerable energy [32]. Secondly, if a very strong light were to be produced, it would readily be spotted by keen eyes of the abundant photopic predators around [32, 33]. The bioluminescent epipelagic species are mainly active at night [34] and use internal counterillumination to deter and avoid dusk-active piscivorous predators [19, 33]. Some exceptions are the headlights of anomalopid and monocentrid species [12, 35] and fishes with photophores, like the batrachoidid *Porichthys*.

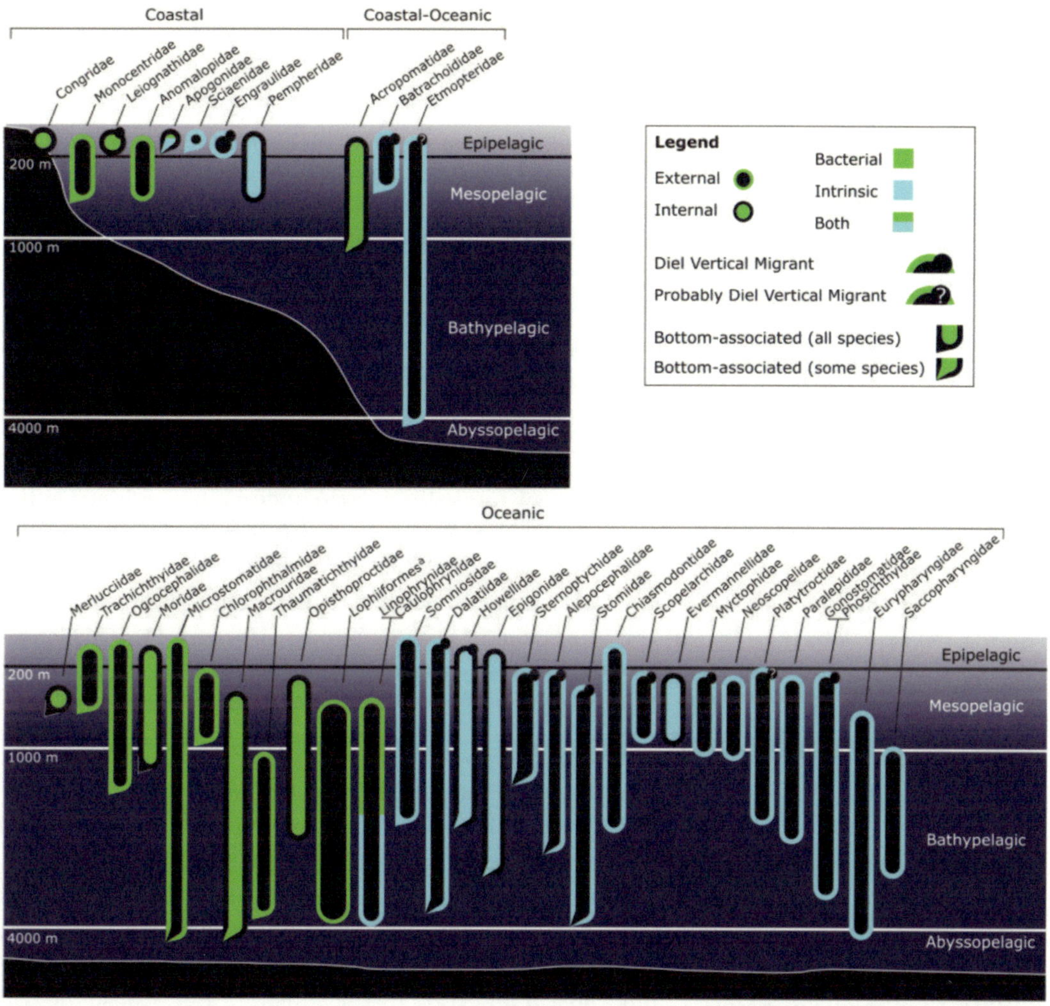

Figure 1. Schematic representation of the spatial distribution of bioluminescent fish families and some ecological details.[a] Lophiiformes families other than Thaumatichthyidae, Linophrynidae, Caulophrynidae and Ogcocephalidae. The data presented only refer to species that are known as bioluminescent. The order of disposition within the coastal, coastal-oceanic and oceanic categories is ecological and not spatial. References used for spatial distribution and diel vertical migrations [10, 13, 15, 26, 40, 65, 79, 95–99] and for bioluminescence [2, 9, 16, 21, 24, 47, 48, 50, 58, 60, 67, 100].

Only dim light from the short wavelength spectra of the blue–green component of sunlight (but no moonlight) reaches the mesopelagic zone [30]. In contrast to the epipelagic zone, the mesopelagic zone is characterized by a uniformity of light in all lateral directions and the highest intensity from above [4, 30]. This is the oceanic 'bioluminescent hotspot' where more than 80% of the species present are bioluminescent [32, 36] and biological light reaches its greatest splendour in diversity and complexity of structures and purposes [37]. In terms of fish biomass, it is the mesopelagic zone that stands out [38]. Myctophids are one of the families of teleosts with a very high number of species, representing the second most abundant taxon of mesopelagic fishes. In terms of sheer numbers, the gonostomatid genera *Vinciguerria* and *Cyclothone* represent the most abundant fauna on the planet [11]. The mesopelagic zone is also the transference zone from an extended origin of light to a point source. The downwelling sunlight is the primary source of light in the upper mesopelagic zone (200–610 m depth) while bioluminescence takes over as the main source of light in the lower mesopelagic zone (610–1000 m depth). Under the essentially monochromatic scotopic light, the use of bioluminescence is far more advantageous for camouflage and vision, be it in the case of predators or prey, than at depths more brightly illuminated by a fuller spectrum of light. In the depths where silhouette distinction is of visual importance and it comes as no surprise that most of the bioluminescent fish species inhabit this realm. The colorations of the fish at greater depths are a reflection of the different light environment. In the upper mesopelagic silver bodies acting as mirrors and assisting in camouflaging are common [30]. The animals inhabiting the lower mesopelagic and bathypelagic tend to have darker bodies, preventing reflections of bioluminescence in these deeper realms [4, 21].

At 1000 m depth, sunlight is no longer strong enough to allow fishes to see in and that depth by definition corresponds to the beginning of the bathypelagic zone [30]. The latter is the aphotic zone where no sunlight is available [30] and complete homogeneity in terms of the light environment occurs in all directions [4, 31]. Here, light from above does not possess much relevance for the impoverished faunal assemblages present and the only source of photic stimulation, in contrast to that of the mesopelagic zone, is bioluminescence [3, 30]. Consequently, there is a drop in the number of bioluminescent species and individuals from that in the mesopelagic zone [2]. Even deeper living abyssal fishes, phylogenetically related to bathypelagic or mesopelagic luminous species, are not bioluminescent (e.g. macrourids) [4, 12].

It is not surprising that the majority of all bioluminescent fishes are oceanic and pelagic [33], as there are no structures blocking the path of the light in this environment, it seems to be optimal for the realization of many of the roles bioluminescence can ultimately be involved in [30, 37]. On the other hand, some benthopelagic species that are not known as vertical migrators possess ventral luminescence (e.g. morid species). These are puzzling cases of bioluminescence and its possible role [2], because emitting ventral light close to the bottom does not fulfil the purpose of camouflage. In fact, it may cause a reflection by the substrate and expose the fish [21]; however, there may be advantages too, which have not been looked into like, for instance, diverting attention and misguiding potential predators to the reflection rather than its originator. Coastal and benthic fishes more often employ bacterial luminescence while the pelagic species possess mostly intrinsic bioluminescence [2, 33, 39].

http://dx.doi.org/10.5772/65385

Diel vertical migrations are common among mesopelagic fishes [40] but are also known from some species of epipelagic [4, 41] and bathypelagic fish [4, 21]. These migrations do not always involve all taxa present [15] or genders in the same population and may vary geographically and seasonally. Generally speaking, the diel vertical migrant approaches surface waters at night to feed on zooplankton or other migrants when diurnally hunting visual predators are less active. The vertical movements of these individuals provide an important source of carbon (and other nutrients) to the deeper layers of the ocean [4].

4. Bioluminescence purposes in fishes

4.1. Light organs and control

Despite the extensive diversity of luminous organs in fishes, the basic structure is rather uniform [37] irrespective of whether we deal with photocytes or bacterial symbionts (**Figure 2**). The photogenic mass is situated at the centre and is associated with an external lens. An internal reflector is enclosed by a pigmented layer. Innervation of blood capillaries is also a common feature [2, 37]. Bacterial light organs usually possess an opening to exterior allowing the release of excess bacterial cells to the environment [42]. Photophores, with the exception of those in neoscopelids, are closed [2, 4].

The majority of the bioluminescent fishes have luminous tissues capable of generating light by themselves, but others maintain a culture of luminous symbiotic bacteria in their light organs [4, 39, 43]. The light organs of the species with symbiotic bacteria possess ducts in which they culture extracellular luminous bacteria [39]. In most cases, these bacteria are species like *Photobacterium*, but *Aliivibrio fischeri* and possibly *Vibrio harveyi* can also be found as symbionts in light organs of some luminous fishes [42]. Intrinsically luminescent fish possess photocytes with *Cypridina* luciferin and coelenterazine. *Cypridina* luciferin is only used by coastal fishes of the families Batrachoididae, Pempheridae and Apogonidae while coelenterazine has been suggested to occur in Stomiiformes, Myctophiformes and Platytroctidae [44]. The remainder of intrinsically bioluminescent fishes may belong to coelenterazine or unknown luciferin types.

A pigmented layer of cells surrounds the light organs internally to absorb stray light [2]. Below the light organ, there usually is a layer of cells with a reflecting material that redirects the light towards the lens [45]. Some species lack reflectors [20, 21], and in others, the reflectors may not be present in all types of luminous tissues, as stomiids [46]. In fishes with indirect bioluminescence, the reflector is located above the ventral diffusive tissues (see **Figure 3**) instead of being inside the light organ [47]. Reflectors affect the emission spectra in some fishes, as in alepocephalids, and myctophids that lack pigmented lenses in front of the light organ [48, 49]. When the light produced by the photogenic chamber is emitted to the outside of the photophores, it passes through the lens [45], whose pigments filter the emitted light and adjust its spectrum to match that of the downwelling light [3, 48]. Some fishes like the monocentrids lack lenses. In the case of *Cleidopus gloriamaris*, a red-orange skin tissue external to the light organ alters the spectrum of the light emission [12].

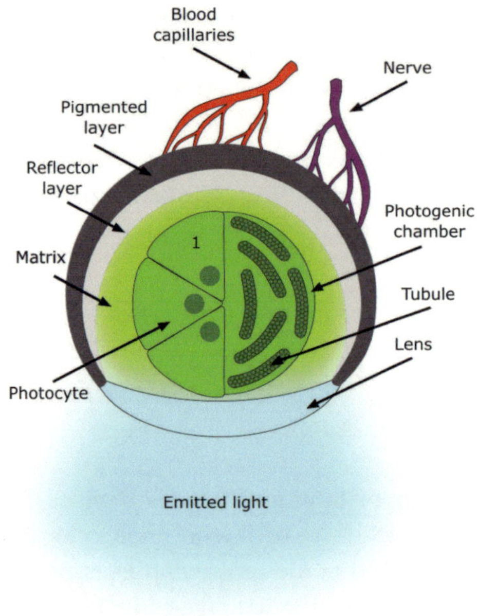

Figure 2. Schematic representation of the general structure of light organs in fishes. External side with light emission and internal side with blood capillaries and nerve supply. Matrix represents the space between the photogenic chamber and other structures. The right half of the photogenic chamber represents an intrinsic light organ with photocytes. The left half of the photogenic chamber represents a symbiotic light organ with transverse section of tubules filled with luminous bacteria.

Fishes with luminous symbiotic bacteria emit light continuously, but may have some control over the light by chromatophores [34] and nutrient supply to the bacteria [34, 43, 45]. Shutters are tissue patches with chromatophores that can partially or totally cover the light organs and are under muscular control [41, 50]. These shutters are different across genera of the same family and may occlude internal organs as in leiognathids and *Siphamia* or may be restricted to external organs as in anomalopids [41, 50, 51]. It is suggested that anglerfishes possess a similar mechanism to control the light emission from the escae [41]. Skin chromophores can be used to control light emissions in bioluminescent fishes with internal bacterial symbiotic organs (leiognathids [50], pempherids [47], acropomatids [19], evermannelids [2]) or externally (Gadiformes [52], Trachichtidae [12]), and bacterial symbionts in the lower jaw organs as in monocentrids [34]. There are some exceptions like *Opisthoproctus*, which may regulate light with ventral scale movements [2] and thaumatichthyids [41] and *Cleidopus gloriamaris* [12] that have light organs inside the mouth closing it to block the light emission. Leiognathids use branchiostegals to control their light emissions [53], and according to the observations of one of the authors (Yuichi Oba), the pigmented pelvic fins of *Chlorophthalmus* may serve the same purpose. The regulation of bacterial luciferase by oxygen and ion supply from the blood has been stated as a possibility, but there are no clear conclusions on this [39]. Observations by Meyer-Rochow [43] on *Anomalops katoptron* have shown that their bacterial light organs become successively dimmer when the fish starves.

http://dx.doi.org/10.5772/65385

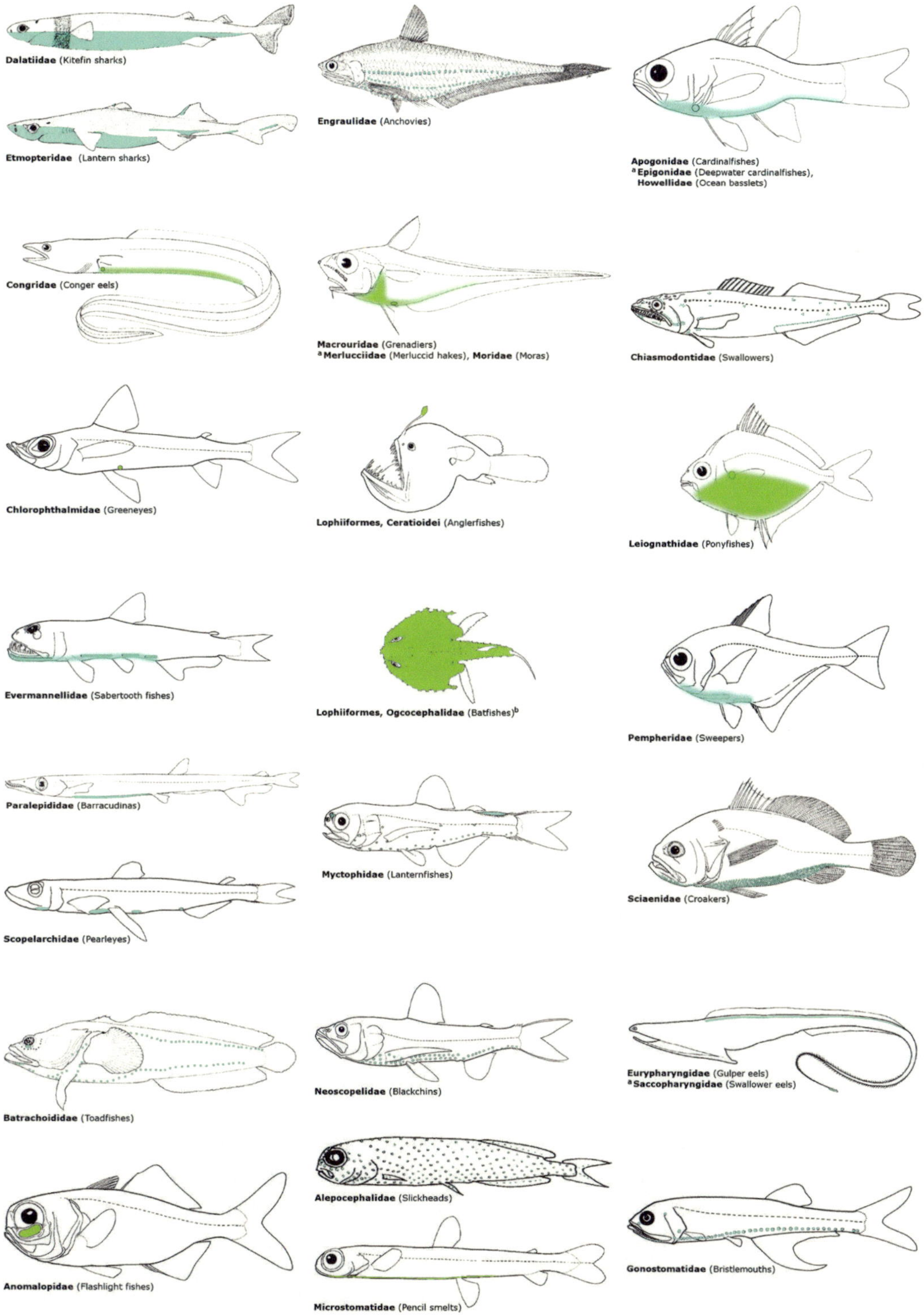

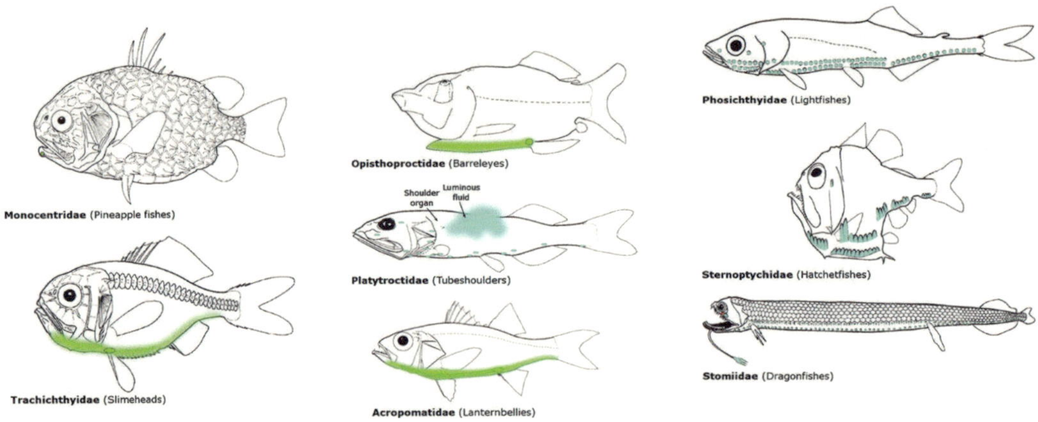

Figure 3. Light organs in fishes per family. Intrinsic bioluminescence in blue and symbiotic in green. The light organs are represented in coloured bands or circles delimited by a black outline. Blur coloured forms represent internal bioluminescence through diffusive tissues. [a]Used in cases that bioluminescent emissions and organs are very similar between families of the same order.[b] Crane [101] also found luminescence in skin samples of the ventral surface of the fish. Images are representative for the families and are not on scale. Same references used for bioluminescence in **Figure 1** plus [10, 13, 15, 95]. Source of drawings: Food and Agriculture Organization of the United Nations [15, 26, 96, 102–105]. Reproduced with permission.

Neural light emission control of intrinsic bioluminescence occurs in sharks [54], Stomiiformes [46, 55], myctophids, *Porichthys* [55] as well as alepocephalids [49] and proceeds through neurons in scopelarchids [56] and other coastal fishes besides *Porichthys* [33]. Neurons may be able to individually regulate light outputs of single photophores as in myctophids [55]. However, blood supply has also been suggested to be involved, at least in the control of the linophrynid's intrinsically luminous barbels [41].

Exceptions are platytroctids, which excrete luminous liquids; a process that invokes muscular control [2, 41]. In some anglerfishes, the liquid may not actually be secreted by the Escae, but from small symbiotic luminous bulbs (the so-called caruncles) that in the family Ceratiidae are dorso-posteriorly positioned from the escae [14].

Luminous organs are predominantly ventral [21, 37] (**Figure 3**) but some light tissues can be dorsal, as in the escae of anglerfishes and photophores in dalatiid sharks. Some fishes have buccal light organs like in the apogonids *Siphamia* spp. [18], the myctophid genus *Neoscopelus* [20] and thaumatichthyids and *Cleidopus*. Light organs are present in or on the heads of some fish especially deep-sea species like alepocephalids [57], platytroctids [13], myctophids, stomiids, and the chiasmodontid *Pseudoscopelus* [58], but also the shallow-water anomalopids and monocentrids [12]. Apart from these organs, there are other types of luminous tissues in fishes. Besides the 'normal' photophores, i.e. primary photophores, some species also possess secondary photophores that are smaller and of less complex structure [57]. These small photophores are then, rarely with some degree of preference, distributed all over the body of the fish, cf. platytroctids, alepocephalids, myctophids and stomiids [13, 15, 57]. Myctophids possess caudal glands that usually present sexual dimorphism [2] and glandular light organs can also be found in the Stomiiformes [2, 46] and dalatiid species [8]. Luminous tissue with a similar structure to photophores can be found in the body of myctophids [2] and stomiids [46] and barbels, while luminous mucus clinging to teeth was seen in some stomiids [20] and anglerfishes [2]. Some studies suggest the existence of bioluminescent skin in *Himantolophus azurlucens* and *Cryptopsaras couesii* [41].

The intensity of light emission in fishes can vary individually depending on the type of organ, depth and function (see further). Nevertheless, it seems that intensity is higher in coastal and shallow rather than mesopelagic, deep water species. Light spectra of the emitted lights usually peak in the blue-green, but are to some extent variable mostly in relation to depth. Coastal and epipelagic fishes produce light principally in the green and meso- and bathypelagic in the blue [3, 4, 33]. A rare case is the red light emission of some stomiids. Regarding angular distribution, bioluminescence starts to exceed downwelling light in the lower mesopelagic zone. At that depth, luminescence signals change from a ventral direction to a point-light scenario with bioluminescence emitted in all directions. The angle of the light emitted by an individual depends on the position of the light organ and its reflector in the body of the fish [2, 48], on the kind of transmission tissue involved and the body position of the light-emitting fish in the water column [37, 50]. Some shallow-water fishes like anomalopids and leiognathids can control the angle of the emitted lights [41, 50]. The light emitted also depends on the trophic level and nutritive state of the fish [43] and whether the light-emitting individual is looking at a conspecific side by side in a school, or, for example, prey above or a predator below.

4.2. Functions of light organs and likely roles of the emitted lights

A bioluminescent signal is defined by its intensity, spectral peak(s), temporal and spatial characteristics [3]. To what extent the degree of polarization, if any, plays a role has never been examined. The same luminous tissue may serve multiple kinds of functions, and the locations of the light organs are generally related to the roles the emitted lights are expected to play. The light generated can be used by a fish to communicate with its congeners, attract prey or avoid and startle predators. The range of functions of biological light is less complex in the coastal species than that in the deep water species, but even in the former, the emitted light has a variety of different biological roles.

4.2.1. Intraspecific communication

Fishes use light signals for intraspecific recognition, schooling and mating. Some fishes possess species-specific structures that purely on account of their placement and position may also assist specific recognition, as the photophore patterns in myctophids apparently do [59]. As aggregations of bioluminescent fishes help confusing predators, light emission may also assist school-forming species of etmopterids [10, 54], chlorophthalmids and macrourids [60].

A role in reproduction seems to be the part of the bioluminescence repertoire of at least some luminescent fishes including leiognathids, with their complex system of controlled light emission, and stomids [2, 3]. Light organs are sexually dimorphic in some species of *Gazza*,

Secutor and *Leiognathus,* and courtship behaviours using light signals have been observed in these fishes [61].

4.2.2. *Interspecific communication*

Some fishes use luminous organs located in or on the head to illuminate their surroundings, in search of prey and to detect predators. The bright headlights of anomalopids illuminate their surroundings, help them to avoid predators [62] and allow them to spot zooplankton [35]. Luminescence may be used for illumination in alepocephalids and macrourids in search of prey on the seafloor [30].

Many fish seem to attract their prey with bioluminescent lures. The escae of anglerfishes have been suggested to mimic faecal pellets with luminous bacteria [32]. Monocentrids have been assumed to attract prey with their light organs [53], but the cephalic photophores of *Diaphus* may be used to stun preys with bright flashes [1]. *Malacosteus niger* has small greenish light-emitting spots near its mouth and larger red light-emitting patches below its eyes. It is believed that the small greenish lights attract prey and the red light, invisible to most deep water organisms, is used by the fish to visually detect the approach of the unsuspecting prey in order to seize it when sufficiently close [2, 30, 48, 63].

Bioluminescent fishes can also use their lights in various ways to distract predators and avoid to be seen by them. Counterillumination is widespread and particularly common in the mesopelagic species, which emit light ventrally to match the downwelling light from above. In that way these often strongly laterally compressed species become camouflaged and almost invisible as silhouettes to the eyes of a predator below [37, 50]. This notion of the use of bioluminescence is strongly supported by biological and ecological facts. Photophores tend to be larger in the species of mesopelagic fishes occurring at shallower depths [2, 4], but the intensity of the ventral light emission from some fishes may change according to the light they find themselves in as from leiognathids [50] and myctophids [64]. Extraocular photophores in Stomiiformes, myctophids [2] and sharks [65] are located in optimal positions for comparisons of the downwelling sunlight with their own light in order to adjust it, a scenario that has also been suggested for apogonids [51]. Counterillumination appears to be of importance to the vast majority of luminous shallow water [33, 35] and mesopelagic teleosts [4, 21, 57] and sharks [66]. Whether the control of the matching luminescence involves eyes and ocular feedback or whether the light organs and cells within them are able to independently and directly regulate photic output of the light organ are questions yet to be answered.

Aposematism is a likely reason for the bioluminescent dorsal spines in etmopterids [67]. Aposematism is one of the proposed functions for bioluminescence in Porichthys as these species possess venomous dorsal spines [2]. When a possible predator is close by and ready to attack, other methods have to be used for luminous fishes to escape. One possible strategy for prey to evade a predator's attack is to emit a brief and blindingly bright flash of light right before changing its swimming direction. This has been suggested for the post-orbital photophores of anomalopids [62] and stomiids [4]. As most of the predators of these fishes have eyes adapted to dim light, very bright flashes may have a temporary effect of rendering them incapable of vision [32, 33]. In addition to the species already mentioned, this survival strategy also seems to be employed in cases of the brief and bright flashes of leiognathids [53], the cephalic photophores in *Diaphus* [1] and the caudal organs of various myctophids [2]. A very similar defence has been proposed in cases in which a bioluminescent liquid is secreted into the water, acting as a luminous smokescreen and confusing the predator while the prey escapes [4]. Luminous liquids or mucoid secretions are produced in the 'shoulder organ' of platytroctids [57], the gill slits of the epigonid *Florenciella* [2], the anal gland of the macrourid *Malacocephalus* [21] and the escae and caruncles of some anglerfishes [21, 41].

4.2.3. *Unreported functions*

For the bioluminescence in several families of fish, no particular reason for the biological light is known (**Table 1**). Nevertheless, most of these unreported cases deserve some attention. As for the majority of other pelagic species, counterillumination is probably the most important function in Somniosidae, Evermannelidae, Microstomatidae, Phosichthyidae, Epigonidae and Howellidae as well as in some other families of the same order. Eurypharyngidae may use their luminescence as a lure to attract food items as with members of Saccopharyngidae. The fact that bioluminescence characteristics exhibit considerable similarities within the families of the order Gadiformes may be used as evidence that functions suggested for Macrouridae can apply equally to Moridae and Merlucciidae.

5. Vision in bioluminescent fishes

Light enters the eye through the cornea (**Figure 4**), passes the pupil and is refracted by the lens and focussed on the retina. Photons are absorbed by pigments in the photoreceptor cells, transformed into synaptic signals, the latter being transmitted through interneurons to the retinal ganglion cells [30, 68, 69]. The axons of these cells are grouped together and form the optic nerve which passes the visual information to the optic areas in the brain [70]. The basic structure of the eye in fish conforms to that of other vertebrates, but there are features that differ, depending on the life style of the fish involved (predator or prey) and on the photic environment [3], i.e. essentially the epipelagic and mesopelagic zones, in which the fish eyes are supposed to operate [30]. The most obvious differences in the structure and function of the fish eye accompany differences in the light intensities prevailing along the depth gradient of the oceanic zones. The majority of the bioluminescent species inhabit deeper waters and their vision is adapted to dim light conditions [30]. The luminescent crepuscular or nocturnal epipelagic fishes [33, 34, 39] have photoreceptors that, too, are mainly adapted to function under low light conditions [68, 70, 71].

5.1. Eye (structure, size and position)

There are no bioluminescent fishes with degenerated eyes and all possess binocular vision [3]. Epipelagic [4, 33] and mesopelagic fishes [30] inhabit environments that can be reached by sunlight and they usually possess large laterally positioned eyes. In contrast, the aphotic bathypelagic species often have smaller eyes [4, 30]. The majority of the bioluminescent fishes have typical single lens, camera-type eyes, but some mesopelagic fishes possess tubular eyes. The latter are cylindrical eyes located at the top of the head and on account of their position allow the fish to distinguish silhouettes against downwelling sunlight. In these eyes, their binocular overlap provides high quantum capture, and accessory retinae and additional visual structures allow them to also detect a certain amount of light coming from lateral and ventral regions [68].

5.2. Lens, pupils and gaps

Fish lenses present little spherical aberration and a gradient of refractive index with the highest value in the centre guarantees that the focus of the light is sharp [4]. Lens transparency helps sensitivity [3], but lens pigments may influence the colour of the light reaching the retina [36]. Large pupils corresponding to wide apertures allow the admission of more photons, increasing

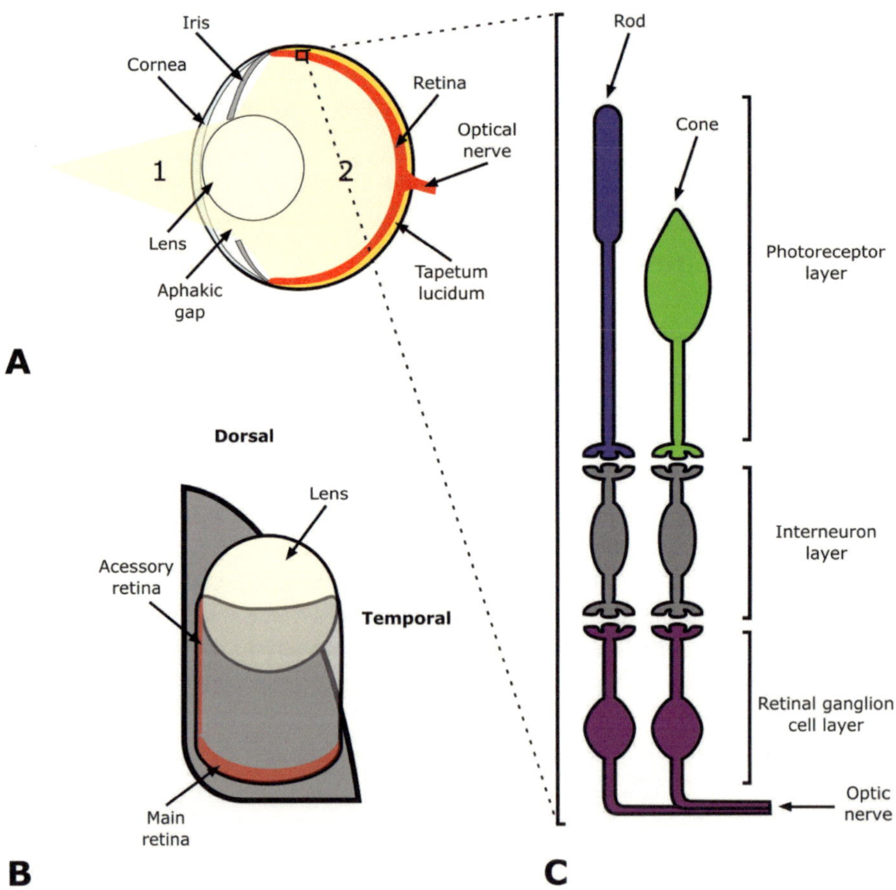

Figure 4. Schematic representations of the eye structure in fishes. (A) Camera-eye representing light entering the eye (1) and refracted by the lens focused on the retina (2). (B) Front-view of a left tubular eye in the head of a fish. (C) Basic retinal structure by cell layers.

visual sensitivity [30]. Aphakic gaps can enhance sensitivity by allowing light capture from specific oblique areas of the visual field [3, 4]. Such gaps are more commonly rostrally placed and extend binocular frontal vision in the lateral eyes. Light passing through gaps may get focussed in specialized retinal areas [30, 31, 72].

5.3. Tapeta

Animals that live in dim environments often possess *tapeta lucida,* which consist of a reflective layer positioned behind the photoreceptors acting as a mirror [30, 71]. This layer reflects the photons that are not captured by the photoreceptors, sending them back to the photoreceptors [68] and doubling quantum capture success; thus improving the sensitivity of the eye [31]. The spectral reflection peak of the tapetal regions in some fish species is thought to coincide with the absorption peaks of the visual pigments in the photoreceptors of the fish, but in some cases specific tapetal areas reflect spectrally different light thereby fulfilling specific visual tasks [36, 72].

5.4. Retina

Photoreceptor cells are responsible for the perception of light and their anatomical and ultrastructural make up, organization, dimensions, density, and distribution, and visual pigment is what has to match the behavioural visual tasks required of them by their owners in the specific photic habitats [69]. Like other vertebrates, fishes have two anatomically distinguishable types of photoreceptors. Cones, of which physiologically different two, three or more kinds with non-identical peak spectral sensitivities may be present, perceive different colours under conditions of bright light. Contrary to cones, rods are monochromatic photoreceptors, but provide the fish with high sensitive vision under low levels of light [36]. Cones occur in some bioluminescent fishes but are often reduced in total number and/or size even in the shallow-water species [60, 68, 70, 71, 73, 74]. Deep-sea fishes have mostly pure rod retinas of high cell density, for there is little need for them to be able to distinguish colours. For them the priority is to possess high sensitivity in a blue-light dominated nearly monochromatic dim, almost dark environment. High sensitivity of the rods is achieved by longer and wider cells. Sensitivity is also enhanced when rods are arranged in groups or are arranged in multi-layers called banks [30].

The differences in function between rods and cones are due to different kinds of photopigments in the visual cells. Rods possess either rhodopsin, a visual pigment with maximum absorbance at short wavelength (chemically a retinaldehyde, retinal 1 or vitamin A_1-based compound) or porphyropsin (i.e. chemically a 3,4-dehydroretinal, retinal 2 or vitamin A_2-based compound) [36]. Deep-sea fishes usually have high concentrations of rhodopsin giving the fish broad sensitivity to blue-green wavelengths [69]. However, some deep-sea fish have rod types containing different wavelength-sensitive rhodopsins or in addition may also have porphyropsins, the latter with longer-wavelength peak absorbances. The variety of receptor types and the development of banked retinas to increase absolute sensitivity can provide deep-water fish with the potential to make use of a wide range of photic stimuli [36, 75].

The density of retinal ganglion cells and, in particular, the ratios between receptors to ganglion cells define visual acuity and sensitivity. A high ratio of rods to retinal ganglion cells is indicative of high absolute sensitivity, while a low ratio would favour acuity. Most of the deep-water fishes possess unspecialized retinae and retinal cell ratios typical of sensitivity improvement [30, 31]. However, even so slight regional enhancements of acuity may be found in some species [68, 76].

Retinal specializations are regions of concentric increases in visual cell densities that enhance the visual sensitivity (rods) or acuity (retinal ganglion cells). These specializations seem to be related to visual behavioural tasks [72, 76]. In bioluminescent fishes, *areae centralis* and foveas can usually be identified as areas with higher visual acuity than the rest of the retina [30, 68], but it is possible that such specialized retinal regions change during the lifetime of a fish [77].

6. Evolution of bioluminescence and vision

The eyes of bioluminescent fishes evolved as adaptations to a dim light environment, in which bioluminescence provided important visual stimuli to perceive predators, prey items and mates. The fact that the bioluminescence in these species shares optical properties with the light pervading the environment the fish inhabit indicates the role selective pressures must have played [4]. Nevertheless, one cannot discount the idea that light emission and reception together represent a mutual and progressive adaptation [3].

In fishes, the adaptation of rhodopsin pigments to dim-light vision occurred in 12 separate occasions, suggesting that dim-light vision was a subject to considerable adaptive evolution [5]. In deep-sea fishes, these pigments were already shifted and adapted to the same visual environment as today in the Early Triassic. This shift was an adaptation to maximize the quantum capture of the dim sunlight and the biologically generated light in the deeper water of the sea. Along the way, convergent evolution seemed to have occurred with regard to rhodopsins of deep-sea fishes for adaptation to specific light regimes. This would explain the variation of wavelength sensitivities in these fishes, some adapted to specific bioluminescence peaks [75, 78]. Diversity in the pigmentation of the lenses of deep-water fishes suggests that it too has been the subject of separate evolutionary events a number of times [36]. Much later than vision, and assumed to have happened from the Early Cretaceous to the Late Cenozoic, bioluminescence appears to have evolved in teleosts [17]. A total of 27 independent occurrences are regarded to have left representatives in the marine teleosts assemblage, appearing first in the Stomiiformes [17]. Bioluminescence in the three closely related families containing luminescent species of sharks is also thought to have evolved from a common ancestor during the Cretaceous [9, 10, 17]. Bioluminescent patterns could have contributed to speciation in fishes [9, 11].

6.1. Bioluminescence evolution

An exposure to different light regimes powerfully influences the type of bioluminescence characteristics in the species of different habitats. This can express itself through different

http://dx.doi.org/10.5772/65385

spectral emission peaks and intensities of the biologically generated lights matching the photic environment of a species' habitat. The angle and direction of the emitted light are also subject to evolutionary pressures as most of luminescence glows ventrally in epipelagic and mesopelagic zones [2, 21, 33], where traces of sunlight from above are still discernible. No light from above, on the other hand, reaches the bathypelagic zone where the bioluminescent species exhibit a reduction in the number and sizes of photophores [2]. A good example of this tendency is the bathypelagic myctophids of the genus *Taaningichthys*, in which only the deepest living species, *T. paurolychnus* [79], lacks photophores (but still possesses caudal glands) [15].

6.1.1. Functions

Bioluminescence plays major roles in the ecology of the luminescent fish species. The evolutionary history of bioluminescence in fishes shows that the ability to generate light arose dozens of times in unrelated clades [17] and yet, the basic structure of the light organs is nearly the same in all cases, independent of substrate, position and functions of the organ [20, 37]: counterillumination seems an excellent example. Fishes developed different ways to hide their silhouette from predators, using their ventral light emission. Some species make use of rows of photophores for that purpose, whereas others use indirect luminescence, but each case arose repeatedly and independently during evolution [17, 21].

An example of different structures employed in counterillumination can be seen in shallow-water species. In that environment, all the fishes use an indirect form of bioluminescence for camouflage with the exception of *Porichthys*, sciaenids and *C. dussumieri* that have photophores for that purpose [2, 33]. Different photogenic origins (i.e. symbiotic and intrinsic) may have led to the evolution of counterillumination, and some excellent examples are those that involve indirect bioluminescence. Some shallow (e.g. pempherids) and deep-water (e.g. *Cocorella atrata*) fishes use intrinsic light organs for indirect ventral bioluminescence, but the majority of species with this kind of light emission are employing symbiotic bacteria [2, 19, 39]. This divergence of photogenic origins can also be observed within apogonids [2]. Ventral luminescence in diel vertical migrants must also have come under selective pressures to conceal the silhouettes of the migrants from predators, when ascending into shallower and more illuminated waters [21, 37]. It is also conceivable that light signals in interspecific schools of diel vertical migrants play a role in keeping individuals together [3]. Although we focussed on counterillumination, there are other less well studied examples of structures of photogenic as well as non-photogenic origins serving identical functions in bioluminescent fishes. What immediately comes to mind are the symbiotic lures in anomalopids [53] and ceratiids [14] and the intrinsic barbels of stomiids [4] and photophores in sharks [8].

6.1.2. Acquisition

Whether we deal with cases of intrinsic or symbiotic bioluminescence in fishes, the acquisition of the bioluminescence must have appeared at some point in the evolutionary history. It is still not clear how and when this happened, but based on some widely accepted suggestions the process behind it is thought to have involved the higher availability of luminous bacteria in coastal areas as a cause for the widespread use of symbiotic bacteria in the bioluminescence of fishes inhabiting seashore-near habitats [80]. Similar kinds of constraints could have acted upon the *Cypridina* luciferin containing species of intrinsic luminous fish. These fishes are coastal species [33, 44] and the ostracods, which produce *Cypridina* luciferin, *Vargula* and *Cypridina* are benthic species and seem to inhabit mainly coastal and shallow waters [81–83]. It is well established that fishes like apogonids, pempherids and *Porichthys* have obtained the *Cypridina* luciferin from ostracods they ingested [18, 47, 84]. As there are no records that fish can synthesize coelenterazine, dietary acquisition seems a plausible scenario. This suggestion is supported by the widespread occurrence of coelenterazine also in luminous mid-water fishes and the fact that this substance is most abundant in the tissues of the digestive tracts of these fishes [85].

On the other hand, we have the symbiotic luminous fishes, in which light-producing bacteria are responsible for the bioluminescent capacity of the host. In return, the host provides protection, a place to grow, nutrition and oxygen to the bacteria [42]. This symbiotic relation could have started with facultative luminous bacteria in the gut of fishes (e.g. leiognathids), evolving into the obligatory symbiosis we now witness in anglerfishes and anomalopids [39, 41]. Hosts are ecologically dependent on the bacterial light and 'host family-bacterial specificity' seems to be the trend [42]. However, contrary to what had been believed for decades, nowadays coevolution between facultative symbiont and the host fish species is regarded as an unlikely scenario [41] as different fish host the same bacterial species and different species of bacteria can be co-symbionts of fish species. Moreover, evolutionary divergence in the bacterial species is unrelated to that of the host species [86]. Nevertheless, the absence of non-luminous bacteria in the organs indicates that some selection must have taken place [42] and associations could have arisen (or been bolstered) through environmental factors like the depth that the metamorphosing larvae inhabit. Support for this notion comes from *Photobacterium leiognathi* and *Photobacterium mandapamensis* found in organs of fishes that inhabit warmer and shallower waters and *Photobacterium kishitanii* that is found in deeper and colder waters [39, 42, 86].

6.2. Visual adaptation to bioluminescence

Vision in fishes evolved primarily as an adaptation to light intensity changes to extended or point sources in the near and far field of vision and secondarily to species-specific ecological needs [30, 31]. Bioluminescence has, to a large extent, affected the evolution of the optical system in luminescent fishes, which is obvious when one compares the eyes of fishes inhabiting the mesopelagic and bathypelagic realms. In the mesopelagic zone, animals use counterillumination to match the dim background light [36] and under such conditions fish evolved certain adaptations like tubular eyes [68], yellow lenses and retinae with banked receptor cells. In terms of their function, such eyes had to be sensitive, able to distinguish small spectral discrepancies from background light and to detect camouflaged prey [36]. For bathypelagic fishes, on the other hand, the only light sources available to them would have been of bioluminescent origins, restricting the range of their visual adaptational possibilities [31]. The bathypelagic species therefore tend to have larger aphakic gaps than mesopelagic species even within the same genus [3, 68]. As food is scarce at these greater depths, fishes living there evolved less energetically expensive small eyes [4]. Nevertheless, these fishes do possess eyes with visual acuities capable of distinguishing point-source bioluminescence. Moreover, the eyes of the bathypelagic bioluminescent species show overall fewer regressed features than the eyes of the non-luminescent bathypelagic species do, indicating that the former rely more on vision than the latter. There are even some bottom-dwelling deep-sea fishes that have severely degenerated eyes ,which are likely to detect no more than the presence or absence of bioluminescence of perhaps invertebrates in the seabed [31].

The types of visual pigments [5] and the balance between visual acuity and sensitivity are the results of millennia of selective pressures stemming from the diverse light regimes and visual tasks fishes had to contend with [4]. Visual pigments evolved primarily to light from the sun, perhaps to guide specimens to darker regions and shelters away from the light or to seek illuminated areas, but in deep-sea fishes photopigments seem to be much more attuned to bioluminescence spectra [4, 6, 30, 36]. The eyes of most of the deep-water species have high concentrations of rhodopsin giving them a broad sensitivity to the blue-green range of the spectrum [69], comprising wavelengths that seem ideally suitable for perceiving bioluminescence emissions of prey, predators and conspecifics [6]. Potential colour vision in some deep-sea fishes may be an adaptation to identify species by differences in their bioluminescence hues [69, 75].

7. Ontogeny of bioluminescence and vision

7.1. Metamorphosis of eye and luminous tissues

During larval metamorphosis in fishes, the eyes change in shape, size, structure and even location [27]. The best examples among bioluminescent species are the tubular eyes, which are formed from laterally placed camera-type eyes in the larvae that become tubular in shape and migrate dorsally during metamorphosis. The lens remains unchanged but the retina suffers many alterations [87]. Retinae in larvae of the deep-water species, whether or not luminescent, are poorly specialized and vision, generally, is much less developed than in adults [88]. Almost all fish larvae have pure-cone retinae that are totally or partially substituted by rods during metamorphosis. The only exceptions in bioluminescent fish larvae are Saccopharyngiformes and *Lumiconger arafura,* in which pure rod retinae are the rule. In retinae with rod multilayers, the number of banks increases throughout development [68, 87, 88].

Photophores are formed during metamorphosis of a fish [22, 27, 87] and some are then already functional [2]. The luminous organs of symbiotic fishes also develop during larval metamorphosis [39]. This is the case of other luminous tissues too, as for example, the caruncles of ceratiids [41] and the caudal glands of myctophids [27]. Furthermore, the barbels of stomiids seem to arise during metamorphosis [27] and continue to elaborate throughout the fish's development [4].

The acquisition of the photogenic substrate in fishes is not well known. The luminous organs of symbiotic fishes start to develop prior to their being colonized by bacteria. Colonization occurs during metamorphosis and it seems that each new generation of fish needs to acquire its symbiont afresh from the environment. The larvae of the shallow-water species are spatially close to adults, which suggests that bacteria may be released from the adult light organ and transferred to the young [39]. This does not seem to apply to the deep-sea species as the larvae are planktonic and hundreds of metres away from the adults. In these cases, it was suggested that larvae are infected when they migrate to coastal areas where luminous bacteria are more abundant [42, 80]. For ceratioids, the secretory cells that are common only in young fish may liberate pheromones to attract the right bacterial symbiotic species [41]. In intrinsic species, maternal transference of *Cypridina* luciferin was observed in *Porichthys* [84]. The same mechanism was suggested to apply to other teleosts and etmopterid sharks [89, 90].

7.2. Ecology and ontogeny: vision and bioluminescence in young fishes

Most symbiotic and intrinsically bioluminescent fishes have epipelagic larvae that will inhabit mid- or deep-waters as adults [4, 12, 15, 22, 26, 39]. During this ontogenic vertical migration to deeper waters, the visual habitat changes to an increasingly dim and monochromatic light scenario. These changes in the light environment require the modifications of the fish's photoreceptive as well as luminescent structures (described above) to coincide with the larvae undergoing metamorphosis. The migration to deeper waters and the metamorphic changes previous to the habitat change are believed to be completed in a relatively short time [87]. The development of rods and the changes in photoreceptor arrangements prepare the metamorphosing larvae to a life as a juvenile or adult in an environment where superior visual sensitivity is paramount. The transformations provide the young fish with visual capabilities for recognizing conspecifics, forming schools, engaging in predation and tracking and evading predator advances. In multi-layered retinae, a greater development of banks appears to coincide with the greater depths juvenile fish seek to spend their lives in as adults. In shark species, a metamorphosis similar to that seen in teleost is not present and the ontogenesis of photogenic organs or eyes in bioluminescent sharks has not been studied in much detail [68, 87].

In teleosts, the development of luminous organs is also affected by the changes in the photic environment and the behaviour of the young fishes. In some Stomiiformes and myctophids, the early development of cephalic photophores suggests that they are used to lure crustacean prey at night [22]. The more common scenario is that in the species whose larvae occur in dimly illuminated waters, photophores develop on the body first [2, 22, 91] and then gradually become functional as the migration to deeper water proceeds [22]. Larvae tend to develop photophores earlier in the species that inhabit deeper water as adults. For juveniles, light emission changes during growth and maturation [22]; ventral luminescence develops rapidly during that phase of the life of many teleosts [2, 4, 12, 52]. Ventral photophores also appear to be more developed in juvenile etmopterid sharks that inhabit shallower depths than the adults [8]. Considered together, these observations highlight the importance of counterillumination in young fishes while they migrate to deeper waters [2, 22].

Not only do photophores and light organs change in relation to a fish's developmental stage, the role or roles bioluminescence is to play in the life of a fish may also change with age. This has been suggested for etmopterid sharks as ventral patterns of photophores used for mating and schooling are more developed in adults [8, 92]. Photophores of young platytroctids are horizontally placed on the body and are probably used to illuminate prey, while distracting predators. These photophores, however, do acquire more vertical positions as the ontogenic development proceeds [13] and this very likely implies a similar explanation to that seen in scopelarchids in which luminescence is limited to the adults [2]. In both cases, the use of the ventral lights only in adults is strong evidence for its function as counterillumination devices to help camouflage silhouettes of larger adult bodies.

8. Visual ecology

Considering bioluminescence as a means for communication and vision may help us understand its origins in fishes. Luminous species exhibit adaptations for a wide range of visual tasks [30] as light organs are multi-purpose and in most cases their predators and preys are also bioluminescent [2, 30]. Most bioluminescent signals are of rather low intensity, but bioluminescent fishes frequently possess rod-dominated retinae, conferring to them high visual absolute sensitivity [68, 70, 71] allowing them to perceive even very dim lights.

8.1. Bioluminescence detection

The optical characteristics, i.e. 'quality and quantity' of the light emitted by the vast majority of mid and deep-water luminescent species of fish relate to the downwelling light of their habitats. Vision accompanies this tendency. All of the luminescent species, whether shallow or deep-water fishes, possess eyes with at least some visual overlap, i.e. binocular vision, and that enhances sensitivity to and detection of distant point sources in dimly lit environments [68]. Bioluminescent fishes, moreover, frequently exhibit large pupils, which help them to detect bioluminescent flashes against a wider background [30]. The visual pigments present in the eyes of deep-sea bioluminescent fishes seem to be perfectly adapted to the bioluminescence spectra they encounter [2, 6, 30, 36].

Additionally to these more general adaptations, bioluminescent fishes evolved numerous visual particularities in order to improve detection of bioluminescence signals. Large eyes in epipelagic and in particular mesopelagic species admit greater amounts of photons, banked photoreceptors further improve the photon yield and efficient tapeta permit an at least two-fold photon catch. Such adaptations are critically important if sensitivity improvement in order to use bioluminescence is the goal. Different spatial relation of aphakic gaps, retinal specializations and tapeta are particularly well developed among the myctophid species [72] and luminous sharks [65]. Most deep-sea fishes can detect bioluminescence signals up to 30 m away [6, 48]. Some are even able to see bioluminescence up to 51 m as in bathypelagic fishes [21]. Bioluminescent bathypelagic fishes possess small eyes with wide pupils, rostral aphakic gaps and acute foveas. Although these eyes are less sensitive than those of the mesopelagic species, their resolving power is better and they seem perfectly adapted to spot point sources of biological lights in their environment. Contrary to non-luminescent bathypelagic fishes, the eyes of the luminous species are not widely separated, aiding them in distances estimation [31, 68].

8.2. Intraspecific communication

In many bioluminescent fish species, their luminescence seems to play a role in communication between conspecifics. The eyes of these species are by necessity optimally adapted to their own bioluminescence. In shallow-water species, anomalopids seem to have eyes and retinae adapted to their own lights [74] and the visual pigments of *Porichthys* are also a good match to the emission spectrum in this species [73].

In deeper species, the eyes of *Chlorophthalmus*, with yellow tainted lenses and a specialized retina with some cones seem specifically adapted to their own blue-green light emissions [60]. Myctophids also possess visual pigments adapted to the spectra of their own light emissions [6]. Most male anglerfishes possess relatively large laterally positioned eyes with aphakic gaps [14] and long rods for heightened absolute sensitivity, adaptations that may be linked to the use of the female's esca to attract the attention of a male [3]. The genera *Malacosteus, Photostomias* and *Aristostomias* possess sub-ocular photophores, which emit red-light used as a 'private waveband'. These species evolved long-wavelength pigments in their rods and possessed tapeta and lenses that allow them to see their own red light [36].

Dalatiid sharks do not school, contrary to etmopterids [10]. In bioluminescent sharks, only the etmopterids seem to use their light emissions for intraspecific purposes, specifically the lateral patterns of the photophores [8, 92]. These patterns are more nasal and temporal in *Etmopterus lucifer* and seem to be detected by specific sensitive and acute zones in the nasal and temporal areas of the retina of this species [65].

8.3. Hunting

The use of bioluminescence in the context of food procurement in luminescent fishes is common. Photic lures to attract prey are one of the better known examples of luminous hunting devices. In order to be seen by and attract prey, the light tissue of the predator needs to be brighter than the background and has to be located in a visible place of the predator body, preferably the head and near the mouth [21]. Indeed, the most luminous structures of fishes that are likely to act as lures are easily spotted and emit an intense light, as the escae of anglerfishes [41], the barbels of stomiids [4] and the head organs of anomalopids [33], monocentrids [12] and *Diaphus* [1]. Most of their prey organisms possess highly sensitive eyes [30], so that the perception by them of the lights emitted by the photic lures is not an issue.

Some luminous fishes use their light to illuminate their surroundings, helping them in the search and seizure of prey, as suggested for anomalopids [35, 62] and *Diaphus* [1]. The same is possible for the benthopelagic deep-sea fishes, as alepocephalids and macrourids, that have uncommonly well-developed eyes, which might serve them to search for prey with their ventrally aimed luminescence [31].

It is common for bioluminescent fishes to prey on other luminous organisms. A good indication of this is the acquisition of *Cypridina* luciferin by fish through the consumption of luminous ostracods [18, 33] and the high amount of coelenterazine in the digestive tissues of bioluminescent fishes [85]. The eyes of these predators, exhibit characteristics that help them perceive the bioluminescence of their prey. The highly sensitive retinae of fishes like myctophids, ensures that light from prey reaching them over relative long distances, i.e. several metres, is registered [72, 76]. Luminous sharks have retinal specializations, aphakic gaps and tapeta that also seem to be used for the detection of luminescent prey. In the specific case of the dalatiid *Squaliolus aliae*, the shortwave sensitive pigments in its retina seem to be more adapted to prey emitting blue-luminescence [65].

The fact that the prey of luminous fishes is frequently bioluminescent can create problems for the predator as most prey may employ counterillumination for concealment [4]. Bioluminescent fishes therefore appear to have evolved visual mechanisms allowing them to distinguish light emissions of prey from the downwelling surrounding lights. Most of the predators seeking to capture fish that employ counterillumination have large eyes that can disrupt the prey's counterillumination at close range [32]. Other species have evolved tubular eyes which possess acute retinal areas [30] perfect to detect at greater distances even silhouettes camouflaged through counterillumination [21, 46] and diverticula and accessory retinae to detect lateral luminescent stimuli [3, 31]. Some predators use yellow eye lenses to cut-off blue spectra, and to recognize the minimal green mismatches revealing the luminous silhouettes of prey [36]. Aphakic gaps 'lined' with retinal specializations of photoreceptor and ganglion cells as in myctophids may further help detecting counterilluminated prey [72, 76]. The enhancement of binocular resolution by foveas in deep-sea fish may also aid in breaking luminescent camouflages [68]. Different visual pigments and banks in rods that potentiate colour vision in deep-sea fishes are probably involved in breaking counterillumination [36, 75] and the translucent skin area above the eyes in etmopterid sharks may filter the spectra of luminescent prey, thereby compromising the camouflage of the latter [65].

8.4. Predator avoidance

One of the major functions of bioluminescence is defence [4, 33]. It is therefore not surprising to see such a vast number of different self-protective mechanisms in luminescent fishes. The trick is to avoid being seen by predators and in the case that has already happened, to deceive the predator and 'to go on the run'. The method of defence as well as vision of the predator and prey is always involved. We already stated the possibility for the predators to be luminous and that the eyes of bioluminescent fishes serving as prey are sufficiently sensitive to detect the predator. Curiously, except for the purpose of camouflage, smaller species tend to produce faster and shorter flashes than the larger predators [93]. This seems a strategy to reduce the risk to be accepted by a predator. A predator cannot afford the effort and risk of checking out every single flash it encounters, so that the predator would probably use the size and duration of a flash it encounters in deciding whether pursuit is worthwhile and the emitter of the luminescence is large or small [32].

The spectrum of a luminescent species' light used in counterillumination is not always a perfect match of the downwelling light [30]. This is not a problem for the deep-sea species as most of their predators do not possess colour vision [4, 48] and intensity rather than spectral matching is most important and apparently within the capability of most species that employ counter-illumination concealment [4, 30, 32, 64]. On the other hand, epipelagic predators of bioluminescent fish have eyes more capable than those of deeper waters [94] to disrupt counterillumination [33]. The majority of coastal luminescent fishes exhibit internal light organs providing them a kind of disruptive luminescence that is more effective in turbid and dynamic near-shore waters than the uniform glow [33, 50] of deep-water species that live in more homogeneous and optically transparent photic environment [4].

Since both nocturnal predators of shallow-water bioluminescent fishes [94] and deep water predators [30] possess highly sensitive eyes, luminescent 'smokescreens' and very bright, brief flashes can temporarily confuse or even blind a predator and allow potential prey to flee [33].

Bioluminescence is not always straightforward and what seems to be an advantage can become a disadvantage under certain conditions. The 'private waveband' of the red light in stomiids seems an outstanding evolutionary trait for a predator, but some prey like scopelarchids [69], gonostomatids [75] and myctophids [6] have retinae with additional long-wave sensitive photopigments enabling the potential prey to see the stomiids red light. Fishes with 'head lamps' to illuminate the surroundings may be coastal and nocturnal or midwater inhabitants. Their bright emissions may be an advantage to pick up prey, but they are also a disadvantage as predators can easily spot these fishes and be attracted to them [12, 32, 33]. Nevertheless, the risks of that happening appear to be reduced as these species feed in schools from which predators usually find it more difficult to focus on an individual fish and seize it [12, 33, 94]. Moreover, the luminescent anomalopid also can rapidly and repeatedly occlude their lights [35, 41, 43].

9. Final remarks

9.1. Methodology

To understand the inter-relationship between vision and bioluminescence in luminescent species, detailed information on eyes and light organs of these animals is paramount. Spectral properties and intensities of the emitted lights, development, organization, structure and function of photophores as well as photoreceptors have to be studied and behavioural correlates need to be identified. None of that is easy; accessibility is difficult, measurements of the light emissions in luminescent fishes can be a tricky undertaking (especially for deep-sea teleosts and sharks) and when caught, these animals are more than often moribund and in most cases have to be stimulated to glow. This is possible using electricity and/or certain chemicals [34, 55]. The intensity and spectra of such induced light emissions can be measured using a spectrometer and a photo detector, but the data may not exactly reveal their properties in the way the fish use their luminescence in their natural environment. The angle of the light emissions depends on the light organ's position on the body and the internal angle of the organ, both analysable to a fair degree of accuracy using histological techniques [46]. Chemical analyses of the chemical compounds involved in the generation of the biological light would help understand the underlying mechanism involved, but control of the light emissions,

http://dx.doi.org/10.5772/65385

whether direct by the photophores themselves, through nervous signals or hormonal mediators requires living, healthy specimens.

Anatomical details of the eye, cell topography, photoreceptor cell types, cell ultrastructures and distributions as well as certain visual parameters can be gleaned from histological investigations, but while functional parameters like sensitivity, acuity, angular acceptance, etc. can to some extent be deduced from the anatomical details, electrophysiological techniques or the use of microspectrophotometry must be the methods of choice when it comes to functional questions. Spectral sensitivity peaks of the visual pigments obtained spectrophotometrically or electrophysiologically ought to be connected with chemical analyses of the photopigments involved and opsin gene analyses [36]. Chemical and optical studies like refractive indices, focal lengths, etc. on the dioptric structures of the fish eye ought to complement the other investigations so that predictions can be made on the way a fish uses its eyes.

Evolutional studies of vision receive support from investigations on the mutagenesis of the visual pigments [5] and genetics of nuclear and mitochondrial gene fragments can also be used in studies on phylogenetic relationships between species [17]. Often lacking first-hand direct observational evidence of vision/luminescence interactions in the natural environment, the next best approach would involve realistic analyses of the underwater transmission of bioluminescence in specific cases and ecological modelling, based on the information available [48].

9.2. Future prospects

Communication by light is a subject of considerable interest to ethologists and ecologists, chemists, geneticists, anatomists, physiologist, opticians and vision researchers. Biotechnologists, too, become increasingly interested in bioluminescence generally and communication by light in particular. For the comparative zoologist it is important to realize that behaviour exhibited by individual luminescent fish in captivity may not be the same as that in the fish's natural environment. This is even more important to keep in mind when interpreting the induced light emissions obtained from nearly dead deep-sea species. Moreover, vision and bioluminescence are only part of a bigger picture. There is evidence that pineal photoreceptors support light perception and are even involved in light output regulation in deep-sea fish [64]. Therefore in analyses of the ecological role, i.e. overall biological significance that bioluminescence plays, we must not ignore the fact that other sense organs like, for instance the lateral line system, semicircular canals, otoliths, chemo, electro and magnetoreceptors, etc. may also be involved [4].

Regarding the eco-ethological roles of the relationship between bioluminescence and vision, some questions have been answered but many more remain to be solved. Remotely operated underwater vehicles (ROVs) can be expected to help in this task and should provide new *in vivo* insights into the various eco-ethological roles of bioluminescence while advances in genetic techniques can be expected to shed further light onto the phylogeny and evolution of these 'brightest creatures of the animal world'.

Acknowledgements

The authors are grateful to the Food and Agriculture Organization of the United Nations for the authorization of the drawings in **Figure 4** and Dr. Hayato Tanaka for the information on luminous ostracods.

Author details

José Paitio[1], Yuichi Oba[1*] and Victor Benno Meyer-Rochow[2,3]

*Address all correspondence to: yoba@isc.chubu.ac.jp

1 Deparment of Environmental Biology, Chubu University, Kasugai, Japan

2 Department of Genetics and Physiology, Oulu University, Finland

3 Research Institute of Luminous Organisms, Hachijojima, Japan

References

[1] Haddock SHD, Moline MA, Case JF. Bioluminescence in the sea. Annual Review of Marine Science. 2010;2:443–493. DOI: 10.1146/annurev-marine-120308-081028.

[2] Herring PJ, Morin JG. Bioluminescence in fishes. In: Herring PJ, editor. Bioluminescence in Action. London: Academic Press; 1978. pp.273–329.

[3] Nicol JA. Bioluminescence and vision. In: Herring PJ, editor. Bioluminescence in Action. London: Academic Press; 1978. pp. 367–408.

[4] Herring P. The Biology of the Deep Ocean. New York: Oxford University Press Inc.; 2002. 314 p.

[5] Yokoyama S. Evolution of dim-light and color vision pigments. The Annual Review of Genomics and Human Genetics. 2008;9:259–282. DOI: 10.1146/annurev.genom.9.081307.164228.

[6] Turner JR, Whit EM, Collins MA, Partridge JC, Douglas RH. Vision in lanternfish (Myctophidae): Adaptations for viewing bioluminescence in the deep-sea. Deep-Sea Research I. 2009;25:1003–1017.DOI: 10.1016/j.dsr.2009.01.007.

[7] WoRMS Editoral Board. World Register of Marine Species [Internet]. [Updated: 2016]. Available from: www.marinespecies.org [Accessed: 20/06/2016].

[8] Claes JM, Mallefet J. Bioluminescence of sharks: First synthesis. In: Meyer-Rochow VB, editor. Bioluminescence in Focus - A Collection of Illuminating Essays. Kerala: Research Signpost; 2009. p. 51–65.

[9] Straube N, Li C, Claes JM, Corrigan S, Naylor GJP. Molecular phylogeny of Squaliformes and first occurrence of bioluminescence in sharks. BioMed Central Evolutionary Biology. 2015;15:162. DOI: 10.1186/s12862-015-0446-6.

[10] Reif W-E. Function of scales and photophores in mesopelagic luminescent sharks. Acta Zoologica. 1985;66(2):111–118.

[11] Davis MP, Holcroft NI, Wiley EO, Sparks JS, Smith WL. Species-specific bioluminescence facilitates speciation in the deep sea. Marine Biology. 2014;161:1140–1148. DOI: 10.1007/s00227-014-2406x.

[12] Herring PJ. Aspects of the bioluminescence of fishes. In: Barnes H, Barnes M, editors. Oceanography and Marine Biology. An Annual Review. Volume 20. Aberdeen: Aberdeen University Press; 1982. pp. 472–541.

[13] Matsui T, Rosenblatt RH. Review of the deep-sea fish family Platytroctidae (Pisces: Salmoniformes). Bulletin of the Scripps Institution of Oceanography. 1987;19:1–159.

[14] Pietsch TW. Oceanic Anglerfishes. Extraordinary Diversity in the Deep Sea. Berkeley: University of California Press; 2009. 557 p.

[15] Carpenter KE, Niem VH, editors. FAO Species Identification Guide for Fishery Purposes. The Living Marine Resources of the Western Central Pacific. Volume 3: Batoid Fishes, Chimaeras and Bony Fishes Part 1 (Elopidae to Linophrynidae). Rome: Food and Agriculture Organization of the United Nations; 1999. pp.1407–2068

[16] Castle PHJ, Paxton JR. A new genus of luminescent eel (Pisces: Congridae) from the Arafura Sea, Northern Australia. Copeia. 1984;1:72–81.

[17] Davis MP, Sparks JS, Smith WL. Repeated and widespread evolution of bioluminescence in marine fishes. PLoS One. 2016;11(6):e0155154. DOI: 10.1381/journal.pone.0155154.

[18] Thacker CE, Roje DM. Phylogeny of cardinalfishes (Teleostei: Gobiiformes: Apogonidae). Molecular Phylogenetics and Evolution. 2009;52:735–746. DOI: 10.1016/j.ympev.2009.05.017.

[19] Haneda Y. Luminous organs of fish which emit light indirectly. Pacific Science. 1950;4:214–227.

[20] Harvey EN. Pisces. In: Harvey EN, editor. Bioluminescence. New York: Academic Press; 1952. pp. 494–553.

[21] McAllister DE. The significance of ventral bioluminescence in fishes. Journal of the Fisheries Research Board of Canada. 1967;24(3):538–554.

[22] Suntsov AV, Widder EA, Sutton TT. Bioluminescence. In: Finn RN, Kapoor BG, editors. Fish Larval Physiology. Enfield: Science Publishers; 2008. pp. 51–88.

[23] Herring P. Systematic distribution of bioluminescence in living organisms. Journal of Bioluminescence and Chemiluminescence. 1987;1:147–163.

[24] Cohen DM, Inada T, Iwamoto T, Scialabba N. FAO Species Catalogue. Gadiform Fishes of the World (Order Gadiformes). An Annotated and Illustrated Catalogue of Cods, Hakes, Grenadiers and Other Gadiform Fishes Known to Date. FAO Fisheries Synopsis. Number 125. Volume 10. Rome: Food and Agriculture Organization of the United Nations; 1990. 442 p.

[25] Ramaiah N, Chandramohan D. Occurrence of Photobacterium leiognathi, as the bait organ symbiont in frogfish Antennarius hispidus. Indian Journal of Marine Sciences. 1991;21:210–211.

[26] Carpenter KE, Niem VH, editors. FAO Species Identification Guide for Fishery Purposes. The Living Marine Resources of the Western Central Pacific. Volume 4: Bony Fishes Part 2 (Mugilidae to Carangidae). Rome: Food and Agriculture Organization of the United Nations; 1999. pp. 2069–2790 .

[27] Moser HG, editor. The Early Stages of Fishes in the California Current Region. Atlas number 33. La Jolla: California Cooperative Oceanic Fishery Investigations; 1996. 1505 p.

[28] McKay RJ. FAO Species Catalogue. Sillaginid Fishes of the World. (Family Sillaginidae). An Annotated and Illustrated Catalogue of Sillago, Smelt or Indo-Pacific Whiting Species Known to Date. FAO Fisheries Synopsis. Number 125. Volume 14. Rome: Food and Agriculture Organization of the United Nations; 1992. 87 p.

[29] Terao A. Photophore in flying fish. Zoological Magazine. 1950;13(40):59–60.

[30] Warrant EJ, Locket NA. Vision in the deep sea. Biological Reviews. 2004;79:671–712. DOI: 10.1017/S1464793103006420.

[31] Warrant EJ, Collin SP, Locket NA. Eye design and vision in deep-sea fishes. In: Collin SP, Marshall NJ, editors. Sensory Processing in Aquatic Environments. New York: Springer-Verlag; 2003. pp. 303–320.

[32] Young RE. Oceanic bioluminescence: An overview of general functions. Bulletin of Marine Science. 1983;33(4):829–846.

[33] Morin JG. Coastal bioluminescence: Patterns and functions. Bulletin of Marine Science. 1983;33(4):787–817.

[34] Harvey, EN. The luminous organs of fishes. In: Brown ME, editor. The Physiology of Fishes. Volume II. Behaviour. New York: Academic Press; 1957. pp. 346–366.

[35] Meyer-Rochow VB. Some observations on spawning and fecundity in the luminescent fish *Photoblepharon palpebratus*. Marine Biology. 1976;37:325–328.

[36] Douglas RH, Partridge JC, Marshall NJ. The eyes of deep-sea fish I: Lens pigmentation, tapeta and visual pigments. Progress in Retinal and Eye Research. 1998;17(4):597–636.

[37] Clarke WD. Function of bioluminescence in mesopelagic organisms. Nature. 1963;198:1244–1246.

[38] Sutton TT. Vertical ecology of the pelagic ocean: Classical patterns and new perspectives. Journal of Fish Biology. 2013;83:1508–1527. DOI: 10.1111/jfb.12263.

[39] Haygood MG. Light organ symbioses in fishes. Critical Reviews in Microbiology. 1993;19(4):191–216.

[40] Salvanes AGV, Kristoffersen JB. Mesopelagic fishes. In: Steele JH, Thorpe SA, Turekian KK, editors. Encyclopedia of Ocean Sciences. Volume 3. 2nd ed. San Diego: Academic Press; 2001. pp. 748–754.

[41] Karplus I. The associations between fishes and luminescent bacteria. In: Karplus I, editor. Symbiosis in Fishes: The Biology of Interspecific Partnerships. New Jersey: John Wiley and Sons; 2014. pp. 7–57.

[42] Urbanczyk H, Ast JC, Dunlap PV. Phylogeny, genomics, and symbiosis of Photobacterium. Federation of European Microbiological Societies. 2011;35(2):324–342. DOI: 10.1111/j.1574-6976.2010.00250.x.

[43] Meyer-Rochow VB. Loss of bioluminescence in *Anomalops katoptron* due to starvation. Experientia. 1976;32:1175–1176.

[44] Thompson EM, Rees J-F. Origins of luciferins: Ecology of bioluminescence in marine fishes. In: Hochachka PW, Mommsen TP, editors. Biochemistry and Molecular Biology of Fishes. Amsterdam: Elsevier Science; 1995. pp. 435–466.

[45] Cavallaro M, Mammola CL, Verdiglione R. Structural and ultrastructural comparison of photophores of two species of deep-sea fishes: *Argyropelecus hemigymnus* and *Maurolicus muelleri*. Journal of Fish Biology. 2004;64:1552–1567. DOI: 10.1111/j. 905-8649.2004.00410.x.

[46] O'Day WT. Luminescent silhouetting in stomiatoid fishes. Contributions in Science from the Natural History Museum of Los Angeles. 1973;246:1–8.

[47] Haneda Y, Johnson FH. The photogenic organs of *Parapriacanthus beryciformes* franz and other fish with the indirect type of luminescent system. Journal of Morphology. 1962;110(2):187–198.

[48] Denton EJ, Herring PJ, Widder EA, Latz MF, Case JF. The roles of filters in the photophores of oceanic animals and their relation to vision in the oceanic environment. Proceedings of the Royal Society of London B: Biological Sciences. 1985;225:63–97.

[49] Best ACG, Bone Q. On the integument and photophores of the alepocephalid fishes Xenodermichthys and Photostylus. Journal of the Marine Biological Association of the United Kingdom. 1976;56:227–236.

[50] McFall-Ngai M, Morin J. Camouflage by disruptive illumination in leiognathids, a family of shallow-water, bioluminescent fishes. Journal of Experimental Biology. 1991;156:119–138.

[51] Dunlap PV, Nakamura M. Functional morphology of the luminescence system of Siphamia versicolor (Perciformes: Apogonidae), a bacterially luminous coral reef fish. Journal of Morphology. 2011;272:897–909. DOI: 10.1002/jmor.10956.

[52] Haneda Y. The luminescence of some deep-sea fishes of the families Gadidae and Macrouridae. Pacific Science. 1951;5(4):372–378.

[53] McFall-Ngai MJ, Dunlap PV. Three new modes of luminescence in the leiognathid fish Gazza minuta: Discrete projected luminescence, ventral body flash, and buccal luminescence. Marine Biology. 1983;73:227–238.

[54] Claes JM, Mallefet J. Comparative control of luminescence in sharks: New insights from the slendertail lanternshark (*Etmopterus molleri*). Journal of Experimental Marine Biology and Ecology. 2015;467:87–94. DOI: 10.1016/j.jembe.2015.03.008.

[55] Baguet F. Excitation and control of isolated photophores of luminous fishes. Progression in Neurobiology. 1975;5(2):97–125.

[56] Case JF, Strause LG. Neurally controlled luminescent systems. In: Herrin PJ, editor. Bioluminescence in Action. London: Academic Press; 1978. pp. 331–366.

[57] Sazonov YI. Morphology and significance of the luminous organs in alepocephaloid fishes. Biosystematics and Ecology Series. 1996;11:156–163.

[58] Tatsuta N, Imamura H, Nakaya K, Kawai T, Abe T, Sakaoka K, Takagi S, Yabe M. Taxonomy of mesopelagic fishes collected around the Ogasawara islands by the T/S Oshoro-Maru. Memoirs of the Faculty of Fisheries Sciences, Hokkaido University. 2014;56(1):1–64.

[59] Nicol JA. Spectral composition of the light of the lanternfish *Myctophum punctatum*. Journal of Marine Biological Association of the United Kingdom. 1960;40:27–32.

[60] Somiya, H. Bacterial bioluminescence in chlorophthalmid deep-sea fish: a possible interrelation between the light organ and the eyes. Experientia. 1977;33:906–909.

[61] McFall-Ngai MJ, Dunlap PV. External and internal sexual dimorphism in leiognathid fishes: Morphological evidence for sex-specific bioluminescent signaling. Journal of Morphology. 1984;182:71–83.

[62] Morin JG, Harrington A, Nealson K, Badlwin TO, Hastings JW. Light for all reasons: Versatility in behavioural repertoire of the flashlight fish. Science. 1975;190:74–76.

[63] Meyer-Rochow VB. Light of my life – Messages in the dark. Biologist. 2001;48(4):163–167.

[64] Young RE, Roper CFE, Walters JF. Eyes and extraocular photoreception in midwater cephalopods and fishes: Their roles in detecting downwelling light for counterillumination. Marine Biology. 1979;51:381–390.

[65] Claes JM, Partridge JC, Hart NS, Garza-Gisholt E, Ho H-C, Mallefet J, Collin SP. Photon hunting in the twilight zone: Visual features of mesopelagic bioluminescent sharks. PLoS One. 2014;9(8):e104213. DOI: 10.1381/journal.pone.0104213.

[66] Claes JM, Ho H-C, Mallefet J. Control of luminescence from pygmy sharks (*Squaliolus aliae*) photophores. The Journal of Experimental Biology. 2012;215:1691–1699. DOI: 10.1242/jeb.066704.

[67] Claes JM, Dean MN, Nilsson D-E, Hart NS, Mallefet J. A deepwater fish with "lightsabers" – a dorsal spine-associated luminescence in a counterilluminating lanternshark. Scientific Reports. 2013;3:1308. DOI: 10.1039/srep01308.

[68] Wagner H-J, Fröhlich E, Negishi K, Colin SP. The eyes of deep-sea fish II. Functional morphology of the retina. Progress in Retinal and Eye Research. 1998;4:637–685.

[69] Douglas RH, Hunt DM, Bowmaker JK. Spectral sensitivity tuning in the deep-sea. In: Collin SP, Marshall NJ, editors. Sensory Processing in Aquatic Environments. New York: Springer-Verlag; 2003. pp. 323–342.

[70] Ali MA. Retinas of Fishes: An Atlas. Berlin: Springer-Verlag; 1976. 284 p. DOI: 10.1007/978-3-642-66435-9.

[71] Warrant E. Vision in the dimmest habitats on earth. Journal of Comparative Physiology A: Neuroethology, Sensory, Neural and Behavioural Physiology. 2004;190:765–789. DOI: 10.1007/s00359-004-0546-z.

[72] de Busserolles F, Fitzpatrick JL, Marshall NJ, Collin SP. The influence of photoreceptor size and distribution on optical sensitivity in the eyes of lanternfishes (Myctophidae). PLoS One. 2014;9(6):e99957. DOI: 10.1381/journal.pone.0099957.

[73] Fernandez HR, Tsuji FI. Photopigment and spectral sensitivity in the bioluminescent fish *Porichthys notatus*. Marine Biology. 1976;34:101–107.

[74] Meyer-Rochow VB, Barburina V, Smirnov S. Histological observations on the eye of the two luminescent fishes *Photoblepharon palpebratus* (Boddaert) and *Anomalops katoptron* (Blkr). Zoologischer Anzeiger – A Journal of Comparative Zoology. 1982;209(1/2):65–72.

[75] Douglas RH, Partridge JC. On the visual pigments of deep-sea fish. Journal of Fish Biology. 1997;50:68–85.

[76] de Busserolles F, Marshal NJ, Collin SP. Retinal ganglion cell distribution and spatial resolving power in deep-sea lanternfishes (Myctophidae). Brain, Behaviour and Evolution. 2014;84(4):262–276. DOI: 10.1159/000365960.

[77] Miyazaki T, Iwami T, Meyer-Rochow VB . The position of the retinal area centralis changes with age in *Champsocephalus gunnari* (Channichthyidae), a predatory fish from coastal Antarctica. Polar Biology. 2011;34:1117–1123. DOI: 10.1007/s00300-011-0969-2.

[78] Hunt DM, Dulai KS, Partridge JC, Cottrill P, Bowmaker JK. The molecular basis for spectral tuning of rod visual pigments in the deep-sea fish. The Journal of Experimental Biology. 2001;204:3333–3344.

[79] Froese R, Pauly D. Fishbase [Internet]. [Updated: 2015]. Available from: fishbase.org [Accessed: 15/06/2016]

[80] Dunlap PV, Davis KM, Tomiyama S, Fujino M, Fukui A. Developmental and microbiological analysis of the inception of bioluminescent symbiosis in the marine fish *Nuchequula nuchalis* (Perciformes: Leiognathidae). Applied and Environmental Microbiology. 2008;74(24):7471–7481. DOI: 10.1128/AEM.01619-08.

[81] Tsuji FI, Lynch III RV, Haneda Y. Studies on the bioluminescence of the marine ostracod crustacean Cypridina serrata. Biological Bulletin. 1970;139(2):386–401.

[82] Cohen AC, Morin JG. Six new luminescent ostracodes of the genus Vargula (Myodocopida: Cypridinidae) from the San Blas region of Panama. Journal of Crustacean Biology. 1989;9(2):297–340.

[83] Morin JG. Firefleas of the sea: Luminescent signaling in marine ostracode crustaceans. The Florida Entomologist. 1986;69(1):105–121.

[84] Mensinger AF, Case JF. Bioluminescence maintenance in juvenile *Porichthys notatus*. Biological Bulletin. 1991;181:181–188.

[85] Shimomura O, Inoue S, Johnson FH, Haneda Y. Widespread occurrence of coelenterazine in marine bioluminescence. Comparative Biochemistry and Physiology – Part B. Biochemistry & Molecular Biology. 1980;65:135–138.

[86] Dunlap PV, Ast JC, Kimura S, Fukui A, Yoshino T, Endo H. Phylogenetic analysis of host-symbiont specificity and codivergence in bioluminescent symbioses. Cladistics. 2007;23:507–532. DOI: 10.1111/j.1096-0031.2007.00157.x.

[87] Evans BI, Fernald RD. Metamorphosis and fish vision. Journal of Neurobiology. 1990;21(7):1038–1052.

[88] Evans BI, Browman HI. Variation in the development of the fish retina. American Fisheries Society Symposium. 2004;10:146–166.

[89] Claes JM, Mallefet J. Early development of bioluminescence suggests camouflage by counter-illumination in the velvet belly lantern shark *Etmopterus spinax* (Squaloidea: Etmopteridae). Journal of Fish Biology. 2008;73:1337–1350. DOI: 10.1111/j.1095-8649.2008.02006.x.

[90] Mallefet J, Shimomura O. Presence of coelenterazine in mesopelagic fishes from the Strait of Messina. Marine Biology. 1995;124:381–385.

[91] Badcock J, Larcombe RA. The sequence of photophore development in *Xenodermichthys copei*. Journal of the Marine Biological Association of the United Kingdom. 1980;60:277–294.

[92] Claes JM, Mallefet J. Ontogeny of photophores pattern in the velvet belly lantern shark, Etmopterus spinax. Zoology. 2009;112:433–441. DOI: 10.1016/j.zool.2009.02.003.

[93] Mensinger AF, Case JF. Luminescent properties of deep-sea fish. Journal of Experimental Marine Biology and Ecology. 1990;144:1–15.

[94] Munz FW, McFarland WN. Evolutionary adaptations of fishes to the photic environment. In: Crescitelli F, editor. Handbook of Sensory Physiology, Volume VII/5.The Visual System in Vertebrates. Berlin: Springer-Verlag; 1977. pp. 193–274. DOI: 10.1007/978-3-642-66468-7

[95] Merret NR, Badcock J, Herring PJ. The status of Benthalbella infans (Pisces: Myctophoidei), its development, bioluminescence, general biology and distribution in the eastern North Atlantic. Journal of Zoology, London. 1973;170:1–48.

[96] Carpenter KE, Niem VH, editors. FAO Species Identification Guide for Fishery Purposes. The Living Marine Resources of the Western Central Pacific. Volume 5: Bony Fishes Part 3 (Menidae to Pomacentridae). Rome: Food and Agriculture Organization of the United Nations; 2001. pp. 2791–3390 .

[97] Fernandez I, Devaraj M. Dynamics of the gold-spotted grenadier anchovy (*Coilia dussumieri*) stock along the northwest coast of India. Indian Journal of Fisheries. 1996;43(1):27–39.

[98] Roe HSJ, Badcock J. The diel migrations and distributions within a mesopelagic community in the North East Atlantic. 5. Vertical migrations and feeding of fish. Progress in Oceanography. 1984;13:399–424.

[99] Badcock J. The vertical distribution of mesopelagic fishes collected on the SOND Cruise. Journal of the Marine Biological Association in the United Kingdom. 1970;50:1001–1044.

[100] Poulsen JY. A new species of pencil smelt *Nansenia boreacrassicauda* (Microstomatidae, Argentiniformes) from the North Atlantic Ocean. Zootaxa. 2015;4030(3):517–532. DOI: 10.11646/zootaxa.4020.3.6.

[101] Crane Jr JM. Bioluminescence in the batfish *Dibranchus atlanticus*. Copeia. 1968;2:410–411.

[102] Carpenter KE, Niem VH, editors. FAO Species Identification Guide for Fishery Purposes. The Living Marine Resources of the Western Central Pacific. Volume 2: Cephalopods, crustaceans, holothurians and sharks. Rome: Food and Agriculture Organization of the United Nations; 1998. pp. 687–1406.

[103] Carpenter KE, Niem VH, editors. FAO Species Identification Guide for Fishery Purposes. The Living Marine Resources of the Western Central Pacific. Volume 6: Bony

Fishes Part 4 (Labridae to Latimeriidae), Estuarine Crocodiles, Sea Turtles, Sea Snakes and Marine Mammals. Rome: Food and Agriculture Organization of the United Nations; 2001. pp. 3391–4218 p.

[104] Carpenter KE, editor. FAO Species Identification Guide for Fishery Purposes. The Living Marine Resources of the Western Central Atlantic. Volume 2: Bony Fishes Part 1 (Acipenseridae to Grammatidae). Rome: Food and Agriculture Organization of the United Nation and American Society of Ichthyologists and Herpetologists. Special Publication Number 5; 2002. pp. 601–1374.

[105] Carpenter KE, editor. FAO Species Identification Guide for Fishery Purposes. The Living Marine Resources of the Western Central Atlantic. Volume 3: Bony Fishes Part 2 (Opistognathidae to Molidae), sea turtles and marine mammals. Rome: Food and Agriculture Organization of the United Nation and American Society of Ichthyologists and Herpetologists. Special Publication Number 5; 2002. pp. 1375–2127 .

[106] Sasaki K. Phylogeny of the family Sciaenidae, with notes on its zoogeography (Teleostei, Perciformes). Memoirs of the Faculty of Fisheries Hokkaido University. 1989;36(1–2):1–138.

Chapter 13

Bioluminescence Microscopy: Design and Applications

Hirobumi Suzuki, May-Maw-Thet,
Yoko Hatta-Ohashi, Ryutaro Akiyoshi and
Taro Hayashi

Additional information is available at the end of the chapter

http://dx.doi.org/10.5772/65048

Abstract

Bioluminescence imaging by microscopy is performed using an ultra-low-light imaging camera. Although imaging devices such as sensor and camera have been greatly improved over time, such improvements have not been attained commercially which are available for microscopes now. We previously optimized the optical system of a microscope for bioluminescence imaging using a short-focal-length imaging lens and evaluated this system with a conventional color charge-coupled device camera. Here, we describe the concept of bioluminescence microscope design using a short-focal-length imaging lens and some representative applications, including intracellular calcium imaging, imaging of clock gene promoter assays, and three-dimensional reconstruction of *Drosophila* larva. This system facilitates the acquisition of bioluminescence images of single live cells using luciferase, which is similar to fluorescence microscopy using a fluorescent protein.

Keywords: bioluminescence microscopy, short-focal-length imaging lens, single live-cell analysis, three-dimensional imaging

1. Introduction

Aequorin is a calcium-specific light-emitting protein extracted from the jelly fish *Aequorea* [1]. Additionally, aequorin-injected eggs of the medaka (*Oryzias latipes*), a fresh water fish, showed a dramatic increase in free calcium during fertilization, as determined by measuring light from the eggs using a photomultiplier tube [2]. Notably, bioluminescence microscopy with an image intensifying system using a vidicon camera was performed in 1978 to show the spatial distribution of the free calcium in the egg [3]. This system revealed a spreading wave of high free calcium (calcium wave) during fertilization from the animal pole, as discussed with the fertilization wave of cortical changes in eggs observed by light microscopy [4]. Although the potential of low-light imaging has been recognized in physiology and developmental biology, this technique was not commonly used at that time due to a lack of commercially available instrumentation. Later, advances in detector and digital imaging processing systems facilitated the commercial production of appropriate instrumentation and made it possible for low-light imaging to be carried out using a silicon-intensifier target (SIT) tube camera or a high-sensitivity cooled charge-coupled device (CCD) camera.

Since the cloning of firefly luciferase in the late 1980s, luciferase has been used as a reporter enzyme to assay the activity of a particular gene promoter using the photon-counting luminometer method [5–7]. Additionally, bioluminescence microscopy of promoter activity in single cells has been performed using ultra-low-light imaging cameras, such as liquid nitrogen-cooled CCD cameras, photon-counting CCD cameras, or image-intensifying CCD cameras [8–16]. However, temporal and spatial resolution was not enough for the observation of cellular biological events and for the detection of single cells compared with that of conventional CCD cameras. Therefore, satisfactory analysis has not been achieved at the single-cell level by bioluminescence microscopy.

Recently, electron-multiplying CCD (EM-CCD) camera, which yields higher sensitivity and image quality, was commercially released and subsequently used for bioluminescence microscopy [17–19]. Although the image sensor of ultra-low-light imaging cameras has been greatly improved over time, such improvements have not been made commercially available for microscopes.

In our previous studies, we optimized an optical system using a short-focal-length imaging lens for bioluminescence microscopy and performed bioluminescence imaging of single live cells expressing the luciferase gene using a conventional CCD camera [20, 21]. This system is commercially available now and has been widely used for gene expression analysis in chronobiology [22–28], neurobiology [29, 30], developmental biology [31], medical research [32–35], signal transduction analysis [36–38], molecular interaction [39–41], and radiation biology [42, 43]. Accordingly, in this study, we describe the concept of bioluminescence microscopy adopting a short-focal-length imaging lens and present several representative applications, including a three-dimensional analysis, to demonstrate the advantages of the short-focal-length imaging lens system.

2. Microscope design

Bioluminescence microscopy is based on the detection of light emitted by living cells expressing a luciferase gene or other luminescence-related gene. Conventional microscopes are inefficient at transmitting light from the sample to the detector, necessitating long exposure times. We designed a new type of microscope for ultra-low-light imaging based on modifications to the imaging lens, vignetting, and effective field area.

http://dx.doi.org/10.5772/65048

2.1. Imaging lens

Figure 1 shows a diagram of an inverted microscope equipped with an infinity-corrected optical system. Light from a sample is collected by an objective lens (OB), and the sample image is created by an imaging lens (tube lens; IM) on a CCD chip. Generally, the degree of brightness (I) of an image is directly proportional to the square of the numerical aperture (NA) of the OB and is inversely proportional to the square of magnification (M) of the image, namely as $I \propto (NA/M)^2$. Therefore, a higher NA and lower M yield much brighter images. However, it is difficult to obtain both conditions. Because higher NA OB yields higher M (shorter focal length), or lower M OB yields lower NA. Thus, high NA and low M are mutual trade-offs. On the other hand, the value of NA/M is the same as the NA of the IM, geometrically denoted as NA′. Therefore, a microscope with a high NA′ (short-focal-length imaging lens) makes it possible to achieve a higher NA and lower M without further improvement of the objective lens. This was the concept on which we based the design of the bioluminescence microscope [20, 21].

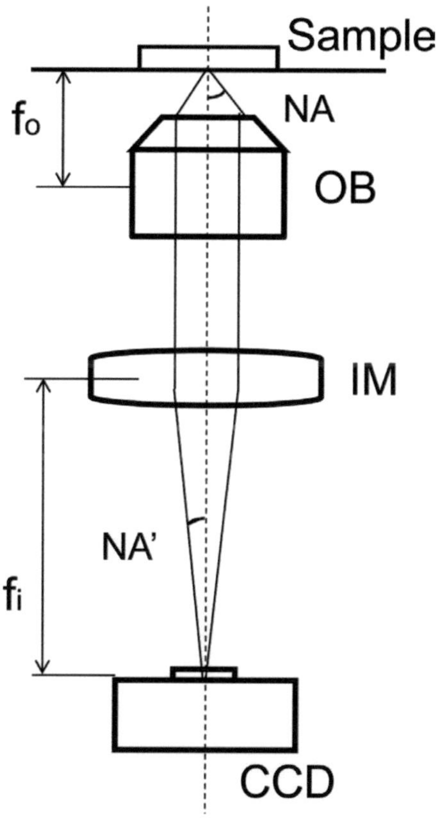

Figure 1. A diagram of an inverted microscope with an infinity-corrected optical system. OB, objective lens; IM, imaging lens; NA, numerical aperture of the objective lens; NA′, numerical aperture of the imaging lens; F_o, focal distance of the objective lens; F_i, focal distance of the imaging lens.

Figure 2 shows one example of the I condition to capture bioluminescence images of live cells using a microscope. The X-axis indicates the I value of the *in vivo* macro-imaging system, OV100 (Olympus, Tokyo, Japan), using a 0.8× objective lens with an NA varying from 0.05 to 0.25 (I = 0.004–0.098). HeLa cells transiently expressing the Luc+ luciferase gene (pGL3 control vector; Promega, Madison, WI, USA) in Hanks' balanced salt solution (HBSS; Invitrogen, Carlsbad, CA, USA) containing 1 mM D-luciferin, potassium salt (Promega) at room temperature were imaged using a CCD camera (ST-7; SBIG, Ottawa, Canada) for astronomical imaging. The exposure time was 1 min, and the cooling temperature was −20°C. The Y-axis indicates the normalized luminescence intensity of the entire area of the image captured (**Figure 2A**). As shown in the graph (**Figure 2B**), luminescence images could be captured at I values of greater than 0.02, although the M was lower for single-cell imaging. Therefore, we designed an IM to achieve an I value of 0.02 with a higher M using a conventional OB.

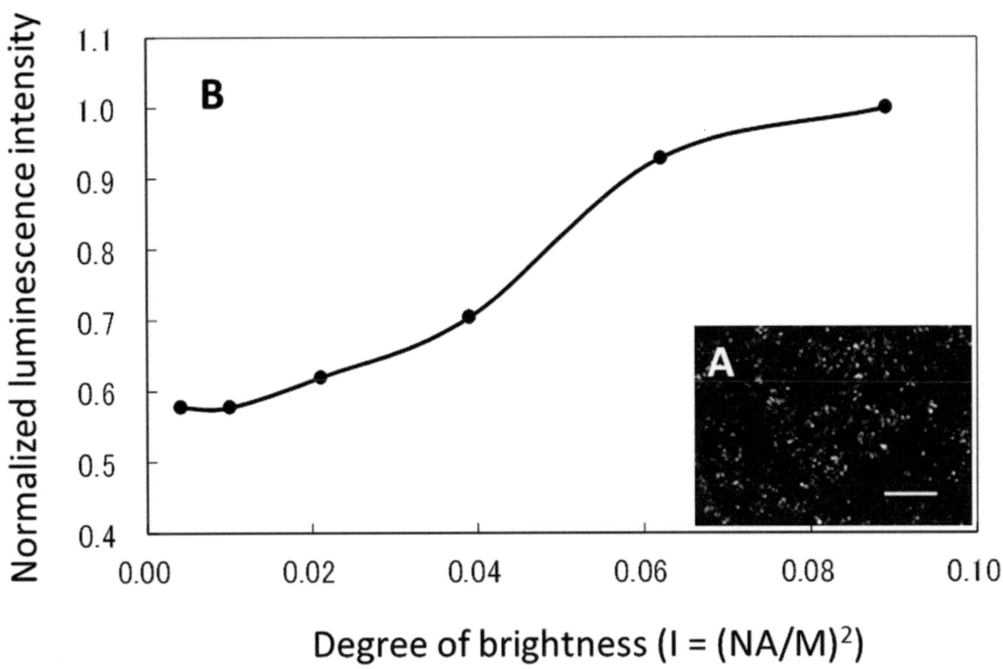

Figure 2. The I condition to capture bioluminescence images of HeLa cells using an OV100 *in vivo* macro-imaging system with a 0.8× objective lens and an NA ranging from 0.05 to 0.25 (I = 0.004 to 0.098). (A) HeLa cells expressing the Luc+ luciferase gene in HBSS containing 1 mM D-luciferin. Scale bars, 1000 μm and (B) normalized luminescence intensity of the image (A) against degree of brightness I.

2.2. Vignetting

Figure 3 shows diagrams of light passing from an object in an infinity-corrected optical system. In this system, light from object runs parallel between the OB and IM (**Figure 3A** and **B**). Therefore, this system is suitable for several observations because several optical elements (such as mirror units for fluorescence observations or polarizing filters) can be inserted

http://dx.doi.org/10.5772/65048

between the OB and IM without light-pass correction for image formation. However, light from peripheral vision (ray "a") is vignetted by the imaging lens, when the distance between the OB and IM becomes longer (**Figure 3C**). Therefore, vignetting can be avoided by shortening the distance between the OB and IM.

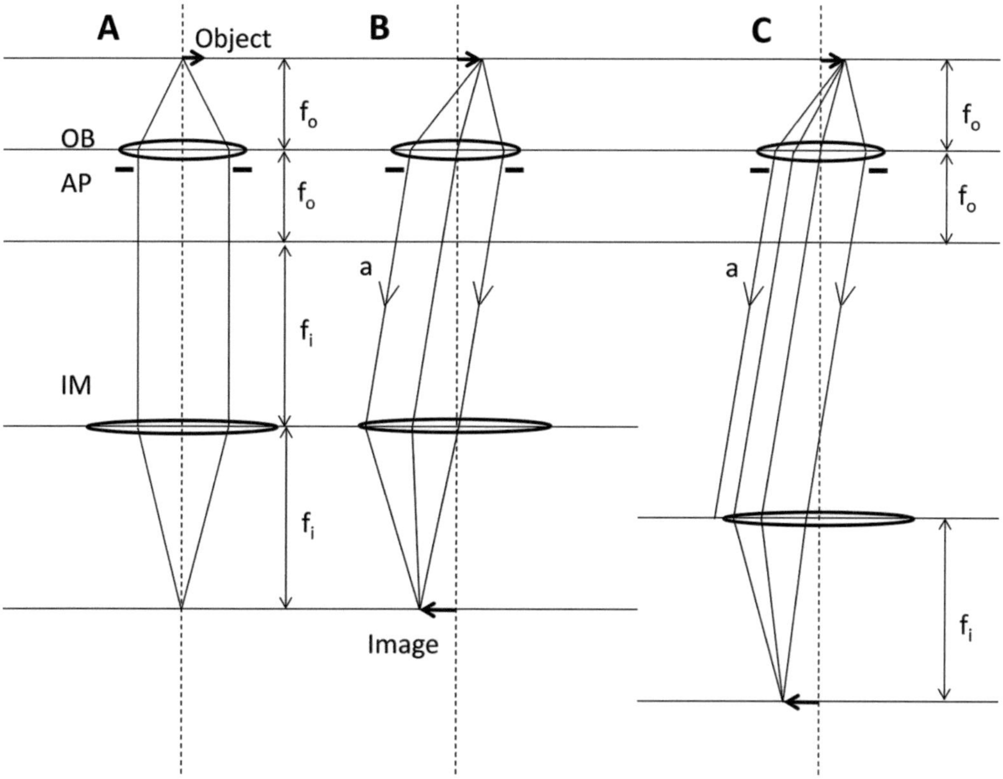

Figure 3. Diagrams of light passing from an object in an infinity-corrected optical system, illustrating the vignetting of light from peripheral vision by the imaging lens. (A) Light passing through the central axis. Light flux was restricted by the aperture of the objective lens; (B) light passing through peripheral vision; and (C) light passing through peripheral vision when the distance between the objective and the imaging lens was greater than that in B. Light from peripheral vision (ray "a") was vignetted by the imaging lens. OB, objective lens; AP, aperture; IM, imaging lens; f_o, focal distance of the objective lens; f_i, focal distance of the imaging lens.

2.3. Effective field area

Figure 4 shows an effective field diagram on a CCD chip. Generally, the area of the CCD chip is smaller than the effective field area because peripheral vision is affected by several optical aberrations. Images on the CCD chip only show part of the light collected by the OB. If all light is collected on the CCD chip as an image by reducing the magnification using an intermediate tube lens or modified IM, the light intensity of each pixel becomes greater. Thus, image quality is sacrificed to obtain brighter images.

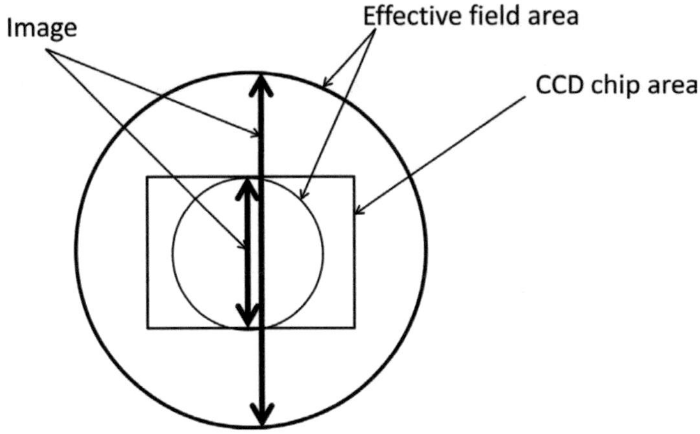

Figure 4. An effective field diagram on a CCD chip. Generally, the area of a CCD chip is smaller than the effective field area, and an image on a CCD chip is only some of light collected by the objective lens. If all light is collected on the CCD chip as an image by reducing the magnification using an intermediate tube lens or modified imaging lens, the light intensity of a pixel of the CCD chip increases.

2.4. The bioluminescence microscope, LV200

Figure 5 shows the inverted bioluminescence microscope used in our studies (Luminoview LV200; Olympus, Tokyo, Japan). A halogen lamp was used as the source of transmitted bright-field light. The light was directed to a sample through a condenser lens with a glass fiber. A short focal-length imaging lens (f = 36 mm, NA = 0.2) was customized based on the condition of I to capture dim bioluminescence images in this system. Normal OBs are available for observation. Using the IM, total magnification was reduced to one-fifth of the magnification of the OB because the focal distance of the IM is fixed (180 mm; Olympus) in a conventional microscope body. The distance between the OB and IM was set at 17 mm to avoid vignetting. A stage-top incubator with temperature and CO_2 gas controllers (MI-IBC-IF; Tokai Hit Co., Shizuoka, Japan) was added to the sample stage. The observation area was covered with a dark box [21, 44].

To evaluate the performance of LV200, bioluminescence images of U2OS cell lines stably expressing CBG99, CBR, and Luc2 beetle luciferase (Promega) were captured using an UPlanFLN 40× oil objective lens (NA = 1.30, I = 0.026) and DP70 color CCD camera (Olympus). Cells were cultured on 35-mm glass-bottomed dishes in Dulbecco's modified Eagle's medium (DMEM; Invitrogen) containing 10% fetal bovine serum and 1 mM beetle D-luciferin at 37°C with 5% CO_2. Binning of the CCD camera was 1 × 1 (1360 × 1024 pixels), International Organization for Standardization gain was 1600, and exposure time was 2 min. **Figure 6A** shows the bioluminescence images of cells expressing CBG99, CBR, and Luc2, captured within 2 min using a conventional color CCD camera as green, red, and orange color, respectively. Notably, bioluminescence images could not be captured under the same conditions (stable cell lines, OB, CCD camera, and exposure time) using a conventional inverted microscope (IX70; Olympus; NA = 1.30, I = 0.001), although a 10-min exposure time was required to obtain images

http://dx.doi.org/10.5772/65048

for the beetle luciferase-expressing cell line (**Figure 6B**) [21]. Despite the use of blank image subtraction, 10 min is the upper limit of exposure time for the DP70 color CCD camera due to intense background evaluation [21]. Thus, bioluminescence images of cells expressing the luciferase gene can be captured using an LV200 microscope with a 40× OB and color CCD camera. In this case, the M of the image was reduced by a power of 8 owing to the short-focal-length IM, and the I value was 0.026. To equalize the I value between the LV200 and IX70 microscopes, a low M and high NA OB (e.g., 8×, NA 1.3) is required for IX70. However, an OB with such a high NA cannot be purchased commercially.

Figure 5. Bioluminescence microscope, LV200. A stage-top incubator with temperature and CO_2 gas controllers was added to the sample stage. The observation area was covered with a dark box.

To show the spatial resolution of the bioluminescence images acquired using LV200, organelle-targeted images were captured using an UPlanFLN 100× oil objective lens (Olympus) and ImagEM EM-CCD camera (C9100-13; Hamamatsu Photonics, Shizuoka, Japan). The binning of the EM-CCD camera was 1 × 1 (512 × 512 pixels), EM-gain was 1024, and exposure time was 300 ms to 1 s. NanoLuc luciferase (Promega), which is 150-fold brighter than beetle luciferase [37], was used as a tag for organelle localization, similar to a fluorescent protein. Before substrate addition (12.5 μM furimazine; Promega), cells were washed with culture medium three times.

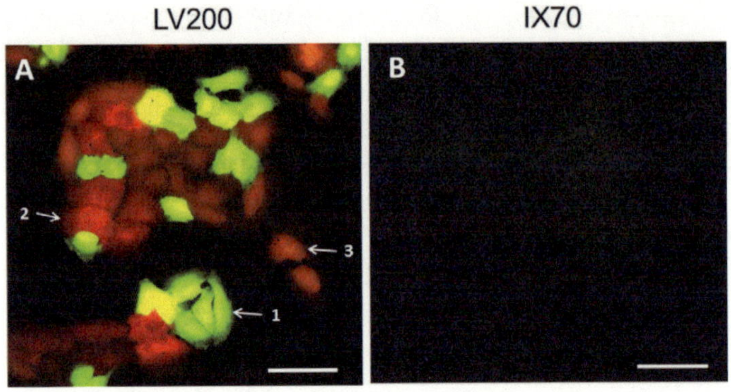

Figure 6. Bioluminescence images of U2OS cells expressing beetle luciferase CBG99 (arrow 1), CBR (arrow 2), and Luc2 (arrow 3) at 37°C captured by LV200 and IX70 microscopes with a UPlanFLN 40× oil objective lens and DP70 color CCD camera. The exposure times were 2 and 10 min for LV200 (M = 8, I = 0.026) and IX70 (M = 40, I = 0.001), respectively. D-Luciferin, 1 mM. Scale bars, 100 μm (A) and 20 μm (B). This figure was quoted and modified from Ref. [21] with Wiley's open access terms and conditions.

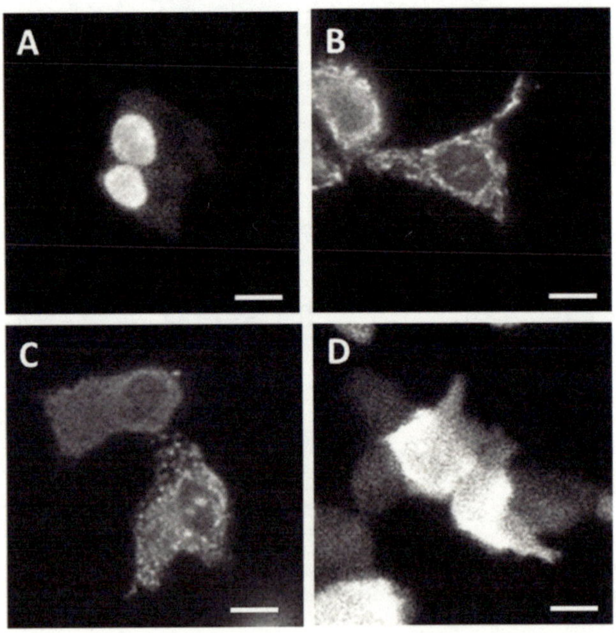

Figure 7. Bioluminescence images of NanoLuc fused with NLS (A), CoxVIII (B), calreticulin (C), and no targeting sequence (D) in U2OS cells at 37°C. Images were captured using an LV200 microscope with an UPlanFLN 100× oil objective lens and an ImagEM EM-CCD camera. Exposure time, 300 ms (A, D), 500 ms (B), and 1 s (C). Furimazine, 12.5 μM. Scale bars, 20 μm. This figure was quoted from Ref. [21] with Wiley's open access terms and conditions.

Figure 7 shows bioluminescence images of NanoLuc fused with nuclear localization sequence (NLS) (**Figure 7A**), mitochondrial targeting sequence (subunit VIII of human cytochrome C oxidase, CoxVIII) (**Figure 7B**), endoplasmic reticulum resident protein, calreticulin with KDEL retrieval sequence (**Figure 7C**), or no targeting sequence (**Figure 7D**) in U2OS cells [21]. The NanoLuc-NLS accumulated in the nucleus of the cell, and the CoxVIII-NanoLuc and calreticulin-NanoLuc-KDEL appeared in a meshwork pattern in the cytoplasm. Thus, the nucleus and cytoplasm were discriminated clearly, and mitochondria and endoplasmic reticulum were recognized in the cytoplasm.

3. Applications

As examples of bioluminescence microscopy using our system (LV200), we introduce three applications: (1) calcium imaging of single cells, (2) imaging of clock gene promoter assays, and (3) three-dimensional imaging of *Drosophila* larva.

3.1. Intracellular Ca^{2+} imaging using obelin

Obelin is a calcium-specific bioluminescent protein similar to aequorin; using obelin, intracellular calcium was imaged by ATP and ionomycine (A23187) stimulation for calcium release from intracellular membranes (mitochondria and endoplasmic reticulum) and inflow from outside of the cell, respectively. The apoobelin gene [45] was inserted into the mammalian expression vector, pCDNA3.1 (Invitrogen), and transfected into HeLa cells. HeLa cells transiently expressing apoobelin were incubated in DMEM containing 60 μM coelenterazine (Promega) for 4 h to reconstitute obelin. The cells were stimulated with 500 μM ATP, and bioluminescence images were captured using an LV200 microscope with an UplanApo 20× OB (NA = 0.70; Olympus) and an iXon EM-CCD camera (DU-8971; Andor Technology, Belfast, UK). Binning of the CCD was 2 × 2, EM gain was maximum, and the exposure time was 25 s with a 30-s interval. The cells were restimulated by 10 mM ionomycine at 20 min after ATP stimulation.

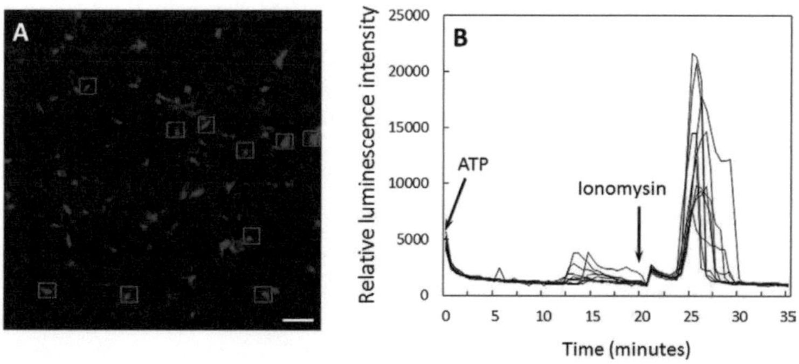

Figure 8. Bioluminescence image of intracellular calcium levels in HeLa cells using the photoprotein obelin (A). Time course of light intensity in single cells (B). Images were captured using an LV200 microscope with an UPlanApo 20× objective lens and an ImagEM EM-CCD camera. Cells were stimulated with 500 μM ATP and 1 mM ionomycine. Coelenterazine, 60 μM. Exposure time, 25 s. Scale bar, 200 μm.

Figure 8A shows pseudocolor-coded bioluminescence images of intracellular calcium in HeLa cells at 8 min after ionomysin stimulation. **Figure 8B** shows a time course of the intracellular calcium response for 10 single cells using time-laps image analysis software TiLIA [46]. Calcium responses in each cell varied temporally, were broad in intensity at around 15 min after ATP stimulation, and were uniform and greater in intensity after ionomysin stimulation [47, 48]. Using this imaging system, ATP-induced calcium oscillation in HEK-293 cells was confirmed using a bioluminescent calcium sensor constructed by aequorin and GFP with 1 s exposure time using a bioluminescence resonance energy transfer (BRET) system [36].

3.2. Imaging of clock gene promoter assays

The circadian rhythm is monitored by measuring the promoter activity of clock genes from individual cells as a cellular clock. However, it is impossible to resolve whether loss of circadian rhythm following stimulation is caused by dis-periodicity or dis-synchronicity in individual cells using a luminometer because the luminometer captures total luminescence from the cell population. Bioluminescence microscopy can provide clear single-cell analyses of promoter activity.

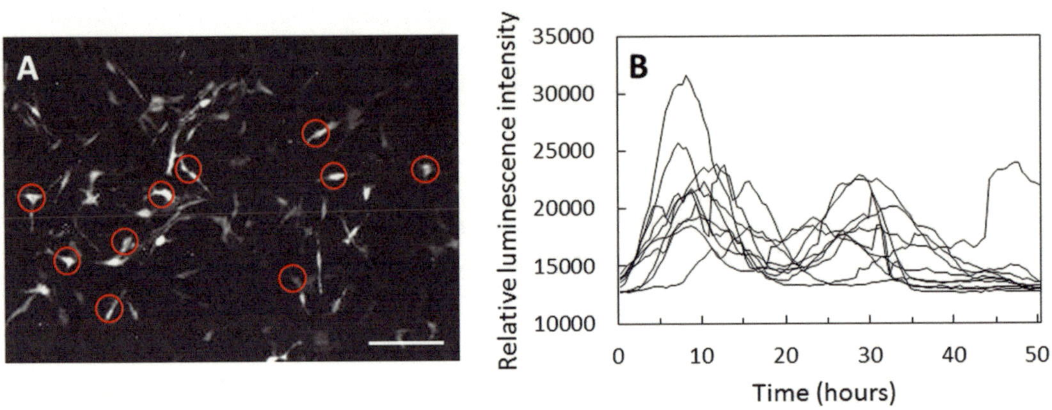

Figure 9. Bioluminescence image of *Per2* gene expression in NIH 3T3 cells using a *Luc+* luciferase promoter vector (A). Time course of light intensity of the single cells (B). Images were captured using an LV200 microscope with an UPlanApo 20× objective lens and a DP30 CCD camera. Cells were stimulated with 1 mM dexamethasone. D-Luciferin, 500 μM. Exposure time, 5 min. Scale bar, 200 μm.

Figure 9 shows an example of imaging of *Per2* clock gene promoter activity in cultured cells. The promoter region of the *Per2* clock gene in mice was inserted into the luciferase promoter vector, pGL3 (Promega), and the vector was transfected into NIH3T3 cells. Cells were cultured in CO_2-independent DMEM (Invitrogen) containing 500 μM D-luciferin. Bioluminescence images were captured using an LV200 microscope with an UPlanApo 20× OB (NA = 0.70) and DP30BW CCD camera (Olympus) at 37°C. The binning of the CCD was 1 × 1 (1024 × 1024 pixels), the cooling temperature of the sensor chip was 5°C, and the exposure time was 5 min with 30-min interval for 28 h. As shown in **Figure 9A**, bioluminescence images of single cells expressing the *Per2* gene were captured clearly using a conventional CCD camera. **Figure 9B**

shows a time course of *Per2* promoter activity in 10 cells selected appropriately for 48 h using TiLIA [46]; this time course allowed us to analyze synchronicity among cells [49]. Using this imaging system, Ukai et al. [23] produced photoresponsive mammalian cells by introducing the photoreceptor melanopsin and monitored the effects of photoperturbation on the state of the cellular clock. They observed that a critical light pulse drove cellular clocks into singularity behavior and proved that loss of the circadian rhythm of a cellular clock may be caused by dis-synchronization of individual cells underlying singularity behavior by single-cell analysis.

3.3. Three-dimensional imaging of *Drosophila larva*

Because our bioluminescence microscope system utilizes a short-focal-length imaging lens, the magnification is lower and the focal depth is shallower than those of conventional microscopy systems using the same OB. Therefore, depth of field is also shallower. This is convenient for three-dimensional image reconstruction by light sectioning.

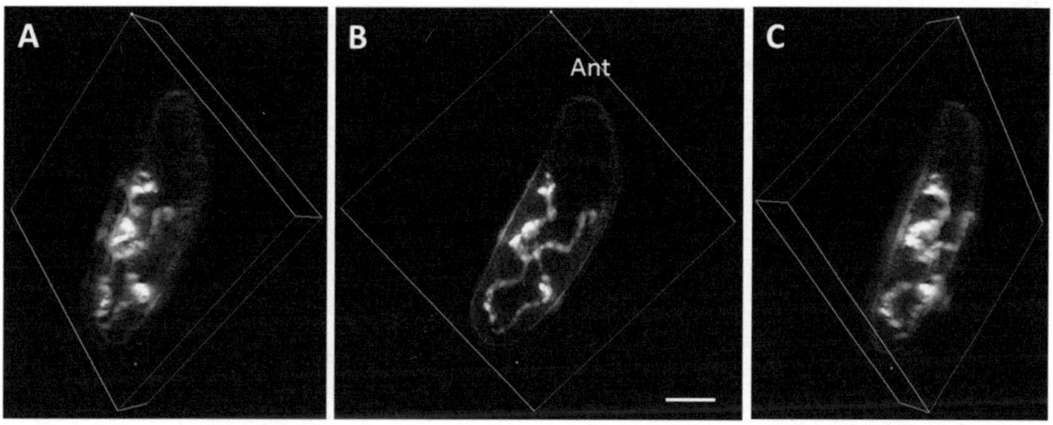

Figure 10. Three-dimensional bioluminescence image of armadillo promoter activity of insular larva for the transgenic *Drosophila melanogaster* reconstructs from 11 sectionalized images (100 μm depth). Expression of *armadillo* was observed in the midgut from a tissue depth of 100 μm. The larva was immersed in 3 mM D-luciferin for 5 min before image acquisition. A and C, slant view. B, front view. Scale bar, 100 μm. Ant, anterior part of the body.

Accordingly, we constructed transgenic *Drosophila melanogaster* carrying an armadillo (a member of the segment polarity gene) promoter and luciferase fusion gene [31]. **Figure 10** shows three-dimensional images of armadillo promoter activity from insular larva of transgenic flies reconstructed from 11 sectionalized images (front and slant views). The larva was immersed in 3 mM D-luciferin for 5 min before image acquisition. Bioluminescence images of the anesthetized larva with cold treatment were captured using an LV200 microscope with an UPlanFLN 60× OB (NA = 0.90) and iXon EM-CCD camera. The binning of the CCD was 1 × 1; EM gain was maximum; and exposure time was 30 s. Eleven sectionalized images were obtained with 10-μm steps from top to bottom (0–100 μm) of the larva. After deconvolution, three-dimensional images were reconstructed using CelSens Dimension image analysis software (Olympus). As shown in the **Figure 10**, the expression of the *armadillo* was observed in the midgut from a tissue depth of 100 μm, although it needed clearing treatment of kidney tissue for imaging of 100–200 mm depth by confocal fluorescence microscopy [50].

4. Conclusion

In this study, we presented the concept of bioluminescence microscopy using a short-focal-length IM. This system facilitates the acquisition of bioluminescence images of single live cells using luciferase, similar to fluorescence microscopy using a fluorescent protein, although M is lower than that of conventional microscopy. Furthermore, this method is applicable for studies of cellular activity at the single cell level, including analyses of signal transduction, gene expression, and embryogenesis.

As bioluminescence microscopy requires no excitation light, it leads to substantive differences from fluorescence microscopy. Bioluminescence observation lacks the phototoxicity and background autofluorescence problems associated with fluorescence observation and permits the long-term, non-lethal observation of living cells such as embryonic stem cells, iPS cells, and embryos. **Table 1** summarized the substantive differences between the fluorescence and bioluminescence microscopy. Thus, bioluminescence microscopy is a powerful tool in cellular biology and complements fluorescence microscopy.

	Fluorescence	**Bioluminescence**
Excitation energy	Photon	Chemical reaction
Auto-fluorescence	Affected	None
Phototoxicity	Affected	None
Long-term observation	Acceptable	Excellent
Observation of photosensitive cell	Acceptable	Excellent
Brightness of image	Excellent	Acceptable
Spatiotemporal resolution	Excellent	Acceptable

Table 1. Substantive differences between fluorescence and bioluminescence microscopy.

Acknowledgements

We thank Drs. ES Vysotski (Russian Academy of Science) and Y Nakajima (National Institute of Advanced Industrial Science and Technology, Dr. Ohmiya's Lab) for providing plasmid vectors, apoobelin, and the Per2 reporter. We also thank our colleagues at Olympus Corporation for technical assistance and discussion during the development of bioluminescence microscope, LV200.

http://dx.doi.org/10.5772/65048

Author details

Hirobumi Suzuki[1*], May-Maw-Thet[1,2], Yoko Hatta-Ohashi[1], Ryutaro Akiyoshi[1] and Taro Hayashi[1]

*Address all correspondence to: hirobumi2_suzuki@ot.olympus.co.jp

1 Evaluation Technology Department 1, Olympus Corporation, Hachioji, Tokyo, Japan

2 Department of Experimental Animal Model for Human Disease, Graduate School of Medical and Dental Sciences, Tokyo Medical and Dental University, Tokyo, Japan

References

[1] Shimomura O, Johnson FH, Saiga Y. Extraction, purification and properties of aequorin, a bioluminescent protein from hydromedusan *Aequorea*. J. Cell. Comp. Physiol. 1962; 59:223–239.

[2] Ridgway EB, Gilkey JC, Jaffe LF. Free calcium increases explosively in activating medaka eggs. Proc. Nat. Acad. Sci. U. S. A. 1977; 74:623–627.

[3] Gilkey JC, Jaffe LF, Ridgway EB, Reynolds GT. A free calcium wave traverses the activating egg of the medaka, *Oryzias latipes*. J. Cell Biol. 1978; 76:448–466.

[4] Yamamoto T. Physiology of fertilization in fish eggs. Int. Rev. Cytol. 1961; 12:361–405.

[5] de Wet JR, Wood KV, DeLuca M, Helinski DR, Subramani S. Firefly luciferase gene: structure and expression in mammalian cells. Mol. Cell Biol. 1987; 7:725–735.

[6] Brasier AR, Tate JE, Habener JF. Optimized use of the firefly luciferase assay as a reporter gene in mammalian cell lines. BioTechniques. 1989; 7:1116–1122.

[7] Alam J, Cook JL. Reporter genes: application to the study of mammalian gene transcription. Anal. Biochem. 1990; 188:245–254.

[8] Frawley LS, Faught WJ, Nicholson J, Moomaw B. Real time measurement of gene expression in living endocrine cells. Endocrinology. 1994; 135:468–471.

[9] Thompson EM, Adenot P, Tsuji FI, Renard JP. Real time imaging of transcriptional activity in live mouse preimplantation embryos using a secreted luciferase. Proc. Natl. Acad. Sci. U. S. A. 1995; 92:1317–1321.

[10] White MRH, Masuko M, Amet L, Elliott G, Braddock M, Kingsman AJ, Kingsman SM. Real-time analysis of the transcriptional regulation of HIV and hCMV promoters in single mammalian cells. J. Cell Sci. 1995; 108:441–455.

[11] Castaño JP, Kineman RD, Frawley LS. Dynamic monitoring and quantification of gene expression in single, living cells: a molecular basis for secretory cell heterogeneity. Mol. Endocrinol. 1965; 10:599–605.

[12] Kennedy HJ, Viollet B, Rafiq I, Kahn A, Rutter GA. Upstream stimulatory factor-2 (USF2) activity is required for glucose stimulation of L-pyruvate kinase promoter activity in single living islet β-cells. J. Biol. Chem. 1997; 272:20636–20640.

[13] Takasuka N, White MRH, Wood CD, Robertson WR, Davis JRE. Dynamic changes in prolactin promoter activation in individual living lactotrophic cells. Endocrinology. 1998; 139:1361–1368.

[14] Maire E, Lelièvre E, Brau D, Lyons A, Woodward M, Fafeur V, Vandenbunder B. Development of an ultralow–light–level luminescence image analysis system for dynamic measurements of transcriptional activity in living and migrating cells. Anal. Biochem. 2000; 280:118–127.

[15] Welsh DK, Yoo SH, Liu AC, Takahashi JS, Kay SA. Bioluminescence imaging of individual fibroblasts reveals persistent, independently phased circadian rhythms of clock gene expression. Curr. Biol. 2004; 14:2289–2295.

[16] Masamizu Y, Ohtsuka T, Takashima Y, Nagahara H, Takenaka Y, Yoshikawa K, Okamura H, Kageyama R. Real-time imaging of the somite segmentation clock: revelation of unstable oscillators in the individual presomitic mesoderm cells. Proc. Natl. Acad. Sci. U. S. A. 2006; 103:1313–1318.

[17] Hoshino H, Nakajima Y, Ohmiya Y. Luciferase-YFP fusion tag with enhanced emission for single-cell luminescence imaging. Nat. Methods. 2007; 4:637–639.

[18] Kwon HJ, Enomoto T, Shimogawara M, Yasuda K, Nakajima Y, Ohmiya Y. Bioluminescence imaging of dual gene expression at the single-cell level. BioTechniques. 2010; 48:460–462.

[19] Suzuki T, Kondo C, Kanamori T, Inouye S. Video rate bioluminescence imaging of secretory proteins in living cells: localization, secretory frequency, and quantification. Anal. Biochem. 2011; 415:182–189.

[20] Suzuki H, Dosaka S, Ohashi-Hatta Y, Sugiyama T. Luminescence microscope for reporter assay of single live cells. In: Proceedings of the 14th International Symposium on Bioluminescence and Chemiluminescence Chemistry, Biology, and Applications; 15–19 Oct 2006; San Deago. Singapore: World Scientific; 2007. pp. 53–56.

[21] Ogoh K, Akiyoshi R, May-Maw-Thet, Sugiyama T, Dosaka S, Hatta-Ohashi Y, Suzuki H. Bioluminescence microscopy using a short focal-length imaging lens. J. Microsc. 2014; 253:191–197.

[22] Sato TK, Yamada RG, Ukai H, Baggs JE, Miraglia LJ, Kobayashi TJ, Welsh DK, Kay SA, Ueda HR, Hogenesch JB. Feedback repression is required for mammalian circadian clock function. Nat. Genet. 2006; 38:312–319.

[23] Ukai H, Kobayashi TJ, Nagano M, Masumoto K, Sujino M, Kondo T, Yagita K, Shigeyoshi Y, Ueda HR. Melanopsin-dependent photo-perturbation reveals desynchronization underlying the singularity of mammalian circadian clocks. Nat. Cell Biol. 2007; 9:1327–1334.

[24] Akashi M, Hayasaka N, Yamazaki S, Node K. Mitogen-activated protein kinase is a functional component of the autonomous circadian system in the suprachiasmatic nucleus. J. Neurosci. 2008; 28:4619–4623.

[25] Dibner C, Sage D, Unser M, Bauer C, d'Eysmond T, Naef F, Schibler U. Circadian gene expression is resilient to large fluctuations in overall transcription rates. EMBO J. 2008; 28:123–134.

[26] Fukuda H, Tokuda I, Hashimoto S, Hayasaka N. Quantitative analysis of phase wave of gene expression in the mammalian central circadian clock network. PLoS One. 2011; 6:e23568, 1–8.

[27] Yagita K, Horie K, Koinuma S, Nakamura W, Yamanaka I, Urasaki A, Shigeyoshi Y, Kawakami K, Shimada S, Takeda J, Uchiyama Y. Development of the circadian oscillator during differentiation of mouse embryonic stem cells in vitro. Proc. Natl. Acad. Sci. U. S. A. 2010; 107:3846–3851.

[28] Edwards MD, Brancaccio M, Chesham JE, Maywood ES, Hastings MH. Rhythmic expression of cryptochrome induces the circadian clock of arrhythmic suprachiasmatic nuclei through arginine vasopressin signaling. Proc. Natl. Acad. Sci. U. S. A. 2016; 113:2732–2737.

[29] Asai S, Takamura K, Suzuki H, Setou M. Single-cell imaging of *c-fos* expression in rat primary hippocampal cells using a luminescence microscope. Neurosci. Lett. 2008; 434:289–292.

[30] Chang E, Pohling C, Natarajan A, Witney TH, Kaur J, Xu L, et al. AshwaMAX and Withaferin A inhibits gliomas in cellular and murine orthotopic models. J. Neurooncol. 2016; 126:253–264.

[31] Akiyoshi R, Kaneuch T, Aigaki T, Suzuki H. Bioluminescence imaging to track real-time armadillo promoter activity in live *Drosophila* embryos. Anal. Bioanal. Chem. 2014; 406:5703–5713.

[32] Sramek C, Mackanos M, Spitler R, Leung LS, Nomoto H, Contag CH, Palanker D. 2011. Non-damaging retinal phototherapy: dynamic range of heat shock protein expression. Invest. Opthalmol. Vis. Sci. 2011; 52:1780–1787.

[33] Horibe T, Torisawa A, Akiyoshi R, Hatta-Ohashi Y, Suzuki H, Kawakami K. Transfection efficiency of normal and cancer cell lines and monitoring of

promoter activity by single-cell bioluminescence imaging. Luminescence. 2014; 29:96–100.

[34] Horibe T, Torisawa A, Kurihara R, Akiyoshi R, Hatta-Ohashi Y, Suzuki H, Kawakami K. Monitoring Bip promoter activation during cancer cell growth by bioluminescence imaging at the single-cell level. Integr. Cancer Sci. Therap. 2015; 2:291–299.

[35] Kikuchi O, Ohashi S, Horibe T, Kohno M, Nakai Y, Miyamoto S, Chiba T, Muto M, Kawakami K. Novel EGFR-targeted strategy with hybrid peptide against oesophageal squamous cell carcinoma. Sci. Rep. 2016; 6:22452, 1–12.

[36] Rogers KL, Martin JR, Renaud O, Karplus E, Nicola MA, Nguyen M, Picaud S, Shorte SL, Brûlet P. Electron-multiplying charge-coupled detector-based bioluminescence recording of single cell Ca^{2+}. J. Biomed. Optics. 2008; 13:031211, 1–10.

[37] Hall MP, Unch J, Binkowski BF, Valley MP, Butler BL, Wood MG, Otto P, Zimmerman K, Vidugiris G, Machleidt T, Robers MB, Benink HA, Eggers CT, Slater MR, Meisenheimer PL, Klaubert DH, Fan F, Encell LP, Wood KV. Engineered luciferase reporter from a deep sea shrimp utilizing a novel imidazopyrazinone substrate. ACS Chem. Biol. 2012; 7:1848–1857.

[38] Sugiyama T, Suzuki H, Takahashi T. Light-induced rapid Ca^{2+} response and MAPK phosphorylation in the cells heterologously expressing human OPN5. Sci. Rep. 2014; 4:5352, 1–10.

[39] Binkowski B, Fan F, Wood K. Engineered luciferases for molecular sensing in living cells. Curr. Opin. Biotechnol. 2009; 20:14–18.

[40] Cosby N. Enabling kinetic studies of GPCRs. Live-cell, non-lytic biosensor for GPCR profiling. Screening. 2009; 4:28–29.

[41] Compan V, Pierredon S, Vanderperre B, Krznar P, Marchiq I, Zamboni N, Pouyssegur J, Martinou JC. Monitoring mitochondrial pyruvate carrier activity in real time using a BRET-based biosensor: investigation of the Warburg effect. Mol Cell. 2015; 59:491–501.

[42] Pratx G, Chen K, Sun C, Martin L, Carpenter CM, Olcott PD, Xing L. Radioluminescence microscopy: measuring the heterogeneous uptake of radiotracers in single living cells. PLoS One. 2012; 7:e46285, 1–9.

[43] Pratx G, Chen K, Sun C, Axente M, Sasportas L, Carpenter C, Xing L. High-resolution radioluminescence microscopy of 18F-FDG uptake by reconstructing the β-ionization track. J. Nucl. Med. 2013; 54:1841–1846.

[44] Goda K, Hatta-Ohashi Y, Akiyoshi R, Sugiyama T, Sakai I, Takahashi T, Suzuki H. Combining fluorescence and bioluminescence microscopy. Microsc. Res. Tech. 2015; 78:715–722.

[45] Marcova SV, Vysotski ES, Blinks JR, Burakova LP, Wang BC, Lee J. Obelin from the bioluminescent marine hydroid *Obelia geniculate*: cloning, expression, and comparison

of some properties with those of other Ca^{2+}-regulated photoproteins. Biochemistry. 2002; 41:2227–2236.

[46] Konno J, Hatta-Ohashi Y, Akiyoshi R, Thancharoen A, Silalom S, Sakchoowong W, Yiu V, Ohba N, Suzuki H. TiLIA: a software package for image analysis of firefly flash patterns. Ecol. Evol. 2016; 6:3026–3031.

[47] May-May-Thet, Sugiyama T, Suzuki H. Bioluminescence imaging of intracellular calcium dynamics by the photoprotein obelin. In: Proceedings of the 15th International Symposium on Bioluminescence and Chemiluminescence Light Emission: Biology and Scientific Applications; 13–17 May 2008; China. Singapore: World Scientific; 2009. pp. 359–362.

[48] Olympus. Intracellular calcium imaging using photoprotein, obelin. 2009. Available from: http://bioimaging.jp/appli/007/pdf/app_lv200.pdf [Accessed: 2016-06-06]

[49] Olympus. Imaging of clock gene promoter assay of culture cells. 2009. Available from: http://bioimaging.jp/appli/004/pdf/app_lv200.pdf [Accessed: 2016-06-06]

[50] Clendenon SG, Young PA, Ferkowicz M, Phillips C, Dunn KW. Deep tissue fluorescent imaging in scattering specimens using confocal microscopy. 2011; Microsc. Microanal. 17:614–617.

Chapter 14

Bioluminescence of the Black Sea Ctenophores-Aliens as an Index of their Physiological State

Tokarev Yuriy Nikolaevich and
Mashukova Olga Vladimirovna

Additional information is available at the end of the chapter

http://dx.doi.org/10.5772/65063

Abstract

Three experiment series on the ctenophores *Mnemiopsis leidyi* and *Beroe ovata* bioluminescence variability investigation were conducted: (1) depending on ctenophores size and ontogeny stage; (2) depending on temperature conditions and (3) depending on season. The ctenophores luminescence was registered using the laboratory complex "Svet" by methods of mechanical and chemical stimulation. Ctenophores light-emission characteristics are changing in the process of ontogenesis and rising proportionally to the organism mass growth. Seasonal dynamics of the ctenophore-aliens light-emission characteristics has been revealed: the highest indices of *M. leidyi* and *B. ovata* bioluminescence are observed in the summer period and minimal indices for both species were registered in the winter-spring period. Environment temperature affects considerably at the amplitude-temporal characteristics of the ctenophores light-emission. The bioluminescence reaction optimum for *M. leidyi* is achieved under the temperature of 26 ± 1°C, and for *B. ovate*—under the temperature of 22 ± 1°C, while its minimum for both ctenophores was registered under the temperature of 10 ± 1°C. Thus, results of the investigations have detected the opportunity to use ctenophores *M. leidyi* and *B. ovata* light-emission characteristics as an index for their physiological state estimation.

Keywords: light-emission characteristics, ecological-physiological indices, *Mnemiopsis leidyi*, *Beroe ovata*, the Black Sea

1. Introduction

Bioluminescence as a manifestation of an organism life activity in a form of electric-magnetic radiation in the visible region of spectrum is the most important ecological factor of marine environment [1]. Quite recently, they considered that microplankton—bacteria and dinoflagellates—makes the main contribution into formation of the Black Sea bioluminescence field [2–5]. But for a number of the World ocean regions, another fraction of plankton community makes the major contribution into bioluminescence field formation, in particular jelly-fish macroplankton [3, 6, 7]. For instance, ctenophores *Mnemiopsis leidyi* and *Beroe ovata*, which quite recently inhabited the Black Sea are luminescent organisms, bioluminescent intensity of which exceeds hundreds of thousands to million times light-emission of the majority of the microplankton representatives.

There are totally about 150 species of ctenophores, of them in 46 species, living in wide range of temperatures, bioluminescent ability has been registered reliably [8, 9]. For the past 30 years, the Black Sea ctenophore fauna became considerably more rich: Until 1980, it has been presented by one species of pleurobrachia (*Pleurobrachia pileus* (O.F. Muller, 1776)), from 1980-1990-th two species from genus mnemiopsis (*Mnemiopsis leidyi* A. Agassiz, 1865) and beroe (*Beroe ovata* Mayer, 1912) were added and in 2007 near the Turkish and Bulgarian shores bolinopsis (*Bolinopsis vitrea* (L. Agassiz, 1860)) was also found. Now it is not yet clear whether this species inhabiting Mediterranean sea will be able to naturalize in the Black Sea,, but it was met already in 2010 [10]. That is why our work will be devoted exclusively to the parameters of life activity of only two alien ctenophores: *Mnemiopsis leidyi* and *Beroe ovata.*

Ctenophores—aliens not only reached the list of the Black Sea macroplankton but they also considerably influenced structure dynamics of its ecosystem, thus attracting great attention to them. The climate warming and increasing of the anthropogenic eutrophication led in a number of cases to considerable growth of not only ctenophores populations but jelly-fish as well, which influenced condition of the marine communities and effected human economic activity: fishing nets and water canals were blocked, obstacles for marine bathing were created, and in the Black Sea anchovy fishing sharply decreased with the first flash of the mnemiopsis mass development [11–13].

At present time, there is quite great number of works devoted to physiology and ecology of different ctenophores species, including the Black Sea populations [12, 14–20]. From 1980, they conduct intensive studies of the ctenophores—aliens in the Black Sea: they reveal features of their distribution by the sea regions in connection with depth, temperature and salinity; they also study peculiarities of nutrition, breathing and reproduction. As for mnemiopsis, they revealed effect of the environment temperature on such characteristics as population vertical distribution in pelagial [13, 21–23], reproduction rate [24], metabolism intensity [25, 26] and some peculiarities of luminescence under experimental conditions [27, 28]. The same data were received for beroe as well [28–30].

But such important ecological characteristic of the ctenophores as bioluminescence still remains to be not much studied. In particular, the studies of the light-emission parameters in the Black Sea populations of *M. leidyi* and *B. ovata* by present time were conducted exclusively in the Department of Biophysical Ecology, IBSS NASU (now—IMBR RAS). Such indices of the ctenophores bioluminescent as a change of intensity and duration of the light-emission in ontogenesis are studied not enough, influence of different environment factors on the bioluminescence parameters is studied insufficiently, still unclear is connection between the organism physiological indices and its luminescence. Nevertheless, it is known that on the base of the amplitude-time characteristics of bioluminescence we can make a conclusion about the organism functional condition [27, 28, 31, 32].

In connection with the above mentioned, we consider it to be extremely important to continue investigation of the light-emission in the Black Sea alien ctenophores, to reveal an influence of different factors on them and to evaluate accordance of their functional state with variability of the bioluminescence parameters.

2. Materials and methods

Experimental investigations were conducted in the Biophysical Ecology Department of the A.O. Kovalevsky Institute of Marine Biological Research (IMBR) from 2007 to 2012. Ctenophores with sizes of 35–40 mm (oral-aboral length for *M. leidyi* and total for *B. ovata*) were collected by Judy net in the Crimea coastal zone at the layer of 0–50 m. Not damaged samples without content in the gastrovascular cavity were chosen for experiments. Three experiment series were conducted: (1) depending on ctenophore size and ontogeny stage; (2) depending on temperature conditions and (3) depending on season. The freshly caught animals were left for 2–3 h to adapt to the conditions similar to *in situ*.

The investigation of *B. ovata* bioluminescence parameters in ontogenesis was carried out in September–November 2007–2009 in three experiment series: (1) depending on ctenophore size; (2) depending on their physiological state and (3) on ontogeny stage. *B. ovata* individuals with wet weight from 0.06 to 19.53 g were taken for the first experiment series. Unbroken individuals were placed into 5 l containers with filtered marine water (membrane filters pore diameter is 35 μm) at a temperature of 21 ± 2°C [33]. For estimation of *B. ovata* luminescence variability in relation to reproduction stage specimens were separated into four groups: (1) 50-mm long—just-caught individuals before gonada formation; (2) 50-mm long individuals with mature gonads; (3) ctenophore eggs spawned by the second group; (4) ctenophore larvae grown from eggs of the third group. Adult just-caught *B. ovata* organisms were put in 20 l aquariums with filtered water under temperature 20 ± 2°C with feeding *M. leidyi* (*L* = 40 mm). Experiments of *B. ovata* light signal registration by ontogeny stages were carried out by the methodology [33, 34].

The investigation of *M. leidyi* bioluminescence parameters in ontogenesis was carried out in July–August 2007–2010. Ctenophores of 3–65 mm size (oral-aboral length) were selected from the plankton samples, taken by Judy net in the upper 10 m layer. Ctenophores wet weight was calculated by the volume of displaced water in measuring cylinder with further weighing of each specimen on the microanalytic weighs AN 50 with accuracy up to 0.01 g. In experiments on the ctenophores size influence on their bioluminescence characteristics, fresh-caught specimens were separated to six size groups: (1) 0.0073 ± 0.00036 g; (2) 0.52 ± 0.026 g; (3) 3.69 ± 0.18 g; (4) 12.77 ± 0.63 g; (5) 35.06 ± 1.75 g and (6) 42.03 ± 2.10 g. To avoid photoinhibition of the ctenophores bioluminescence, they were kept before measuring for 2 h in darkness with 24 ± 2°C temperature. Ctenophores were kept in vessels with 3–5 l volume marine water, filtered through the membrane filters with 35 μm pores diameter. Proper filtration of the ctenophores medium was necessary for exclusion of the by-catches of another luminescent organisms (first of all dinoflagellates), which could distort the results of experiments when studying ctenophores bioluminescence at the initial ontogenesis stages (eggs and larvae). To study an influence of the ctenophores reproductive system condition on the characteristics of their bioluminescence, the caught in the sea adult specimens (40 mm length) were divided to three experimental groups: (1) ctenophores freshly caught in the sea (with gonads at early stage of development, which served as a control after 2-h adaptation in the filtrated water); (2) ctenophores with eggs clutches, formed in the laboratory conditions after experimental feeding; (3) specimens after eggs spawning out. Such ctenophores were preliminary kept under conditions analogous to those of the second group during 6 h; during this period, they produced new eggs clutches and they spawned eggs out. The third group ctenophores bioluminescence characteristics were registered directly after their spawning.

It is known that at natural conditions calanoid copepods, dominating in mesozooplankton composition in the second half of the summer season make the main part of the Black Sea mnemiopsis feeding [35]. That is why we used calanoid copepods *Calanipeda aquaducles*, grown in the laboratory for fish cultivation for nutrition of the ctenophores in the experimental conditions. Before measuring bioluminescent characteristics, the second group ctenophores were kept isolated during 5 h in 5 l vessels with concentration of the late copepodite stages of copepods at the level of 60 ex·l^{-1} (with food supply of 300 ex for one specimen of ctenophores). Copepods concentration in the experimental vessels was determined before the beginning of experiment, counting specimens in an aliquota of volume in the Bogorov camera. In 3 h after beginning of exposition, concentration of food was corrected to initial volumes. With such food supply, ctenophores reproduce actively and in 5 h of nutrition they form ready for spawning eggs clutches [18]. This group of ctenophores was lighted directly after formation of clutch in them.

For estimation of variability of luminescence biophysical characteristics in the ctenophores in ontogenesis, they were divided into four groups: (1) freshly caught in the sea specimens of 40 mm length before gonads formation, adapted to the conditions of experiment under complete darkness during 2 h; (2) ctenophores of 40 mm length with matured gonads, formed as a result of experimental nutrition during 5–6 h after catching; (3) eggs, spawned out by the second group ctenophores, 0.40–0.50 mm diameter; (4) developed from the ctenophores eggs larvae, 0.25–0.30 mm diameter. To receive eggs and then larvae, the freshly caught adult ctenophores were isolated in 5 l vessels with filtered water, where they were fed by copepods. Eggs clutched by ctenophores were collected by filtration of all the water volume through 100 μm sieve. Eggs collected on the sieve were washed into 200 ml glass cylinder, and the number of eggs was calculated in all the volume under microscope. Size of eggs and larvae were measured with accuracy of 0.01 mm under microscope. The measurements of the bioluminescence characteristics were conducted in 15–20 specimens of each experimental group and repeated three times. Before light-emission stimulation ctenophores were kept in the filtered marine water with 24 ± 2°C temperature. The given temperature conditions are optimal for quick eggs spawning by ctenophores and further larvae development [25].

For investigation of temperature variability, uni-sized (35–40 mm length) ctenophores were divided in the laboratory into five groups and contained in different temperature conditions: (1) 10 ± 1°C; (2) 16 ± 1°C; (3) 22 ± 1°C; (4) 26 ± 1°C and (5) 30 ± 1°C. *M. leidyi* and *B. ovata* were kept in the temperature-controlled aquariums (50 l) with the filtered marine water, being adapted during 6–8 h to the temperature close to such in the sea in the given period [36].

The main parameters: amplitude, energy and bioluminescence duration of the alien-ctenophore under the different temperature conditions were compared. For research of seasonal dynamics, bioluminescence uniform-sized samples group (40 mm) of ctenophores were taken. The adaptive period before experiments on ctenophore bioluminescence was 2 h. Experiments on ctenophore bioluminescence characteristics registration on the laboratory complex—luminescope "Svet" [31] were conducted after the adaptive period. Special cuvette for mechanical, chemical and electrical stimulation of the plankton organisms, made of transparent organic glass, in which experimental organisms were placed, was set into the luminescope dark chamber. Biophysical characteristics of the ctenophore light-emission were investigated by mechanical and chemical stimulation in our experiments. Mechanical stimulation method, the most adequate to the natural stimuli, chemical stimulation by ethyl alcohol give more prolonged and bright signals with maximal values [32, 33].

3. Results

3.1. Seasonal dynamics of the *Mnemiopsis leidyi* bioluminescence

The studies conducted had revealed in *M. leidyi* bioluminescence intensity considerable seasonal fluctuations for its amplitude characteristic as well as for temporal one (**Figure 1**).

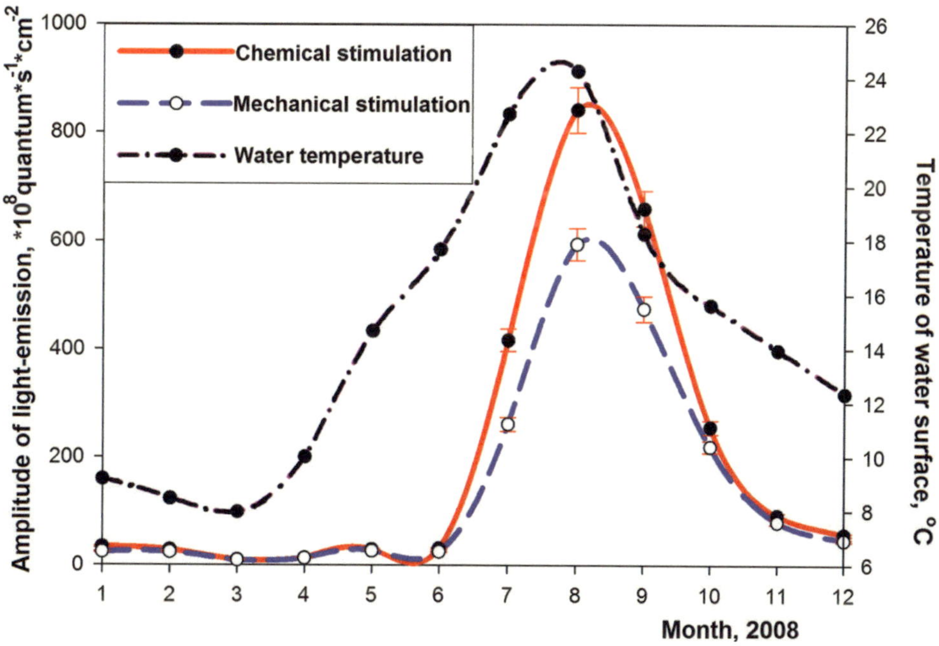

Figure 1. *Mnemiopsis leidyi* light-emission amplitude seasonal dynamics under different stimulation types [31].

Thus, the ctenophores in the winter period gave not very intensive flash with the amplitude until 70.0 ± 3.4·10^8 quantum·s^{-1}·cm^{-2} and duration about 2.0 s. Low bioluminescence values were observed in the spring period with minimum on March (9.93 ± 0.49&·10^8 quantum·s^{-1}·cm^{-2} and 9.21 ± 0.46·10^8 quantum·s^{-1}·cm^{-2}) under the chemical and mechanical stimulation correspondingly [31].

The average luminescence amplitude (260.94 ± 13.04·10^8 quantum·s^{-1}·cm^{-2}) was registered in June. The light-emission characteristics rise with peak in August and make 841.97 ± 42.09·10^8 quantum·s^{-1}·cm^{-2}. It is related with ctenophores reproduction in July–August. *M. leidyi* luminescence intensity under the chemical stimulation is 2–2.5 times greater ($p < 0.05$) than under the mechanical one [31].

The luminescence amplitude of *M. leidyi* decreases almost 11 times in the middle of November, if compared with the summer period. Light-emission energy of ctenophores depending on season changes analogically with their amplitude indices (**Figure 2**) [31].

Thus, minimal energy values of *M. leidyi* were registered in February and maximal in August, making 659.97 ± 32.98·10^8 quantum·cm^{-2} and 393.39 ± 19.66·10^8 quantum·cm^{-2} under the chemical and the mechanical stimulation correspondingly. The *M. leidyi* luminescence

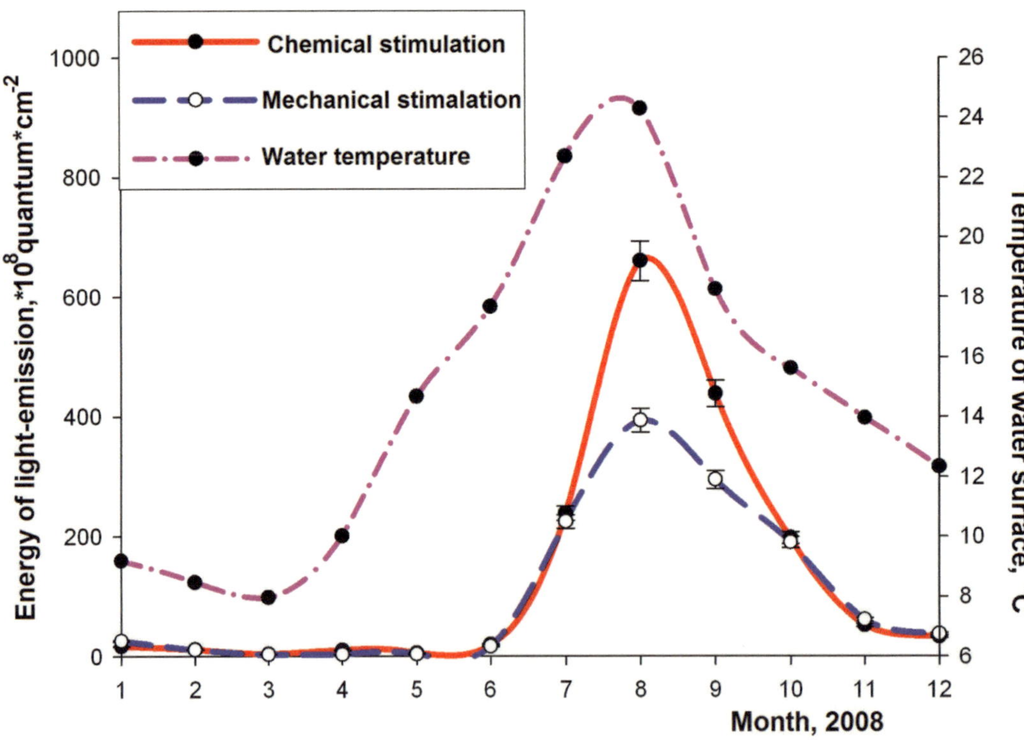

Figure 2. *Mnemiopsis leidyi* light-emission energy seasonal dynamics under different stimulation types [31].

energy reduces during the following period and decreases 12 times in November if compared with July.

M. leidyi light-emission duration changes considerably depending on season (**Figure 3**). Thus, the shortest flashes are registered in February–March, making 0.79–1.32 s and more prolonged luminescence duration is observed in August–September and it achieves 2.77–3.46 s ($p < 0.05$) [31].

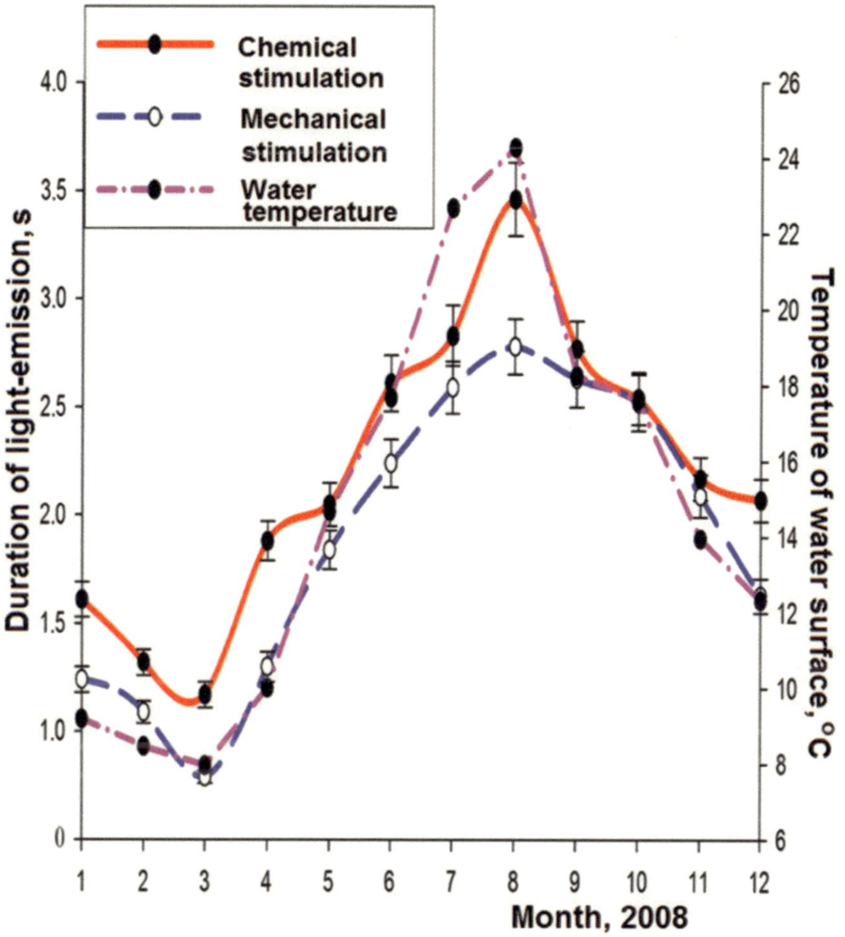

Figure 3. *Mnemiopsis leidyi* light-emission duration seasonal dynamics under different stimulation types [31].

3.2. Seasonal variability of the *B. ovata* bioluminescence characteristics

The typical luminescence signal of *B. ovata* is represented by a number of flashes, superimposing one on another, with several amplitude peaks with sharp increasing background and the same damping decrement.

The beroe luminescence has significant seasonal differences [31]. Thus, several weak signals may be observed for ctenophore, luminous in the winter period (**Figure 4**), followed by the flash of negligible intensity with the greatest amplitude (56.7 ± 2.83·10⁸ quantum·s⁻¹·cm⁻²).

The ctenophore bioluminescence is depressed even more in the spring period, with the minimal values in May: one to two weak signals are observed with the amplitude up to 35.96 ± 1.79·10^8 quantum·s^{-1}·cm^{-2}. The bioluminescence intensity increases up to 537.6 ± 26.88·10^8 quantum·s^{-1}·cm^{-2} which is registered in summer. *B. ovata* maximal bioluminescence is registered in July, their intensity achieves 1382.25 ± 69.11·10^8 quantum·s^{-1}·cm^{-2} and duration up to 2.86 ± 0.14 s. Ctenophore light-emission intensity is 1.5 times higher under the mechanical stimulation than under the chemical one ($p < 0.05$) [31]. *B. ovata* luminescence characteristics decrease up to 98.75 ± 4.93·10^8 quantum·s^{-1}·cm^{-2} in August. The second peak of light-emission intensity is observed in September, achieving 852.56 ± 42.62·10^8 quantum·s^{-1}·cm^{-2}. Luminescence amplitude reduces 15 times by December, if compared with the autumn peak and makes 56.7 ± 2.83·10^8 quantum·s^{-1}·cm^{-2} and 27.01 ± 1.35·10^8 quantum·s^{-1}·cm^{-2} under the mechanical and chemical stimulation correspondingly. Ctenophore luminescence energy seasonal changes are the same (**Figure 5**).

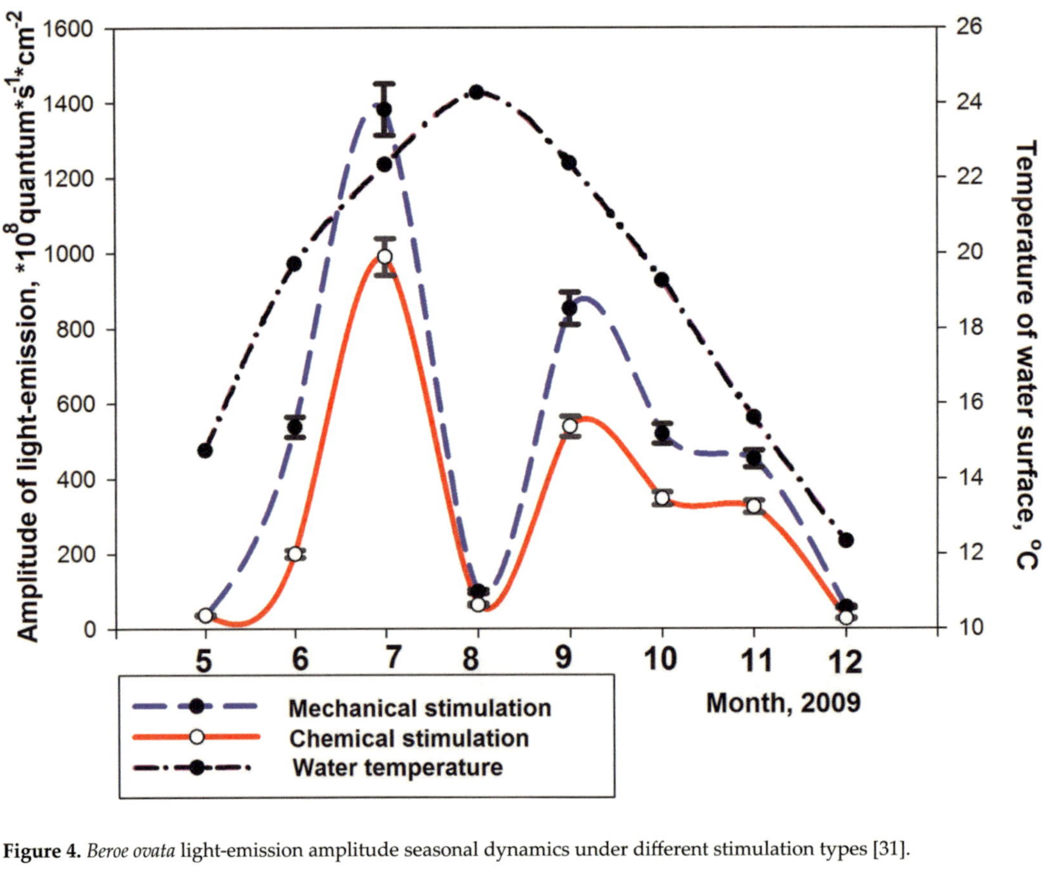

Figure 4. *Beroe ovata* light-emission amplitude seasonal dynamics under different stimulation types [31].

http://dx.doi.org/10.5772/65063

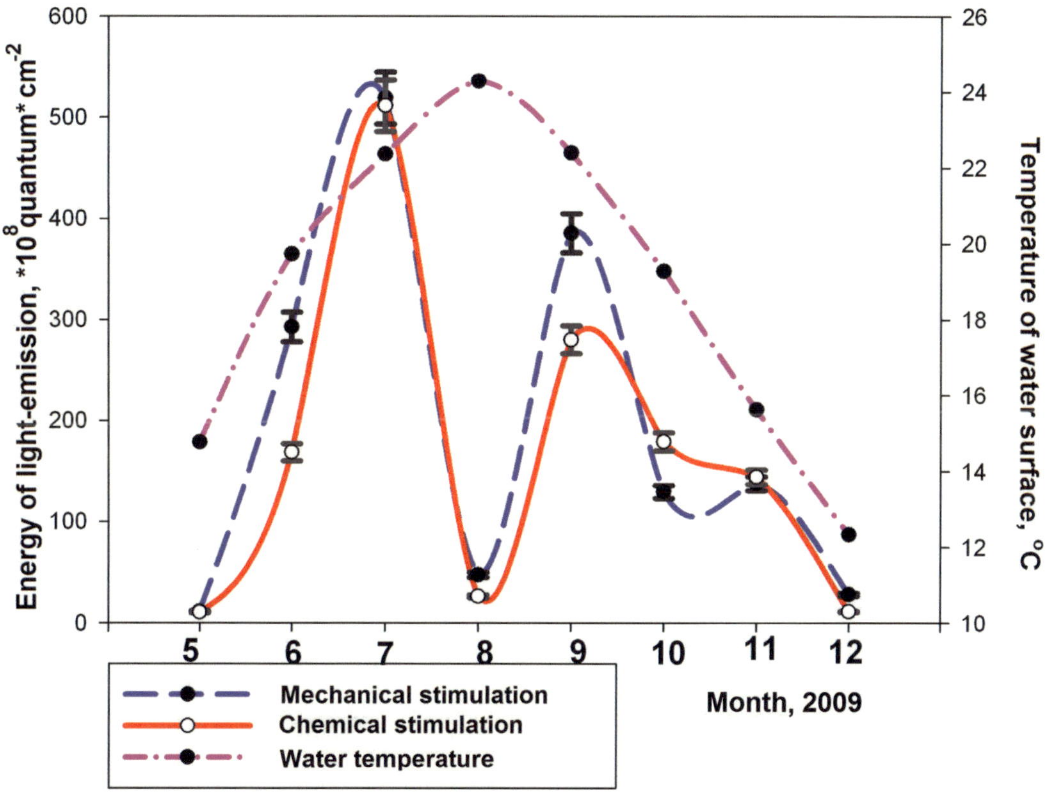

Figure 5. *Beroe ovata* light-emission energy seasonal dynamics under different stimulation types [31].

Low energy values of ctenophores are observed in winter-spring period with minimum in May. *B. ovata* bioluminescence energy is maximal in July, making 518.94 ± 25.94·10^8 quantum·cm^{-2} under the mechanical stimulation and 511.88 ± 25.59·10^8 quantum·cm^{-2} under the chemical one. Decrease of the luminescence energy indices is observed in August, if compared with the previous month. *B. ovata* light-emission amplitude rises again in autumn with maximum in September and decreases 1.5 times if compare with the summer period ($p < 0.05$). Light-emission energy decreases 11 times ($p < 0.05$) by December [31]. Light-emission duration of ctenophore like its intensity in the different seasons change considerably (**Figure 6**).

More prolonged signals are registered in July and September, making 2.54–2.86 s, the shortest luminescent signals of *B. ovata* are observed in May (1.06 s) and in December (0.9 s) ($p < 0.05$). *B. ovata* trophic state like this of *M. leidyi* is depressed in the winter-spring period [31, 37], and it reveals itself in reducing its luminescence amplitude-temporal characteristics. But *B. ovata* nutritive conditions are the most favorable in early autumn period, in September especially [37], which affects the ctenophore bioluminescence activity increase in the given period.

B. ovata, if compared to other jelly-fish, is the species sensitive to the temperature swings more than others [38]. The temperature rise in the Black Sea in May up to 16°C leads to *B. ovata* early

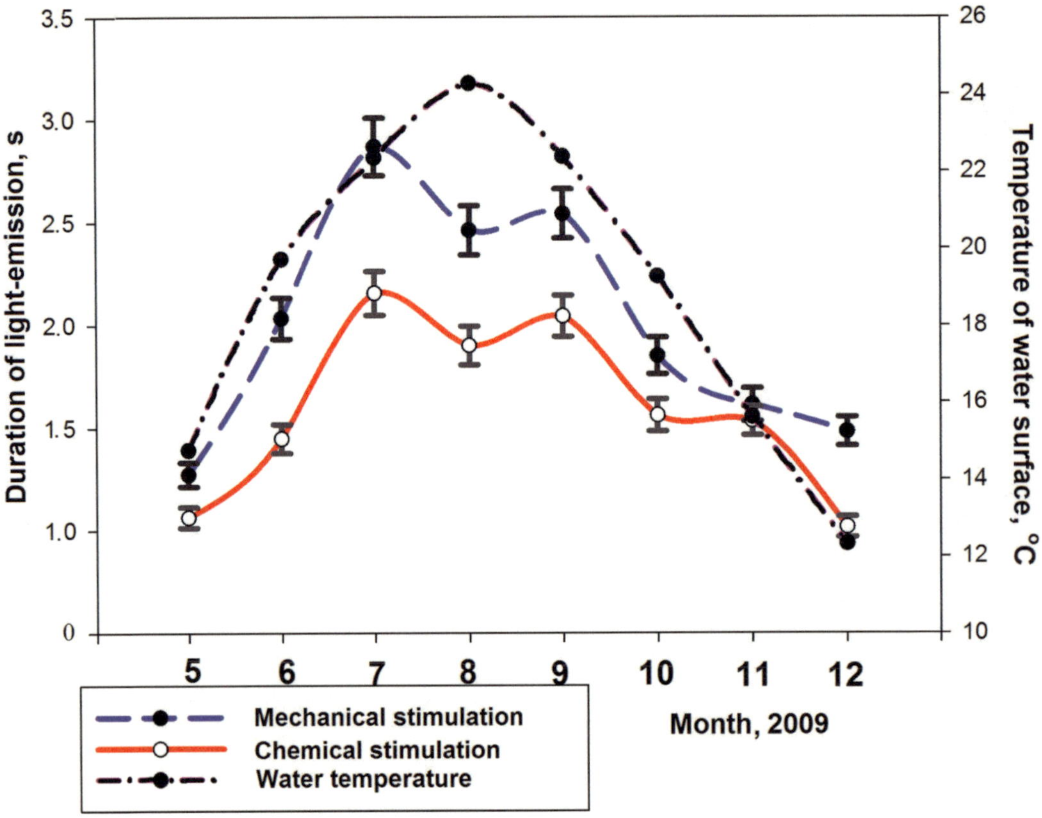

Figure 6. *Beroe ovata* light-emission duration seasonal dynamics under different stimulation types [31].

appearance, but ctenophore luminescence values are low in the spring period. The first ctenophore bioluminescence peak is observed in July. The water temperature rise in August up to 26°C leads to *B. ovata* abundance decrease in the given period. The ctenophore second maximum was found in September while the water temperature falls to 20 ± 2°C. From October till March, the ctenophore state is depressed. Food supplies and mass spawning reduce affect unfavorably the *B. ovata* functional state as well as its bioluminescence indices [31, 39].

Thus, seasonal variability of ctenophores light-emission parameters was established. Our investigations showed that maximal bioluminescence values for mnemiopsis are registered in August, whereas beroe maximal bioluminescence is observed twice—in July and in September. Light-emission minimal values for both ctenophores were observed in the winter-spring period [31].

3.3. Influence of the temperature on the *M. leidyi* bioluminescence

The investigation results have shown considerable changes of the *M. leidyi* bioluminescence intensity, connected with temperature changes (**Table 1**). Thus, maximal indices of the ctenophore signals amplitude were observed under the temperature of 26 ± 1°C. *M. leidyi* bioluminescence intensity under chemical stimulation was 1.5 times higher, than under the mechanical one, making 1432.94 ± 71.64·10^8 quantum·s^{-1}·cm^{-2}.

The temperature increases up to 30°C leads to four times decrease of ctenophore luminescence intensity, making 322.34 ± 16.1·10^8 quantum·s^{-1}·cm^{-2}. *M. leidyi* light-emission intensity decreased two times ($p < 0.05$) comparing with optimum under the temperature decrease down to 22°C. The temperature decreases down to 10 ± 1°C leads to more considerable bioluminescence intensity change, up to its minimal values 17.32 ± 0.83·10^8 under mechanical and 17.93 ± 0.89·10^8 quantum·s^{-1}·cm^{-2} under the chemical stimulation correspondingly [36].

Characteristics of light-emission	Amplitude of light-emission, quantum·s^{-1}·cm^{-2}		Energy of light-emission, quantum·cm^{-2}		Duration of light-emission, s	
Stimulation types	**1**	**2**	**1**	**2**	**1**	**2**
10 ± 1°C	29.52 ± 1.47·10^8	33.52 ± 1.67·10^8	12.47 ± 0.62·10^8	15.51 ± 0.77·10^8	1.82 ± 0.09	1.94 ± 0.097
16 ± 1°C	219.45 ± 10.97·10^8	332.33 ± 16.61·10^8	197.43 ± 9.87·10^8	283.97 ± 14.19·10^8	2.51 ± 0.12	2.70 ± 0.13
22 ± 1°C	545.75 ± 27.28·10^8	632.95 ± 31.64·10^8	407.19 ± 20.35·10^8	417.65 ± 20.388·10^8	2.89 ± 0.14	3.48 ± 0.17
26 ± 1°C	910.81 ± 45.54·10^8	1432.94 ± 71.64·10^8	725.33 ± 36.26·10^8	894.64 ± 44.73·10^8	3.14 ± 0.16	3.53 ± 0.17
30 ± 1°C	322.34 ± 16.12·10^8	488.43 ± 24.42·10^8	294.89 ± 14.74·10^8	265.15 ± 13.25·10^8	2.54 ± 0.12	2.67 ± 0.17

Remark: 1, mechanical stimulation; 2, chemical stimulation.

Table 1. Light-emission characteristics of *M. leidyi* under different temperatures.

M. leidyi light-emission energy changes under different temperatures (**Table 1**). Thus, maximal values of ctenophore luminescence energy were registered under 26°C, making 894.64 ± 44.7·10^8 and 725.33 ± 36.2·10^8 quantum·cm^{-2}—under the chemical and mechanical stimulation correspondingly.

Bioluminescence energy decreases two times ($p < 0.05$) under the temperature of 22°C. *M. leidyi* light-emission energy minimal indices were observed under the temperature of 10°C [36]. Temperature fluctuations affected the *M. leidyi* light-emission duration change with minimal indices under the temperature 10°C, under its rise up to 30°C making 1.94 and 2.67 s correspondingly. The most continuous signals were registered under the temperature of 26°C, 3.54 ± 0.15 s, under the chemical stimulation especially.

3.4. Influence of the temperature on the *B. ovata* bioluminescence

Amplitude and light-emission energy considerable changes, connected with the environment temperature change, were revealed in ctenophore *B. ovata* (**Table 2**). Thus, *B. ovata* flashes in amplitude had the maximal indices under the temperature of 22 ± 1°C regardless the type of stimulation, having achieved 1150 ± 57.51·10^8 quantum·s^{-1}·cm^{-2} under the mechanical and 822.03 ± 41.10·10^8 quantum·s^{-1}·cm^{-2} under the chemical stimulation correspondingly.

Ctenophore reacts with more low light-emission amplitude indices with the temperature rise up to 26°C, but minimal values of the luminescence amplitude are registered under the temperature of 30°C, achieving 49.01 ± 2.4·10^8 under the mechanical stimulation and

Characteristics of light-emission	Amplitude of light-emission, quantum·s^{-1}·cm^{-2}		Energy of light-emission, quantum·cm^{-2}		Duration of light-emission, s	
Stimulation types	**1**	**2**	**1**	**2**	**1**	**2**
10 ± 1°C	4.92 ± 0.24·10^8	3.42 ± 0.16·10^8	2.95 ± 0.12·10^8	1.67 ± 0.08·10^8	1.03 ± 0.05	1.02 ± 0.05
16 ± 1°C	551.14 ± 27.55·10^8	482.89 ± 24.14·10^8	262.22 ± 13.11·10^8	156.12 ± 7.8·10^8	1.91 ± 0.09	1.76 ± 0.08
22 ± 1°C	1150.36 ± 57.51·10^8	822.03 ± 41.10·10^8	530.19 ± 26.51·10^8	482.65 ± 24.13·10^8	3.03 ± 0.15	2.47 ± 0.12
26 ± 1°C	577.06 ± 28.85·10^8	268.81 ± 13.44·10^8	166.97 ± 8.34·10^8	148.63 ± 7.43·10^8	2.12 ± 0.10	2.08 ± 0.10
30 ± 1°C	49.01 ± 2.45·10^8	29.23 ± 1.46·10^8	14.73 ± 0.73·10^8	13.84 ± 0.69·10^8	1.53 ± 0.07	1.49 ± 0.07

Remark: 1, mechanical stimulation; 2, chemical stimulation.

Table 2. Light-emission characteristics of *B. ovata* under different temperatures.

29.23 ± 1.46·10^8 quantum·s^{-1}·cm^{-2} under the chemical one. The *B. ovata* functional state is negative under low temperature also. Thus, temperature decreases down to 10 ± 1°C weakens ctenophore moving activity and lowers their luminescence intensity values: down to 3.42 ± 0.16·10^8 quantum·s^{-1}·cm^{-2} and 4.92 ± 0.22·10^8 quantum·s^{-1}·cm^{-2} under the chemical and mechanical stimulation correspondingly. bioluminescence energy of beroe change as well as its amplitude depending on different temperatures. Thus, maximal values are registered under the temperature of 22°C (530.19 ± 26.5·10^8 quantum·cm^{-2}) and minimal—under the temperature of 10°C, making 2.95 ± 0.12·10^8 quantum·cm^{-2}. *B. ovata* light-emission duration varied considerably under the temperature changes [36].

The shortest bioluminescent signals were observed under the temperature of 10°C, making 1.02 ± 0.05 s, and the most continuous under 22°C, achieving 3.03 ± 0.15 s. Ctenophore light-emission characteristics changes, under different temperature conditions, can be explained, we believe, by these organisms physiological adaptations to the environment temperature oscillations point of view. Indeed, the most intensive *M. leidyi* luminescence is observed under the temperature of 26 ± 1°C, and *B. ovata*—under 22 ± 1°C, which are the most favorable for their functional state. Thus, according to the data of Anninsky with co-authors, *M. leidyi* [25] breeding peak is observed under the temperature of 24–26°C in August, and ctenophore *B. ovata* [12, 19] autumn abundance growth under the temperature 20–22°C. Ctenophores under the temperature 22°C are actively breeding, and their metabolism is considerably higher than under lower temperatures. Ctenophores light-emission amplitude decreases for several orders under the temperature to 10°C can be explained by their populations abundance sharp reduction reduces in the autumn-winter period [23, 36].

Maximal activity of the enzyme-substrate complex, basic for the ctenophores luminescence was observed under the temperature of 30°C *in vivo* [36, 40, 41]. Thus, light-emission amplitude maximum was observed in our investigations under following temperatures: under 26°C for *M. leidyi* and under 22°C for *B. ovata* [36].

3.5. Bioluminescence characteristics changes in the *M. leidyi* ontogenesis

After 5–6 h of experimental feeding ctenophores of 40 mm length produced from 3.0 to 4.5 thousands of viable. The spawning peak was observed at night (23–24 h), which corresponds to the data of other researchers [24]. Duration of development from eggs spawning to larvae getting out in our investigations was of 16–19 h. Typical bioluminescent signals of ctenophores *M. leidyi* under mechanical and chemical stimulations at different stages of ontogenesis are presented in **Figure 7a** and **b**. As one can see in **Figure 7** and **Table 3**, ctenophores luminescence characteristics change considerably depending on the development stage.

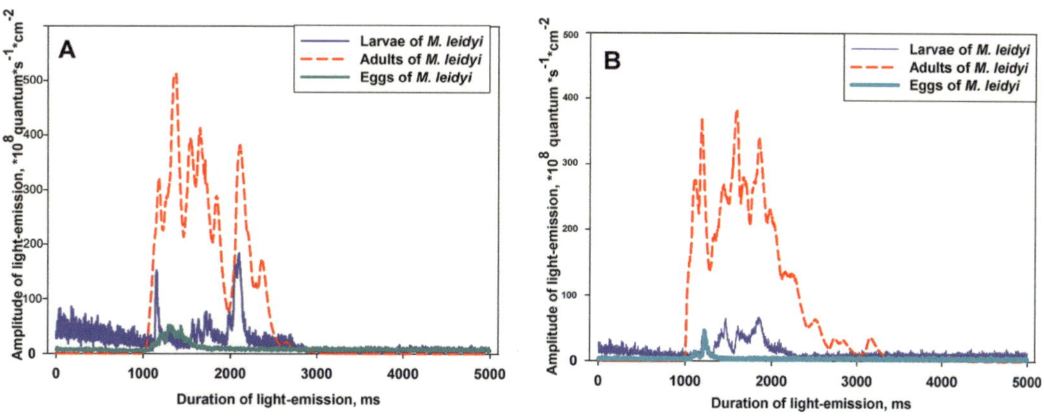

Figure 7. The typical bioluminescence signals of *M. leidyi* at the different ontogenesis stages: (A) under mechanical stimulation and (B) under chemical stimulation.

The most intensive bioluminescence is observed in adult specimens (with matured gonads), in which amplitude-time characteristics reach maximum magnitudes: amplitudes up to

Ontogenesis stages of *M. leidyi*	N	L (mm)	Amplitude of light-emission (quantum·s^{-1}·cm^{-2})		Energy of light-emission (quantum·cm^{-2})		Duration of light-emission, s	
Stimulation types			**1**	**2**	**1**	**2**	**1**	**2**
Just-caught individuals (control)	43	40	(112.16 ± 5.61) ·10^8	(144.18 ± 7.20) ·10^8	(109.68 ± 5.48) ·10^8	(143.36 ± 7.16) ·10^8	2.39 ± 0.12	2.75 ± 0.13
Reproductive ctenophores	38	40	(424.46 ± 21.22) ·10^8	(470.98 ± 23.54) ·10^8	(284.76 ± 14.23) ·10^8	(311.24 ± 15.56) ·10^8	3.28 ± 0.16	3.93 ± 0.19
Ctenophore eggs	25	0.40–0.50	(0.39 ± 0.019) ·10^8	(0.89 ± 0.04) ·10^8	(0.23 ± 0.012) ·10^8	(0.52 ± 0.026) ·10^8	0.45 ± 0.02	0.76 ± 0.03
Ctenophore larvae	30	0.25–0.30	(1.44 ± 0.08) ·10^8	(3.13 ± 0.15) ·10^8	(0.48 ± 0.022) ·10^8	(1.07 ± 0.05) ·10^8	1.33 ± 0.067	1.86 ± 0.11

Remark: 1, mechanical stimulation; 2, chemical stimulation.

Table 3. The bioluminescence characteristics of ctenophore *M. leidyi* at the ontogenesis.

(470.98 ± 23.54)·10^8 quantum·s^{-1}·cm^{-2} and duration of signal—up to 3.93 ± 0.19 s. Light-emission amplitude in the adult specimens three times and signal energy two times ($p < 0.05$) exceeds analogous characteristics of the control group ctenophores.

Luminescence durations in the given ctenophore groups also differ considerably. For example, luminescence duration in the adult specimens for 1.18 s exceeds the same in control. Signal duration in the control group ctenophores three to four times exceeded those in their eggs and larvae. The weakest luminescence was registered in ctenophores eggs (**Table 3**), expressed in low amplitudes (less than 0.39 ± 0.019·10^8 quantum·s^{-1}·cm^{-2}) and light-emission energy (less than 0.23 ± 0.012·10^8 quantum·cm^{-2}), as well as small duration of the bioluminescent signal—up to 0.45 ± 0.02 s. Comparing bioluminescence of the ctenophore eggs and larvae, we stated that the larval stage luminescence amplitude was 3.5 and energy two to three times higher than analogous characteristics of the eggs bioluminescence. Signal durations of ctenophore larvae also two to three times exceeded analogous parameters in eggs ($p < 0.05$). The results of correlation of the light-emission in *M. leidyi*

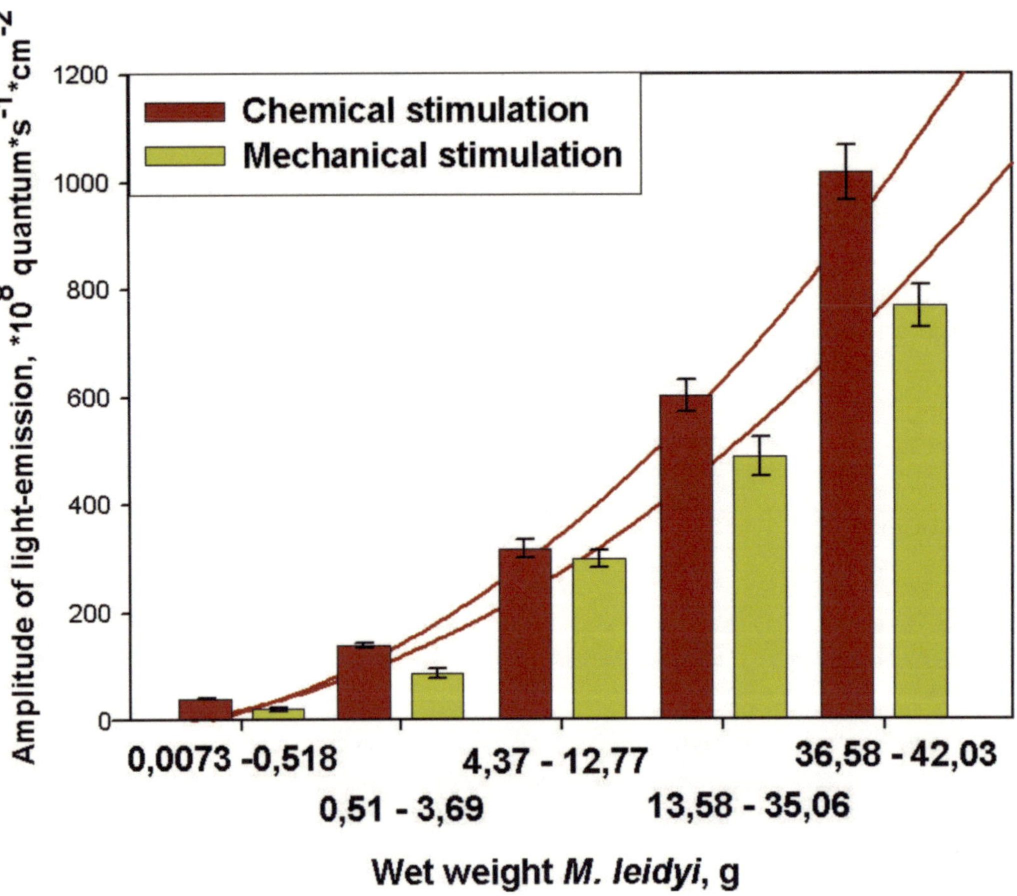

Figure 8. Variability of the bioluminescence amplitude ctenophore *M. leidyi* depending on wet weight of the individuals under mechanical and chemical stimulation.

ctenophores depending on specimen size under the mechanical and chemical stimulation are given in **Figures 8–10**.

It has been revealed that magnitudes of amplitude, energy and duration of the bioluminescent signals in the freshly caught ctenophores depended directly on their size. For example, luminescence intensity in *M. leidyi* with wet weight 0.52 ± 0.026 g makes under mechanical stimulation $1.32 \cdot 10^8$ quantum·s^{-1}·cm^{-2} and under chemical—$3.55 \cdot 10^8$ quantum·s^{-1}·cm^{-2} (**Figure 8**), while in big specimens (with wet weight 42.03 g) its intensity makes $(767.56 \pm 42.21) \cdot 10^8$ quantum·s^{-1}·cm^{-2} under mechanical stimulation and $(1016.93 \pm 50.84) \cdot 10^8$ quantum·s^{-1}·cm^{-2} under chemical one. Analogous situation is observed at the bioluminescence energetic indices (**Figure 9**), which grow with an increase of the organism size (from $(0.89 \pm 0.035) \cdot 10^8$ to $(1004.28 \pm 40.17) \cdot 10^8$ quantum·cm^{-2} under chemical stimulation and from $(0.29 \pm 0.01) \cdot 10^8$ quantum·cm^{-2} to $(868.26 \pm 39.07) \cdot 10^8$ quantum·cm^{-2} under mechanical one). Luminescence duration (**Figure 10**) of less in size organisms (with wet weight 0.0073 ± 0.00036 g) made under mechanical stimulation 0.79 ± 0.03 and under chemical—1.37 ± 0.06 s, but in the second size group (with wet weight 0.52 ± 0.026 g) ctenophores bioluminescence duration under both types of stimulation increased 2–2.5 times. Further on with an increase of the specimen size in groups from 10 to 65 mm ctenophore luminescence duration practically did not change,

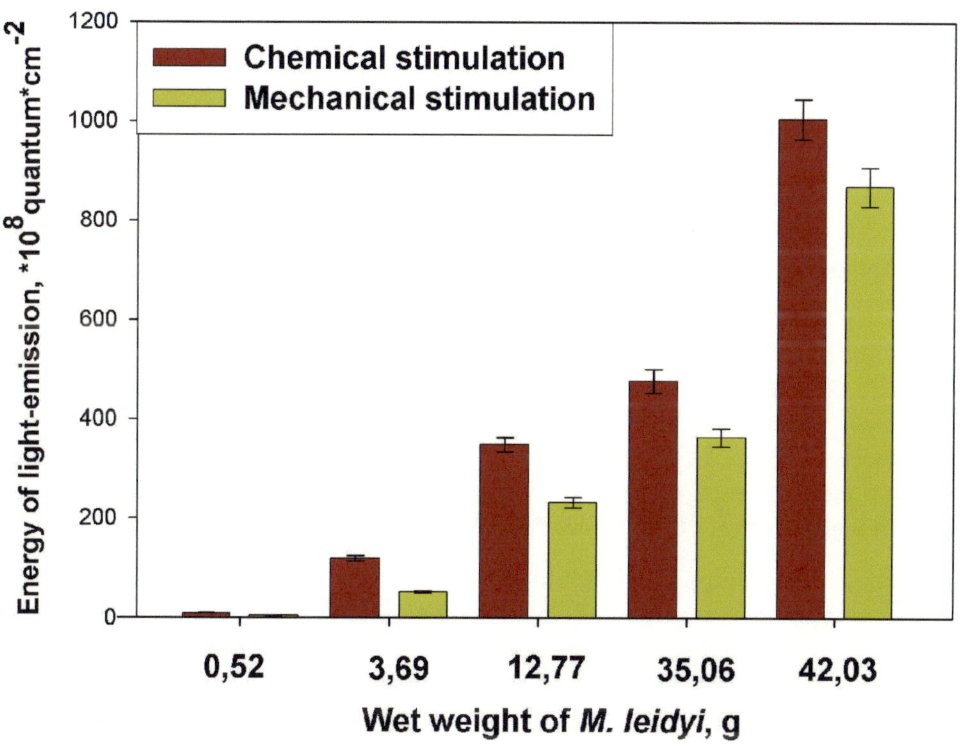

Figure 9. Variability of the bioluminescence energy ctenophore *M. leidyi* depending on wet weight of the individuals under mechanical and chemical stimulation.

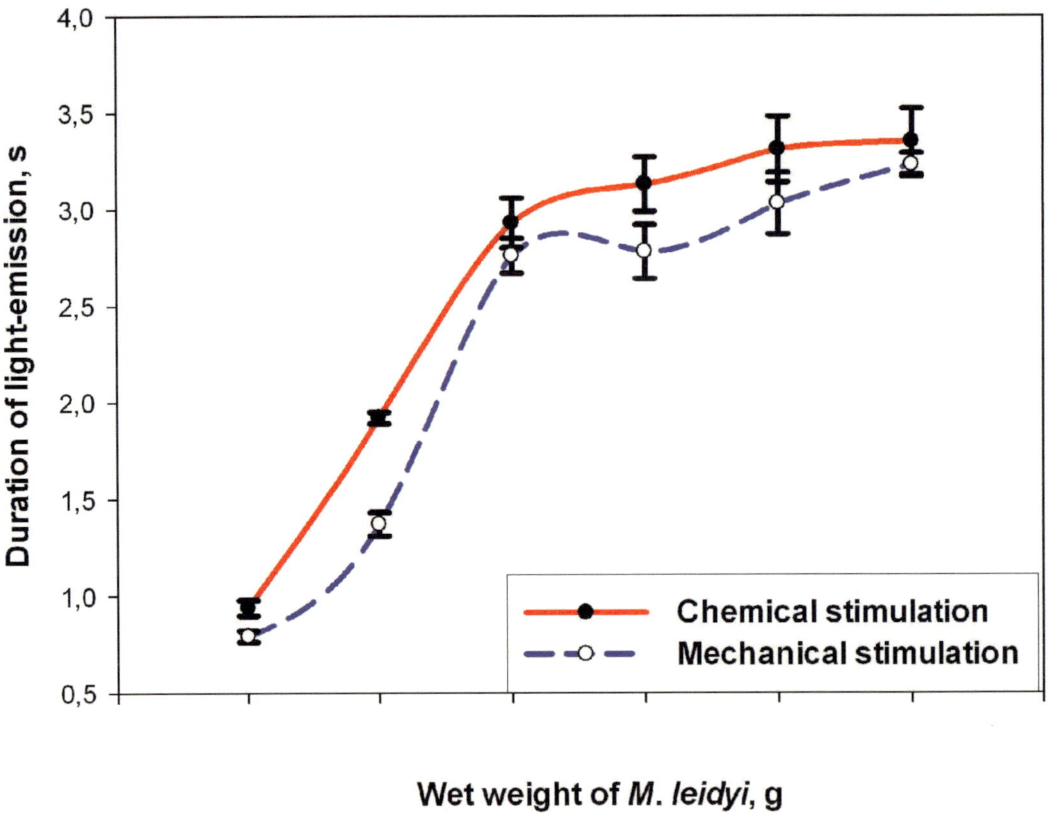

Figure 10. Variability of the bioluminescence duration ctenophore *M. leidyi* depending on wet weight of the individuals under mechanical and chemical stimulation.

reaching in the biggest specimens 3.35 ± 0.16 s under mechanical stimulation and 2.78 ± 0.13 s under chemical one.

Figure 11 represents variability of the biophysical characteristics of *M. leidyi* light-emission depending on the stage of the organism's reproduction. As it can be seen at the illustrative material presented amplitude of light signals appeared to be the most sensitive index of the bioluminescence (**Figure 11**), its maximum magnitudes were registered in a group of ctenophores with eggs clutches, where they two to three times ($p < 0.05$) exceeded luminescent intensity in the freshly caught specimens (control).

Having compared ctenophore bioluminescence after spawning and those in control we found that luminescence amplitude in the control group 14 times exceeded amplitude in the spawned specimens. Light-emission energy in the spawning ctenophores with clutch reached if compared with other groups of organisms maximum magnitudes up to $(139.46 \pm 8.36) \cdot 10^8$ quantum·cm^{-2}, which 1.5 times exceeded analogous indices in specimens from the control group and 53 times ($p < 0.05$) the same indices in the spawned ctenophores, showing the lowest energetic indices to $(2.62 \pm 0.13) \cdot 10^8$ quantum·cm^{-2}. The signals duration in ctenophores

http://dx.doi.org/10.5772/65063

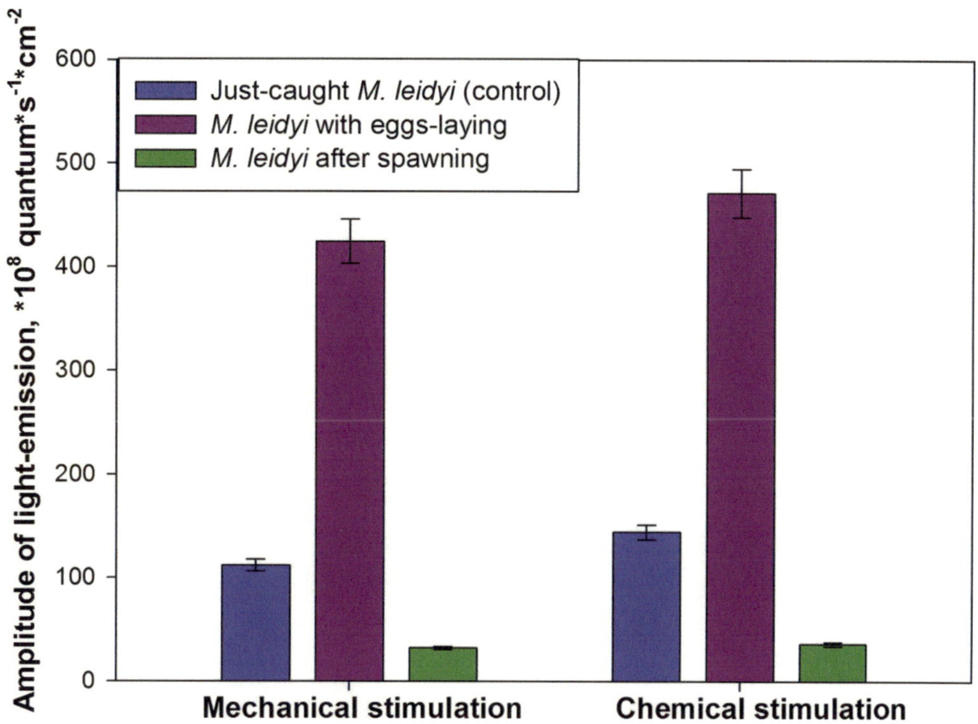

Figure 11. Variability of the bioluminescence amplitude ctenophore *M. leidyi* at the reproduction period.

with clutch and control exemplars practically did not differ, making from 3.28 to 3.62 s, but 3.5 times they exceeded such signals in the spawned specimens, which gave the least time of luminescence to 0.90 ± 0.045 s.

3.6. Bioluminescence variability in the *B. ovata* ontogenesis

Bioluminescence energy values depend on quantity of secret, produced in the time of organism irritation. So with the increase of the ctenophore age and body mass growth, the more is secret content. Thus, luminescence intensity is a function of organism's mass, that is, $A = f(W)$. Amplitude and bioluminescent signal duration of newly caught ctenophores directly depend on dimension, that is, on wet weight of the investigated organism (**Figure 12**) [33].

B. ovata light-emission amplitude of organisms with body mass till 0.06 ± 0.003 g under mechanical stimulation was two times more than the one under chemical stimulation, achieving $(11.39 \pm 0.56) \cdot 10^8$ quantum·s^{-1}·cm^{-2}. Light-emission intensity grows when *B. ovata* body mass increases from 0.06 to 19.53 g, achieving $(925.74 \pm 45.27) \cdot 10^8$ quantum·s^{-1}·cm^{-2}. The shortest luminescent signals (0.46–0.94 s) were produced by small-sized ctenophores (**Figure 13**).

Beroe light-emission duration increased, achieving from 1.44 to 2.37 s, as body mass raised [33]. The organisms with body mass 19.53 ± 0.97 g produce 2–2.5 times more prolonged

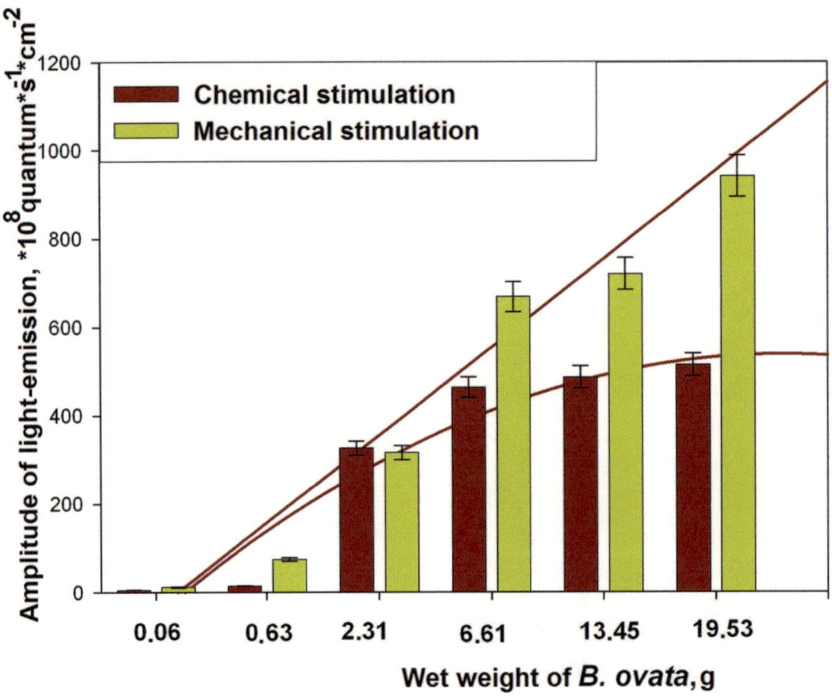

Figure 12. *B. ovata* light-emission amplitude in terms of organism wet weight under mechanical and chemical stimulation [33].

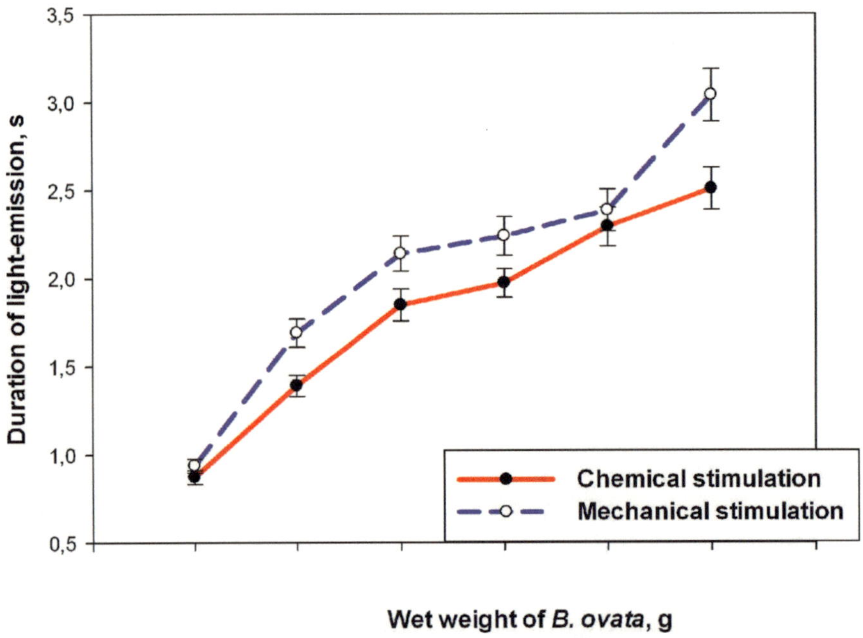

Figure 13. *B. ovata* bioluminescent signal duration depending on body mass wet weight under mechanical and chemical stimulations.

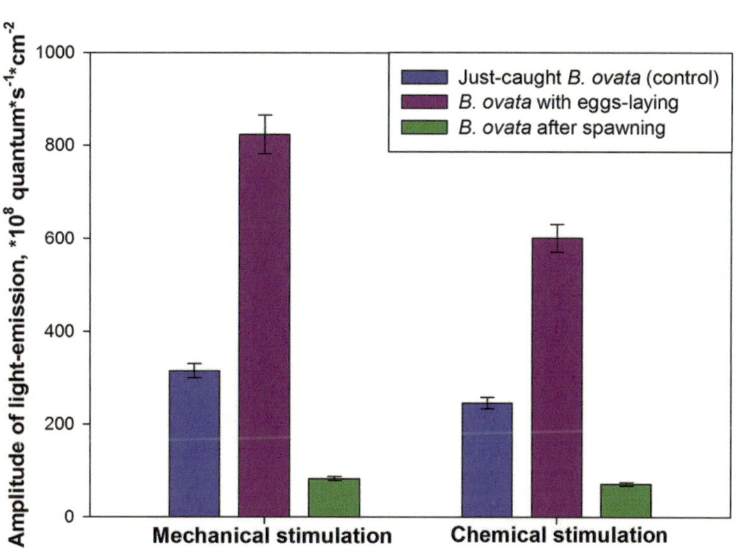

Figure 14. *B. ovata* bioluminescent signal amplitude in reproduction period [33].

light-emission signals than small-sized ctenophores. Experiment's results of the ctenophore reproduction system investigations detected that bioluminescence amplitudes were maximum in ctenophores with egg clutches (**Figure 14**), being two to three times more ($p < 0.05$) than in the control group. However, bioluminescence indices were three to four times more in the control organisms than in the post-spawning group. Light-emission energy of spawning individuals if compared with other groups achieved maximal indices until $(434.41 \pm 21.7) \cdot 10^8$ quantum·cm^{-2} [33].

The post-spawning group gave the lowest energy indices until $(56.77 \pm 2.83) \cdot 10^8$ quantum·cm^{-2}. Light-emission durations in ctenophores with eggs clutches were the same as in the control

Ontogenesis stages of *B. ovata*	L (mm)	Amplitude of light-emission (quantum·s^{-1}·cm^{-2})		Energy of light-emission (quantum·cm^{-2})		Duration of light-emission, s	
Stimulation types		**1**	**2**	**1**	**2**	**1**	**2**
Just-caught individuals (control)	50	$(315.36 \pm 15.76) \cdot 10^8$	$(246.23 \pm 12.31) \cdot 10^8$	$(331.09 \pm 16.55) \cdot 10^8$	$(177.60 \pm 8.88) \cdot 10^8$	2.27 ± 0.11	1.39 ± 0.06
Reproductive ctenophores	50	$(823.91 \pm 41.18) \cdot 10^8$	$(601.72 \pm 30.08) \cdot 10^8$	$(434.41 \pm 21.72) \cdot 10^8$	$(259.75 \pm 12.98) \cdot 10^8$	2.49 ± 0.12	1.86 ± 0.09
Ctenophore eggs	0.80–0.85	$(0.76 \pm 0.03) \cdot 10^8$	$(0.28 \pm 0.01) \cdot 10^8$	$(0.53 \pm 0.02) \cdot 10^8$	$(0.21 \pm 0.01) \cdot 10^8$	0.89 ± 0.04	0.33 ± 0.016
Ctenophore larvae	0.4–0.5	$(6.07 \pm 0.3) \cdot 10^8$	$(2.26 \pm 0.1) \cdot 10^8$	$(3.71 \pm 0.17) \cdot 10^8$	$(1.49 \pm 0.06) \cdot 10^8$	1.64 ± 0.08	1.08 ± 0.05

Remark: 1, mechanical stimulation; 2, chemical stimulation.

Table 4. The bioluminescence characteristics of ctenophore *B. ovata* at the ontogenesis [33].

group. The lowest light-emission time was in the post-spawned group up to 1.51 ± 0.07 s. The *B. ovata* clutch contained from 2.0 till 7.0 thousands of eggs with size up to 0.80–0.85 mm. Free-swimming larvae with body length of 0.4–0.5 mm appeared on the third day after spawning.

Ctenophore eggs have low luminescence indices with intensity peaks up to $(0.76 \pm 0.03) \cdot 10^8$ $quantum \cdot s^{-1} \cdot cm^{-2}$, light-emission energy values—up to $(0.53 \pm 0.02) \cdot 10^8$ $quantum \cdot cm^{-2}$ and short bioluminescent signal—up to 0.89 ± 0.048 s. It was shown also that larvae biolumi-

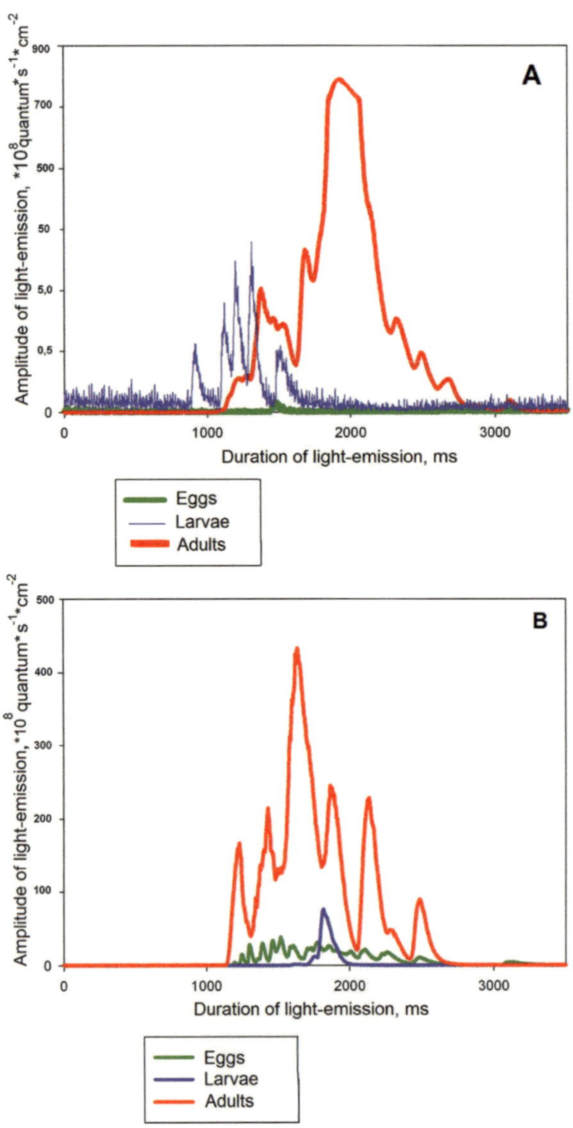

Figure 15. Typical *B. ovata* bioluminescent signals at different ontogeny stages under mechanical (A) and chemical (B) stimulations [33].

nescence intensity was eight times and energy—seven times more than eggs had ($p < 0.05$) (**Table 4, Figure 15**).

The same situation was observed for eggs and larvae light-emission durations. Thus, larvae luminescence duration was two to three times more than the eggs' one. *B. ovata* light-emission amplitude rises with ctenophore growth. Adult individual's light-emission amplitude exceeded the larvae one. Luminescence duration in the control organisms is 0.63 s more than in the larvae. Consequently, present research detected that *B. ovata* light-emission characteristics significantly change in ontogeny, at the reproductive stages and rise proportionally with body mass growth [33].

4. Discussion

M. leidyi and *B. ovata* light-emission seasonal variability can be conditioned by specificities of ctenophores chemical composition seasonal dynamics. Thus, different food supply is the main reason of the organism's biochemical composition changes. Ctenophores physiological state in the winter-spring periods is depressed, that is connected with food concentration deficit [27]. That is why light-emission intensity and energy have the lowest values in these periods. Food conditions of ctenophore are most favorable in the summer, which is connected with rise of glycogen and waxes concentrations in the *M. leidyi* [42]. Other maximal light-emission amplitude values are registered in August as well. Other reason of the bioluminescence seasonal changes is the Black Sea water temperature variability. Thus, water temperature lowering to 8 ± 2°C in winter-spring leads to decrease in amplitude-temporal indices of ctenophores bioluminescence. Low temperatures are unfavorable for the *M. leidyi* vital activity, their motion function and metabolic processes and negatively influence reproduction condition [39, 43, 44]. That is why the ctenophore light-emission characteristics decrease in this period. Ctenophores light-emission seasonal dynamics can be explained by the seasonal changes of their biochemical composition, connected with the food supply [31, 39].

That allows using our experiments results in different variants of the ecological monitoring of the coastal water area. Environment temperature affects considerably the amplitude-temporal characteristics of the Black Sea alien-ctenophore light-emission. It was revealed that bioluminescence reaction optimum for *M. leidyi* is achieved under the temperature of 26 ± 1°C and for *B. ovata*—under the temperature of 22 ± 1°C, while its minimum for both ctenophore species was registered under the temperature of 10 ± 1°C. As it follows from the results of the experiments conducted, bioluminescence is characteristic for *M. leidyi* ctenophores at all the stages of individual development, but with considerable changes in its parameters during ontogenesis. It is necessary to underline that our investigations were conducted at the period of the ctenophores reproduction from July to September, maximum of which in *M. leidyi* takes place in August [43].

But at the period of intensive growth and reproduction, when fodder zooplankton biomass cannot supply needs for support and reproduction of the population, ctenophores are under deficit of food [18, 20]. That is why increase in the mnemiopsis abundance during reproduction is accompanied with a decrease of its average individual mass, first of all due to a decrease of the big size individuals share in the population [33, 39]. During the given investigations, we also observed in the zooplankton samples domination of fry, eggs and larvae of the ctenophores and to a less extent availability of matured specimens.

Undoubtedly flashes intensity depends on a number of photoprotein in photocytes and maturation of the photocytes themselves. Its content increases with age and consequently with an increase if linear sizes and body mass. Thus, light-emission energy is a function of the organism mass, that is, E=f (W). It is also known that trophic factor effects considerably life activity and bioluminescent characteristics of ctenophores [27]. For example, according to our data, freshly caught ctenophores with full stomachs gave eggs and germs averagely in 6 h. But without feeding germs did not develop and they perished, having not reached the larval stage. In the laboratory conditions with satisfactory supply of food ctenophores are close to the conditions of the specimens *in situ* [18, 20]. At the before-spawning period, ctenophore is getting ready for the reproduction, accumulates necessary for this organic substances and contains quite great energetic potential, equal to the sum of its own one and this of eggs. That is why just at the given period, we observe the highest amplitude-energetic parameters of their bioluminescence. Visual observations of the ctenophores behavior at the period of their spawning have shown that individuals after ovulation feel themselves worse, become less mobile, some of them fall to the bottom. Such behavior is identical to this *in situ*, when ctenophores spawning not only influences their moving activity but in some cases also causes specimens death [33]. According to some researchers, organic losses in the ctenophores after spawning can make 6.9% of body. But the full-day losses for exchange in ctenophore with body mass of 25 g at 26°C are estimated as 5.6% of body [43].

In other words, the loss of substance with sex products is quite comparable with losses of an organism for breathing. This points to domination of the generative metabolism strategy in ctenophores and explains accompanying slowing of its growing at the reproductive period [14]. As bioluminescence is closely connected with the breathing chain of organism [3], it is quite understandable that considerable change in the functional condition and metabolism in ctenophores during reproduction are reflected in the observed low indices of the bioluminescence in the spawned individuals if compared with the control. Differences in the ctenophores bioluminescence parameters we revealed at different ontogenesis stages can be also explained by changes in their biochemical composition during their individual development. For instance, according to Finenko and Anninsky, organic substance composition differs considerably in eggs and larvae from the same in the matured specimens. In particular, content of organic substance in *M. leidyi* eggs makes only 0.25 $\mu g \cdot mg^{-1}$, but in the body of two-day larvae of *M. leidyi* of 0.26–0.30 mm size 25.1 ± 8.3 $\mu g \cdot mg^{-1}$ of wet substance [14].

Due to the fact that the organic substances stock provide early survival of larvae and maximum growth rate parallel to minimum exchange more bright lighting of larvae if compared with the ctenophore eggs can be explained, as we think, by great content of organic substance in larvae. Together with this, specific content of organic substance in the ctenophore early larvae is 20–30 times higher than the corresponding magnitudes for adult specimens. Change in number of photocytes in developing individuals can present one more reason of the registered by us variability of the luminescence characteristics on the ctenophores in ontogenesis. For example at early stages of the ctenophores development, when growth of twinkling rowing plates begins in organisms we observe an increase of the photocytes cytological maturity. At more late stages, when embryo begins to feed itself we observe an increase of the photocytes number. And at last with development of the organisms, we register an increase of the photoprotein number in the photocytes tissues of the adult specimens [40].

That is why it is quite explainable that quantum issue of the ctenophores bioluminescence is minimal at early stages of the organism's development and it is maximal at those late. Besides differences in the ctenophores bioluminescence parameters can be conditioned, according to our opinion, by peculiarities of the ctenophores biochemical composition, determined by their dependence on nutrition quantity and spectrum. According to the data of Anninsky et al. [14], concentration of organic substance in the ctenophore body depends considerably on their size. Protein in the ctenophore body is dominating oxidized substrate and its share in the ctenophore organic substance is of 80–85%. Correlation of concentrations of free amino acids and protein is maximal in small individuals with highly active metabolism and minimal in big organisms. There is domination in lipids of fractions, characteristic for the cell membranes: phospholipids make 35.7 ± 9.6% of general lipids. But in bigger organisms, they observe a tendency to increase number of waxes and sterine ethers. For example their content was of 4.0 ± 3.6; 5.5 ± 3.2 и 7.1 ± 4.0% in ctenophores with the size 10–20, 21–30 and 31–50 mm, correspondingly. In carbohydrates, glycogen dominated; its content grew a bit with an increase of ctenophores size and made 25 ± 4; 28 ± 5; and 36 ± 12 $\mu g \cdot g^{-1}$, when body length was 10, 11–20 and 31–50 mm correspondingly [42].

And at last with organisms growing hydration increases and individuals motility decreases. Thus, protein-lipid and carbohydrate exchange effect changes in the ctenophores bioluminescence parameters. But, as it has been already marked with development of organisms' quantity of photoprotein in the ctenophores photocytes and concentration of the substrate of the bioluminescent reaction—luciferin increase, which influence reinforcement of the bioluminescent activity in adult ctenophores [40]. Taking into consideration fermentative nature of the bioluminescent reaction, we can presume that change in the rate of fermentative processes affects duration of the bioluminescent signals. Really maximal bioluminescence is observed in small specimens with higher fermentative activity and shorter signal duration. In adult individuals, we observed decrease in metabolism and connected with this decrease in luciferase fermentative activity, which facilitates more long light-emission [45]. Thus, development of organisms along the way of increasing body hydration and decrease of the active exchange in more big specimens, lowering of their motility and maneuver is compensated by the most important ecological characteristic: less access for predators due to more developed luminescent organs and correspondingly maximal yield of the bioluminescence energy. It gives grounds for supposition that bioluminescence protective function is the most important component in the ctenophores ecology.

Our investigations with *Beroe* larvae were conducted in the period of ctenophore reproduction—from September to November. *B. ovata* spawning peak is observed in October [19, 33, 34]. Juveniles, eggs and larvae predominated in the zooplankton samples from mid-September till October. Similar to *M. leidyi Beroe* prepared for reproduction in prespawning period. Composition of ctenophores changed. Probably, for this reason, high bioluminescent amplitude-energetic parameters of *B. ovata* were observed. Recording beroe behavior in the spawning period has shown that the organisms after fertilization became less mobile. The *B. ovata* eggs clutch *in situ* was similar to our laboratory experiments, but varied from 4500 ± 250 until 28,000 eggs per day. Distinction in clutch sizes was determined by different organisms' sizes, temperature conditions and availability of nutrition budget [14, 33, 46]. Adult *B. ovata* organisms lose large amount of organic material with reproductive products [14]. Ctenophores with body mass 15.4 g loose from 6% to 8% of organic matter per day, from different data, at temperature conditions of 19–21°C [14, 33, 38].

Accordingly, the fact that bioluminescence is closely related to biochemical processes in organism and to its physiological state [28, 47] is well substantiated by our data that the lowest ctenophore bioluminescent parameters are produced by post-spawned individuals at reproduction period. It is revealed by us that ctenophore bioluminescent parameters dissimilarity at different reproductive stages are explained by changes of their biochemical composition in ontogenesis. The eggs and larvae composition of organic matter differed much from the adult individuals [14]. Beroe light-emission parameters changeability in ontogenesis can be related with photocyte quantitative variability of growing individuals and their cytological maturity [40]. We suppose that ctenophore light-emission characteristics changeability with body mass growth can be determined by specialty of their biochemical composition depending on sizes [23].

At the same time, as the organism develops along the way of body growing hydration and active metabolism [14, 23], decrease of great individuals' mobility and maneuverability is compensated by one of highly important qualities: the lowered survival capability due to more developed bioluminescent organs and, consequently, maximum bioluminescent energy discharge.

Acknowledgments

The authors are grateful for the valuable advices during the given work conduction to Khanaichenko A.N., Finenko G.A. the scientists of KIMBR, Russia; to Juk V.F., Belogurova Yu.B. and M.I. Silakov, leading engineers of KIMBR, Russia, for the help in work with laboratory equipment and making the program of its verification.

Author details

Tokarev Yuriy Nikolaevich[1] and Mashukova Olga Vladimirovna[2*]

*Address all correspondence to: olgamashukova@yandex.ru

1 Scientific Supervisor of IMBR, Head of the Biophysical Ecology Department of IMBR, RAS, Sevastopol, Nakhimov Av

2 Senior research scientist, the Biophysical Ecology Department of The Kovalevsky Institute of marine biological research (IMBR), RAS, Sevastopol, Russia

References

[1] Harvey E. Bioluminescence. New York: Academic Press; 1952. 649 p.

[2] Gitelzone I, Levin L, Utyushev R. et al. Bioluminescence in the ocean. St. Petersburg: Gidrometeoizdat; 1992. 283 p.

[3] Tokarev Yu. Basin of the hydrobionts biophysical ecology. Sevastopol: ECOSI-Gidrophizika; 2006. 342 p.

[4] Valiadi M, Iglesias-Rodriguez D. Understanding bioluminescence in dinoflagellates: How far have we come? Mint: Microorganisms. 2013; 1: 3–25.

[5] Marsinko C, Painter S, Martin A, Allen J. A review of the measurement and modeling of dinoflagellate bioluminescence. Mint: Progress in Oceanography. 2013; 109: 117–129.

[6] Widder E. Bioluminescence in the ocean: Origins of biological, chemical, and ecological diversity. Mint: Science. 2010; 328: 704–708.

[7] Pennisi E. Light in the deep. Mint: Science. 2012; 235(3): 1060–1063.

[8] Determinator of fauna of the Black and Azov seas. Kiev: Naukova dumka; 1968. 437 p.

[9] Purcell J. Climate effects on formation of jellyfish and ctenophore blooms: A review. Mint: Journal of the Marine Biological Association UK 2005; 85: 461–476.

[10] Öztürk B, Mihneva V, Shiganova T. First records of *Bolinopsis vitrea* (L. Agassiz, 1860) (Ctenophora: Lobata) in the Black Sea. Mint: Aquatic Invasions. 2011; 6(3): 355–360. Doi:10.3391/ai.2011.6.3.12

[11] Shushkina E, Musaeva E, Anochina L. et al. Role of the jellyfish macroplankton: Sifonophore *Aurelia aurita*, ctenophores *Mnemiopsis leidyi* and *Beroe ovata* in the planktonic community of the Black Sea. Mint: Oceanology. 2000; 40(6): 859 – 861.

[12] Vostokov S, Arashkevich E, Drits A. et al. Ecological and physiological characteristics of the ctenophore *Beroe ovata* in the coastal water of the Black sea: Quantity, biomass, size distribution, hunting behavior, feeding and metabolism. Mint: Okeanologiya. 2001; 41(1): 109–155.

[13] Mutlu E, Bingel F, Gucu A. et al. Distribution of the new invader *Mnemiopsis* sp. and the resident *Aurelia aurita* and *Pleurobrachia pileus* populations in the Black Sea in the years 1991–1993. Mint: ICES Journal of Marine Science. 1994; 51: 407–421.

[14] Anninsky B, Finenko G, Abolmasova G et al. Body organic content *Mnemiopsis leidyi* (Ctenophora: Lobata) and *Beroe ovata* (Ctenophora: Beroida) in early ontogenetic stages. Mint: Biology Exterminating. 2007; 33(6): 457–464.

[15] Zaika V, Sergeeva N. Morphology and development of *Mnemiopsis mccradyi* (Ctenophora, Lobata) in the Black sea. Mint: Zoologicheskii Zhurnal. 1990; 69(2): 5–11.

[16] Minkina N, Pavlova E. Daily changes of respiration intensity of comb jelly *Mnemiopsis leidyi* in the Black sea. Mint: Okeanologiya. 1995; 35 (2): 241–245.

[17] Parkhomenko A. Phosphorus excretion by the Black sea *Mnemiopsis leidyi*. Mint: Ekologya Sea. 2006; 72: 70–76.

[18] Finenko G, Abolmasova G, Romanova Z. Intensity of the nutrition, respiration, and growth *Mnemiopsis mccradyi* in relation to grazing conditions. Mint: Biologya of the Sea. 1995; 20(5): 315–320.

[19] Finenko G, Romanova Z, Abolmasova G. The ctenophore *Beroe ovata*: A recent invader to the Black sea. Mint: Ekologya Sea. 2000; 50: 21–25.

[20] Tsikhon-Lukanina E, Reznichenka O, Lukasheva T. Quantitative aspects of feeding in the Black sea ctenophore *Mnemiopsis leidyi* . Mint: Okeanologiya. 1991; 31(2): 272–276.

[21] Zaika V, Sergeeva N. Daily changes of structure of population of vertical distribution of a comb jelly of *Mnemipsis mccradyi* Mayer (Ctenophora) in the Black Sea. Mint: Gidrobiologichesky Zhurnal. 1991; 27(2): 15–19.

[22] Zaika V, Ivanova N. *Mnemiopsis mccradyi* in the autumn hyponeuston of the Black sea. Mint: Ekologya Sea. 1992; 42: 6–10.

[23] Zaika V. About the connection of *Mnemiopsis leidyi* size structure in the Black Sea with dynamics of its growth and reproduction. Mint: Morskyi Ekolohichnyi Zhurnal. 2005; 4(3): 59–64.

[24] Zaika V, Revkov N. Anatomy of gonads and regime of spawning of ctenophore *Mnemiopsis* sp. in the Black sea. Mint: Zoologicheskii Zhurnal. 1994; 73(3): 5–9.

[25] Anninsky B, Abolmasova G. Temperature, as a factor of ctenophore *Mnemiopsis leidyi* metabolism intensity and mass development in the Black Sea. Mint: Okeanologiya. 2000; 40(5): 63–69.

[26] Finenko G, Abolmasova G, Datsyk N. Effect of food composition and temperature on ctenophore-invader *Mnemiopsis leidyi* A. Agassiz feeding rate *in situ*. Mint: Russian Journal of Biological Invasions. 2013; 4: 78–90.

[27] Tokarev Yu, Mashukova O. Variability of the bioluminescence characteristics of the Black Sea ctenophores-aliens in connection with different conditions of nutrition. Mint: Advances in Bioscience and Biotechnology. 2013; 4: 968–973. http://www.scirp.org/journal/abb/

[28] Tokarev Yu, Mashukova O, Vasilenko V. The bioluminescence of ctenophore-aliens *Mnemiopsis leidyi* and *Beroe ovata* in the Black sea at the mechanical and chemical stimulation. Mint: Ekologya Sea. 2008; 76: 61–66.

[29] Luppova N. *Beroe ovata* Mayer, 1912 (Ctenophora, Athentaculata, Beroidea) in the coastal waters of the north-east of the Black sea. Mint: Ekologya Sea. 2002; 59: 23–25.

[30] Finenko G, Romanova Z, Abolmasova G. et al. Trophic interaction in the Black sea plankton community at the present stage. Mint: Ecologiya Moray. 2006; 71: 50–54.

[31] Mashukova O, Tokarev Yu. Seasonal dynamic of the Black sea ctenophores-aliens. Mint: Natura Montenegrina, Podgorica. 2013; 12(2): 345–355.

[32] Mashukova O, Tokarev Yu. Bioluminescence daily rhythm of ctenophore *Beroe ovata* Mayer, 1912. Proceedings of the Global Congress on ICM: Lessons Learned to Address New Challenges. Mint: Marine Biology Microbiology. 2013; 2: 729–736.

[33] Tokarev Yu, Mashukova O, Sibirtsova E. Bioluminescence characteristics changeability of ctenophore *Beroe ovata* Mayer 1912 (Beroida) in ontogenesis. Mint: Turkish Journal of Fisheries and Aquatic Sciences. 2012; 12: 479–484. http://www.trjfas.org/

[34] Arashkevich E, Anokhina L, Vostokov S. et al. Reproduction Strategy of *Beroe ovata* (Ctenophora, Atentaculata, Beroida): A New Invaser in the Black Sea. Mint: Oceanology. 2001; 41(1): 116–120.

[35] Gubanova A. Mnogoletnie izmeneniya v soobschestve zooplanktona Sevastopol'skoy bukhty. In: Eremeev V, Gaevska A. editors. Sovremennoe sostoyanie pribrezhnykh vod Kryma. IBSS, Sevastopol Sevastopol: ECOSI-Gidrophizika; 2003. p. 83–94.

[36] Mashukova O, Tokarev Yu. Influence of the temperature at the Black Sea ctenophores-aliens bioluminescence characteristics. Mint: Advances in Bioscience and Biotechnology. 2012; 3: 269–273. http://www.scirp.org/journal/abb/

[37] Finenko G, Romanova Z, Abolmasova G. et al. Ctenophores: Invaders and their role in the trophic dynamics of the planktonic community in the coastal regions off the Crimean Coasts of the Black Sea (Sevastopol Bay). Mint: Oceanology. 2006; 46(4): 472–482.

[38] Shiganova T, Bulgakova Y, Volovik S. et al. The new invader *Beroe ovata* Mayer, 1912 and its effect on the ecosystem in the northeastern Black Sea. Mint: Hydrobiologia. 2001; 451: 187–197.

[39] Khoroshilov V. The Black sea ctenophore *Mnemiopsis leidyi* population seasonal dynamics. Mint: Okeanologiya. 1993; 33(4): 558–562.

[40] Shimomura O. Bioluminescence: Chemical principles and methods. World Scientific; 2006. 470 p.

[41] Haddock S, Moline M, Case J. Bioluminescence in the Sea. Mint: Annual Review of Marine Science. 2010; 2: 443–493.

[42] Anninsky B. Seasonal dynamics of the chemical composition of organic matter in the ctenophora *Mnemiopsis leidyi* A. Agassiz in the Black sea. Mint: Okeanologiya. 1995; 35(3): 426–429.

[43] Finenko G, Arashkevitch E, Kideys A. et al. Reproduction characteristics and growth rate of ctenophore *Beroe ovata* larvae in the Caspian and Black Sea waters. Mint: Morskyi Ekologichnyi Zhurnal. 2011; 10(1): 77–88.

[44] Kharchuk I. Anabiosis: Basic definitions and processes, accompanied it. Mint: Ecologiya Moray. 2005; 70: 62–78.

[45] Lapota D. Bioluminescence: Recent advances in oceanic measurements and laboratory applications. InTech Janeza Trdine; 2012; 9: 190 p.

[46] Gucu A. Can overfishing be responsible for the successful establishment of *Mnemiopsis leidyi* in the Black Sea? Mint: Estuarine, Coastal and Shelf Science. 2002; 54: 439–451.

[47] Mashukova O.V., Tokarev Yu.N. Variability of *Mnemiopsis leidyi* A. Agassiz (Ctenophora: Lobata) bioluminescence in relation to regeneration. Mint: Marine Biological Journal. 2016; 1(1): 36–42.